Semiconductor Devices for High-Speed Optoelectronics

Providing an all-inclusive treatment of electronic and optoelectronic devices used in high-speed optical communication systems, this book emphasizes circuit applications, advanced device design solutions, and noise in sources and receivers. Core topics covered include semiconductors and semiconductor optical properties, high-speed circuits and transistors, detectors, sources, and modulators. It discusses in detail both active devices (heterostructure field-effect and bipolar transistors) and passive components (lumped and distributed) for high-speed electronic integrated circuits. It also describes recent advances in high-speed devices for 40 Gbps systems. Introductory elements are provided, making the book open to readers without a specific background in optoelectronics, whilst end-of-chapter review questions and numerical problems enable readers to test their understanding and experiment with realistic data.

Giovanni Ghione is Full Professor of Electronics at Politecnico di Torino, Torino, Italy. His current research activity involves the physics-based and circuit-oriented modeling of high-speed electronic and optoelectronic components, with particular attention to III-N power devices, thermal and noise simulation, electrooptic and electroabsorption modulators, coplanar passive components, and integrated circuits. He is a Fellow of the IEEE and has authored or co-authored over 200 technical papers and four books.

Semiconductor Devices for High-Speed Optoelectronics

GIOVANNI GHIONE

Politecnico di Torino, Italy

CAMBRIDGE UNIVERSITY PRESS
Cambridge, New York, Melbourne, Madrid, Cape Town, Singapore, São Paulo, Delhi

Cambridge University Press
The Edinburgh Building, Cambridge CB2 8RU, UK

Published in the United States of America by Cambridge University Press, New York

www.cambridge.org
Information on this title: www.cambridge.org/9780521763448

First published 2009

Printed in the United Kingdom at the University Press, Cambridge

A catalogue record for this publication is available from the British Library

ISBN 978-0-521-76344-8 hardback

Additional resources for this publication at www.cambridge.org/Ghione

To the memory of my parents

Contents

Preface

The development of high-speed fiber-based optical communication systems that has taken place since the early 1970s can be really considered as a technological wonder. In a few years, key components were devised (such as the semiconductor laser) with the help of novel technological processes (such as epitaxial growth) and found immediate application thanks to the development of low-loss optical fibers. New compound semiconductor alloys (namely, InGaAsP) were ready to provide their potential to emit the right wavelengths needed for long-haul fiber propagation. When electronic repeaters seemed unable to provide a solution to long-haul propagation, fiber amplifiers were developed that allowed for all-optical signal regeneration. And the list could be continued. A miracle of ingenuity from a host of researchers made it possible to assemble this complex puzzle in a few years, thus bringing optoelectronic technology to a consumer electronics level.

Increasing the system capacity by increasing the transmission speed was, of course, a main concern from the early stages of optical system development. While optoelectronic devices behave, on the electronic side, in a rather conventional way up to speeds of the order of 1 Gbps, for larger speeds (up to 40 Gbps and beyond) RF wave propagation has to be accounted for in designing and modeling optoelectronic devices. When speed increases, the distributed interaction between RF and optical waves becomes a useful, sometimes indispensable, ingredient in many optoelectronic devices, like modulators and (to a lesser extent) detectors. Similarly, the electronic circuits that interface light sources, modulators, and detectors should provide broadband operation up to microwave or millimeter-wave frequencies, thus making it mandatory to exploit compound semiconductor electronics (GaAs- or InP-based) or advanced Si-based solutions (like SiGe HBT integrated circuits or nanometer MOS processes).

Increasing speed beyond the 10 Gbps limit by improving device performance, however interesting it is from the research and development side, may in practice be less appealing from the market standpoint. The ultimate destiny of optoelectronic devices (such as sources, modulators, and detectors) optimized for 40 Gbps (or even faster) systems after the post-2000 market downturn still is uncertain, and research in the field has followed alternative paths to the increase of system capacity. At the same time, new application fields have been developed, for instance in the area of integrated all-Si optical signal processing systems, and also for integrated circuit level high-capacity communications. However, the development of high-speed optoelectronic devices has raised a number of stimulating (and probably lasting) design issues. An example is the principle of the distributed interaction between optical and RF waves, which is common to a variety of high-speed components. Another relevant theme is the co-design and the (possibly monolithic) integration of the electronic and optoelectronic components of a system, not to mention the critical aspects concerning device packaging and interconnection in systems operating at 40 Gbps and beyond.

Taking the above into account, it is not surprising that the main purpose of the present book is to provide a kind of unified (or, perhaps, not too widely separated) treatment of high-speed electronics and optoelectronics, starting from compound semiconductor basics, down to high-speed transistors, ICs, detectors, sources and modulators. Part of the material was originally developed for a number of postgraduate and Master courses, and therefore has the ambition (but also the limitation) of providing a treatment starting from the very basics. It is hoped that this justifies both the presence of introductory material on semiconductors and semiconductor optical properties, and a treatment of high-speed electronics starting from a review of transmission lines and scattering parameters. From this standpoint, the text attempts to be as self-contained as possible. Of course, the choice of subjects is somewhat influenced by the author's personal tastes and previous research experience (not to mention the need to keep the page count below 500): some emphasis has been put on noise, again with an attempt to present a self-contained treatment of this rather difficult topic, and many important optoelectronic components have not been included (to mention one, semiconductor optical amplifiers). Yet another innovative subject that is missing is microwave photonics, where of course the RF and microwave and optoelectronic worlds meet. Nevertheless, the text is (in the author's opinion, at least) different enough from the many excellent textbooks on optoelectronics available on the market to justify the attempt to write it.

I wish to thank a number of colleagues (from Politecnico di Torino, unless otherwise stated) for their direct or indirect contribution to this book. Ivo Montrosset provided many useful suggestions on the treatment of optical sources. Incidentally, it was under the guidance of Ivo Montrosset and Carlo Naldi that (then an undergraduate student) I was introduced to the basics of passive and active optoelectronic devices, respectively; this happened, alas, almost 30 years ago. Helpful discussions with Gian Paolo Bava and Pierluigi Debernardi (Consiglio Nazionale delle Ricerche) on laser noise, with Simona Donati Guerrieri on the semiconductor optical properties and with Fabrizio Bonani and Marco Pirola on active and passive high-speed semiconductor electronic devices and circuits are gratefully acknowledged. Michele Goano kindly revised the sections on compound semiconductors and the numerical problems, and provided useful suggestions on III-N semiconductors. Federica Cappelluti prepared many figures (in particular in the section on photodetectors), initially exploited in lecture slides. Finally, Claudio Coriasso (Avago Turin Technology Center, Torino) kindly provided material on integrated electroabsorption modulators (EAL), including some figures. Additionally, I am indebted to a number of ME students who cooperated in research, mainly on lithium niobate modulators; among those, special mention goes to F. Carbonera, D. Frassati, G. Giarola, A. Mela, G. Omegna, L. Terlevich, P. Zandano. A number of PhD students also worked on subjects relevant to the present book: Francesco Bertazzi (now with Politecnico di Torino) on EM modeling of distributed electrooptic structures; Pietro Bianco, on high-speed modulator drivers; Federica Cappelluti, on electroabsorption modulator modeling; Gloria Carvalho, on EAL modeling; Antonello Nespola (now with Istituto Superiore Mario Boella), on the modeling of distributed high-speed photodetectors. Part of the thesis work of Antonello Nespola and Federica Cappelluti was carried out within the framework of a cooperation with UCLA (Professor Ming Wu, now at University of California, Berkeley). Finally, I gratefully recall many helpful discussions with colleagues from the industry: among those, Marina Meliga, Roberto Paoletti, Marco Romagnoli, and Luciano Socci.

Giovanni Ghione
January 2009

1 Semiconductors, alloys, heterostructures

1.1 Introducing semiconductors

Single-crystal semiconductors have a particularly important place in optoelectronics, since they are the starting material for high-quality sources, receivers and amplifiers. Other materials, however, can be relevant to some device classes: polycrystalline or amorphous semiconductors can be exploited in light-emitting diodes (LEDs) and solar cells; dielectrics (also amorphous) are the basis for passive devices (e.g., waveguides and optical fibers); and piezoelectric (ferroelectric) crystals such as lithium niobate are the enabling material for a class of electrooptic (EO) modulators. Moreover, polymers have been recently exploited in the development of active and passive optoelectronic devices, such as emitters, detectors, and waveguides (e.g., fibers). Nevertheless, the peculiar role of single crystal semiconductors justifies the greater attention paid here to this material class with respect to other optoelectronic materials.

From the standpoint of electron properties, semiconductors are an intermediate step between insulators and conductors. The electronic structure of crystals generally includes a set of allowed energy bands, that electrons populate according to the rules of quantum mechanics. The two topmost energy bands are the *valence* and *conduction* band, respectively, see Fig. 1.1. At some energy above the conduction band, we find the *vacuum level*, i.e., the energy of an electron free to leave the crystal. In *insulators*, the valence band (which hosts the electrons participating to the chemical bonds) is separated from the conduction band by a large energy gap E_g, of the order of a few electronvolts (eV). Due to the large gap, an extremely small number of electrons have enough energy to be promoted to the conduction band, where they could take part into electrical conduction. In insulators, therefore, the conductivity is extremely small. In *metals*, on the other hand, the valence and conduction bands overlap (or the energy gap is *negative*), so that all carriers already belong to the conduction band, independent of their energy. Metals therefore have a large conductivity. In *semiconductors*, the energy gap is of the order of 1–2 eV, so that some electrons have enough energy to reach the conduction band, leaving *holes* in the valence band. Holes are pseudo-particles with positive charge, reacting to an external applied electric field and contributing, together with the electrons in the conduction band, to current conduction. In pure (*intrinsic*) semiconductors, therefore, charge transport is *bipolar* (through electrons and holes), and the conductivity is low, exponentially dependent on the gap (the larger the gap, the lower the conductivity). However, impurities can be added (*dopants*) to provide large numbers of electrons to

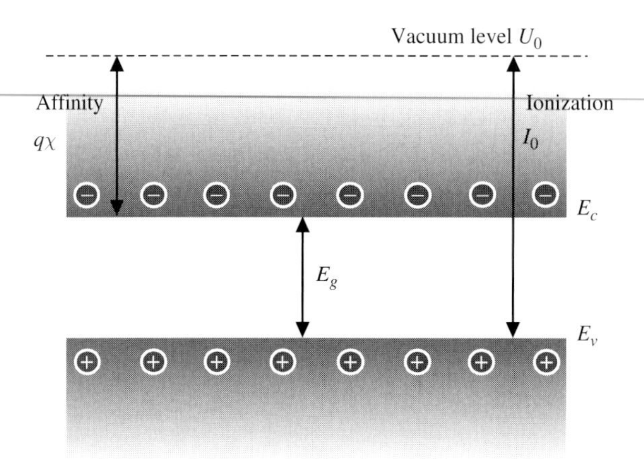

Figure 1.1 Main features of semiconductor bandstructure. E_g is the energy gap; E_c is the conduction band edge; E_v is the valence band edge.

the conduction band (*donors*) or of holes to the valence band (*acceptors*). The resulting doped semiconductors are denoted as n-type and p-type, respectively; their conductivity can be artificially modulated by changing the amount of dopants; moreover, the dual doping option allows for the development of pn junctions, one of the basic building blocks of electronic and optoelectronic devices.

1.2 Semiconductor crystal structure

Crystals are regular, periodic arrangements of atoms in three dimensions. The point set $\underline{r}$ defining the crystal nodes, corresponding to the atomic positions (Bravais lattice) satisfies the condition $\underline{r} = k\underline{a}_1 + l\underline{a}_2 + m\underline{a}_3$, where k, l, m are integer numbers and $\underline{a}_1$, $\underline{a}_2$, $\underline{a}_3$ are the *primitive vectors* denoting the *primitive cell*, see Fig. 1.2. Bravais lattices can be formed so as to fill the entire space only if the angles $\alpha_1, \alpha_2, \alpha_3$ assume values from a discrete set (60°, 90°, 120°, or the complementary value to 360°). According to the relative magnitudes of a_1, a_2, a_3 and to the angles $\alpha_1, \alpha_2, \alpha_3$, 14 basic lattices can be shown to exist, as in Table 1.1. In semiconductors, only two lattices are technologically important at present, i.e. the *cubic* and the *hexagonal*. Most semiconductors are cubic (examples are Si, Ge, GaAs, InP...), but some are hexagonal (SiC, GaN). Both the cubic and the hexagonal structure can be found in carbon (C), where they are the diamond and graphite crystal structures, respectively.

Three kinds of Bravais cubic lattices exist, the simple cubic (sc), the face-centered cubic (fcc) and the body-centered cubic (bcc), see Fig. 1.3. The cubic semiconductor crystal structure can be interpreted as two *shifted* and *compenetrated* fcc Bravais lattices.

Let us consider first an elementary semiconductor (e.g., Si) where all atoms are equal. The relevant cubic lattice is the *diamond lattice*, consisting of two interpenetrating

Table 1.1 The 14 Bravais lattices.

Name	Bravais lattices	Conditions on primitive vectors
Triclinic	1	$a_1 \neq a_2 \neq a_3, \alpha_1 \neq \alpha_2 \neq \alpha_3$
Monoclinic	2	$a_1 \neq a_2 \neq a_3, \alpha_1 = \alpha_2 = 90^\circ \neq \alpha_3$
Orthorhombic	4	$a_1 \neq a_2 \neq a_3, \alpha_1 = \alpha_2 = \alpha_3 = 90^\circ$
Tetragonal	2	$a_1 = a_2 \neq a_3, \alpha_1 = \alpha_2 = \alpha_3 = 90^\circ$
Cubic	3	$a_1 = a_2 = a_3, \alpha_1 = \alpha_2 = \alpha_3 = 90^\circ$
Trigonal	1	$a_1 = a_2 = a_3, \alpha_1 = \alpha_2 = \alpha_3 < 120^\circ \neq 90^\circ$
Hexagonal	1	$a_1 = a_2 \neq a_3, \alpha_1 = \alpha_2 = 90^\circ, \alpha_3 = 120^\circ$

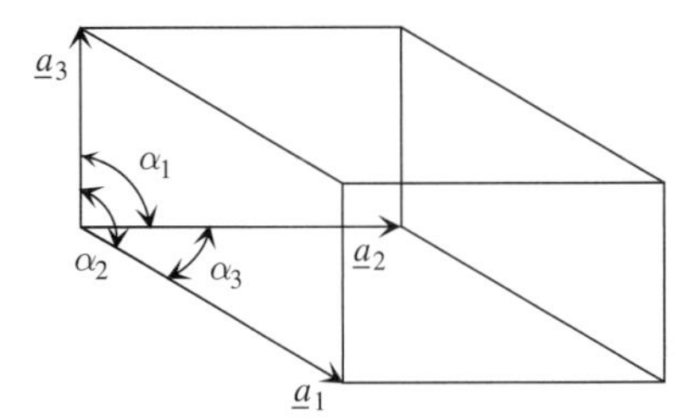

Figure 1.2 Semiconductor crystal structure: definition of the primitive cell.

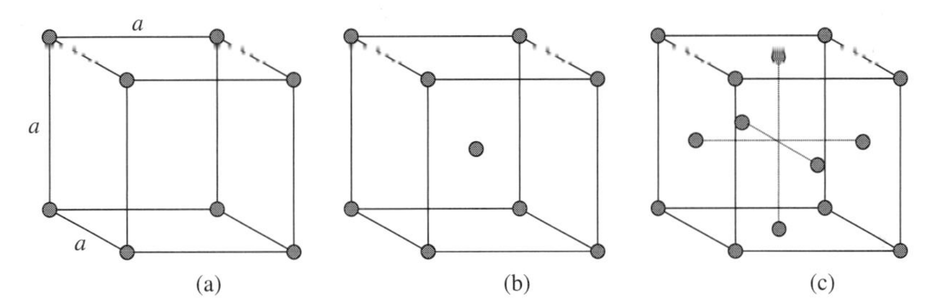

Figure 1.3 Cubic Bravais lattices: (a) simple, (b) body-centered, (c) face-centered.

fcc Bravais lattices, displaced along the body diagonal of the cubic cell by one-quarter the length of the diagonal, see Fig. 1.4. Since the length of the diagonal is $d = a\left|\hat{x} + \hat{y} + \hat{z}\right| = a\sqrt{3}$, the displacement of the second lattice is described by the vector

$$\underline{s} = \frac{a\sqrt{3}}{4} \frac{\hat{x} + \hat{y} + \hat{z}}{\sqrt{3}} = \frac{a}{4}\left(\hat{x} + \hat{y} + \hat{z}\right).$$

1.2.1 The Miller index notation

The Miller indices are a useful notation to denote planes and reference directions within a lattice. The notation (h, k, l), where h, k, l are integers, denotes the set of parallel planes that intercepts the three points $\underline{a}_1/h$, $\underline{a}_2/k$ and $\underline{a}_3/l$, or some multiple thereof, while $[h, k, l]$ in square brackets is the direction orthogonal to plane (h, k, l).

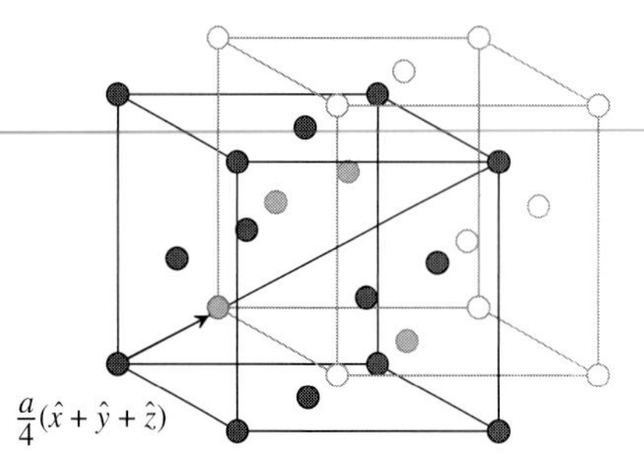

Figure 1.4 The diamond lattice as two cubic face-centered interpenetrating lattices. The pale and dark gray points represent the atoms falling in the basic cell.

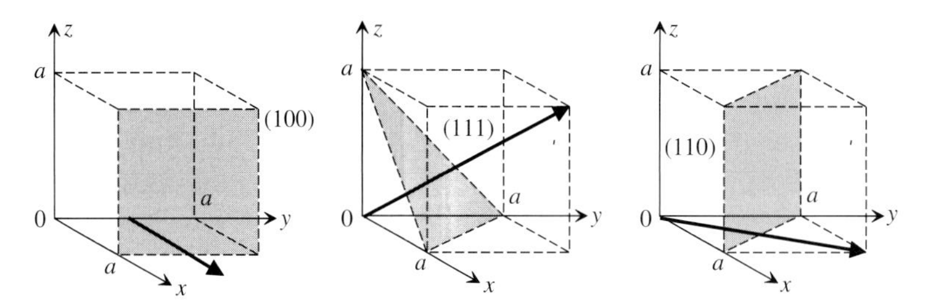

Figure 1.5 Examples of planes and directions according to the Miller notation.

Additionally, $\{h, k, l\}$ is a family of planes with symmetries and $\langle h, k, l\rangle$ is the related direction set. In cubic lattices, the primitive vectors coincide with the Cartesian axes and $a_1 = a_2 = a_3 = a$, where a is the lattice constant; in this case, we simply have $[h, k, l] \equiv h\widehat{x} + k\widehat{y} + l\widehat{z}$ where $\widehat{x}$, $\widehat{y}$ and $\widehat{z}$ are the Cartesian unit vectors.

To derive the Miller indices from the plane intercepts in a cubic lattice, we normalize with respect to the lattice constant (thus obtaining a set of integers (H, K, L)), take the reciprocal (H^{-1}, K^{-1}, L^{-1}) and finally multiply by a minimum common multiplier so as to obtain a set (h, k, l) such as $h : k : l = H^{-1} : K^{-1} : L^{-1}$. Notice that a zero index corresponds to an intercept point at infinity. Examples of important planes and directions are shown in Fig. 1.5.

Example 1.1: Identify the Miller indices of the following planes, intersecting the coordinate axes in points (normalized to the lattice constant): (a) $x = 4$, $y = 2$, $z = 1$; (b) $x = 10$, $y = 5$, $z = \infty$; (c) $x = 3.5$, $y = \infty$, $z = \infty$; (d) $x = -4$, $y = -2$, $z = 1$.

We take the reciprocal of the intercept, and then we multiply by the minimum common multiplier, so as to obtain an integer set with minimum module. In case (a), the reciprocal set is $(1/4, 1/2, 1)$, with minimum common multiplier 4, leading to the Miller indices $(1, 2, 4)$. In case (b), the reciprocals are $(1/10, 1/5, 0)$ with Miller indices $(1, 2, 0)$. In case (c), the plane is orthogonal to the z axis, and the Miller indices simply are $(1, 0, 0)$. Finally, case (d) is similar to case (a) but with negative intercepts; according to the Miller notation we overline the indices rather than using a minus sign; we thus have $(\overline{1}, \overline{2}, 4)$.

1.2.2 The diamond, zinc-blende, and wurtzite semiconductor cells

The cubic diamond cell includes 8 atoms; in fact, if we consider Fig. 1.6, the corner atoms each contribute to eight adjacent cells, so that only $8/8 = 1$ atom belongs to the main cell. The atoms lying on the faces belong half to the main cell, half to the nearby ones, so that only $6/2 = 3$ atoms belong to the main cell. Finally, the other (internal) 4 atoms belong entirely to the cell. Therefore, the total number of atoms in a cell is $1 + 3 + 4 = 8$. In the diamond cell, each atom is connected to the neighbours through a tetrahedral bond. All atoms are the same (C, Si, Ge...) in the diamond lattice, while in the so-called *zinc-blende lattice* the atoms in the two fcc constituent lattices are different (GaAs, InP, SiC...). In particular, the corner and face atoms are metals (e.g., Ga) and the internal atoms are nonmetals (e.g., As), or vice versa.

In the diamond or zinc-blende lattices the Miller indices are conventionally defined with respect to the cubic cell of side a. Due to the symmetry of the tetrahedral atom bonds, planes (100) and (110), etc. have two bonds per side, while planes (111) have three bonds on the one side, two on the other. Moreover, the surface atom density is different, leading, for example, to different etch velocities.

Some semiconductors, such as SiC and GaN, have the hexagonal *wurtzite* crystal structure. Hexagonal lattices admit many *polytypes* according to the stacking of successive atom layers; a large number of polytypes exists, but only a few have interesting semiconductor properties (e.g. 4H and 6H for SiC). The wurtzite cell is shown in Fig. 1.7, including 12 equivalent atoms. In the ideal lattice, one has

$$|\underline{a}_3| = c, \quad |\underline{a}_1| = |\underline{a}_2| = a, \quad \frac{c}{a} = \sqrt{\frac{8}{3}} \approx 1.633.$$

Some properties of semiconductor lattices are shown in Table 1.2.[1] It can be noted that wurtzite-based semiconductors are often anisotropic (uniaxial) and have two dielectric constants, one parallel to the c-axis, the other orthogonal to it.

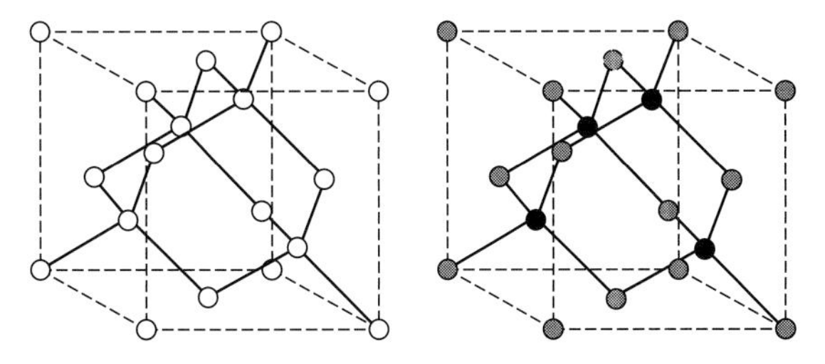

Figure 1.6 The diamond (left) and zinc-blende (right) lattices.

[1] Semiconductor properties are well documented in many textbooks; an excellent online resource is provided by the Ioffe Institute of the Russian Academy of Sciences at the web site [1].

Table 1.2 Properties of some semiconductor lattices: the crystal is D (diamond), ZB (zinc-blende) or W (wurtzite); the gap is D (direct) or I (indirect); $\epsilon_{\parallel}$ is along the c axis, $\epsilon_{\perp}$ is orthogonal to the c axis for wurtzite materials. Permittivities are static to RF. Properties are at 300 K.

Material	Crystal	E_g (eV)	D/I gap	ϵ_r or $\epsilon_{\parallel}$	$\epsilon_{\perp}$	a (Å)	c (Å)	Density, ρ (g/cm^3)
C	D	5.50	I	5.57		3.57		3.51
Si	D	1.12	I	11.9		5.43		2.33
SiC	ZB	2.42	I	9.72		4.36		3.17
Ge	D	0.66	I	16.2		5.66		5.32
GaAs	ZB	1.42	D	13.2		5.68		5.32
GaP	ZB	2.27	I	11.11		5.45		4.14
GaSb	ZB	0.75	D	15.7		6.09		5.61
InP	ZB	1.34	D	12.56		5.87		4.81
InAs	ZB	0.36	D	15.15		6.06		5.67
InSb	ZB	0.23	D	16.8		6.48		5.77
AlP	ZB	2.45	I	9.8		5.46		2.40
AlAs	ZB	2.17	I	10.06		5.66		3.76
AlSb	ZB	1.62	I	12.04		6.13		4.26
CdTe	ZB	1.47	D	10.2		6.48		5.87
GaN	W	3.44	D	10.4	9.5	3.17	5.16	6.09
AlN	W	6.20	D	9.14		3.11	4.98	3.25
InN	W	1.89	D	14.4	13.1	3.54	5.70	6.81
ZnO	W	3.44	D	8.75	7.8	3.25	5.21	5.67

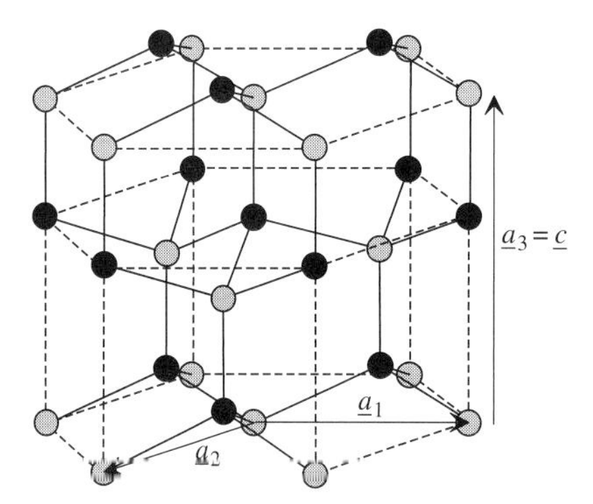

Figure 1.7 The hexagonal wurtzite cell. The c-axis corresponds to the direction of the $\underline{a}_3 = \underline{c}$ vector.

1.2.3 Ferroelectric crystals

Ferroelectric materials have a residual spontaneous dielectric polarization after the applied electric field has been switched off. The behavior of such materials is somewhat similar to that of ferromagnetic materials. Below a transition temperature, called the Curie temperature T_c, ferroelectric materials possess a spontaneous polarization or electric dipole moment. The magnitude of the spontaneous polarization is greatest at temperatures well below the Curie temperature, and approaches zero as the Curie

Table 1.3 Properties of some ferroelectric crystals. KDP stands for potassium dihydrogen phosphate. Data from [2], Ch. 13, Table 2.

Material class	Material	Curie temperature T_C (K)	Spontaneous polarization P_s (μC/cm^2)
KDP	KH_2PO_4	123	4.75
Perovskites	$BaTiO_3$	408	26
Perovskites	$LiNbO_3$	1480	71
Perovskites	$KNbO_3$	708	30

temperature is approached. Ferroelectric materials are inherently piezoelectric; that is, in response to an applied mechanical load, the material will produce an electric charge proportional to the load. Similarly, the material will produce a mechanical deformation in response to an applied voltage. In optoelectronics, ferroelectric materials are particularly important because of the excellent *electrooptic properties*, i.e., the strong variation of the material refractive index with an applied electric field. The crystal structure is often cubic face-centered, and the material is anisotropic and uniaxial. The most important ferroelectric crystal for optical applications is probably lithium niobate, $LiNbO_3$ (LN for short); some other materials (such as barium titanate) belonging to the so-called *perovskite* class are also sometimes used. The crystal structure of perovskites is face-centered cubic. Above the Curie temperature, the crystal is strictly cubic, and positive and negative ions are located in the cell so as to lead to zero dipole moment. Below the Curie temperature, however, a transition takes place whereby positive and negative ions undergo a shift in opposite directions; the crystal structure becomes tetragonal (i.e., the elementary cell height a_3 is different from the basis $a_1 = a_2$) and, due to the charge displacement, a net dipole moment arises. Table 1.3 shows a few properties of ferroelectric crystals, namely the spontaneous polarization P_s and the Curie temperature [2].

1.2.4 Crystal defects

In practice, the crystal lattice is affected by defects, either native (i.e., not involving external atoms) or related to nonnative impurities. Moreover, defects can be point defects (0D), line defects (1D), surface defects (2D), such as dislocations, and volume defects (3D), such as precipitates. Native point defects are *vacancies*, see Fig. 1.8, and *self-interstitials*, while *interstitials* are nonnative atoms placed in the empty space between the already existing lattice atoms. *Substitutional* defects involve an external atom, e.g., a dopant, which replaces one native atom. Typically, dopants act as donors or acceptors only if they are in a substitutional site; if they are in an interstitial site, they are inactive (chemically inactivated).[2]

[2] Dopants can also be electrically inactivated when they are not ionized.

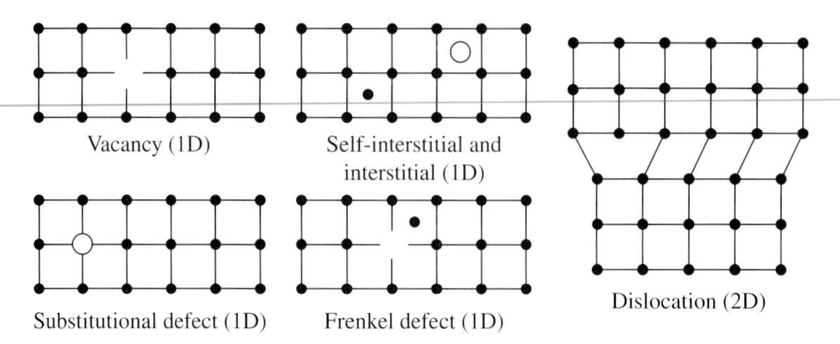

Figure 1.8 Point defects in a crystal (1D) and dislocations (2D).

1.3 Semiconductor electronic properties

1.3.1 The energy–momentum dispersion relation

A crystal is a periodic arrangement of atoms; since each positively charged nucleus induces a spherically symmetric Coulomb potential, superposition yields in total a periodic potential $U(\underline{r})$ such as

$$U(\underline{r}) = U(\underline{r} + \underline{L}),$$

where $\underline{L} = k\underline{a}_1 + l\underline{a}_2 + m\underline{a}_3$. In such a periodic potential, electrons follow the rules of quantum mechanics, i.e., they are described by a set of *wavefunctions* associated with allowed electron states. Allowed states correspond to allowed energy bands, which collapse into energy levels for isolated atoms; allowed bands are separated by forbidden bands. Low-energy electrons are bound to atoms, and only the two topmost allowed bands (the last, being almost full, is the *valence band*; the uppermost, almost empty, is the *conduction band*) take part in carrier transport. As already recalled, the vacuum level U_0 is the minimum energy of an electron free to move in and out of the crystal.

Electrons in a crystal are characterized by an energy–momentum relation $E(\underline{k})$, where the wavevector $\underline{k}$ is related to the electron momentum $\underline{p}$ as $\underline{p} = \hbar\underline{k}$. The *dispersion relation* $E(\underline{k})$ is defined in the $\underline{k}$ space, also called the reciprocal space; it is generally a multivalued function, periodic in the reciprocal space, whose fundamental period is called the *first Brillouin zone* (FBZ). A number of branches of the dispersion relation refer to the valence band, a number to the conduction band; the total number of branches depends on the crystal structure and is quite large (e.g., 12 for the conduction band and 8 for the valence band) in wurtzite semiconductors.

In cubic semiconductors, the FBZ is a solid with six square faces and eight hexagonal faces, as shown in Fig. 1.9. Owing to symmetries, only a portion of the FBZ, called the *irreducible wedge*, actually includes independent information; all the rest can be recovered by symmetry. Important points in the FBZ are the center (Γ point), the X point (center of the square face), and the L point (center of the hexagonal face).

The full details of the dispersion relation are not essential for understanding low-energy phenomena in semiconductors; attention can be restricted to the branches

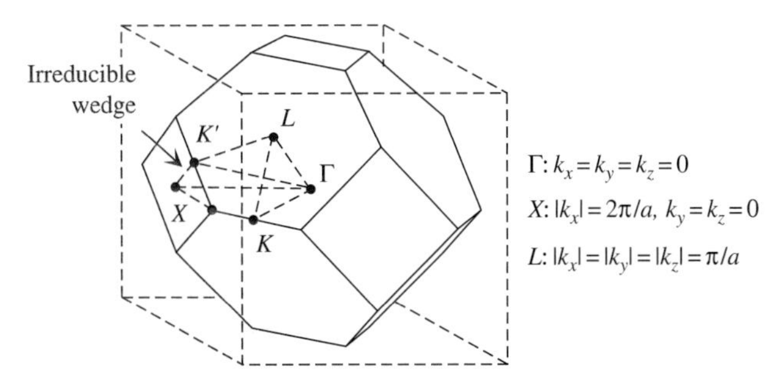

Figure 1.9 The first Brillouin zone (FBZ) in a cubic lattice (lattice constant a).

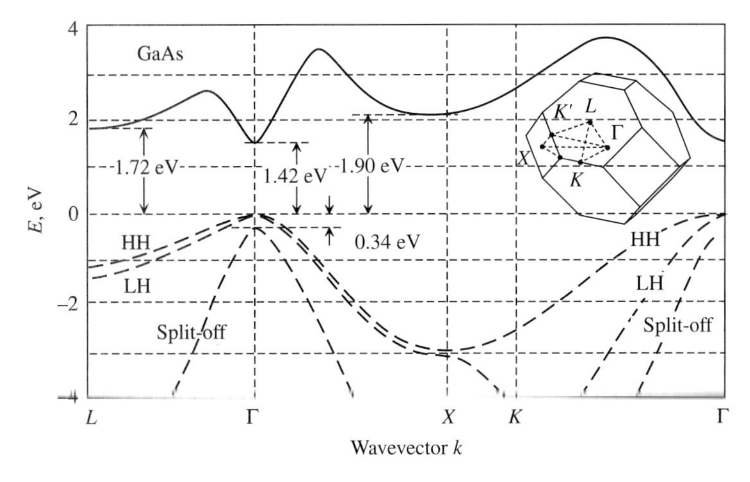

Figure 1.10 Simplified dispersion relation for GaAs.

describing low-energy electrons in the conduction band (around the *conduction band edge* E_c) and high-energy electrons (low-energy holes) in the valence band (around the *valence band edge* E_v). Valence band electrons are more efficiently described in terms of pseudoparticles (the *holes*) related to electrons missing from the valence band. Holes behave as particles with positive charge and potential energy opposite to the electron energy, so that the topmost branches of the dispersion relation (i.e., the branches describing low-energy holes) define the valence band edge.

As a relevant example, let us discuss the dispersion relation for a direct-bandgap semiconductor, GaAs. The term *direct bandgap* refers to the fact that the minimum of the conduction band and the maximum of the valence band (both located in the Γ point) correspond to the same momentum $\hbar \underline{k}$, in this case $\hbar \underline{k} = 0$. The dispersion relation shown in Fig. 1.10 is simplified, in the sense that only the lowest branch of the conduction band is shown, while three branches of the valence band appear, the *heavy hole* (HH), the *light hole* (LH), and the *split-off* band. Light and heavy hole bands are degenerate, i.e., they share the same minimum in the Γ point, and they differ because of the $E(\underline{k})$ curvature near the minimum, which corresponds to a larger or smaller hole effective mass. The split-off band enters some transport and optical processes but can be neglected in a first-order treatment. The conduction band has the lowest minimum at the Γ point, and two secondary minima at the L and X points. The main gap is 1.42 eV, while the secondary gaps are 1.72 eV (L point) and 1.90 eV (X point). Only a section of the dispersion relation is presented, running from the L point to the Γ point (the center of the FBZ), and then from the Γ point to the X point and back to the origin through the K point.

Since electrons and holes have, at least in the absence of an applied field, a Boltzmann energy distribution (i.e., their probability to have energy E is proportional to $\exp(-E/k_B T)$, where $k_B T = 26$ meV at ambient temperature), most electrons and holes can be found close to the conduction band and valence band edges, respectively.

Consider now the lowest minimum of the conduction band or highest maximum in the valence band; the dispersion relation can be approximated (around the Γ point) by a parabola as

$$E_n - E_c \approx \frac{\hbar^2 k^2}{2m_n^*}, \qquad E_v - E_h \approx \frac{\hbar^2 k^2}{2m_h^*},$$

where m_n^* and m_h^* are the electron and hole *effective masses*.[3] Therefore, the electron kinetic energy $E_n - E_c$ or hole kinetic energy $E_v - E_h$ (assuming the valence band edge energy E_v and the conduction band edge energy E_c to be the energy of a hole or of an electron, respectively, at rest) have, approximately, the same expression as the free-space particle kinetic energy, but with an effective mass m_n^* or m_h^* instead of the *in vacuo* inertial mass m_0. If the minimum is not located in the center of the first BZ (as for the conduction band of indirect bandgap semiconductors) the momentum (in a dynamic sense) can be defined "with respect to the minimum," so that the following approximation applies:

$$E_n - E_c \approx \frac{\hbar^2 \left|\underline{k} - \underline{k}_{\min}\right|^2}{2m_n^*}.$$

The effective mass can be evaluated from the inverse of the curvature of the dispersion relation around a minimum or a maximum. In general, the approximating surface can be expressed as

$$E_n - E_c = \frac{\hbar^2 k_a^2}{2m_{na}^*} + \frac{\hbar^2 k_b^2}{2m_{nb}^*} + \frac{\hbar^2 k_c^2}{2m_{nc}^*},$$

which is an ellipsoid; the coordinate system coincides with the principal axes. If the three effective masses are equal, the ellipsoid degenerates into a spherical surface, and we say that the minimum is *spherical*, with *isotropic* effective mass. This typically happens at Γ point minima. In indirect-bandgap semiconductors, the constant-energy

[3] Corrections to the parabolic approximation accounting for nonparaboliticity effects can be introduced (e.g., in the conduction band) through the expression:

$$E_k (1 + \alpha E_k) = \frac{\hbar^2 k^2}{2m_n^*},$$

where E_k is the electron kinetic energy $E_n - E_c$ and α is a nonparabolicity correction factor.

surfaces are rotation ellipsoids, and we can define two effective masses, one transversal m^*_{nt} (common to two principal directions) and one longitudinal m^*_{nl} (along the third principal direction). The electron effective mass increases with E_g, according to the fitting law (see (2.9)):

$$\frac{m^*_n}{m_0} \approx \frac{E_g\big|_{\text{eV}}}{13}.$$

Due to degeneracy, the valence bands have a more complex behavior near the valence band edge, but can anyway be approximated with isotropic masses; however, since the heavy and light hole populations mix, a properly averaged effective mass has to be introduced; the same remark applies for electrons with anisotropic effective mass. The averaging law is related to the application, and is not unique; we can therefore have an effective mass for transport and also (as discussed later) an effective mass for the density of states that follow different averaging criteria. Concerning the density of states mass (denoted with the subscript D), we have for the electrons

$$m^*_{n,D} \triangleq \left(m^*_{na} m^*_{nb} m^*_{nc}\right)^{1/3} M_c^{2/3}.$$

The above expression refers to the general case of ellipsoidal minima with multiplicity M_c (more than one minimum in the FBZ); for a Γ point spherical minimum in the conduction band we have simply

$$m^*_{n,D} = m^*_n,$$

while for the rotation ellipsoid case in Si (where 6 equivalent minima are present in the FBZ) we obtain

$$m^*_{n,D} \triangleq 6^{2/3} (m^*_{nl})^{1/3} (m^*_{nt})^{2/3}.$$

For holes, in the case of degeneracy:

$$m^*_{h,D} \triangleq \left[(m^*_{hh})^{3/2} + (m^*_{lh})^{3/2}\right]^{2/3},$$

while of course $m^*_{h,D}$ reduces to m^*_{hh} or m^*_{lh} if degeneracy is removed (as in a strained quantum well, see Section 1.7). Concerning the effective masses for transport, since on average the electron moves along all three principal directions with the same probability, we have that the transport or conductivity average electron mass is given by

$$\frac{1}{m^*_{n,tr}} = \frac{1}{3m^*_{na}} + \frac{1}{3m_{nb}} + \frac{1}{3m_{nc}},$$

which reduces, for Si, to

$$\frac{1}{m^*_{n,tr}} = \frac{2}{3m^*_{nt}} + \frac{1}{3m_{nl}}.$$

In a spherical minimum (isotropic effective mass) we finally have

$$m^*_{n,tr} = m^*_n.$$

For holes, the situation is more complex, since heavy and light holes exist. It can be shown that the transport hole effective mass is given by a weighted average over the heavy and light holes as (see e.g., [3], Section 8.1.2)

$$\frac{1}{m^*_{h,tr}} = \frac{p_{hh}}{pm^*_{hh}} + \frac{p_{lh}}{pm^*_{lh}},$$

where p_{lh} and p_{hh} are the light and heavy hole densities and $p = p_{lh} + p_{hh}$ is the total hole density. At or near equilibrium, the HH and LH populations are related through the effective densities of states, so that

$$\frac{p_{hh}}{p} = \frac{m^{*\,3/2}_{hh}}{m^{*\,3/2}_{lh} + m^{*\,3/2}_{hh}}, \quad \frac{p_{lh}}{p} = \frac{m^{*\,3/2}_{lh}}{m^{*\,3/2}_{lh} + m^{*\,3/2}_{hh}};$$

it follows that

$$\frac{1}{m^*_{h,tr}} = \frac{m^{*\,1/2}_{hh} + m^{*\,1/2}_{lh}}{m^{*\,3/2}_{hh} + m^{*\,3/2}_{lh}}.$$

For instance, in Si we have $m^*_{hh} = 0.49\, m_0$, $m^*_{lh} = 0.16\, m_0$; thus:

$$\frac{m_0}{m^*_{h,tr}} = \frac{0.49^{1/2} + 0.16^{1/2}}{0.16^{3/2} + 0.49^{3/2}} \rightarrow m^*_{h,tr} = 0.37 m_0.$$

1.3.2 The conduction and valence band wavefunctions

Electrons and holes belonging to the conduction and valence bands are characterized, from the standpoint of quantum mechanics, by a wavefunction. According to the Bloch theorem, wavefunctions in a periodic potential (e.g., a crystal) can be generally expressed as

$$\psi_{\underline{k}}(\underline{r}) = \exp(-\mathrm{j}\underline{k} \cdot \underline{r}) u_{\underline{k}}(\underline{r}), \tag{1.1}$$

where $u_{\underline{k}}(\underline{r})$ is a periodic function in the crystal space, such as $u_{\underline{k}}(\underline{r}) = u_{\underline{k}}(\underline{r} + \underline{L})$, $\underline{L}$ being a linear combination (with integer indices) of the primitive lattice vectors. The functional form of the wavefunction in (1.1), called the *Bloch wave*, ensures that the probability associated with the wavefunction is indeed a periodic function in the crystal space. For $\underline{k} \approx 0$ (e.g., near the Γ point) one has $\psi_{\underline{k}}(\underline{r}) \approx u_0(\underline{r})$, where $u_0(\underline{r})$ follows single-atom-like wavefunctions (s-type or p-type, see Fig. 1.11).

Since the detailed spatial behavior of wavefunctions is relevant to optical properties, we recall that *conduction band wavefunctions are, near the* Γ *point, s-type*, i.e., they have a probability distribution with spherical constant-probability surfaces. On the other hand, *the valence band wavefunctions are p-type*, i.e., they are even with respect to two orthogonal directions and odd with respect to the third, see Fig. 1.11. For instance, p_x is even with respect to the y and z axes and odd with respect to the x axis. The detailed shape of the wavefunctions is much less important than their property of being even in all directions (the s-type wavefunction) or odd with respect to one direction.

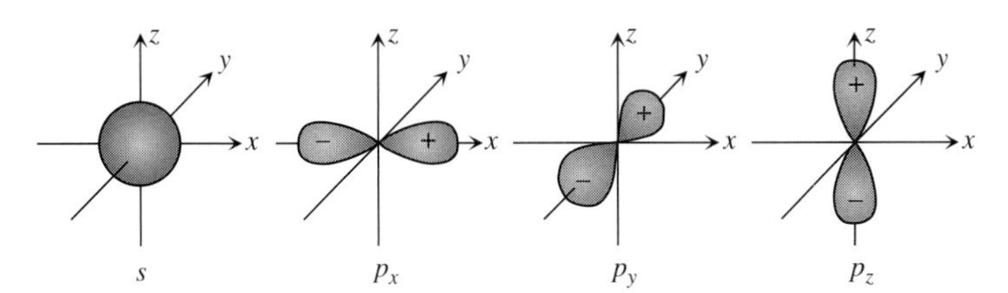

Figure 1.11 Conduction band (s-type) and valence band (p-type) wavefunctions: probability distribution and wavefunction sign (for p-type).

More specifically, it can be shown that heavy and light hole wavefunctions result from a superposition of p-type wavefunctions:

$$\phi_{HH}(x, y, z) = -\frac{1}{\sqrt{2}}\left(p_x \pm \mathrm{j}p_y\right) \tag{1.2}$$

$$\phi_{LH}(x, y, z) = -\frac{1}{\sqrt{6}}\left(p_x \pm \mathrm{j}p_y \mp 2p_z\right), \tag{1.3}$$

where the prefactors are introduced for normalization, see e.g., [4], Section 2.4.

1.3.3 Direct- and indirect-bandgap semiconductors

A simplified version of the dispersion relation, including the main conduction band minima and valence band maxima, is often enough to explain the electronic and optical behavior of a semiconductor. Such an example is shown in Fig. 1.12(a), for GaAs: the coincident maxima and minima in the Γ point make this semiconductor a typical example of *direct-bandgap* material. Direct-bandgap semiconductors are particularly important in optics, because they are able to interact directly with photons; in fact, those can provide an energy of the order of the energy gap, but negligible momentum. To promote an electron from the valence to the conduction band, an energy larger than the gap has to be provided, but, in GaAs, negligible momentum, since the valence band maximum and conduction band minimum are both at $\underline{k} = 0$. Since the interaction involves only one electron and one photon, the interaction probability is high.

Silicon, the most important semiconductor in electronics, is an example of an indirect-bandgap semiconductor, i.e., a material in which the valence band maximum and conduction band minimum occur at different values of $\underline{k}$, see Fig. 1.12(b). In particular, the main conduction band minimum is close to point X but within the FBZ, and six minima exist in the FBZ. The electron energy around such minima can be expressed as a function of the transverse (e.g., orthogonal to (100)) and of the longitudinal (e.g., parallel to (100)) wavenumbers:

$$E_n \approx E_c + \frac{\hbar^2 k_t^2}{2m_{nt}^*} + \frac{\hbar^2 k_l^2}{2m_{nl}^*}.$$

In Si, the electron–photon interaction leading to band-to-band processes requires a substantial amount of momentum, which has to be supplied by a further particle, typically a *lattice vibration* (phonon). The multibody nature of the interaction makes it less

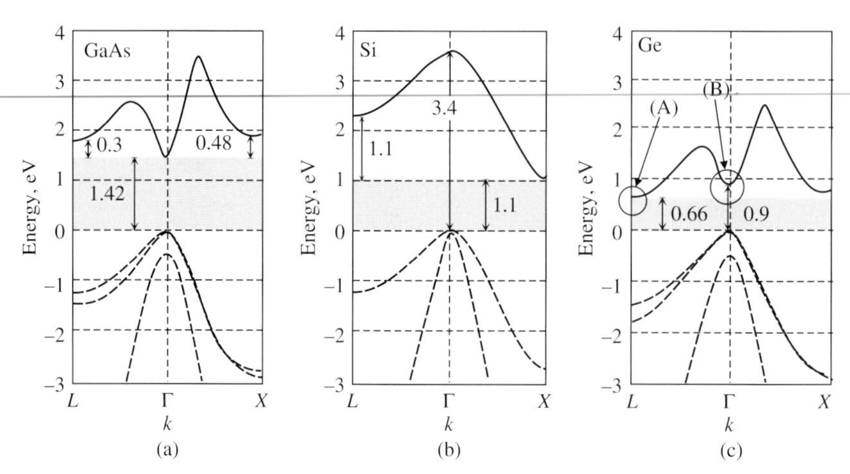

Figure 1.12 Simplified dispersion relation for (a) GaAs, (b) Si, (c) Ge. In Ge, the main conduction band minimum (A) is indirect, and has an impact on transport and low-energy optical properties; the secondary direct minimum (B) influences the optical properties at high photon energy.

probable, and therefore the interaction strength is lower. Typically, direct-bandgap semiconductors are able to absorb and emit light; indirect-bandgap semiconductors absorb light (albeit less efficiently) but are unable to operate as high-efficiency light emitters, particularly in lasers. Germanium (Ge), see Fig. 1.12(c), is an indirect-bandgap semiconductor; the lowest conduction band minimum is at point L, but a direct bandgap exists with a higher energy (0.9 eV) at point Γ. As a result, the main transport properties of Ge are typical of an indirect-bandgap material, but optical properties are influenced by the fact that high-energy photons can excite electrons directly from the valence band to the direct minimum. Some of germanium's optical properties (e.g., the absorption) exhibit both indirect- and direct-bandgap semiconductor features, depending on the photon energy.

In the above materials, the central minima can be characterized by isotropic or quasi-isotropic (as for the valence band) effective masses, while indirect bandgap minima are typically anisotropic and have to be described in terms of a longitudinal and transverse effective mass. A summary of the effective masses and other band properties in Si and GaAs is shown in Table 1.4.

Many III-V semiconductors have a bandstructure similar to GaAs. InP, see Fig. 1.13(a), has a slightly lower bandgap, but a larger difference between the central and the lateral minima. This has important consequences on transport properties, since it increases the electric field at which the electrons are scattered from the central minimum (characterized by high mobility, i.e., high electron velocity with the same applied electric field) to the lateral minima (with low mobility). This ultimately leads to a decrease of the average electron velocity with increasing field, see Fig. 1.14. The maximum velocity is larger in InP than in GaAs, allowing for the development of electron devices (such as transistors) with superior properties in terms of maximum speed. The peak electron velocity (corresponding to the onset of the negative differential mobility region) occurs at a field $\mathcal{E}_m$ related to the energy difference ΔE between the Γ and

Table 1.4 Main band properties of Si and GaAs. The electron mass m_0 is 9.11×10^{-34} kg.

Property	Si	GaAs
Electron effective masses	$m^*_{nl} = 0.98 m_0$	$m^*_n = 0.067 m_0$
	$m^*_{nt} = 0.19 m_0$	
	$m^*_{n,D} = 1.08 m_0$	$m^*_{n,D} = 0.067 m_0$
	$m^*_{n,tr} = 0.26 m_0$	$m^*_{n,tr} = 0.067 m_0$
Hole effective masses	$m^*_{hh} = 0.49 m_0$	$m^*_{hh} = 0.45 m_0$
	$m^*_{lh} = 0.16 m_0$	$m^*_{lh} = 0.08 m_0$
	$m^*_{h,D} = 0.55 m_0$	$m^*_{h,D} = 0.47 m_0$
	$m^*_{h,tr} = 0.37 m_0$	$m^*_{h,tr} = 0.34 m_0$
Energy gap $E_g(T)$, T (K)	$1.17 - \dfrac{4.37 \times 10^{-4} T^2}{636 + T}$	$1.52 - \dfrac{5.4 \times 10^{-4} T^2}{204 + T}$
Electron affinity $q\chi$ (eV)	4.01	4.07

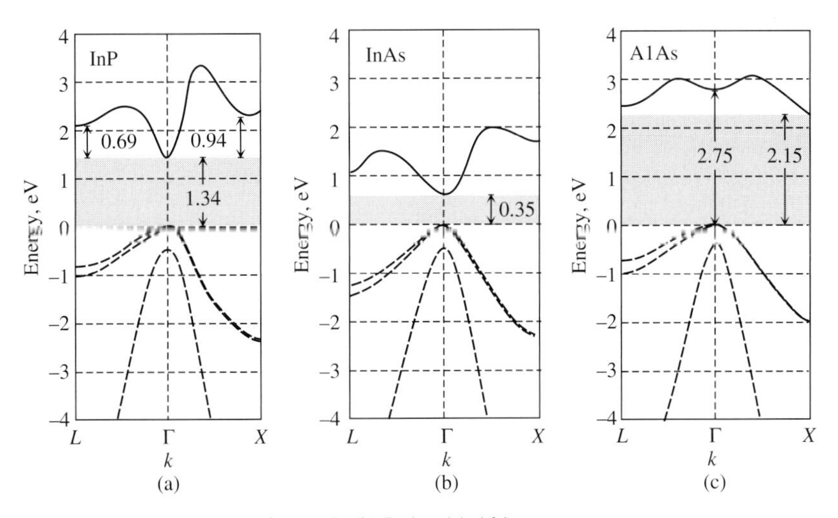

Figure 1.13 Simplified bandstructure of (a) InP, (b) InAs, (c) AlAs.

lateral minima; in GaAs $\Delta E \approx 300$ meV with $\mathcal{E}_m \approx 3.2$ kV/cm while in InP $\Delta E \approx 700$ meV with $\mathcal{E}_m \approx 10$ kV/cm.[4] InAs, see Fig. 1.13(b), has a very similar bandstructure, but with lower energy gap. For certain compound semiconductors, such as AlAs, see Fig. 1.13(c), the central minimum is higher than the lateral minima, thus making the material of indirect-bandgap type. InAs and AlAs are not particularly important per se, but rather as the components of semiconductor alloys. Some additional compound semiconductor properties are listed in Table 1.5, where v_s is the electron high-field saturation velocity (also denoted as $v_{n,\mathrm{sat}}$), $v_{\max}$ is the maximum steady-state electron velocity. Notice that, while the saturation velocity is almost independent of doping, the maximum in the nonmonotonic velocity–field curve of most compound semiconductors

[4] In GaN, on the other hand, $\Delta E \approx 3$ eV, leading to a peak field in excess of 200 kV/cm, see, e.g., [5].

Table 1.5 Band properties of some important compound semiconductors. Mobility data are upper bounds referring to undoped material.

Property	$In_{0.53}Ga_{0.47}As$	GaAs	InP	AlAs	InAs
a (Å)	5.869	5.683	5.869	5.661	6.0584
E_g @300 K (eV)	0.717	1.424	1.34	2.168	0.36
$q\chi$ (eV)		4.07	4.37	3.50	4.90
m_n^*/m_0	0.041	0.067	0.077	0.150	0.027
m_{lh}^*/m_0	0.044	0.08	0.12	0.150	0.023
m_{hh}^*/m_0	0.452	0.45	0.6	0.76	0.60
$\epsilon(0)/\epsilon_0$	13.77	13.18	12.35	10.16	14.6
$\epsilon(\infty)/\epsilon_0$	11.38	10.9	9.52	8.16	12.25
$\mathcal{E}_{br}$ (kV/cm)	3.0	3.2	11		
μ_n (cm^2/Vs)	12000	8500	5500		40000
$v_{\max}$ (10^7 cm/s)	≈ 2.5	≈ 1.7	≈ 2.7		
v_s (10^7 cm/s)	0.7	1			

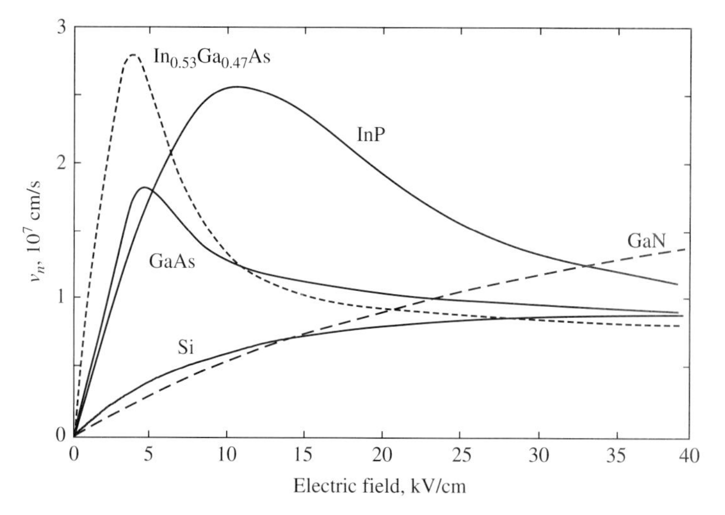

Figure 1.14 Electron drift velocity–field curves of Si, GaAs, InP, GaN, and InGaAs lattice matched to InP. The GaN velocity has a peak toward 200 kV/cm and then saturates with GaAs-like behavior. Adapted from [6], p. 13.

(see Fig. 1.14) depends on the low-field mobility and therefore on doping; the values provided (referring to intrinsic material) are therefore indicative.

Compound semiconductor families are classified according to the chemical nature of the metal and nonmetal components. If the metal component belongs to group III and the nonmetal to group V, we obtain a III-V compound. Examples of III-V compounds are GaAs, InP, GaSb, InAs (direct bandgap) and AlAs, GaP (indirect bandgap). III-V compounds with nitrogen such as GaN, InN, AlN are often referred to as III-N compounds. III-V compounds are probably the most important semiconductors for high-frequency electronics and optoelectronics. II-VI compounds include CdTe, HgTe,

ZnS, CdSe, ZnO (direct bandgap; note that HgTe has a negative bandgap and therefore has a metal rather than semiconductor behavior). IV-IV compounds are SiC and SiGe, both of them of indirect bandgap type. Finally, I-VII semiconductor compounds also exist, such as AgI and CuBr.

1.4 Carrier densities in a semiconductor

1.4.1 Equilibrium electron and hole densities

According to the picture drawn so far, a simplified representation of the semiconductor bandstructure includes two energy bands, the valence and conduction bands, separated by the energy gap E_g. Some electrons have large enough energy to be promoted from the valence to the conduction band, leaving behind positive charges called holes. Both electrons and holes can interact with an external electric field, and with photons or other particles. Further details of the bandstructure are introduced in Fig. 1.1, such as the electron affinity $q\chi$, i.e., the distance between the conduction band edge and the vacuum level U_0, and the ionization I_0, i.e., the distance between the valence band edge and the vacuum level. The electron and hole populations n and p depend on the number of electron and hole states per unit volume in the two bands (density of states N_c and N_v, respectively, both functions of the energy), and on how those states are populated as a function of the energy. According to statistical mechanics, electrons and holes follow at equilibrium the *Fermi–Dirac* distribution,[5] while the out-of-equilibrium distribution can be often approximated, in optoelectronic devices, by the so-called *quasi-Fermi* distribution.

In the effective mass approximation, the density of states (DOS) in a 3D (bulk) semiconductor can be shown to be

$$N_c(E) \equiv g_c(E) = \frac{4\pi}{h^3}(2m^*_{n,D})^{3/2}\sqrt{E - E_c}$$

$$N_v(E) \equiv g_v(E) = \frac{4\pi}{h^3}(2m^*_{h,D})^{3/2}\sqrt{E_v - E},$$

whose behavior is shown in Fig. 1.15. Owing to the effect of heavy holes, the valence band DOS typically is larger than the conduction band DOS.

The Fermi–Dirac distributions describing the electron and hole equilibrium occupation statistics are expressed as

$$f_n(E) = \frac{1}{1 + \exp\left(\dfrac{E - E_F}{k_B T}\right)} \quad (1.4)$$

$$f_h(E) = \frac{1}{1 + \exp\left(\dfrac{E_F - E}{k_B T}\right)}, \quad (1.5)$$

[5] Or by the *Boltzmann* distribution, an approximation of the Fermi–Dirac distribution holding for energies larger than the Fermi energy.

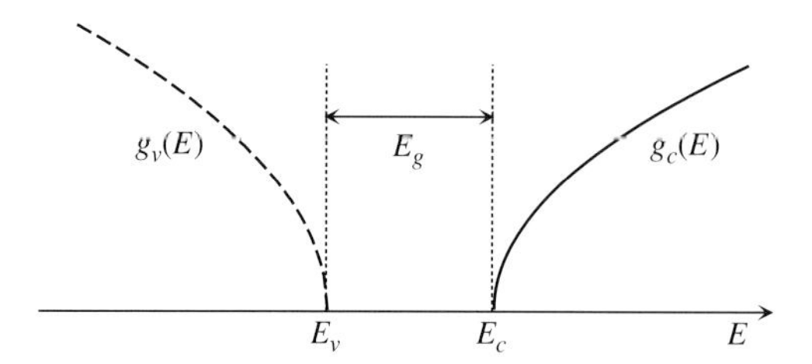

Figure 1.15 Valence (g_v) and conduction band (g_c) density of states in a bulk semiconductor.

where the Fermi level E_F is constant in the whole system. If the Fermi level is within the energy gap (this case corresponds to *nondegenerate semiconductors*) the Boltzmann approximation of the statistics holds:

$$f_n(E) \underset{E \gg E_F}{\approx} \exp\left(\frac{E_F - E}{k_B T}\right), \quad f_h(E) \underset{E \ll E_F}{\approx} \exp\left(\frac{E - E_F}{k_B T}\right).$$

The Boltzmann approximation applies, in fact, if the distance between E and E_F is larger than a few $k_B T$ units. In the *degenerate* case the Fermi level can fall into the conduction or valence bands, and this condition is violated; in such cases, the full Fermi–Dirac statistics has to be used.

The behavior of the two Fermi–Dirac distributions for electrons and holes is shown in Fig. 1.16. Integrating the product between the density of states and the statistical distributions (with the Boltzmann approximation) over all energies (i.e., from E_c to $\approx \infty$ for the conduction band and from $\approx -\infty$ to E_v for the valence band), we have

$$n = \int_{E_c}^{\infty} N_c(E) f_n(E)\, \mathrm{d}E = N_c \exp\left(\frac{E_F - E_c}{k_B T}\right)$$
$$p = \int_{-\infty}^{E_v} N_v(E) f_h(E)\, \mathrm{d}E = N_v \exp\left(\frac{E_v - E_F}{k_B T}\right),$$

where the *effective densities of states* are

$$N_c = 2\frac{(2\pi m^*_{n,D} k_B T)^{3/2}}{h^3}, \quad N_v = 2\frac{(2\pi m^*_{h,D} k_B T)^{3/2}}{h^3}. \tag{1.6}$$

In an *intrinsic* (undoped) semiconductor $p = n = n_i$, where

$$n_i = N_c \exp\left(\frac{E_{Fi} - E_c}{k_B T}\right) = p_i = N_v \exp\left(\frac{E_v - E_{Fi}}{k_B T}\right),$$

from which the intrinsic Fermi level can be derived; the intrinsic Fermi level is located at midgap, with a small (typically negative) correction related to the ratio $N_c/N_v = (m^*_{n,D}/m^*_{h,D})^{3/2}$:

$$E_{Fi} = k_B T \log\sqrt{\frac{N_c}{N_v}} + \frac{E_c + E_v}{2}.$$

Moreover, the intrinsic concentration can be directly related to the energy gap:

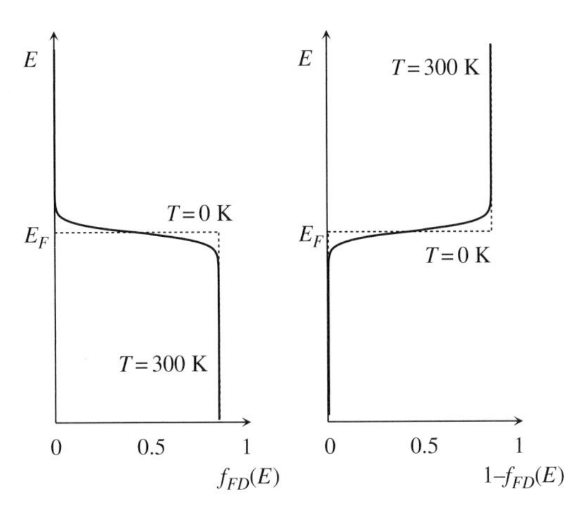

Figure 1.16 Fermi–Dirac distributions for electrons (left) and holes (right).

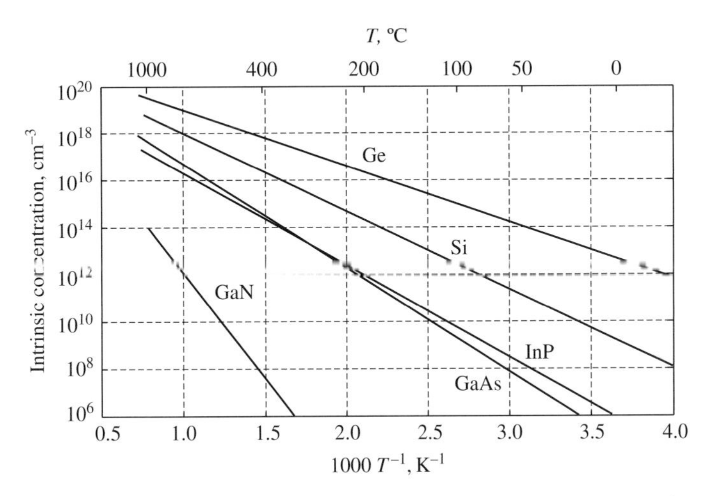

Figure 1.17 Intrinsic concentration for Ge, Si, GaAs, InP and GaN (wurtzite) as a function of the lattice temperature. Data from [1].

$$n_i p_i = n_i^2 = N_c N_v \exp\left(-\frac{E_g}{k_B T}\right). \tag{1.7}$$

The intrinsic concentration as a function of the temperature for Si, Ge, GaAs, InP and GaN is shown in Fig. 1.17. With increasing T, the intrinsic concentration increases exponentially; this is one of the main limitations in high-temperature semiconductor operation, since when the intrinsic concentration is of the order of the doping, doping becomes ineffective.

In equilibrium conditions, the product of the concentrations n and p does not depend on the position of the Fermi level, and is equal to the square of the intrinsic concentration (*mass action law*):

$$np = n_i^2. \tag{1.8}$$

1.4.2 Electron and hole densities in doped semiconductors

The mass action law also holds for doped semiconductors. A semiconductor can be doped with a donor (density N_D), an element able to provide an additional electron when substituting an atom of the native semiconductor lattice. Examples of donors in Si are As and P (both belonging to group V, and therefore with an extra electron in the outer shell vs. Si). The additional electron is weakly bound to the donor (ionization energy into the conduction band of the order of 10 meV for shallow donors) and therefore can easily be ionized and enter the conduction band, thus participating in conduction.[6] In this case, the semiconductor is called n-type. Semiconductors can also be doped with acceptors (concentration N_A). For instance, Si atoms have 4 electrons in the outermost shell; acceptors (e.g., B, a group III element) have 3 electrons in the outermost shell (i.e., one electron less than the substituted native atom) and can therefore attract an electron from the valence band, leaving behind a hole (again with a ionization energy of the order of 10 meV). The semiconductor in this case is called p-type.

If donors and acceptors are fully ionized one has, also taking into account the mass action law (1.8):

$$n \approx N_D^+ \approx N_D, \qquad p \approx n_i^2/N_D \quad n\text{-type semiconductor}$$
$$p \approx N_A^- \approx N_A, \qquad n \approx n_i^2/N_A \quad p\text{-type semiconductor.}$$

In a doped semiconductor, the carrier concentration evolves with temperature according to a three-region behavior; the relevant intervals are the freeze-out, the saturation, and the intrinsic range.

At extremely low temperature, most carriers do not have enough energy to ionize into the conduction band, and the carrier population decreases with T well below the value $n \approx N_D$ (*freeze-out range*). The intermediate range (called the *saturation range*), corresponding to normal device operation, begins at a temperature such as $(3/2)k_B T \approx 20$ meV, i.e., $T \approx 150$ K (this is just an indicative value, since the donor or acceptor ionization energy depends on the doping and semiconductor materials), and ends at a temperature such as $n_i(T) \approx N_D$ (in n-type Si with $N_D = 10^{15}$ cm^{-3} this corresponds to $T \approx 200^\circ$C). In the saturation range, $n \approx N_D$ or $p \approx N_A$; the maximum operating temperature increases with increasing gap. Finally, above the saturation range we find the *intrinsic range*: at high temperature the intrinsic concentration becomes large enough to flood the semiconductor with electrons not originating from the donors (or holes not originating from the acceptors).

[6] A donor or acceptor introduces an isolated energy level in the forbidden band. Shallow donors have an energy level E_D close to the conduction band edge (typically a few meV), while for shallow acceptors the energy level E_A is close to the valence band. Deep donors and acceptors have energy levels close to the center of the gap and act more as electron or hole *traps* (or *recombination centers*) than as dopants, since their ionization (or electrical activation) is low. Ionized dopants follow electron- or hole-like Fermi statistics: donors are almost 100% activated if the Fermi level is *below* the donor level, while acceptors are almost 100% activated if the Fermi level is *above* the acceptor level. This implies, for example, that a deep donor is not ionized in an n-type semiconductor, and even the activation of shallow donors ultimately drops for extremely large n-type doping, since for increasing donor concentration the Fermi level finally becomes larger than the donor level.

From the expressions for the electron and hole densities, the Fermi level can easily be evaluated. In n-type semiconductors, the Fermi level increases vs. E_{Fi}, becoming closer to the conduction band edge, while for p-type semiconductors the Fermi level decreases and becomes closer to the valence band edge. For very high doping (e.g., in excess of 10^{19} cm^{-3}), donors and acceptors cannot be assumed to be 100% ionized (or electrically activated) any longer, but their ionization is related to the very position of the Fermi level and typically decreases, as already remarked, when the Fermi level becomes larger than the donor or smaller than the acceptor energy level.

In a degenerate semiconductor, the Fermi level (or the quasi-Fermi level out of equilibrium) is very close to the conduction or valence band edges or even falls within one of the two bands. Typically, a semiconductor cannot be made degenerate by doping, but degeneracy is a condition that can be achieved out of equilibrium (e.g., in a direct-bias *pn* junction under high carrier injection).

1.4.3 Nonequilibrium electron and hole densities

To address the out-of-equilibrium statistics in a simplified way, we note that deviations from thermodynamic equilibrium may imply two quite different consequences: disequilibrium between the electron and the hole populations, and disequilibrium in carrier populations due to an applied (electric) field.

In equilibrium, the electron and hole populations follow the mass action law, any deviation from this being compensated for by generation–recombination (GR) processes whereby electron–hole (e-h) pairs are generated or disappear by recombination. The excess charge n' or p' (with respect to equilibrium) is removed according to the time behavior

$$n'(t) = n'(0) \exp(-t/\tau_n),$$

with a characteristic time (called the excess lifetime, τ_n or τ_h for electrons and holes, respectively) whose order of magnitude can range from a few milliseconds to nanoseconds according to the restoring mechanism. Recombination processes basically involve an exchange of energy and momentum with other particles, e.g., phonons (lattice vibrations, corresponding to the so-called thermal GR process), photons (radiative GR), other electrons and holes (Auger recombination and impact generation). If the carrier population deviation with respect to equilibrium is *maintained by an external cause* (e.g., a photon flux leading to radiative generation of e-h pairs) the resulting out-of-equilibrium condition can be characterized by a slightly modified form of the equilibrium probability distribution (called the *quasi-Fermi distribution*).

A second nonequilibrium situation derives from the effect of an applied electric field. While the average carrier velocity is zero at equilibrium, and therefore the *carrier distribution* in the velocity space is symmetrical with respect to the origin, application of an electric field leads to an increase of the average velocity and to a nonsymmetrical velocity distribution. For very large fields, the change in shape of the distribution with respect to the equilibrium may become dramatic and a simple quasi-Fermi approach will not be sufficient. However, this form of extreme field–carrier disequilibrium is not essential in the analysis of most optoelectronic devices, and therefore a simplified discussion based on the static carrier velocity–field properties will suffice.

To describe electron–hole imbalance with respect to the equilibrium, we therefore introduce the so called quasi-Fermi statistics, where the single Fermi level is replaced by two separate *quasi-Fermi levels* E_{Fn} and E_{Fh} according to the following formulae:

$$f_n(E, E_{Fn}) = \frac{1}{1 + \exp\left(\dfrac{E - E_{Fn}}{k_B T}\right)} \underset{E \gg E_{Fn}}{\approx} \exp\left(\frac{E_{Fn} - E}{k_B T}\right) \tag{1.9}$$

$$f_h(E, E_{Fh}) = \frac{1}{1 + \exp\left(\dfrac{E_{Fh} - E}{k_B T}\right)} \underset{E \ll E_{Fh}}{\approx} \exp\left(\frac{E - E_{Fh}}{k_B T}\right), \tag{1.10}$$

where the relevant Boltzmann approximations have also been introduced. Within the Boltzmann approximation the carrier densities become

$$n = N_c \exp\left(\frac{E_{Fn} - E_c}{k_B T}\right), \quad p = N_v \exp\left(\frac{E_v - E_{Fh}}{k_B T}\right),$$

while the mass action law can be modified to allow for a difference in the two quasi-Fermi levels (in equilibrium $E_{Fn} = E_{Fh} = E_F$):

$$np = n_i^2 \exp\left(\frac{E_{Fn} - E_{Fh}}{k_B T}\right). \tag{1.11}$$

In particular,

$$np > n_i^2 \quad \text{for } E_{Fn} > E_{Fh} \text{ (carrier injection)}$$
$$np < n_i^2 \quad \text{for } E_{Fn} < E_{Fh} \text{ (carrier depletion).}$$

In the degenerate case, the Boltzmann approximation is invalid and we have to express the charge density with the help of special functions (the Fermi–Dirac integrals):

$$n = \frac{2}{\sqrt{\pi}} N_c \mathcal{F}_{1/2}\left(\frac{E_{Fn} - E_c}{k_B T}\right), \quad p = \frac{2}{\sqrt{\pi}} N_v \mathcal{F}_{1/2}\left(\frac{E_v - E_{Fh}}{k_B T}\right).$$

The computation of the Fermi–Dirac integral can be performed through suitable analytical approximations; an example is given by the Joyce–Dixon (inverse) formulae:

$$E_{Fn} \approx E_c + k_B T \left[\log \frac{n}{N_c} + \frac{1}{\sqrt{8}} \frac{n}{N_c}\right] \tag{1.12}$$

$$E_{Fh} \approx E_v - k_B T \left[\log \frac{p}{N_v} + \frac{1}{\sqrt{8}} \frac{p}{N_v}\right]. \tag{1.13}$$

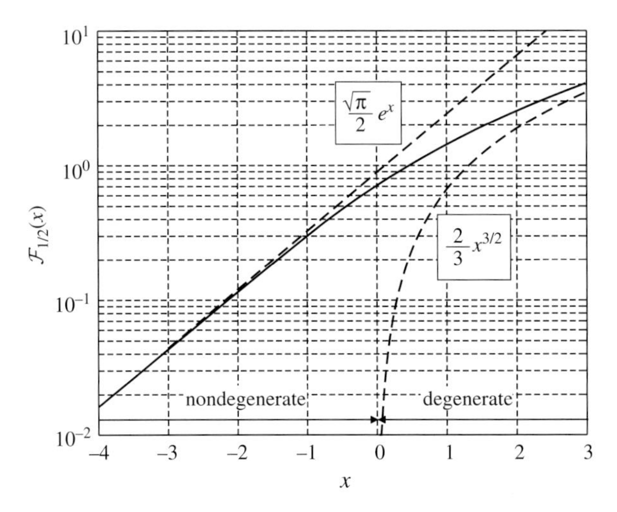

Figure 1.18 Behavior of the Fermi–Dirac integral ($\mathcal{F}_{1/2}$) in the degenerate and nondegenerate ranges.

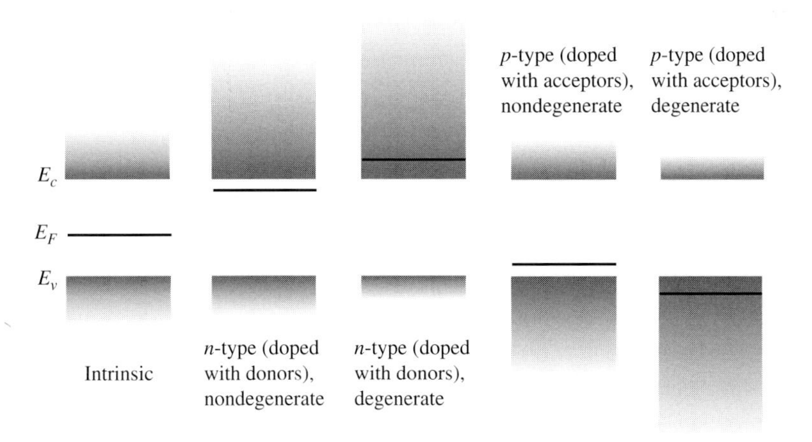

Figure 1.19 Examples of the position of the Fermi level in several semiconductors at equilibrium. In practice, semiconductors cannot be made degenerate by doping.

The overall behavior of the Fermi integral in the two ranges (nondegenerate and degenerate) is shown in Fig. 1.18. For extreme degeneration, the following polynomial approximation holds:

$$n \approx \frac{\sqrt{2} m_{n,D}^{3/2}}{\pi^2 \hbar^3} \frac{2}{3} (E_{Fn} - E_c)^{3/2}, \quad p \approx \frac{\sqrt{2} m_{h,D}^{3/2}}{\pi^2 \hbar^3} \frac{2}{3} (E_v - E_{Fh})^{3/2}.$$

A summary of some possible equilibrium bandstructures is shown in Fig. 1.19; notice that the n-type and p-type degenerate cases are purely theoretical, since increasing the doping level beyond a certain level makes $E_F > E_D$ or $E_F < E_A$, thus decreasing the donor or acceptor activation. This implies that the degenerate condition cannot practically be obtained at equilibrium. Finally, Fig. 1.20 concerns examples out of equilibrium; degeneracy arises in these cases from the high-injection condition.

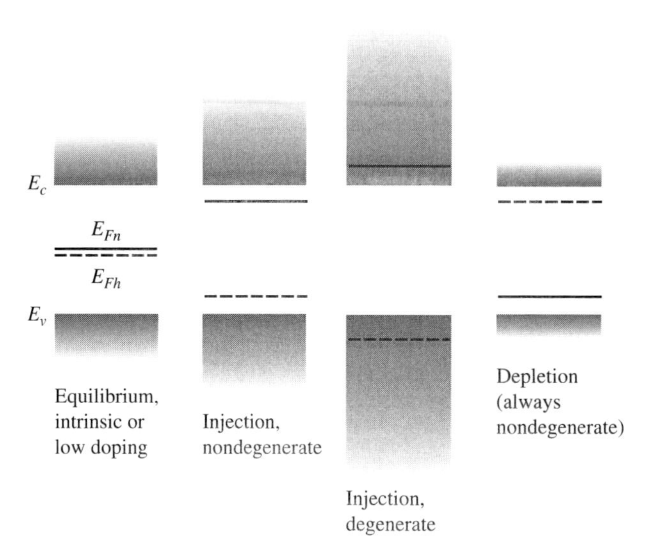

Figure 1.20 Examples of transition from the equilibrium (left) to the out-of-equilibrium bandstructure for degenerate and nondegenerate semiconductors.

1.5 Heterostructures

Crystals with different lattice constants grown on top of each other by epitaxial techniques are affected by interface defects called *misfit dislocations*. Such defects operate as electron or hole traps, and therefore the resulting structure is unsuited to the development of electron devices. However, if the lattice mismatch between the substrate and the heteroepitaxial overlayer is low or zero, an ideal or almost ideal crystal can be grown, made of two different materials. The resulting structure is called a *heterostructure*, and, since the electronic properties of the two layers are different, we also refer to it as a *heterojunction*. The material discontinuity arising in the heterojunction leads to important electronic and optical properties, such as confinement of carriers (related to the discontinuity of the conduction or valence bands) and confinement of radiation (due to the bandgap discontinuity and to the related refractive index step).

Heterostructures can be lattice-matched (if the two sides have the same lattice constant) or affected by a slight mismatch (indicatively, the maximum mismatch is of the order of 1%), which induces tensile or compressive strain. In this case, we talk about *pseudomorphic* or *strained* heterostructures, see Fig. 1.21. A small amount of strain in the heterostructure can be beneficial to the development of electronic or optoelectronic devices, since it leads to additional degrees of freedom in the band structure engineering, and in many cases allows for an improvement of the material transport or optical properties.

A double heterojunction made with a thin semiconductor layer (the thickness should be typically of the order of 100 nm) sandwiched between two layers (e.g., AlGaAs/GaAs/AlGaAs) creates a potential well in the conduction and/or valence band and is often referred to as a quantum well (QW). A succession of weakly interacting

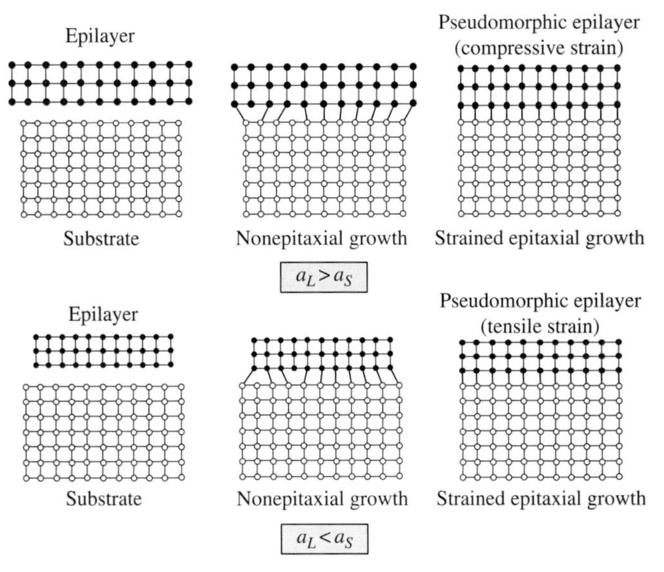

Figure 1.21 Pseudomorphic or strained growth. Above, the epilayer lattice constant is larger than that of the substrate: nonepitaxial growth with interface misfit dislocations and strained epitaxy. Below, the epilayer lattice constant is smaller than that of the substrate.

quantum wells is called a multi quantum well (MQW); if the MQW has many layers, with significant overlapping between the wavefunctions of adjacent wells, we finally obtain a superlattice (SL). The artificial periodicity imposed by the superlattice over the natural periodicity of the crystal introduces important modifications in the electronic properties.

1.6 Semiconductor alloys

Heterostructures are largely based on semiconductor alloys. The idea behind alloys is to create semiconductors having intermediate properties with respect to already existing "natural" semiconductors. Among such properties are the lattice constant a and the energy gap E_g. In several material systems, both a and E_g approximately follow a linear law with respect to the individual component parameters. The motivation to tailor the lattice constant is of course to achieve lattice matching to the substrate; tailoring the energy gap gives the possibility to change the emitted photon energy, thus generating practically important wavelengths, such as the 1.3 or 1.55 μm wavelengths needed for long-haul fiber communications (since they correspond to minimum fiber dispersion and absorption, respectively, see Fig. 1.22). Examples are alloys made of two components and three elements (called *ternary alloys*: e.g., AlGaAs, alloy of GaAs and AlAs) and alloys made of four components and elements (called *quaternary alloys*, e.g., InGaAsP, alloy of InAs, InP, GaAs, GaP). By proper selection of the alloy composition, semiconductor alloys emitting the right wavelength and matched to the right substrate can be generated.

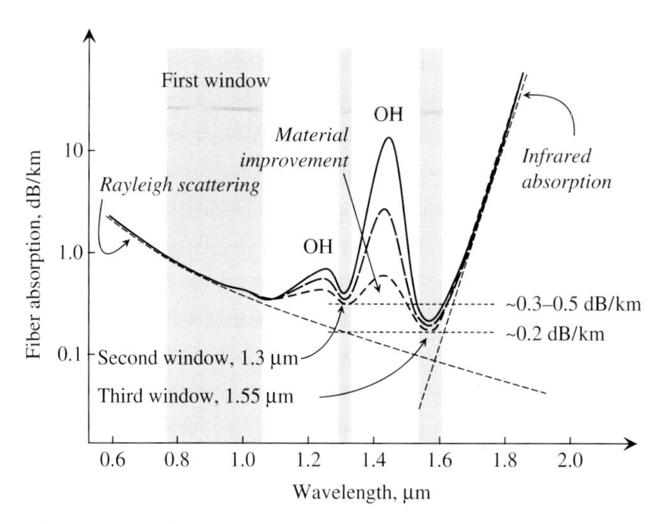

Figure 1.22 Absorption profile of a glass optical fiber.

In order to quantitatively define an alloy, we have to consider that compound semiconductors (CS) are polar compounds with a metal M combined with a nonmetal N in the form MN. Two different CSs sharing the same metal or nonmetal give rise to a ternary alloy or compound:

$$(M_1N)_x\,(M_2N)_{1-x} = M_{1x}M_{2(1-x)}N, \qquad \text{e.g., } \mathrm{Al}_x\mathrm{Ga}_{1-x}\mathrm{As}$$
$$(MN_1)_y\,(MN_2)_{1-y} = MN_{1y}N_{2(1-y)}, \qquad \text{e.g., } \mathrm{GaAs}_y\mathrm{P}_{1-y},$$

where x and $1-x$ denote the mole fraction of the two metal components, and y and $1-y$ denote the mole fraction of nonmetal components. Four different CSs sharing two metal and two nonmetal components yield a quaternary alloy or compound. In the following formulae, M and m are the metal components, N and n are the nonmetal components, and $\alpha+\beta+\gamma=1$:

$$(MN)_\alpha\,(Mn)_\beta\,(mN)_\gamma\,(mn)_{1-\alpha-\beta-\gamma} = M_{\alpha+\beta}m_{1-\alpha-\beta}N_{\alpha+\gamma}n_{1-\alpha-\gamma}$$
$$= M_xm_{1-x}N_yn_{1-y} \quad (\text{e.g., } \mathrm{In}_x\mathrm{Ga}_{1-x}\mathrm{As}_y\mathrm{P}_{1-y}).$$

Most alloy properties can be derived from the component properties through (global or piecewise) linear interpolation (Vegard law), often with second-order corrections (Abeles law); examples are the lattice constant, the energy gap, the inverse of the effective masses, and, in general, the bandstructure and related quantities. Varying the composition of a ternary alloy (one degree of freedom) changes the gap and related wavelength, but, at the same time, the lattice constant; in some cases (AlGaAs) the two components (AlAs and GaAs) are already matched, so that alloys with arbitrary Al content are lattice matched to the substrate (GaAs).

On the other hand, varying the composition of a quaternary alloy (two degrees of freedom) independently changes both the gap and the lattice constant, so as to allow for lattice matching to a specific substrate, e.g., InGaAsP on InP.

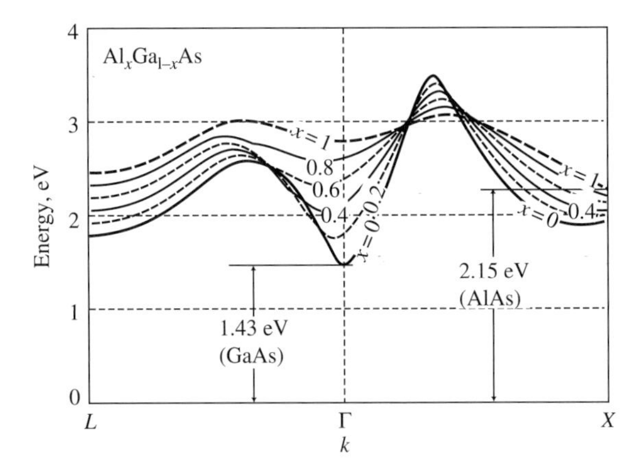

Figure 1.23 Evolution of the bandstructure of AlGaAs changing the Al content from 0 to 1.

The Vegard or Abeles laws must be applied with care in some cases. As an example, consider the $Al_xGa_{1-x}As$ alloy and call P an alloy parameter, such as the energy gap. The Vegard law can be written as:

$$P(x) = (1 - x)P_{\text{GaAs}} + xP_{\text{AlAs}};$$

by inspection, this yields a linear interpolation between the two constituent parameters. However, this law fails to accurately reproduce the behavior of the AlGaAs energy gap because GaAs is direct bandgap, and AlAs is indirect. To clarify this point, let us consider the simplified bandstructure of the alloy as shown in Fig. 1.23. We clearly see that the main and secondary (X point) minima have the same level for $x = 0.45$; for larger Al mole fraction, the material becomes indirect bandgap. Since the composition dependence is different for the energy levels of the Γ and X minima, a unique Vegard law fails to approximate the gap for any alloy composition, and a piecewise approximation is required:

$$E_g \approx 1.414 + 1.247x, \quad x < 0.45$$
$$E_g \approx 1.985 + 1.147(x - 0.45)^2, \quad x > 0.45.$$

The same problems arise in the InGaAsP alloy, since GaP is indirect bandgap; thus, a global Vegard approximation of the kind

$$P_{\text{InGaAsP}} = (1 - x)(1 - y)P_{\text{GaAs}} + (1 - x)yP_{\text{GaP}} + xyP_{\text{InP}} + x(1 - y)P_{\text{InAs}}$$

(by inspection, the approximation is bilinear and yields the correct values for the four semiconductor components) may be slightly inaccurate.

1.6.1 The substrate issue

Electronic and optoelectronic devices require to be grown on a suitable (typically, semiconductor) substrate. In practice, the only semiconductor substrates readily available are those that can be grown into monocrystal ingots through Czochralsky or Bridgman techniques – i.e., in order of decreasing quality and increasing cost, Si, GaAs, InP, SiC, and a few others (GaP, GaSb, CdTe). Devices are to be grown so as to be either lattice matched to the substrate, or slightly (e.g., 1%) mismatched (pseudomorphic approach). The use of graded buffer layers allows us to exploit mismatched substrates, since it distributes the lattice mismatch over a larger thickness. This approach is often referred to as the *metamorphic* approach; it is sometimes exploited both in electronic and in optoelectronic devices. Metamorphic devices often used to have reliability problems related to the migration of defects in graded buffer layers; however, high-quality metamorphic field-effect transistors with an InP active region on a GaAs substrate have recently been developed with success.

1.6.2 Important compound semiconductor alloys

Alloys are often represented as a straight or curved segment (for ternary alloys) or quadrilateral area (for quaternary alloys) in a plane where the x coordinate is the lattice constant and the y coordinate is the energy gap; see Fig. 1.24. The segment extremes and the vertices of the quadrilateral are the semiconductor components. In Fig. 1.24 some important alloys are reported:

- AlGaAs, lattice-matched for any composition to GaAs, direct bandgap up to an Al mole content of 0.45.
- InGaAsP, which can be matched either to GaAs or to InP substrates; InP substrate matching includes the possibility of emitting 1.55 or 1.3 μm wavelengths;[7] the alloy is direct bandgap, apart from around the GaP corner, whose gap is indirect.
- InAlAs, which can be lattice matched to InP with composition $Al_{0.48}In_{0.52}As$.
- InGaAs, a ternary alloy matched to InP with composition $Ga_{0.47}In_{0.53}As$; it is a subset of the quaternary alloy InGaAsP.
- InGaAsSb, the antimonide family, a possible material for long-wavelength devices, but with a rather underdeveloped technology vs. InGaAsP.
- HgCdTe, a ternary alloy particularly relevant to far infrared (FIR) detection owing to the very small bandgap achievable.
- SiGe, an indirect bandgap alloy important for electronic applications (heterojunction bipolar transistors) but also (to a certain extent) for detectors and electroabsorption modulators;
- III-N alloys, such as AlGaN and InGaN, with applications in short-wavelength sources (blue lasers) but also in RF and microwave power transistors. AlGaN can be grown by pseudomorphic epitaxy on a GaN virtual substrate; GaN has in turn no native substrate so far, but can be grown on SiC, sapphire (Al_2O_3) or Si. The InGaN alloy is exploited in optoelectronic devices such as blue lasers and LEDs, besides being able to cover much of the visible spectrum.[8]

[7] InGaAsP lattice-matched to InP can emit approximately between 0.92 and 1.65 μm.
[8] The InN gap is controversial, and probably is much smaller than the previously accepted value around 2 eV. The nitride data in Fig. 1.24 are from [8], Fig. 3.

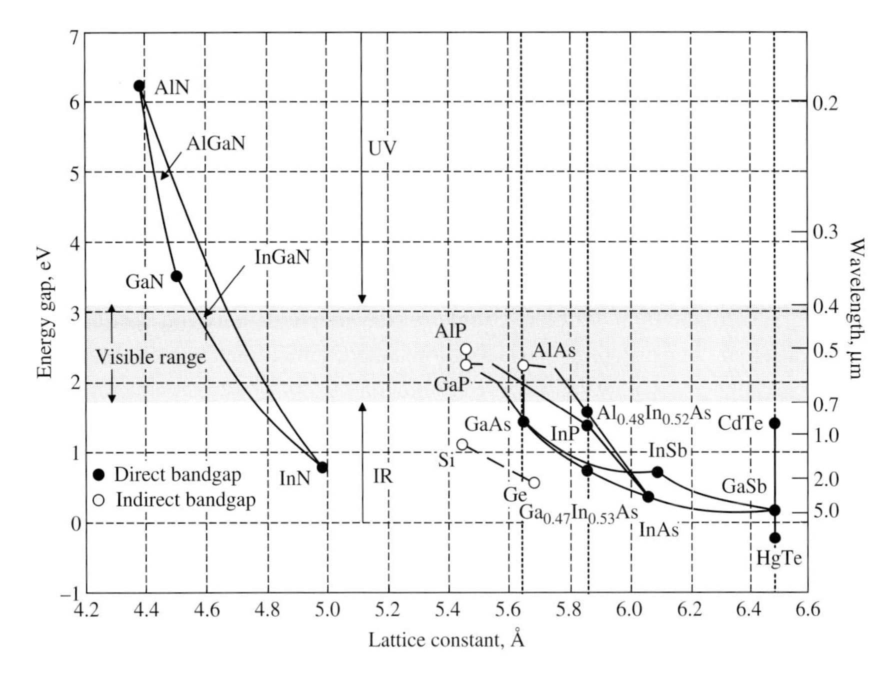

Figure 1.24 Some important alloys in the lattice constant–energy gap plane. In order of increasing gap and decreasing lattice constant: HgCdTe, InGaAsSb, InGaAsP, SiGe, AlGaAs, AlGaP, InGaN, AlGaN. For the widegap (wurtzite) nitrides (GaN, InN, AlN) the horizontal axis reports the equivalent cubic lattice constant. The InGaAsP, AlGaAs and InGaAsSb data are from [7]; the GaN, AlN and InN data are from [8], Fig. 3.

Since GaN, AlN, and InN have the wurtzite (hexagonal) crystal structure, an equivalent lattice constant $a_{C,eq}$ has to be defined for comparison with cubic crystals, so as to make the volume of the wurtzite cell V_H (per atom) equal to the volume of a cubic cell (per atom); taking into account that the wurtzite cell has 12 equivalent atoms, while the cubic cell has 8 equivalent atoms, we must impose:

$$\frac{1}{12}V_H = \frac{1}{12}\frac{3\sqrt{3}}{2}ca_H^2 = \frac{1}{8}a_{C,eq}^3 \rightarrow a_{C,eq} = \left(\sqrt{3}ca_H^2\right)^{1/3},$$

where V_H is the volume of the wurtzite cell prism of sides a and c. For GaN $a_H = 0.317$ nm, $c = 0.516$ nm; it follows that

$$a_{C,eq} = \left(\sqrt{3}ca_H^2\right)^{1/3} = \left(\sqrt{3}\cdot 0.516\cdot 0.317^2\right)^{1/3} = 0.448 \text{ nm}.$$

1.7 Bandstructure engineering: heterojunctions and quantum wells

Although the bandstructure of a semiconductor depends on the lattice constant a, which is affected by the operating temperature and pressure, significant variations in the bandstructure parameters cannot be obtained in practice. Nevertheless, semiconductor alloys enable us to generate new, "artificial" semiconductors with band properties intermediate with respect to the components. A more radical change in the bandstructure occurs when heterojunctions are introduced so as to form quantized structures. A deep variation in the density of states follows, with important consequences in terms of optical properties (as we shall discuss later, the absorption profile as a function of the photon energy mimics the density of states). Moreover, strain in heterostructures allows for further degrees of freedom, like controlling the degeneracy between heavy and light hole subbands.

Heterojunctions are ideal, single-crystal junctions between semiconductors having different bandstructures. As already recalled, lattice-matched or strained (pseudomorphic) junctions between different semiconductors or semiconductor alloys allow for photon confinement (through the difference in refractive indices), carrier confinement (through potential wells in conduction or valence bands), and quantized structures such as superlattices, quantum wells, quantum dots, and quantum wires. An example of a heterostructure band diagram is shown in Fig. 1.25, where the band disalignment derives from application of the *affinity rule* (i.e., the conduction band discontinuity is the affinity difference, the valence band discontinuity is the difference in ionizations). In many practical cases, however, band disalignments are dominated by interfacial effects and do not follow the affinity rule exactly; for instance, in the AlGaAs-GaAs heterostructure one has

$$|\Delta E_c| \approx 0.65\Delta E_g, \quad |\Delta E_v| \approx 0.35\Delta E_g. \tag{1.14}$$

More specifically, the valence and conduction band discontinuities as a function of the Al fraction are (in eV) [1]:

$$|\Delta E_v| = 0.46x$$

$$|\Delta E_c| = \begin{cases} 0.79x, & x < 0.41 \\ 0.475 - 0.335x + 0.143x^2, & x > 0.41. \end{cases}$$

According to the material parameters, several band alignments are possible, as shown in Fig. 1.26; however, the most important situation in practice is the Type I band alignment in which the energy gap of the narrowgap material is included in the gap of the widegap material.

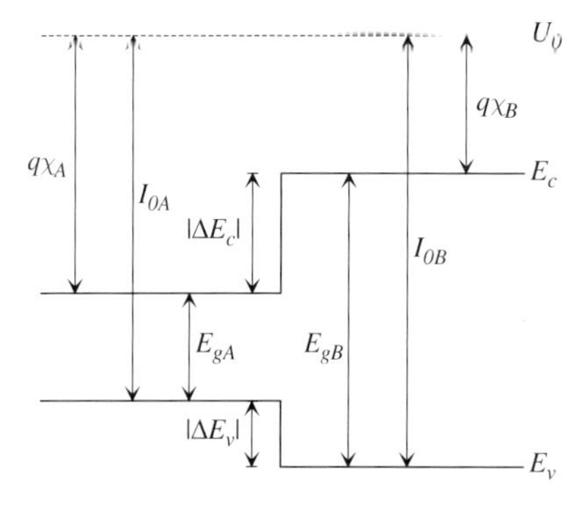

Figure 1.25 Heterostructure band alignment through application of the affinity rule to two materials having different bandstructures.

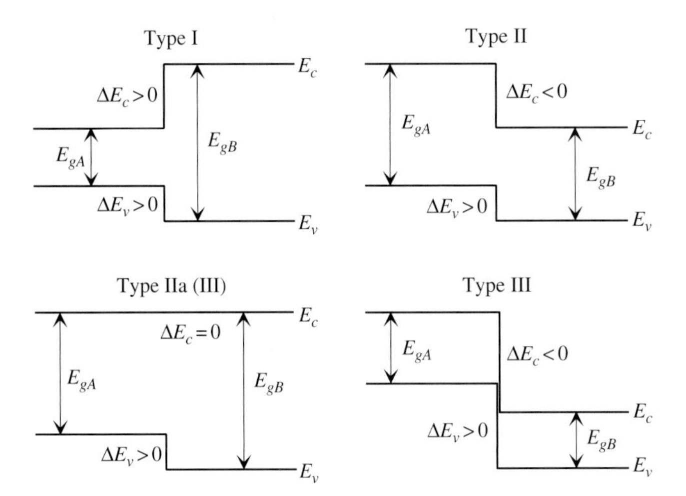

Figure 1.26 Classification of heterostructures according to band alignment; $\Delta E_j = E_{jB} - E_{jA}$.

Heterojunctions can be made with two n-type or p-type materials (*homotype heterojunctions*) so as to form a pn junction (*heterotype heterojunctions*). Often, the widegap material is conventionally denoted as N or P according to the type, the narrowgap material as n or p. According to this convention, a heterotype heterojunction is, for example, Np or nP and a narrowgap intrinsic layer sandwiched between two widegap doped semiconductors is NiP.

Single or double heterostructures can create potential wells in the conduction and/or valence bands, which can confine carriers so as to create conducting channels (with application to electron devices, such as field-effect transistors), and regions where confined carriers achieve high density and are able to recombine radiatively. In the second case, the emitted radiation is confined by the refractive index step associated with the heterostructure (the refractive index is larger in narrowgap materials). An example of this concept is reported in Fig. 1.27, a NiP structure in direct bias that may operate like the active region of a light-emitting diode or a semiconductor laser.

Carriers trapped by the potential well introduced by a double heterostructure are confined in the direction orthogonal to the well, but are free to move in the two other directions (i.e., parallel to the heterojunction). However, if the potential well is very narrow the allowed energy levels of the confined electrons and holes will be quantized. The resulting structure, called a quantum well (QW), has a different bandstructure vs. bulk, where sets of energy subbands appear (see Fig. 1.28). Also the density of states is strongly affected.

The quantum behavior of carriers in narrow (conduction or valence band) potential wells originated by heterojunctions between widegap and narrowgap semiconductors can be analyzed by applying the Schrödinger equation to the relevant particles (electron or holes) described in turn by a 3D effective mass approximation. Solution of the Schrödinger equation enables us to evaluate the energy levels and subbands, given the well potential profile. In a rectangular geometry, we start from bulk (3D motion possible,

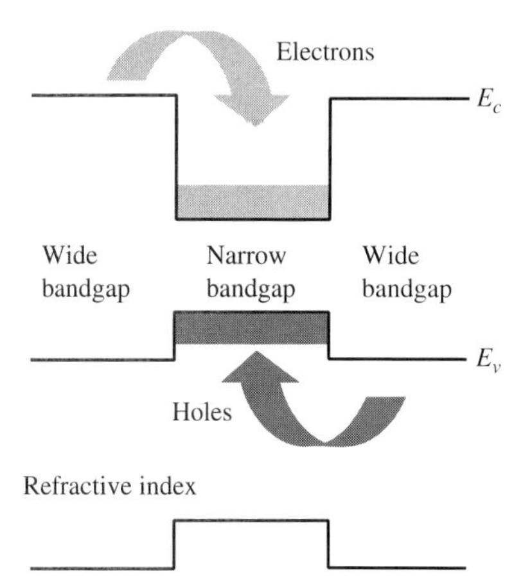

Figure 1.27 Example of carrier and light confinement in a NiP double heterostructure in direct bias.

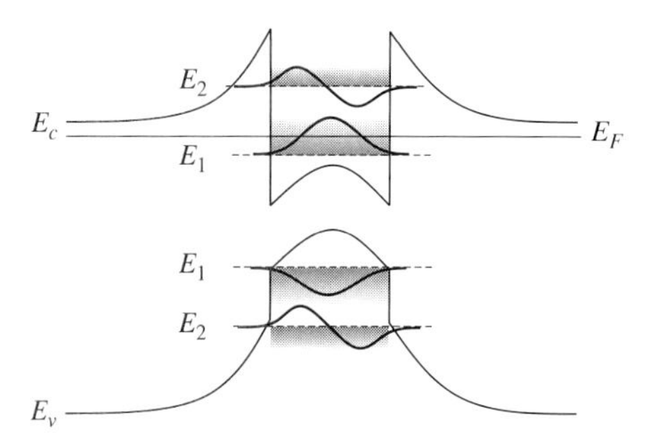

Figure 1.28 Qualitative example of quantization in a quantum well: in the conduction and valence bands, subbands arise with minimum energies corresponding to the levels $E_1, E_2, \ldots$; the total electron and hole wavefunction is given by the product of the 3D (bulk) wavefunction and of the envelope wavefunction shown. For simplicity, only the heavy hole subbands are shown.

no confinement) and obtain, by progressively restricting the degrees of freedom of the particle, the so-called *reduced dimensionality structures* corresponding to:

1. Confinement in one direction (x): particles are confined along x by a potential well but are free to move along y and z (**quantum well**).
2. Confinement in two directions (x and y): particles are confined along x and y, but they are free to move along z (**quantum wire**).
3. Confinement in three directions (x, y, z): particles are entirely confined and cannot move (**quantum dot**).

For a quantum well defined in a Cartesian space, the particle (electron or hole) wavefunction can be expressed, in the effective mass approximation, as the product of three factors:[9]

$$\psi(x, y, z) = \phi(x) \exp\left(-jk_y y\right) \exp\left(-jk_z z\right). \tag{1.15}$$

[9] The effective mass approximation implies that, in the absence of the well, the particle is completely free, i.e., it behaves like a free electron with proper effective mass. The approximate total electron wavefunction of a confined electron is therefore the product of the electron Bloch function and the envelope wavefunction.

For a particle at rest ($k_y = 0$, $k_z = 0$) the wavefunction reduces to $\phi(x)$, called the *envelope wavefunction*, to be derived from the solution of the 1D Schrödinger equation with the well potential profile $U(x)$. Notice that, independent on the kinetic energy of the particle, the probability (corresponding to the particle distribution in space) is always:

$$|\psi(x, y, z)|^2 = |\phi(x)|^2 ,$$

i.e. the particle density only depends on the shape of the envelope wavefunction $\phi(x)$, and in particular only depends on the x coordinate. The envelope wavefunction therefore provides a picture of the particle distribution in the well. From the allowed energies E_i we can express the total particle energy as

$$E = E_k + E_i = E_i + \frac{\hbar^2}{2m^*}(k_y^2 + k_z^2),$$

where E_k is the particle kinetic energy. This expression holds for all values of i; therefore the quantum well introduces in the conduction or valence bands a subband structure; in each subband the dispersion relation is, at least approximately, parabolic, but, contrarily to the 3D case, the particle motion occurs in the plane (y, z) only. Particles therefore form a *2D gas*, i.e. a population able to move in a 2D space only. We suppose that the effective mass is isotropic and (for simplicity) independent of the subband considered.

For a quantum wire (QWire), we have a similar situation: the particle wavefunction can be written in the form

$$\psi(x, y, z) = \phi(x, y) \exp(-jk_z z) ,$$

where $\phi(x, y)$ satisfies a 2D Schrödinger equation with proper energies E_{ij}. The particle energy can therefore be written as

$$E = E_k + E_{ij} = E_{ij} + \frac{\hbar^2}{2m^*}k_z^2,$$

i.e., the dispersion relation is parabolic in k_z, starting from one of the allowed energies E_{ij}. Finally, a quantum dot only exhibits finite-energy states with energy E_{ijk}, in which the particle cannot move along any spatial direction (but can undergo radiative recombination and intersubband transitions).

1.7.1 Carrier density and density of states in a quantum well

We want now to evaluate the sheet density n_s (cm^{-2}) of the electrons trapped in a conduction band QW (a similar treatment would apply to holes in a valence band QW), on the basis of the system Fermi level E_F. As usual, we can exploit the product of the density of states in the 2D system (QW) multiplied by the Fermi–Dirac distribution f_n, see (1.4):

$$\mathrm{d}n_s(E) = g_{2D}(E) f_n(E) \mathrm{d}E,$$

where $g_{2D}(E)$ is the QW density of states (per unit surface). Assuming that the QW energy levels are E_l, $l = 1, 2, 3...$ and that each energy level has an associated wavefunction

$$\psi_l(x, y, z) = \phi_l(x) \exp(-jk_y y) \exp(-jk_z z),$$

we immediately notice that the carrier distribution depends only on x according to the probability $|\phi_l(x)|^2$. We suppose that the wavefunction is normalized with respect to the total concentration per unit surface of the electrons belonging to each subband (having minimum energy E_l). The total density of states can be obtained by summing the densities relative to each subband:

$$g_{2D}(E) = \sum_l g_l(E).$$

To evaluate g_l, we suppose that the 2D electron gas is enclosed in a (large) potential well with infinitely high barriers placed in $y = 0$, $y = L_y$ and $z = 0$, $z = L_z$, with $L_y, L_z \gg W$, where W is the QW thickness. Quantization in the transversal plane only allows for the (positive) wavenumbers

$$k_y = m\frac{\pi}{L_y}, \qquad k_z = n\frac{\pi}{L_z},$$

where m and n are integer numbers. Let us evaluate the number of states $N(E)$ having kinetic energy $E_k < E$; those are the states included in the circle

$$E_k = \frac{\hbar^2}{2m^*}(k_y^2 + k_z^2),$$

where an isotropic effective mass has been introduced. The number of allowed states with $E < E_k$ can be derived, taking into account that only positive wavevectors are considered, by dividing the area of one quarter of the circle with radius $\sqrt{2m^* E_k}/\hbar$ of the (k_x, k_y) plane by the area associated with each state, i.e., $\pi^2/(L_x L_y)$. With a further factor 2 accounting for the spin, we finally obtain

$$N(E) = N(E_l + E_k) = 2 \times \frac{1}{4}\pi \frac{2m^* E_k}{\hbar^2} \frac{L_x L_y}{\pi^2} = \frac{m^* (E - E_l)}{\hbar^2} \frac{L_x L_y}{\pi},$$

for $E > E_l$ ($N(E) = 0$ for $E < E_l$), with corresponding surface state density (we divide by the total laterally confined 2D gas area $L_y L_z$):

$$g_l(E) = \frac{1}{L_y L_z} \frac{\mathrm{d}N(E)}{\mathrm{d}E} = \frac{m^*}{\pi\hbar^2} = \frac{4\pi m^*}{h^2},$$

for $E > E_l$; for $E < E_l$, $g_l(E) = 0$. Introducing the Heaviside step function $u(E)$, ($u = 0$ for $E < 0$, $u = 1$ for $E > 0$) we can express the total density of states of the 2D electron gas as the following staircase function:

$$g_{2D}(E) = \sum_{l=1}^{\infty} \frac{4\pi m^*}{h^2} u(E - E_l) = \sum_{l=1}^{\infty} \frac{m^*}{\pi\hbar^2} u(E - E_l),$$

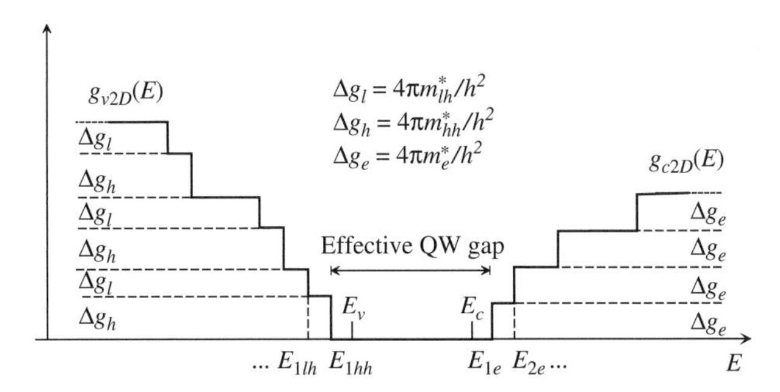

Figure 1.29 Staircase-shaped density of states in a QW for the valence and conduction bands as a function of energy.

where E_l are the allowed energies obtained by solving the Schrödinger equation in direction x. Integrating on all energies we finally obtain the sheet carrier density for electrons:

$$n_s = \int_{E_c}^{\infty} g_{2D}(E) f_{FD}(E)\,\mathrm{d}E = \frac{4\pi m_n^*}{h^2} \sum_l \int_{E_l}^{\infty} \frac{\mathrm{d}E}{1+\exp\left(\frac{E-E_F}{k_B T}\right)}$$
$$= \frac{4\pi m_n^* k_B T}{h^2} \sum_l \log\left[1+\exp\left(\frac{E_F - E_l}{k_B T}\right)\right].$$

Contrary to the 3D case, we do not have to invoke the Boltzmann approximation to obtain a closed-form expression in terms of elementary functions. The above expression is therefore valid for both degenerate and nondegenerate cases.

A sketch of the QW DOS showing the electron and the hole density of states is shown in Fig. 1.29. In the valence band the DOS actually follows a composite staircase behavior including both heavy hole and light hole states:

$$g_{2D,v}(E) = \sum_{l=1}^{\infty} \frac{4\pi m_{hh}^*}{h^2} u(E - E_l) + \sum_{k=1}^{\infty} \frac{4\pi m_{lh}^*}{h^2} u(E - E_k);$$

the state ordering with increasing hole energy is HH_1, LH_1, HH_2, LH_2, etc. In both staircases the position of steps is variable, but the step size is the same for each kind of particle (electrons, heavy holes, light holes). In general, the fundamental level for heavy and light holes will be different ($E_{1lh} \neq E_{1hh}$), thus removing the degeneracy between HH and LH typical of 3D systems.

Example 1.2: A double AlGaAs-GaAs-AlGaAs heterojunction generates a potential well (QW) both in the conduction and in the valence band. Evaluate the energy levels and dispersion relations for electrons and holes assuming for simplicity that the potential barrier has infinite height. Assume the potential well is located between $z = 0$ and $z = W$.

In a QW the wavefunction is factorized; assuming that the particle is able to move in the directions x and y, while it is quantized in z, we can express the total wavefunction as

$$\psi = \phi(z)\psi_x(x)\psi_y(y).$$

Substituting in the 3D Schrödinger equation we have (for the moment we use a generic effective mass m^*)

$$\frac{\hbar^2}{2m^*}\left[\phi\psi_y\frac{\mathrm{d}^2\psi_x(x)}{\mathrm{d}x^2} + \phi\psi_x\frac{\mathrm{d}^2\psi_y(y)}{\mathrm{d}y^2} + \psi_x\psi_y\frac{\mathrm{d}^2\phi(z)}{\mathrm{d}z^2}\right] + [E - U(z)]\,\phi(z)\psi_x(x)\psi_y(y) = 0$$

i.e.,

$$\underbrace{\frac{1}{\psi_x}\frac{\mathrm{d}^2\psi_x}{\mathrm{d}x^2}}_{-k_x^2} + \underbrace{\frac{1}{\psi_y}\frac{\mathrm{d}^2\psi_y}{\mathrm{d}y^2}}_{-k_y^2} + \frac{1}{\phi}\frac{\mathrm{d}^2\phi}{\mathrm{d}z^2} + \frac{2m^*}{\hbar^2}[E - U] = 0,$$

from which we can derive, by separation of variables, the factorized 1D equations:

$$\frac{\hbar^2}{2m^*}\frac{\mathrm{d}^2\phi(z)}{\mathrm{d}z^2} + \left[E - \frac{\hbar^2}{2m^*}(k_x^2 + k_y^2) - U(z)\right]\phi(z) = 0 \tag{1.16}$$

$$\frac{\mathrm{d}^2\psi_x(x)}{\mathrm{d}x^2} + k_x^2\psi_x(x) = 0 \tag{1.17}$$

$$\frac{\mathrm{d}^2\psi_y(y)}{\mathrm{d}y^2} + k_y^2\psi_y(y) = 0. \tag{1.18}$$

Equation (1.16) can also be written as

$$\frac{\hbar^2}{2m^*}\frac{\mathrm{d}^2\phi(z)}{\mathrm{d}z^2} + \left[E' - U(z)\right]\phi(z), \quad E' = E - \frac{\hbar^2}{2m^*}(k_x^2 + k_y^2). \tag{1.19}$$

From the solution of (1.17) and (1.18) we obtain

$$\psi_x(x) = A_x\exp(-\mathrm{j}k_x x), \quad \psi_y(y) = A_y\exp(-\mathrm{j}k_y y),$$

where A_x and A_y are arbitrary constants, while from (1.19), assuming that the well is located between $z = 0$ and $z = W$, and that therefore (due to the infinite height of the well) $\phi(0) = \phi(W) = 0$, we obtain

$$\phi(z) = A_z\sin(k_z z), \quad k_z = \sqrt{\frac{2m^*E'}{\hbar^2}}.$$

In order to have $A_z\sin(k_z W) = 0$ the following quantization condition must be enforced:

$$k_z W = \sqrt{\frac{2m^*E'}{\hbar^2}}W = n\pi, \quad n = 1, 2, ...,$$

i.e.,

$$E'_n = \frac{n^2\hbar^2\pi^2}{2m^* W^2},$$

or

$$E_n = E'_n + \frac{\hbar^2}{2m^*}(k_x^2 + k_y^2) = \frac{n^2\hbar^2\pi^2}{2m_n^* W^2} + \frac{\hbar^2}{2m_n^*}(k_x^2 + k_y^2).$$

In the above expression the conduction band edge E_c is taken as the reference energy (since we assumed $U = 0$ in the well). The total wavefunction is thus

$$\psi = \phi(z)\psi_x(x)\psi_y(y) = A\sin(k_z z)\exp(-jk_x x)\exp(-jk_y y),$$

which can be normalized, for example, by imposing that the total probability is unity:

$$\int_0^W |\psi|^2 \, dz = A^2 \int_0^W \sin^2(k_z z)\, dz = A^2 \int_0^W \left(\frac{1}{2} - \frac{1}{2}\cos 2k_z z\right) dz$$
$$= A^2 \frac{W}{2} + A^2 \frac{1}{4k_z} \sin 2k_z z \Big|_0^W = A^2 \frac{W}{2} = 1 \rightarrow A = \sqrt{\frac{2}{W}},$$

i.e.,

$$\psi(x, y, z) = \sqrt{\frac{2}{W}} \sin(k_z z)\exp(-jk_x x)\exp(-jk_y y).$$

Wavefunctions in the well therefore are either *even* or *odd* with respect to the center of the well, the fundamental state (minimum energy) corresponding to an even probability distribution with respect to the well center. From the expression for E', specializing for electrons (reference energy E_c) and heavy or light holes (reference energy E_v and $E_h = -E_e$), we obtain ($k_T^2 = k_x^2 + k_y^2$, k_T transverse wavevector)

$$E_e^{(n)}(k_T) = E_c + \frac{n^2\hbar^2\pi^2}{2m_n^* W^2} + \frac{\hbar^2 k_T^2}{2m_n^*} = E_{ne} + \frac{\hbar^2 k_T^2}{2m_n^*}$$
$$E_{hh}^{(m)}(k_T) = E_v - \frac{m^2\hbar^2\pi^2}{2m_{hh}^* W^2} - \frac{\hbar^2 k_T^2}{2m_{hh}^*} = E_{mhh} - \frac{\hbar^2 k_T^2}{2m_{hh}^*}$$
$$E_{lh}^{(l)}(k_T) = E_v - \frac{l^2\hbar^2\pi^2}{2m_{lh}^* W^2} - \frac{\hbar^2 k_T^2}{2m_{lh}^*} = E_{llh} - \frac{\hbar^2 k_T^2}{2m_{lh}^*}.$$

Assembling the parabolic dispersion relation for the valence and conduction bands, we finally have the situation shown in Fig. 1.30. Note that $E_{lh}^{(1)}(0) \equiv E_{1lh} < E_{hh}^{(1)}(0) \equiv E_{1hh}$ due to the larger effective mass of heavy holes; thus, in an (unstrained) QW heavy holes are the fundamental valence band level.

In practice, the valence band dispersion relation is more involved than shown in Fig. 1.30 due to HH and LH coupling (note that the light and heavy hole dispersion relations would cross at high energy).

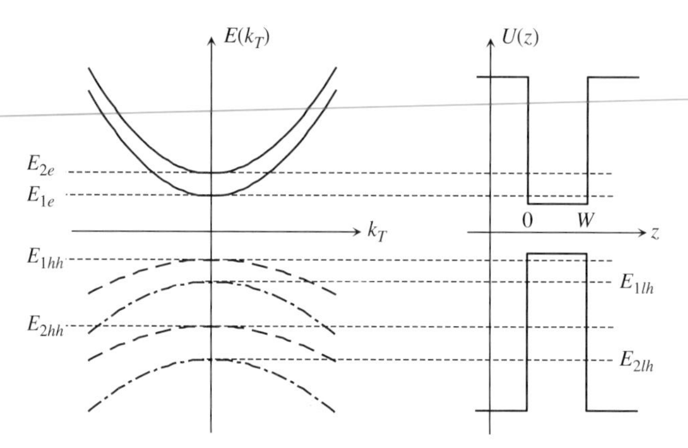

Figure 1.30 Approximate dispersion relation (left) for a QW with high (ideally, infinite) potential barriers (right).

1.7.2 Carrier density and density of states in a quantum wire

The density of states $n_l(E)$ of a quantum wire can be evaluated in a similar way. In the 1D case, we can write

$$\mathrm{d}n_l(E) = g_{1D}(E) f_n(E) \mathrm{d}E,$$

where $g_{1D}(E)$ is the number of states per unit length and unit energy interval. The energy levels E_{ij} of the wire, obtained through a solution of the 2D quantum problem, can be conveniently ordered through a single integer index l as E_l, $l = 1, 2, 3...$ such as $E_l < E_{l+1}$; each energy level has the associated wavefunction $\psi_l(x, y, z)$:

$$\psi_l(x, y, z) = \phi_l(x, y) \exp(-\mathrm{j}k_z z)$$

with envelope wavefunction $\phi_l(x, y)$. The carrier distribution in the plane orthogonal to the wire direction follows the probability $|\phi_l(x, y)|^2$, which we assume as normalized with respect to the total concentration of the electrons in subband l. Also in this case, the total state density can be obtained by summing contributions from all individual subbands:

$$g_{1D}(E) = \sum_l g_l(E).$$

To evaluate g_l we artificially limit the QW through potential barriers located in $x = 0$, $x = L_z$. Quantization allows for the positive wavenumbers:

$$k_z = n\frac{\pi}{L_z}.$$

The number of states $N(E)$ with kinetic energy $E_k < E$ can be evaluated by considering the states in the interval $[0, E_k]$, where

$$E_k = \frac{\hbar^2}{2m^*}k_z^2.$$

Since the corresponding k_z value is $\sqrt{2m^* E_k}/\hbar$, the number of allowed states having energy $< E_k$ can be obtained from the ratio of $\sqrt{2m^* E_k}/\hbar$ and the length associated with each state, which is π/L_z; also accounting for the spin we finally obtain

$$N(E) = N(E_l + E_k) = 2 \times \frac{\sqrt{2m^* E_k}}{\hbar} \frac{L_z}{\pi} = \frac{\sqrt{8m^*(E - E_l)}}{\hbar} \frac{L_z}{\pi}, \qquad E > E_l;$$

the corresponding density of states per unit length will be

$$g_l(E) = \frac{1}{L_z} \frac{\mathrm{d}N(E)}{\mathrm{d}E} = \frac{2\sqrt{2m^*}}{h} \frac{1}{\sqrt{E - E_l}}, \qquad E > E_l$$

while $g_l(E) = 0$, $E < E_l$. We can thus express the total state density as a kind of staircase function with singular (but integrable) steps:

$$g_{2D}(E) = \sum_{l=1}^{\infty} \frac{2\sqrt{2m^*}}{h} \frac{1}{\sqrt{E - E_l}} u(E - E_l).$$

The electron density can be finally recovered as

$$n_f = \int_{E_c}^{\infty} g_{1D}(E) f_{FD}(E) \, \mathrm{d}E = \frac{2\sqrt{2m_n^*}}{h} \sum_l \int_{E_l}^{\infty} \frac{(E - E_l)^{-1/2} \mathrm{d}E}{1 + \exp\left(\dfrac{E - E_F}{k_B T}\right)}.$$

Note that the integral cannot be expressed in an elementary way.

An overall picture of the 3D and 2D DOS is shown in Fig. 1.31, while the 1D DOS is shown in Fig. 1.32, left. For the zero-dimensionality (0D) quantum dot (QD or QDot), only discrete energy levels exist rather than subbands and the DOS simply

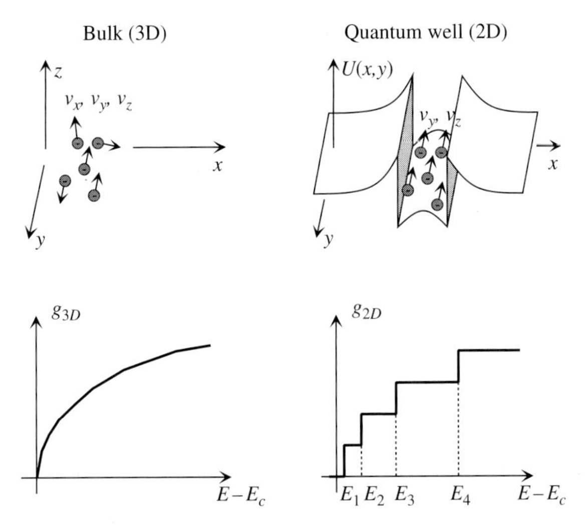

Figure 1.31 Above: degrees of freedom for electrons in bulk (left) or in a quantum well along x, with potential profile U. Below: density of states for bulk (left) and QW (right).

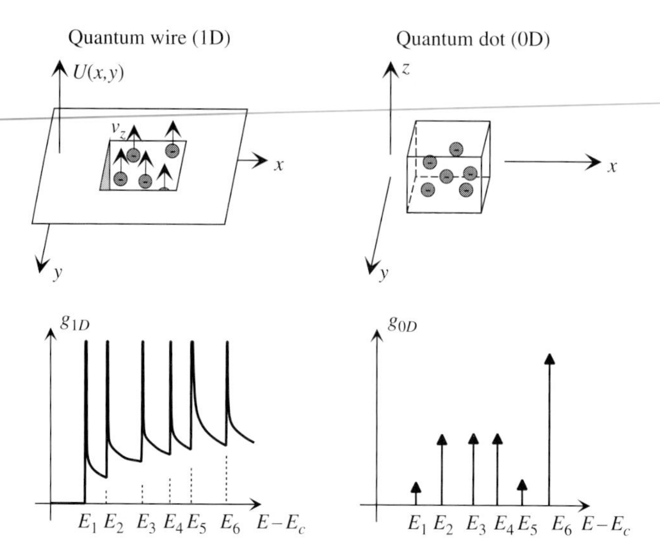

Figure 1.32 Above: degrees of freedom for electrons in a quantum wire (left) with potential profile U, allowing motion along z only, or in a quantum dot (right), no motion possible. Below: QWire (left) and QDot density of states (right).

is a summation of Dirac delta functions of the energy located at each energy level; see Fig. 1.32, right.[10]

1.7.3 Superlattices

Molecular beam epitaxy (MBE) allows several, almost monoatomic, layers to be grown in a controlled and orderly way. The resulting structure is a MQW or, for a large number of (coupled) wells, a superlattice. From the electronic standpoint, N coupled QWs cause the N-fold splitting of the system energy levels, finally leading to subbands, in much the same way as coupled atoms merge their individual energy levels into crystal energy bands.[11] An example of such a subband structure is shown in Fig. 1.33. Superlattices are therefore a kind of artificial 3D medium allowing for new features – e.g., low-energy transitions between subbands can be exploited to absorb (emit) long-wavelength IR. Superlattices can also be obtained by a periodic arrangement of QWires or QDs.

1.7.4 Effect of strain on bandstructure

Tensile or compressive strain changes the semiconductor bandstructure in several ways. A first consequence concerns the lattice constant becoming larger (tensile isotropic

[10] The QD density of states is purely indicative, since the position and area of each δ function depends on the dot shape. In real structures, the density of states has finite peaks due to linewidth-broadening mechanisms.

[11] The basic differences between a MQW and a superlattice is the well number (typically small or large) and above all in the coupling between the wavefunctions of neighboring wells. If the coupling is weak almost isolated energy levels are generated rather than subbands.

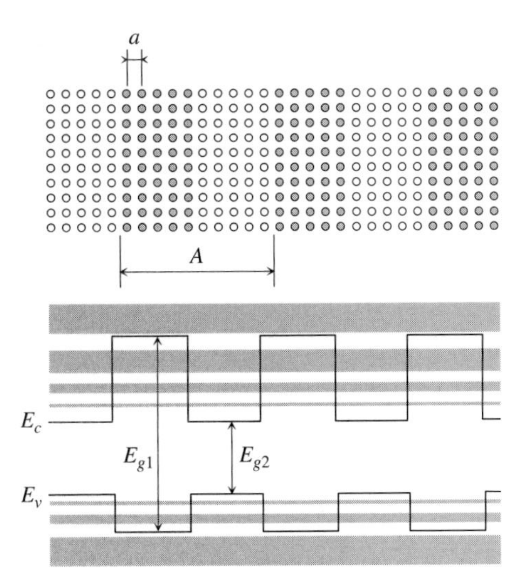

Figure 1.33 Superlattice (SL) and SL bandstructure with subbands in the valence and conduction bands.

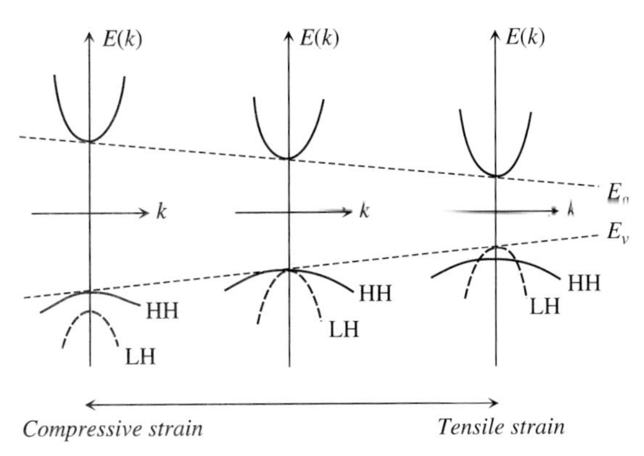

Figure 1.34 Effect of strain on bandstructure in a 3D (bulk) semiconductor. In the figure $m^*_{hh} \approx 10m^*_{lh}$ and $m^*_{lh} \approx m^*_n$.

strain) or smaller (compressive isotropic strain); since the energy gap is inversely proportional to the lattice constant (at least in a direct bandgap material: in indirect bandgap semiconductors local trends can mask this simple effect), E_g increases with compressive strain and decreases with tensile strain. Moreover, strain eliminates the degeneracy of HH and LH states, leading (in 3D) to heavier (compressive uniaxial strain) or lighter (tensile uniaxial strain) holes. Such effects are shown in Fig. 1.34.

In a QW, the situation is even more complex, since the equivalent gap (i.e., the distance between the fundamental levels in the conduction and valence bands) changes. However, degeneracy is usually already removed without strain (the first heavy hole subband being typically on top, as already remarked).

Proper tensile strain can bring back degeneracy, which, in some optoelectronic devices (e.g., in electroabsorption modulators or detectors) allows for polarization-independent operation, see p. 407. The first light hole subband can be also brought on top by tensile strain, thus reducing the hole density of states effective mass. This makes it easier, at least in theory, to achieve population inversion and therefore positive gain, with advantages in QW laser threshold condition, see p. 287.

1.8 Semiconductor transport and generation–recombination

1.8.1 Drift and diffusion

Under the application of an external electric field, electrons and holes experience a driving force that increases their average velocity. Electrons and holes under equilibrium condition (no applied field) at a temperature T have a Fermi–Dirac (Boltzmann) energy distribution, with average (kinetic) energy $E_{\text{ave}} \approx 3k_BT/2$ (39 meV at 300 K). This means that, although the velocity or momentum distribution is symmetric with respect to the origin (i.e., carriers have *zero* ensemble average velocity), the root mean square (r.m.s.) carrier velocity (also called the thermal velocity) is extremely high (of the order of 10^7 cm/s). In the presence of an applied field, the ensemble average velocity assumes a value proportional to the electric field:

$$\underline{v}_{n,\text{ave}} = -\mu_n \underline{\mathcal{E}}, \quad \underline{v}_{h,\text{ave}} = \mu_h \underline{\mathcal{E}},$$

where μ_n and μ_h are the electron and hole *mobilities*, measured in $\text{cm}^2\,\text{V}^{-1}\text{s}^{-1}$. The low-field mobility depends on the interaction with lattice vibrations (phonons), impurities, etc., and typically decreases with increasing doping and increasing temperature (at least, at ambient temperature and above). For very large fields (values around 10 kV/cm, depending on the semiconductor) the average velocity saturates:

$$v_{n,\text{ave}} \to v_{n,\text{sat}}, \quad v_{h,\text{ave}} \to v_{h,\text{sat}},$$

where the saturation velocities have magnitude around 10^7 cm/s. The motion of electrons and holes due to the application of an electric field is called the *drift motion* and gives rise to the *drift (conduction) current density*:

$$\underline{J}_{n,\text{dr}} = -qn\underline{v}_{n,\text{ave}} = qn\mu_n\underline{\mathcal{E}} \tag{1.20}$$

$$\underline{J}_{h,\text{dr}} = qp\underline{v}_{h,\text{ave}} = qp\mu_h\underline{\mathcal{E}}. \tag{1.21}$$

As already remarked, the average carrier velocity is generally limited by scattering with phonons and impurities. However, for extremely small time (picoseconds) or space ($\ll$ 1 μm, also depending on material) scales, electrons and holes can experience *ballistic motion*, i.e., motion unaffected by collisions. In such conditions, the average carrier velocity can be substantially higher in the presence of strong electric fields than the static saturation velocity. Such effect is called *velocity overshoot* and plays a role in increasing the speed of transistors such as nanometer-gate field-effect transistors.

Electrons and holes also exhibit *diffusion* due to concentration gradients, yielding the *diffusion current densities*

$$\underline{J}_{n,\mathrm{d}} = qD_n\nabla n, \quad \underline{J}_{h,\mathrm{d}} = -qD_h\nabla p,$$

where D_n and D_h are the electron and hole diffusivities, respectively. At or near equilibrium the diffusivities and mobilities follow the Einstein relation $D_\alpha = (k_BT/q)\mu_\alpha$, $\alpha = n, h$.

The electron velocity–field curves of a few semiconductors are shown in Fig. 1.14. In compound semiconductors such as GaAs, InP, and InGaAs (lattice matched to InP) the electron velocity–field curve is nonmonotonic: at low electric field, electrons are mainly in the Γ minimum of the conduction band, but with increasing field they are ultimately scattered into the indirect-bandgap minima, where their velocity is lower. As a result, the average velocity decreases. This happens for electric fields of the order of 3 kV/cm in GaAs where the energy difference is about 300 meV, but for larger fields in InP where a larger energy difference is involved. The saturation velocity is, however, similar in all semiconductors, including Si, which has a monotonically increasing velocity with a much lower initial (low field) mobility. The hole mobility in compound semiconductors is similar to (or even worse than) the hole mobility in Si, and therefore n-type (or npn) transistors are preferred for high-speed applications.

1.8.2 Generation and recombination

Generation–recombination (GR) of carriers is described by generation and recombination rates for electrons and holes (G_n for the number of electrons generated per unit time and volume, R_n for the number of electrons recombining per unit time and volume, and similarly for holes) and by the electron and hole net recombination rates:

$$U_n = R_n - G_n, \quad U_h = R_h - G_h.$$

In many conditions, moreover, $U_n = U_h$. This always happens in DC stationary conditions, and whenever the GR process is band-to-band, i.e., involves direct transitions between the valence and conduction bands. In such a case, generation of an e-h pair immediately causes the increase of the electron and hole populations. However, GR can take place through intermediate traps or recombination centers, acting as electron or hole reservoirs; in such cases, the instantaneous net recombination rates of electrons and holes can be different in time-varying conditions. Since the net recombination rate should vanish in equilibrium, in many cases we can write

$$U_n = r_n(pn - n_i^2),$$

where the recombination term is proportional to pn and the generation term to n_i^2. From a physical standpoint, the pn dependence suggests that recombination involves a collision process whose probability increases with increasing electron and hole densities. Very often the recombination rate $r_n pn$ can be conveniently expressed in terms of a *lifetime* τ. For definiteness, let us consider the electron (excess) density n; the excess electron lifetime is defined by the rate equation

$$\frac{dn'}{dt} \approx -\frac{n'}{\tau_n}, \tag{1.22}$$

where in principle the electron lifetime τ_n depends inversely on the carrier population ($\tau_n = 1/r_n p$). However, τ_n is constant for excess minority carriers (electrons) in a p-type semiconductor (where $p \approx N_A$; same for the dual case of an n-type semiconductor with excess holes). The lifetime definition implies (at least, if τ_n is constant) an excess electron population exponentially decreasing with time,

$$n'(t) = n'(0)\exp(-t/\tau_n),$$

and is defined as the average time between the carrier (electron–hole pair) creation and the carrier recombination, i.e.,

$$\langle t \rangle = \frac{\int_0^\infty t n'(t)\, dt}{\int_0^\infty n'(t)\, dt} = \frac{\int_0^\infty t \exp(-t/\tau_n)\, dt}{\int_0^\infty \exp(-t/\tau_n)\, dt} = \frac{\tau_n^2}{\tau_n} = \tau_n.$$

As already recalled, GR mechanisms can be phonon-assisted or *thermal*, photon-assisted or *radiative* (optical), and, finally, assisted by other electrons or holes. Moreover, generation and recombination can occur through interband transitions (*direct* mechanisms), or through *indirect* mechanism assisted by intermediate trap levels in the forbidden band. In direct-bandgap semiconductors, direct optical GR is typically the dominant mechanism, whereas in indirect-bandgap semiconductors trap-assisted GR can be a stronger competitor to the weaker optical GR.[12]

The spatial evolution of excess carrier densities (for definiteness, we consider excess electrons with density n') in a region with negligible electric field is dominated by the carrier GR and diffusion; solution of the continuity equation for the diffusion current in the presence of carrier recombination modeled according to the lifetime approximation (1.22) yields for $n'(x)$ the exponential solution

$$n'(x) = A\exp(-x/L_n) + B\exp(x/L_n),$$

where A and B are determined from the boundary conditions and L_n is the (excess) *electron diffusion length.* A similar solution holds for excess holes, with L_h the (excess) *hole diffusion length.* From the continuity equation we obtain

$$L_\alpha = \sqrt{D_\alpha \tau_\alpha}, \quad \alpha = n, h.$$

1.8.3 Trap-assisted (Shockley–Read–Hall) recombination

Consider a semiconductor with a trap density N_t, and suppose that traps introduce, in the forbidden gap, a discrete energy level E_t. Thermal carrier transitions from the valence to the conduction bands are made easier by the trap level, since two successive transitions with $\Delta E \approx E_g/2$ (if the trap level is at midgap) are much more probable than a single transition with $\Delta E = E_g$. A detailed analysis shows that, in stationary conditions, the net trap-assisted recombination rate can be expressed as

[12] In indirect-bandgap semiconductors the optical GR is also phonon-assisted, and therefore (partly) thermal.

$$U^{SRH} = \frac{np - n_i^2}{\tau_{h0}^{SRH}(n + n_1) + \tau_{n0}^{SRH}(p + p_1)} \tag{1.23}$$

where

$$\tau_{h0}^{SRH} = \frac{1}{r_{ch}^{SRH} N_t}, \quad \tau_{n0}^{SRH} = \frac{1}{r_{cn}^{SRH} N_t}.$$

The parameters r_{ch}^{SRH} and r_{cn}^{SRH} are the trap capture coefficients for electrons and holes, while

$$p_1 = n_i g \exp\left(\frac{E_{Fi} - E_t}{k_B T_0}\right), \quad n_1 = n_i \frac{1}{g} \exp\left(-\frac{E_{Fi} - E_t}{k_B T_0}\right).$$

Trap-assisted GR is called Shockley–Read–Hall (SRH) generation–recombination. The coefficient g is an nondimensional parameter, called the trap degeneracy factor, and E_{Fi} is the intrinsic Fermi level, close to midgap. Expression (1.23) can be simplified under some important conditions.

Consider first a doped semiconductor, e.g., n-type, and suppose $r_{ch}^{SRH} \approx r_{cn}^{SRH} = r_c^{SRH}$. In low-injection conditions (excess electrons concentration negligible vs. the equilibrium electron concentration, $n' \ll N_D$; excess hole concentration $p' \gg n_i^2/N_D$) we have $np - n_i^2 \approx np'$; therefore:

$$U^{SRH} \approx \frac{p'}{\tau_0^{SRH}}\left[1 + \frac{2n_i}{n}\cosh\left(\frac{E_{Fi} - E_t}{k_B T_0}\right)\right]^{-1},$$

where

$$\tau_0^{SRH} = \frac{1}{r_c^{SRH} N_t}. \tag{1.24}$$

Since $n_i \ll n$, the minimum lifetime is obtained for a trap energy near midgap; such a minimum lifetime will be given by (1.24). The lifetime is independent of semiconductor doping and decreases for increasing trap density. Midgap traps are often called *recombination centers*, owing to their ability to cause recombination in a semiconductor and to reduce the thermal lifetime vs. the intrinsic material. Similar remarks apply to the electron lifetime in a p-doped semiconductor. A second important case concerns the role of SRH recombination as a *competitor to radiative recombination* in high-injection conditions. Suppose now that for quasi-neutrality $n \approx p$ and assume for simplicity again $r_{ch}^{SRH} \approx r_{cn}^{SRH} = r_c^{SRH}$; we have

$$U^{SRH} = \frac{p}{2\tau_0^{SRH}}\left[1 + \frac{n_i}{n}\cosh\left(\frac{E_{Fi} - E_t}{k_B T_0}\right)\right]^{-1}.$$

For traps near the midgap, the hole (or electron) lifetime in high injection is therefore

$$\tau_{0hi}^{SRH} = 2\tau_0^{SRH} = \frac{2}{r_c^{SRH} N_t}.$$

Finally, if the quasi-neutrality condition does not hold but $np \gg n_i^2$ and $E_t \approx E_{Fi}$, we also have, with the condition $r_{ch}^{SRH} \approx r_{cn}^{SRH} = r_c^{SRH}$, that

$$U^{SRH} \approx \frac{np}{\tau_0^{SRH}(n+p)}.$$

The capture coefficients can be evaluated as the product of two parameters, the *thermal velocity* v_{th} and the electron or hole trap *cross section* σ_n or σ_h:

$$r_{cn}^{SRH} = v_{th}\sigma_n, \quad r_{ch}^{SRH} = v_{th}\sigma_h.$$

Trap cross sections are of the order of 10^{-15} cm^2 in Si for electrons and holes while, for III-V materials, $\sigma_n \approx 10^{-14}$ cm^2, $\sigma_h \approx 10^{-13}$ cm^2. Finally, the thermal velocity is $v_{th} = \sqrt{3k_BT/m^*}$, with order of magnitude $v_{th} \approx 10^7$ cm/s.

Trap-assisted recombination is a strong competitor of radiative recombination in indirect-gap semiconductors. In fact, consider that in Si we have

$$r_c^{SRH} \approx 10^5 \text{ m/s} \cdot 10^{-17} \text{ m}^2 = 10^{-12} \text{ m}^3\text{/s},$$

so that, for a middle-gap trap density as low as $N_t = 10^{14}$ cm^{-3}, one obtains

$$\tau_0^{SRH} = \frac{1}{10^{-12} \cdot 10^{20}} = 10 \text{ ns},$$

a value close to the radiative lifetime in indirect-bandgap semiconductors. The situation is much more favorable in direct-bandgap semiconductors, where the radiative lifetime is lower.

1.8.4 Auger recombination and generation by impact ionization

The electron- or hole-assisted recombination is called *Auger recombination*, and the related rate is proportional to p^2n or pn^2, implying proportionality not only with respect to the colliding populations (electrons and holes) but also to the population of the energy suppliers. Due to this dependence, the Auger recombination is important (and is indeed an unwanted competitor of the radiative recombination) in high-injection devices like semiconductor lasers. The inverse process of the Auger recombination is the *generation by impact ionization*. In high-field conditions (i.e., for fields of the order of 100 kV/cm), electrons and holes gather enough energy from the electric field between two successive scattering events (i.e., collisions with phonons, impurities or – less important – other carriers) to be able to interact with another electron and promote it to the conduction band. Each electron or hole is therefore able to generate, over a certain length, a number of electron–hole pairs, that undergo in turn the same process (energy increase, scattering and e-h pair generation). The resulting chain can lead to diverging current, i.e., to *avalanche breakdown* in the semiconductor. Avalanche breakdown occurs for electric fields of the order of the breakdown field, which increases exponentially with the material gap; see Fig. 1.35.

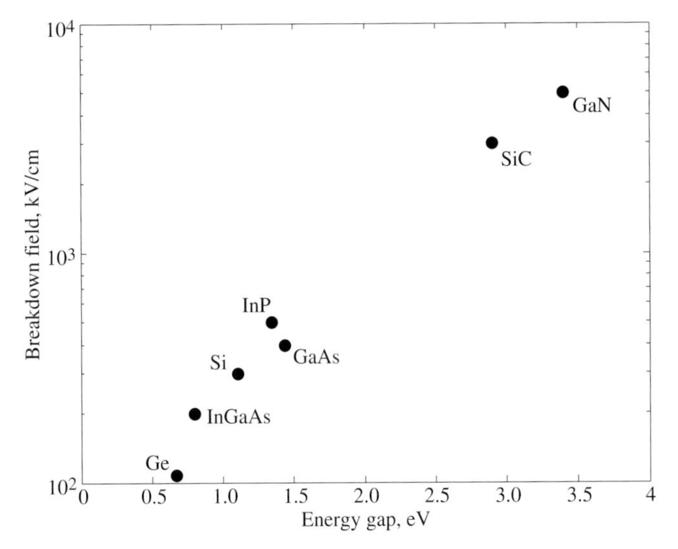

Figure 1.35 Avalanche breakdown fields for some important semiconductors. Insulators (such as C, SiO_2, Si_3N_4) have breakdown fields in excess of 10^4 kV/cm.

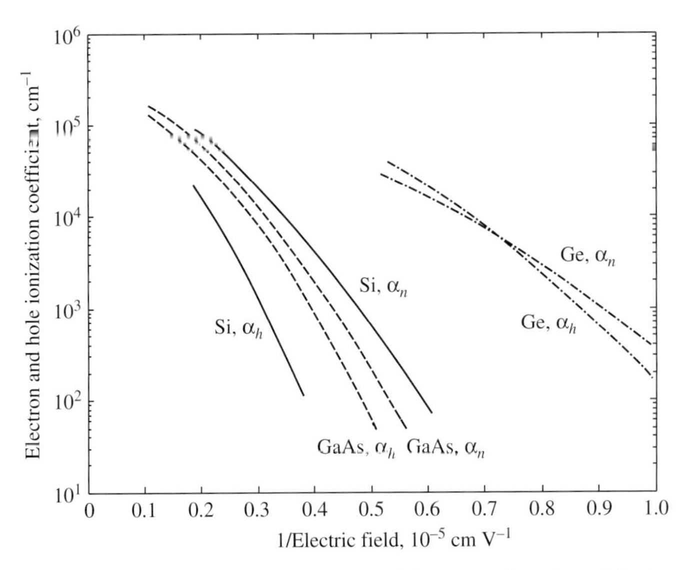

Figure 1.36 Electron and hole impact ionization coefficients as a function of the inverse of the electric field for Si, Ge, GaAs. Adapted from [9], Figs. 2, 4 and 6.

The impact ionization phenomenon can be described by the following carrier generation model:

$$\nabla \cdot \underline{J}_n = -qG_n - qG_h, \quad \nabla \cdot \underline{J}_h = qG_n + qG_h,$$

where

$$G_\alpha = \frac{1}{q}\alpha_\alpha(\mathcal{E})J_\alpha, \quad \alpha = n, h. \tag{1.25}$$

The impact ionization coefficients α_n and α_h (or α and β, dimension cm^{-1}) show a strong increase with the electric field; see Fig. 1.36 [9]. In Si, $\alpha_n \gg \alpha_h$, while $\alpha_n \approx \alpha_h$ in GaAs and Ge. The condition of different ionization coefficients will be shown to represent an optimum condition in low-noise avalanche photodiodes, which exploit the avalanche generation of electron–hole pairs to amplify the photocurrent.

1.9 Questions and problems

1.9.1 Questions

1. Describe the crystal structure of Si and of GaAs. Explain the difference between the diamond and zinc-blende lattice cells.
2. What is the *wurtzite* crystal structure? Quote at least one important wurtzite semiconductor.
3. Explain why ferroelectric crystals already have, with zero applied electric field, a net dipole moment. What is the Curie temperature?
4. Quote at least one important ferroelectric material exploited in electrooptic modulators.
5. What is the behavior of the semiconductor energy gap E_g of direct-bandgap semiconductors vs. the lattice constant a?
6. Give examples of point defects in a crystal. Explain what kind of defect is a dopant atom.
7. Qualitatively sketch the equilibrium band diagram of a n-type AlGaAs – intrinsic GaAs heterostructure. Assume an Al fraction $x = 0.3$. Explain why the heterojunction generates a dipole charge layer (ionized donors on the AlGaAs side, free electrons on the GaAs side). (Hint: the conduction band and valence band discontinuities are given by Eq. (1.14); assume that the difference in AlGaAs and GaAs affinities is $\approx \Delta E_c$ and impose a constant Fermi level.)
8. To grow a heterostructure, is *perfect* lattice matching needed? Explain.
9. Two materials (with different energy gaps and different doping) form a heterostructure. Suppose that the initially isolated materials are ideally connected, and that electrons and holes move to reach the equilibrium condition. Explain whether the carrier motion takes place:
   (a) in opposite directions, always;
   (b) in directions depending on the E_c and E_v discontinuity;
   (c) always from the wide-gap to the narrowgap material.
   (Hint: before equilibrium the local Fermi levels act as quasi-Fermi levels for electrons and holes, whose gradient is the driving force of the two carriers, having opposite charge. When the two materials are joined and before equilibrium the two quasi-Fermi levels have a jump, implying the same driving force for electrons and holes.)
10. What is a *strained* or *pseudomorphic* heterostructure?
11. Justify the importance of the InGaAsP alloy in optoelectronic devices.

12. HgTe has a negative energy gap and therefore metal-like electronic behavior. Explain this statement.
13. A semiconductor has a direct bandgap of 2 eV and an indirect bandgap of 1.5 eV. Justify this statement and define the semiconductor E_g.
14. Sketch the bandstructures of GaAs, InP, and AlAs.
15. Explain the transition between GaAs (direct bandgap) and AlAs (indirect bandgap) when changing the alloy Al content.
16. What are the Γ, X and L points in the Brillouin zone for a cubic crystal? Explain the meaning of the *irreducible wedge* in the FBZ.
17. In a QW made with an heteroepitaxial layer (double heterojunction) the heavy and light hole degeneracy in the Γ point is typically removed. Explain this remark and suggest some consequences for the material density of states and transport properties. Describe the effect of strain on the removal of the hole degeneracy.
18. Sketch the effect of strain on the semiconductor bandstructure for a bulk direct-bandgap semiconductor.
19. The energy–momentum relation of a semiconductor is approximated by a second-order surface (a parabola in the variable k for a spherical minimum) near the conduction band minimum. Justify this approximation and explain the meaning of the *electron effective mass* in this context. What is the nonparabolic correction to the dispersion relation?
20. Explain the meaning of *isotropic* and *anisotropic* effective mass for the electrons.
21. Sketch the electron velocity–field curve for Si and for GaAs. Explain the origin of negative differential mobility in GaAs.
22. Explain why the onset of the negative differential mobility region occurs at higher electric fields in InP when compared with GaAs.
23. Sketch the density of states in 3D (bulk), 2D (quantum well), 1D (quantum wire), 0D (quantum dot).
24. What are the degrees of freedom of a carrier in a quantum well, a quantum wire, a quantum dot?
25. Explain why a quantum well allows for carrier and photon confinement.
26. Explain how carriers can be injected and removed from a quantum dot, taking into account that carrier motion is not allowed in the dot.
27. A quantum well locally modifies the electronic structure of the bulk semiconductor. Justify this remark.
28. What is a superlattice, and what features has the superlattice bandstructure?
29. Explain the difference between a *degenerate* and a *nondegenerate* semiconductor.
30. Is the mass action law $pn = n_i^2$ always true, or only in a nondegenerate semiconductor?
31. Describe the behavior of the carrier density in a doped semiconductor as a function of the lattice temperature.
32. Explain the physics behind thermal, radiative and Auger GR mechanisms in a semiconductor.
33. The impact ionization coefficients in a semiconductor increase with increasing applied electric field. Justify this fact.

34. Explain why the avalanche breakdown field is larger in a wide-gap semiconductor than in a narrow-gap semiconductor.

1.9.2 Problems

1. Sketch the planes denoted, according to the Miller notation, as (101), (001), $(\bar{1}\bar{1}\bar{1})$.
2. Consider a Si crystal that has been cut 1 degree off the (001) direction toward the (110) direction. Due to the misalignment, steps form on the surface. Assuming that the step height is a, where a is the lattice constant, what is the step width?
3. In an out-of-equilibrium semiconductor, where $N_c = 10^{18}$ cm^{-3}, $N_v = 10^{19}$ cm^{-3}, evaluate the position of the quasi-Fermi levels with respect to the conduction and valence band edges, assuming $n \approx p \approx 2 \times 10^{18}$ cm^{-3}. (Hint: use the Joyce–Dixon approximation.)
4. Assume that the defect density in Si ranges from 10^{13} cm^{-3} (intrinsic material) to 10^{19} cm^{-3} (highly doped material). What is the average distance between impurity atoms in the two cases? For comparison, evaluate the number of Si atoms per unit volume.
5. Evaluate the composition range of InGaAsP achieving lattice matching to InP (use the Vegard law). Repeat the computation for InGaAs. For InGaAsP, evaluate the alloy lattice-matched to InP and able to emit 1.3 μm wavelength. Exploit the following values for the semiconductor parameters:

$$\text{GaAs: } a_{\text{GaAs}} = 5.68 \text{ Å}, E_{g\text{GaAs}} = 1.42 \text{ eV}$$

$$\text{GaP: } a_{\text{GaP}} = 5.45 \text{ Å}, E_{g\text{GaP}} = 2.27 \text{ eV}$$

$$\text{InP: } a_{\text{InP}} = 5.87 \text{ Å}, E_{g\text{InP}} = 1.34 \text{ eV}$$

$$\text{InAs: } a_{\text{InAs}} = 6.06 \text{ Å}, E_{g\text{InAs}} = 0.36\text{eV}.$$

6. A QW with infinite potential discontinuity is realized in the GaAs conduction band. The quantum well thickness is $d = 20$ nm. Evaluate the first two energy levels of the QW and the electron sheet concentration, assuming that the electron Fermi level is 20 meV above the first energy level. Try to repeat the computation assuming a finite conduction band well.
7. A quantum dot is made with a cubic potential box with infinitely high walls. The box side is 5 nm. Exploiting for the electron an effective mass of $0.1m_0$, sketch the density of states of the QD. Consider an array of quantum dots with 20 nm spacing between the centers.
8. A deep submicrometer CMOS has a gate length of 40 nm. How many Si atoms can be found along the gate length?
9. Calculate the cell density per unit volume in Si, GaAs, and GaN. (Hint: for GaN, exploit the equivalent cubic cell.)
10. In GaAs, assume that the electron lifetime, for an (excess) electron concentration $n = 10^{17}$ cm^{-3}, is $\tau_n = 1$ μs at 300 K. Evaluate the net recombination rate U_n.
11. The momentum of an electron in the Γ valley of GaAs is $|\underline{k}| = 0.1/a$, where a is the lattice constant. What is the electron energy from the conduction band

edge E_c? What is the electron semiclassical velocity? Can the electron be easily scattered into the lowest secondary valley, located at the L point (assume $\Delta E = E_{cL} - E_{c\Gamma} = 300$ meV)? Supposing that the electron energy is ΔE from the conduction band edge, what would be its momentum? Compare the result with the momentum difference between the L and Γ minima. Repeat the problem exploiting the nonparabolic correction $E_k (1 + \alpha E_k) = (\hbar k)^2 / 2m_n^*$ with $\alpha = 0.67$ eV^{-1}.

12. In a GaAs sample the radiative lifetime is 0.5 ns. Considering a trap defect at midgap with cross section $\sigma_n \approx \sigma_h \approx \sigma = 5 \times 10^{-15}$ m^2 (assume for the thermal velocity at 300 K $v_{th} = 10^7$ cm/s), evaluate the trap concentration for which the nonradiative minority carrier lifetime is the same as the radiative lifetime $\tau_o \approx 0.5$ ns (a) at 300 K, (b) at 77 K. Evaluate in such conditions the electron and hole diffusion lengths. Assume $\mu_n = 4000$ cm^2 V^{-1} s^{-1}, $\mu_h = 600$ cm^2 V^{-1} s^{-1}.
13. An electron is injected into a high-field semiconductor sample with length $W = 10$ μm. The electron ionization coefficient is $\alpha_n = 10^4$ cm^{-1}. Evaluate the number of carriers directly generated by the carrier through impact ionization while crossing the sample, and the total number of carriers generated in a direct or secondary way. Assume that the hole impact ionization is zero.
14. In a semiconductor sample $\alpha_n = \alpha_h = \alpha$. Derive the breakdown condition, i.e., the condition in which the current diverges over a length W. Assume that an electron current is injected in W as $J_n(W)$ and that the hole current in 0 is $J_h(0) = 0$. (Hint: for a full discussion, see Section 4.11.1.)

2 Semiconductor optical properties

2.1 Modeling the interaction between EM waves and the semiconductor

The interaction between an electromagnetic (EM) wave and a semiconductor can be analyzed at several levels. EM waves can be described as classical fields (i.e., in terms of the electric and magnetic field) or, according to quantum mechanics, as a set of *photons* traveling with the speed of light. The crystal's response to the EM wave may be modeled in terms of macroscopic parameters such as the permittivity and permeability (the *macroscopic picture* of the interaction) or with reference to the microscopic interaction mechanisms between carriers and photons, leading to photon absorption or emission (spontaneous and stimulated) and, correspondingly, to electron–hole (e-h) pair generation or recombination (the *microscopic picture*). Moreover, we can look at the interaction between the EM wave and the semiconductor from two closely related viewpoints: the *EM wave standpoint*, yielding the wave absorption and gain, and the *semiconductor standpoint*, leading to the e-h generation and recombination rates.

Figure 2.1 presents a summary of the EM spectrum with decreasing energy and increasing wavelength. Two frequency (energy) bands are of main interest for high-speed optoelectronic applications.

The first band concerns the *semiconductor sources*, whose energy ranges from the near infrared to the UV (roughly 3 octaves, from 0.5 eV to 4 eV). Semiconductor energy gaps cover the same interval. The second band is the frequency range corresponding to *radiofrequencies* and *microwaves* (up to 40 GHz) or even *millimeter waves*, up to 300 GHz, where the spectra of the modulating and/or detected electrical signals carried by an optical transmission system are found. Electronic devices and circuits for high-speed optoelectronic systems operate in this range.

The visible spectrum (from 700 to 400 nm, see Table 2.1), is relevant to several applications, also involving, as might be expected, human vision, but in fact most telecom systems work at lower frequencies, around 800 nm ("first window" of optical fibers) or around 1300 and 1550 nm ("second window" and "third window" of optical fibers), see Fig. 1.22. As already noted, the second and third windows correspond to minimum dispersion and attenuation, respectively.

Table 2.1 The visible spectrum and the colors.

Color	Wavelength (nm)	Frequency (10^3 THz)	Energy (eV)
Red (limit)	700	4.29	1.77
Red	650	4.62	1.91
Orange	600	5.00	2.06
Yellow	580	5.16	2.14
Green	550	5.45	2.25
Cyan	500	5.99	2.48
Blue	450	6.66	2.75
Violet (limit)	400	7.50	3.10

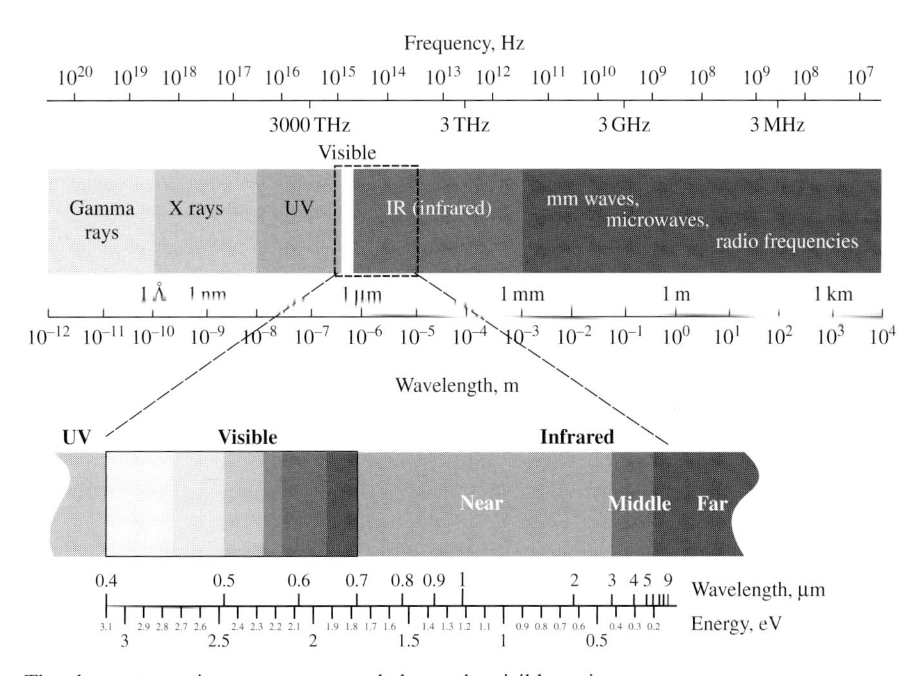

Figure 2.1 The electromagnetic spectrum, expanded near the visible region.

2.2 The macroscopic view: permittivities and permeabilities

We recall now a few macroscopic properties of the interaction between the EM wave and matter. Assuming the electric field $\underline{\mathcal{E}}$ (V/m) and the magnetic field $\underline{\mathcal{H}}$ (A/m) as causes, the effects produced in the interacting material are the dielectric displacement vector $\underline{\mathcal{D}}$ (C/m^2) and the magnetic induction $\underline{\mathcal{B}}$ (tesla, T), respectively. In a linear, *memoriless* (or *nondispersive*) isotropic medium the instantaneous value of the cause is directly proportional to the instantaneous value of the effect, so that[1]

$$\underline{\mathcal{D}}(t) = \epsilon \underline{\mathcal{E}}(t), \quad \underline{\mathcal{B}}(t) = \mu \underline{\mathcal{H}}(t),$$

where ϵ is the dielectric permittivity (farad per meter, F/m) and μ is the magnetic permeability (henry per meter, H/m). Vacuum (or *free space*) is a particularly simple example of a linear, nondispersive medium, with parameters

$$\epsilon = \epsilon_0 = 8.85 \times 10^{-12} \text{ F/m}, \quad \mu = \mu_0 = 4\pi \times 10^{-7} \text{ H/m}.$$

The permittivity and permeability can in turn be expressed in terms of the *in vacuo* parameter as

$$\epsilon = \epsilon_r \epsilon_0, \quad \mu = \mu_r \mu_0,$$

where ϵ_r is the relative permittivity and μ_r is the relative permeability. In all cases considered here, $\mu_r = 1$, i.e., the medium is nonmagnetic. The material *refractive index* (or index of refraction) is $n_r = \sqrt{\epsilon_r}$.

For time-varying harmonic fields (i.e., having time dependence $\exp(j\omega t)$) we associate to the electrical fields the complex phasors $\underline{D}(\omega)$ and $\underline{E}(\omega)$ such that

$$\underline{\mathcal{E}}(t) = \sqrt{2}\,\text{Re}\left[\underline{E}(\omega)\exp(j\omega t)\right], \quad \underline{\mathcal{D}}(t) = \sqrt{2}\,\text{Re}\left[\underline{D}(\omega)\exp(j\omega t)\right].$$

The two phasors are related by a complex, frequency-dependent dielectric permittivity as

$$\underline{D}(\omega) = \epsilon(\omega)\underline{E}(\omega) = (\epsilon' - j\epsilon'')\underline{E}(\omega),$$

where the imaginary part is related to material losses. In a *dispersive* material, the complex dielectric permittivity varies with frequency. With the complex permittivity we can associate a complex refractive index $n_r = \sqrt{\epsilon_r}$, whose imaginary part is associated with losses:

$$n_r(\omega) = n_r'(\omega) - jn_r''(\omega).$$

In general, EM waves interact with all the electric charges or dipoles in a medium, including (with decreasing mass) molecules, atoms, and electrons. With increasing frequency, each of those interactions ceases in turn to be effective at a certain cutoff frequency, above which the relevant contribution to the dielectric response vanishes. Moreover, close to each cutoff frequency losses have a peak. The resulting overall behavior is qualitatively as shown in Fig. 2.2.

Since the permittivity describes the frequency-domain response of a causal system (where the input is the electric field and the output the dielectric displacement field),

[1] Given an independent variable x (the cause) and a dependent variable y (the effect) we say that the relation between x and y is *memoriless* ($y(t) = f(x(t))$) when the instantaneous value of y depends only on the instantaneous value of x. If this does not hold true, the relation is called *dispersive* or *with memory* and formally we write $y(t) = f(x(t), \mathrm{d}/\mathrm{d}t)$.

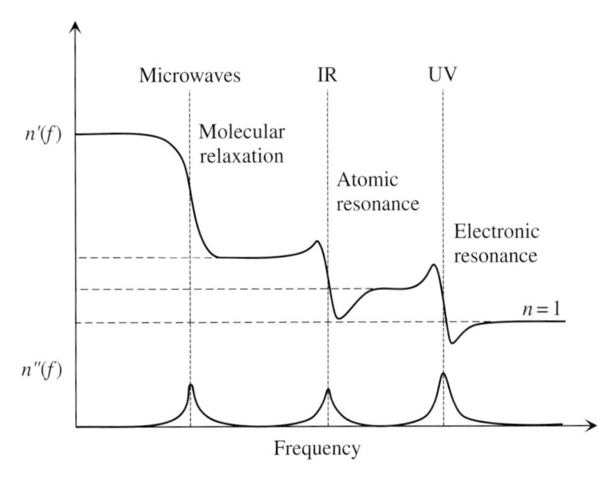

Figure 2.2 Qualitative behavior of the real and imaginary part of the refractive index as a function of the EM wave frequency.

it has to satisfy to a set of integral relations between the real and imaginary parts of $\epsilon(\omega) = \epsilon'(\omega) - \mathrm{j}\epsilon''(\omega)$, known as the *Kramers–Kronig relations*:

$$\epsilon'(\omega) = \epsilon_0 + \frac{2}{\pi} \cdot \mathcal{P} \int_0^\infty \frac{\Omega\epsilon''(\Omega)}{\Omega^2 - \omega^2} \, \mathrm{d}\Omega$$

$$\epsilon''(\omega) = -\frac{2\omega}{\pi} \cdot \mathcal{P} \int_0^\infty \frac{\epsilon'(\Omega) - \epsilon_0}{\Omega^2 - \omega^2} \, \mathrm{d}\Omega,$$

where $\mathcal{P}$ denotes the principal part. Since the real and imaginary parts of the permittivity are related, the real part can be derived from measurements of the imaginary part and vice versa. Moreover, a variation of the real part (due, e.g., to the electrooptic effect, see Section 6.3.1) implies a variation of the imaginary part (i.e., of the absorption), and vice versa. This has a significant impact on the spurious frequency modulation (chirp) of electroabsorption modulators (EAMs) and of directly modulated lasers, see Section 6.8.3 and 5.12.3.

Example 2.1: Suppose that the real part of the dielectric permittivity has a step behavior:

$$\epsilon'(\omega) = \begin{cases} \epsilon, & \omega \leq \omega_0 \\ \epsilon_0 & \omega > \omega_0. \end{cases}$$

Derive the corresponding behavior of the imaginary part from the Kramers–Kronig relations.

Since

$$\epsilon'(\omega) - \epsilon_0 = \begin{cases} \epsilon - \epsilon_0, & \omega \leq \omega_0 \\ 0 & \omega > \omega_0, \end{cases}$$

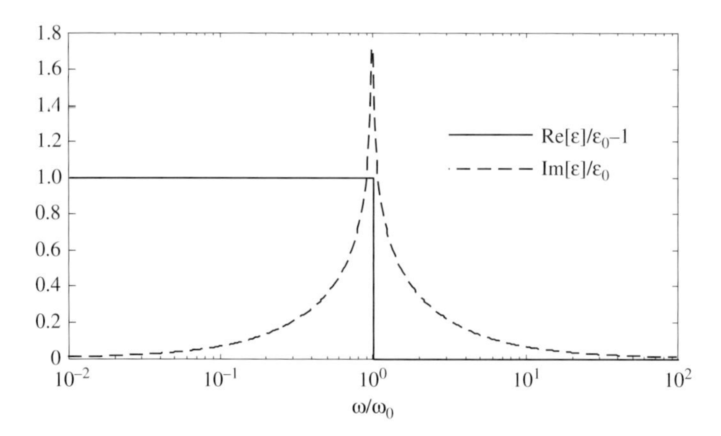

Figure 2.3 Imaginary part of permittivity in the presence of a step in the real part.

the Kramers–Kronig integral becomes

$$
\begin{aligned}
\epsilon''(\omega) &= -\frac{2\omega}{\pi} \cdot \mathcal{P} \int_0^{\infty} \frac{\epsilon'(\Omega) - \epsilon_0}{\Omega^2 - \omega^2} \, d\Omega = -\frac{2\omega}{\pi} \cdot \mathcal{P} \int_0^{\omega_0} \frac{\epsilon - \epsilon_0}{\Omega^2 - \omega^2} \, d\Omega \\
&= -\frac{2\omega}{\pi} (\epsilon - \epsilon_0) \lim_{\xi \to 0} \left[\int_0^{\omega - \xi} \frac{\epsilon - \epsilon_0}{\Omega^2 - \omega^2} \, d\Omega + \int_{\omega - \xi}^{\omega_0} \frac{\epsilon - \epsilon_0}{\Omega^2 - \omega^2} \, d\Omega \right] \\
&= -\frac{\epsilon - \epsilon_0}{\pi} \lim_{\xi \to 0} \left(\ln \left| \frac{-\xi}{2\omega} \right| - \ln \left| \frac{0 - \omega}{0 + \omega} \right| + \ln \left| \frac{\omega_0 - \omega}{\omega_0 + \omega} \right| - \ln \left| \frac{\xi}{2\omega} \right| \right) \\
&= \frac{\epsilon - \epsilon_0}{\pi} \ln \left| \frac{\omega_0 + \omega}{\omega_0 - \omega} \right| .
\end{aligned}
$$

An example of the resulting behavior is shown in Fig. 2.3 for the normalized value $\epsilon = 2\epsilon_0$ at low frequency; removing the step discontinuity by modeling the permittivity variation through a continuous function, the singularity in the imaginary part peak also disappears. In general, a variation of the real part corresponds to a peak in material losses.

In a lossy medium, the phasor associated with the electric field of a *plane wave* (i.e., a wave whose constant phase surfaces are planes, often used as a local approximation for a generic EM wave) propagating in the z direction is

$$
\underline{E} = \underline{E}_0 e^{-jkz} = \left[\underline{E}_0 e^{-\bar{\alpha} z} \right] e^{-j\beta z},
$$

since

$$
-jkz = -j\omega n_r \sqrt{\epsilon_0 \mu_0} z = j\frac{\omega}{c_0} \left(n_r' - jn_r'' \right) z = -j\beta z - \bar{\alpha} z,
$$

where $k = \beta - \mathrm{j}\bar{\alpha}$ (or $\gamma = \bar{\alpha} + \mathrm{j}\beta$) is the *complex propagation constant*; β is the *propagation constant*, and $\bar{\alpha}$ is the *field attenuation*:

$$\beta = \frac{\omega}{c_0} n'_r = \frac{2\pi}{\lambda} = \frac{2\pi}{\lambda_0} n'_r \quad \text{(rad/m)}$$
$$\bar{\alpha} = \frac{\omega}{c_0} n''_r \quad (\text{Np/m or m}^{-1}),$$

and λ is the wavelength, λ_0 is the free space wavelength, c_0 is the light velocity *in vacuo*.[2] Therefore, the magnitude of a harmonic electric field decreases exponentially in the direction of propagation, while its phase decreases linearly with distance.

From the field magnitude $|E(z)| = |\underline{E}_0| \exp(-\bar{\alpha}z)$, we obtain that the Poynting vector decreases as

$$\underline{S} = \frac{|\underline{E}(z)|^2}{Z_0}\widehat{z} = \frac{|\underline{E}_0|^2 \exp(-2\bar{\alpha}z)}{Z_0}\widehat{z},$$

where $Z_0 = \sqrt{\mu/\epsilon}$ is the medium impedance. The component of $\underline{S}$ along the propagation direction is the power density associated with the optical wave, P_{op} (W/m^2); one has, therefore,

$$P_{op}(z) = \underline{S} \cdot \widehat{z} = P_{op}(0) \exp(-2\bar{\alpha}z) \equiv P_{op}(0) \exp(-\alpha z),$$

where $\alpha = 2\bar{\alpha}$ is the *absorption*. The absorption coefficient α is always associated with the same symbol as the attenuation (denoted *only in this chapter* as $\bar{\alpha}$). Absorption and attenuation can be expressed in *natural units* (m^{-1} or Np/m, Neper per meter) and in this case they differ by a factor 2. However in decimal *log units* one has

$$\bar{\alpha}z|_{\text{dB}} = -20\log_{10}\frac{|\underline{E}(z)|}{|\underline{E}_0|} = 20\log_{10}\mathrm{e}^{\bar{\alpha}z} = 8.686\bar{\alpha}z \rightarrow \bar{\alpha}|_{\text{dB/m}} = 8.686\,\bar{\alpha}|_{\text{Np/m}}$$
$$\alpha z|_{\text{dB}} = -10\log_{10}\frac{P_{op}(z)}{P_{op}(0)} = 10\log_{10}\mathrm{e}^{\alpha z} = 4.343\alpha z \rightarrow \alpha|_{\text{dB/m}} = 8.686\,\bar{\alpha}|_{\text{Np/m}},$$

i.e., attenuation and absorption in dB/m are expressed *by the same number*. Remember that the value in dB/m must not be used in connection with $\exp(-\alpha z)$ or $\exp(-\overline{\alpha}z)$ expressions; the following conversion formulae can be used ($\overline{\alpha}$ is the attenuation, α the absorption):

$$\overline{\alpha}|_{\text{dB/m}} = 8.6859\,\overline{\alpha}|_{\text{Np/m}}$$
$$\alpha|_{\text{dB/m}} = 4.3429\,\alpha|_{\text{Np/m}} = 8.6859\,\overline{\alpha}|_{\text{Np/m}}$$
$$\overline{\alpha}|_{\text{Np/m}} = 0.1151\,\overline{\alpha}|_{\text{dB/m}} = 11.51\,\overline{\alpha}|_{\text{dB/cm}}$$
$$\alpha|_{\text{Np/m}} = 0.2303\,\alpha|_{\text{dB/m}} = 0.2303\,\overline{\alpha}|_{\text{dB/m}} = 23.03\,\alpha|_{\text{dB/cm}}.$$

[2] In discussing the microscopic interaction between photons and electrons the photon energy is, however, always related to the *free space wavelength*, which for simplicity will be denoted in that context as λ.

2.2.1 Isotropic vs. anisotropic media

The $\mathcal{D}$ and $\mathcal{E}$ fields are parallel only in *isotropic* materials. In an *anisotropic* medium, $\underline{\mathcal{D}}$ and $\underline{\mathcal{E}}$ will, in general, have different directions. This can be modeled by introducing the permittivity as a matrix, i.e., as the Cartesian representation of a tensor:

$$\underline{D} = \boldsymbol{\epsilon} \cdot \underline{E} \rightarrow \begin{pmatrix} D_x \\ D_y \\ D_z \end{pmatrix} = \begin{pmatrix} \epsilon_{xx} & \epsilon_{xy} & \epsilon_{xz} \\ \epsilon_{yx} & \epsilon_{yy} & \epsilon_{yz} \\ \epsilon_{zx} & \epsilon_{zy} & \epsilon_{zz} \end{pmatrix} \begin{pmatrix} E_x \\ E_y \\ E_z \end{pmatrix}.$$

In all media considered, the permittivity matrix is symmetrical, i.e., $\epsilon_{ij} = \epsilon_{ji}$. It can be shown that a proper choice of the rectangular reference system (the *principal axes*) leads to a *diagonal* permittivity matrix. The permittivity tensor is diagonal in an *isotropic medium* for any orientation of the Cartesian axes; moreover, all diagonal values are equal, i.e., $\boldsymbol{\epsilon} = \epsilon \boldsymbol{I}$, where $\boldsymbol{I}$ is the identity matrix. In an *anisotropic medium* the representation of the permittivity tensor in the principal axes can assume different features, according to whether all diagonal values are different, or two are equal and one is different. We have three possible cases:

1. All diagonal values are different, i.e.

$$\boldsymbol{\epsilon} = \begin{pmatrix} \epsilon_{xx} & 0 & 0 \\ 0 & \epsilon_{yy} & 0 \\ 0 & 0 & \epsilon_{zz} \end{pmatrix}.$$

   In this case, the medium is *biaxial*.

2. Two diagonal values are equal: assume for instance that $\epsilon_{xx} = \epsilon_{yy} = \epsilon_o$ (the *ordinary permittivity*), while $\epsilon_{zz} = \epsilon_e$ (the *extraordinary permittivity*). In this case, we have

$$\boldsymbol{\epsilon} = \begin{pmatrix} \epsilon_o & 0 & 0 \\ 0 & \epsilon_o & 0 \\ 0 & 0 & \epsilon_e \end{pmatrix}.$$

   The z axis is the *optical* or *extraordinary axis*, while any direction in the plane orthogonal to z is the *ordinary axis*. Therefore, the medium is isotropic with respect to rotations in the plane orthogonal to the optical axis. In this case, the medium is *uniaxial*.

3. All diagonal values are equal:

$$\boldsymbol{\epsilon} = \begin{pmatrix} \epsilon & 0 & 0 \\ 0 & \epsilon & 0 \\ 0 & 0 & \epsilon \end{pmatrix} = \epsilon \begin{pmatrix} 1 & 0 & 0 \\ 0 & 1 & 0 \\ 0 & 0 & 1 \end{pmatrix} = \epsilon \boldsymbol{I}.$$

   In this case, the medium is *isotropic*. All cubic semiconductors are isotropic.

A summary of the cases considered is shown in Fig. 2.4.

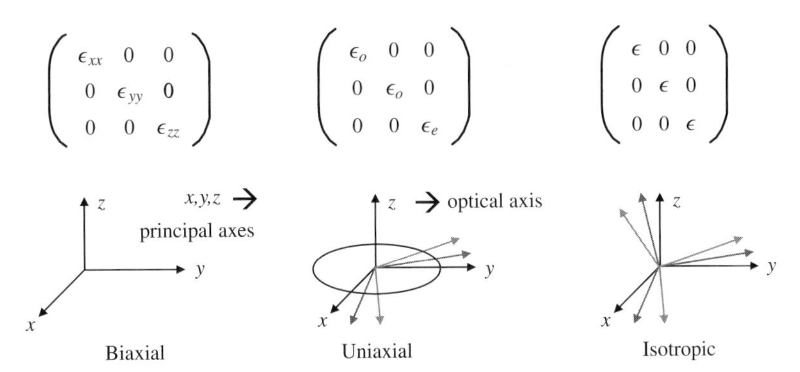

Figure 2.4 Permittivity matrix in the principal reference frame (above) and optical axes (below) in a biaxial, uniaxial and isotropic medium. In the isotropic case the principal axes are arbitrary.

2.3 The microscopic view: EM wave–semiconductor interaction

At a microscopic level, an EM wave with frequency f is interpreted as a collection of photons of energy $E_{ph} = hf = \hbar\omega$. Photons are zero-mass particles, traveling with speed c_0/n_r. Photons have momentum $\underline{p} = \hbar\underline{k}$, where $\underline{k}$ is the propagating wave wavevector;[3] the magnitude of the momentum is $2\pi/\lambda$. Always at a microscopic level, semiconductors are containers of charged particles (electrons and holes) that interact with the EM wave photons. The interaction can be visualized as a collision or scattering process; the possibility of interaction is quite obvious, since charged particles in motion are subject to the *Coulomb force* (EM wave electric field) and to the *Lorentz force* (magnetic field of the EM wave). However, the semiconductor response is peculiar when compared with the response of a generic dielectric material due to the presence of band-to-band processes. In fact, the useful semiconductor response is dominated by the ability of radiation to cause band-to-band carrier transitions with corresponding *emission* or *absorption* of a photon.

A useful relation exists between the EM wave photon energy and the related wavelength; since $\lambda f = c_0$ one has

$$E_{ph} = hf = \frac{hc_0}{\lambda} = \frac{4.136 \times 10^{-15}\ \text{eV s} \cdot 2.998 \times 10^{8}\ \text{m/s}}{\lambda|_{\mu\text{m}} \times 10^{-6}\ \text{m}} = \frac{1.24}{\lambda|_{\mu\text{m}}}\ \text{eV}. \quad (2.1)$$

Therefore, wavelengths of the order of 1 μm are associated with energies of the order of 1 eV, which in fact correspond to "typical" semiconductor bandgaps.

In the EM wave–semiconductor interaction, three cases are possible according to the value of E_{ph} vs. the energy gap E_g:

1. $E_{ph} < E_g$, as in RF, microwaves, and far infrared (FIR): the interaction is weak and does not involve band-to-band processes, but only the dielectric response and interband processes (e.g., the so-called free electron/hole absorption);

[3] For brevity the wavevector k is often referred to as *momentum*, in particular when discussing particles.

2. $E_{ph} \approx E_g$ and $E_{ph} > E_g$, as in near infrared (NIR), visible light, and ultraviolet (UV): light interacts strongly through band-to-band processes leading to the *generation–recombination* of e-h pairs and, correspondingly, to the *absorption–emission* of photons;
3. $E_{ph} \gg E_g$, as for X rays: high-energy ionizing interactions take place, i.e., each photon causes the generation of a high-energy e-h pair, which generates a large number of e-h pairs through avalanche processes. This case is exploited in high-energy particle and radiation detectors, but it is outside the scope of our discussion.

Optical processes leading to band-to-band transitions involve at least one photon and one e-h pair. At a microscopic level, there are three possible fundamental processes:

1. *Photon absorption* (and *e-h pair generation*): the photon energy (momentum conservation is discussed later) is supplied to a valence band electron, which is promoted to the conduction band, leaving a free hole in the valence band. Because of the absorption process, the EM wave decreases its amplitude and power.
2. *Photon stimulated emission* (and *e-h pair recombination*): a photon stimulates the emission of a second photon with the same frequency and wavevector; the e-h pair recombines to provide the photon energy. The emitted photon is coherent with the stimulating EM wave, i.e., it increases the amplitude of the EM field and the EM wave power through a gain process.
3. *Photon spontaneous emission* (and *e-h pair recombination*): a photon is emitted spontaneously; the e-h pair recombines to provide the photon energy. Since the emitted photon is incoherent, the process does not imply the amplification of an already existing wave, but rather the excitation of an EM field with a possibly broad frequency spectrum (if many photons are incoherently emitted in a specific bandwidth).

The three processes are shown in Fig. 2.5, Fig. 2.6, and Fig. 2.7, respectively. Note that absorption and stimulated emission are in fact the *same process* with time reversal.

The rules of the photon–electron interaction are established by quantum mechanics through the so-called *perturbation theory.* According to the perturbation theory, the interaction must satisfy two sets of rules. The *first set* is quite obvious and coincides with

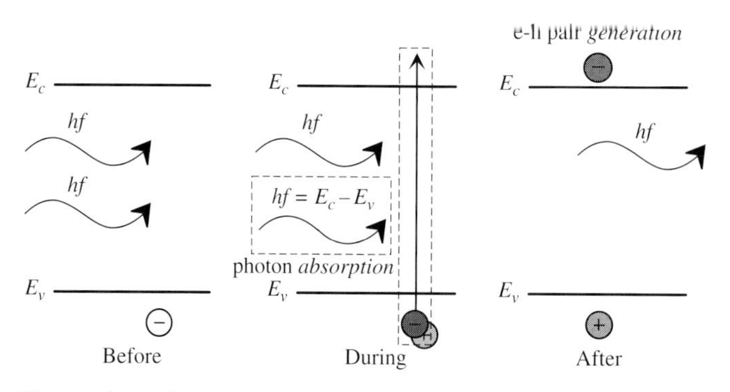

Figure 2.5 Photon absorption.

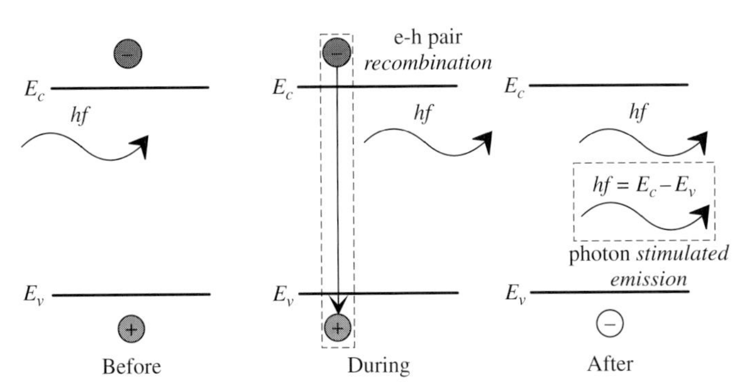

Figure 2.6 Photon stimulated emission.

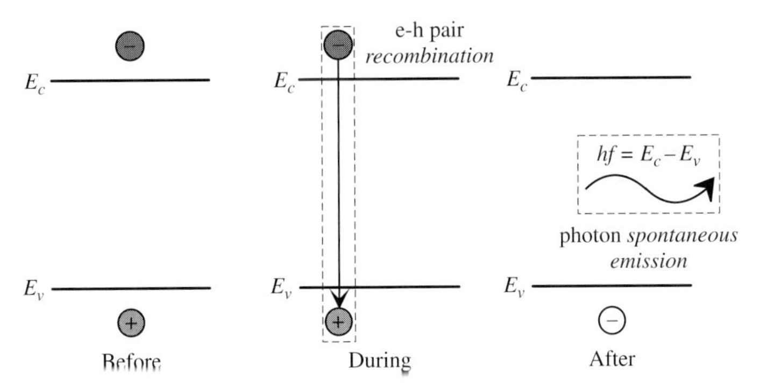

Figure 2.7 Photon spontaneous emission.

classical collision rules, i.e., the conservation of the total energy and momentum from before to after the interaction. The *second set* is less obvious and consists of the so called *selection rules,* according to which some interactions are in fact not allowed (i.e., their probability, or interaction strength, is zero), although they satisfy energy and momentum conservation. Selection rules derive from a detailed quantum-mechanical treatment of the system before and after the interaction, including the particle wavefunctions in the initial and final states (e.g., the electron wavefunctions in the valence and conduction bands).

2.3.1 Energy and momentum conservation

Let us discuss first energy and momentum (or wavevector) conservation, which can be generally expressed as

$$\sum_i E_{i,\text{before}} = \sum_i E_{i,\text{after}}, \quad \sum_i \underline{k}_{i,\text{before}} = \sum_i \underline{k}_{i,\text{after}},$$

where the summation extends to all particles involved in the interaction in the initial and final states (namely, electrons and photons). Holes are not explicitly taken into account, as long as their energy and momentum do not vary in the process. Moreover, energies and momenta are evaluated with respect to an absolute reference, not a local reference to the conduction band minimum or valence band maximum.

To make explicit examples of energy and momentum conservation we refer to a few important cases:

1. Absorption of a photon through a *direct process*, i.e., a process involving only electrons and photons. One has

$$E_f = E_i + E_{ph}, \quad \underline{k}_f = \underline{k}_i + \underline{k}_{ph},$$

where E_f is the final energy of the electron in the conduction band, E_i is the initial energy of the electron in the valence band, E_{ph} is the energy of the absorbed photon (similarly for momenta). We can equivalently write

$$\Delta E_{fi} = E_f - E_i = E_{ph} = hf, \quad \Delta\underline{k}_{fi} = \underline{k}_f - \underline{k}_i = \underline{k}_{ph} = \frac{2\pi}{\lambda}\widehat{k}_{ph},$$

implying that the changes in energy and momentum of the electron are supplied by the absorbed photon.

2. Spontaneous or stimulated emission of a photon through a *direct process*, i.e., a process involving only electrons and photons. One has:

$$E_f + E_{ph} = E_i, \quad \underline{k}_f + \underline{k}_{ph} = \underline{k}_i,$$

where E_f is the final energy of the electron in the valence band, E_i is the initial energy of the electron in the conduction band, E_{ph} is the energy of the emitted photon (similarly for momenta). We can equivalently write

$$\Delta E_{if} = E_i - E_f = E_{ph} = hf, \quad \Delta\underline{k}_{if} = \underline{k}_i - \underline{k}_f = \underline{k}_{ph} = \frac{2\pi}{\lambda}\widehat{k}_{ph},$$

implying that the changes in energy and momentum of the electron are supplied to the emitted photon.

Direct processes typically involve a photon energy of the order of the energy gap E_g, which photons in the NIR and visible ranges are able to provide, see (2.1). However, the amount of momentum a photon can supply is indeed very small if the wavelength is in the micrometer range. To clarify this point, remember that electron momenta are defined in the reciprocal space, whose fundamental cell is the first Brillouin zone (FBZ). We can assume as a *large* electron momentum one corresponding to the distance between the center of the FBZ (the Γ point, where the maximum of the valence band is found) and the periphery of the FBZ (close to the minimum of the conduction band in indirect-gap semiconductors). Such a distance, see Fig. 1.9, is of the order of

$$k_{e,\max} \approx \frac{\pi}{a},$$

where a is the lattice constant, of the order of 0.5 nm. Since the photon wavelength is of the order of 1 μm we conclude that

$$\frac{k_{e,\max}}{k_{ph}} \approx \frac{\lambda}{2a} \approx 10^3,$$

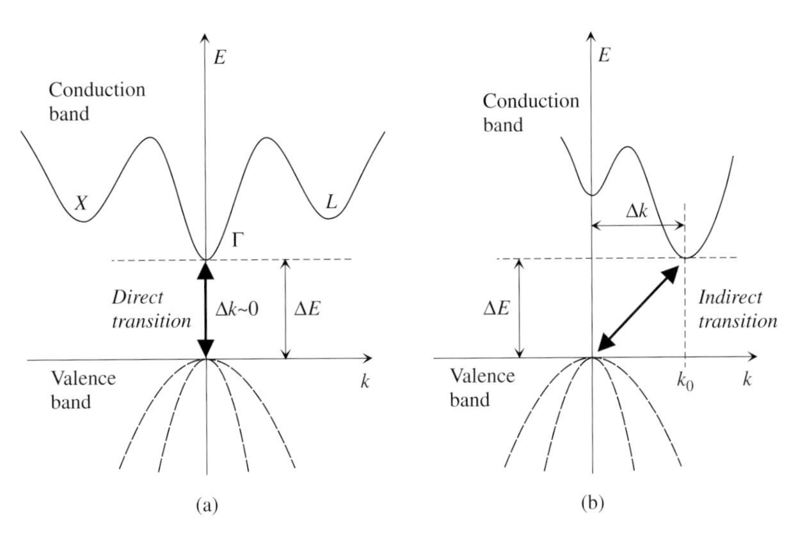

Figure 2.8 Simplified bandstructure of (a) a direct-bandgap semiconductor (GaAs, InP, ...) showing radiative transitions involving negligible momentum difference; (b) an indirect-bandgap semiconductor (Si, SiC, [Ge]), showing radiative transitions involving a large momentum difference.

or $k_{ph} \ll k_{e,\max}$. Thus, the momentum difference between the initial and final states caused, in a direct transition, by the photon momentum, is altogether negligible. Direct transitions are possible in direct-bandgap semiconductors, where electron and hole states are available near the point Γ with negligible momentum difference; see Fig. 2.8(a).

In indirect bandgap semiconductors, such as Si, radiative transitions involve a large momentum variation, which is indeed of the order of π/a, see Fig. 2.8(b). Such a momentum cannot be provided by photons, and the transition is possible only by involving other particles, the *phonons*, having low energy but comparatively large momentum. Phonons are quantized elastic waves carrying mechanical energy (in the form of sound or heat) through the crystal.

In the presence of one or more phonons, the possible interaction mechanisms multiply, since a photon can be emitted or absorbed through the help of an emitted or absorbed phonon. Let us denote as $E_\phi = hF$ the phonon energy (F is the phonon frequency), as $k_\phi = 2\pi/\Lambda$ the phonon momentum (Λ is the phonon wavelength). We have four possible cases:

1. Absorption of a photon through an *indirect process*, i.e., a process involving electrons and photons but also phonons. The phonon can be absorbed or emitted.
   a. Phonon absorption:

$$E_f = E_i + E_{ph} + E_\phi, \quad \underline{k}_f = \underline{k}_i + \underline{k}_{ph} + \underline{k}_\phi,$$

   where E_f is the final energy of the electron in the conduction band, E_i is the initial energy of the electron in the valence band, E_{ph} is the energy of the absorbed photon and E_ϕ is the energy of the absorbed phonon (similarly for momenta).

b. Phonon emission:

$$E_f + E_\phi = E_i + E_{ph}, \quad \underline{k}_f + \underline{k}_\phi = \underline{k}_i + \underline{k}_{ph},$$

where E_f is the final energy of the electron in the conduction band, E_ϕ is the energy of the emitted phonon, E_i is the initial energy of the electron in the valence band, E_{ph} is the energy of the absorbed photon (similarly for momenta).

2. Emission of a photon through an *indirect process*, i.e., a process involving electrons and photons but also phonons. The phonon can be absorbed or emitted.

a. Phonon absorption:

$$E_f + E_{ph} = E_i + E_\phi, \quad \underline{k}_f + \underline{k}_{ph} = \underline{k}_i + \underline{k}_\phi,$$

where E_f is the final energy of the electron in the conduction band, E_{ph} is the energy of the emitted photon, E_i is the initial energy of the electron in the valence band, and E_ϕ is the energy of the absorbed phonon (similarly for momenta).

b. Phonon emission:

$$E_f + E_{ph} + E_\phi = E_i, \quad \underline{k}_f + \underline{k}_{ph} + \underline{k}_\phi = \underline{k}_i,$$

where E_f is the final energy of the electron in the conduction band, E_{ph} is the energy of the emitted photon, E_ϕ is the energy of the emitted phonon, E_i is the initial energy of the electron in the valence band (similarly for momenta).

Since indirect interactions are many-body processes, their probability of occurrence is smaller than for direct processes. In particular, the emission efficiency of an indirect-bandgap semiconductor is low enough to make (at least at present) the realization of devices based on stimulated emission, like lasers, extremely critical (indirect-bandgap light-emitting diodes, LEDs, working with spontaneous emission, are feasible but have low efficiencies). On the other hand, devices based on the absorption process, like photodetectors, can still be realized through indirect-bandgap semiconductors, albeit with inferior performance with respect to direct-bandgap materials.

Concentrating on the indirect (phonon-assisted) photon absorption process, we can express the energy or momentum difference between the final (conduction band) electron state and the initial (valence band) electron state as

$$\Delta E_{fi} = E_{ph} \pm E_\phi \approx E_{ph}, \quad \Delta \underline{k}_{if} = \underline{k}_{ph} \pm \underline{k}_\phi \approx \pm \underline{k}_\phi,$$

where the upper sign corresponds to phonon emission and the lower sign to phonon absorption. Phonon absorption slightly decreases the minimum energy a photon must have in order to be able to promote an electron from the valence band to the conduction band, i.e., it slightly decreases the absorption threshold energy with respect to E_g. On the other hand, phonon emission slightly increases the absorption threshold energy with respect to E_g.

If we compare a direct and an indirect transition, see Fig. 2.9, we see that the momentum (the wavevector k) that the phonon should provide is typically of the order of π/a, since, as already mentioned, indirect conduction band minima are located at or near the boundary of the FBZ (while the valence band maximum is always located in point Γ).

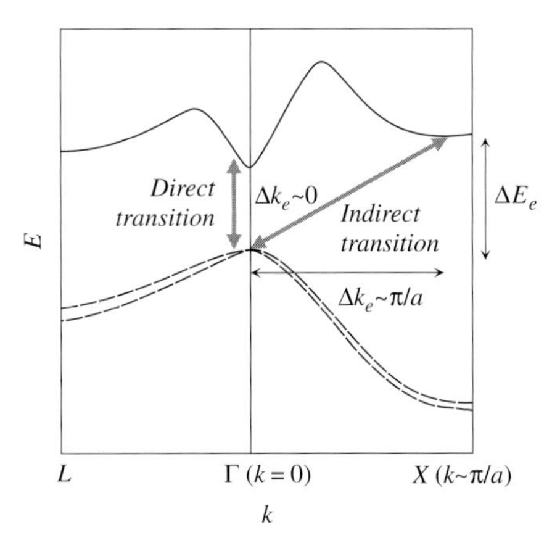

Figure 2.9 Direct and indirect optical transitions in a semiconductor: momentum and energy variations vs. the bandstructure parameters.

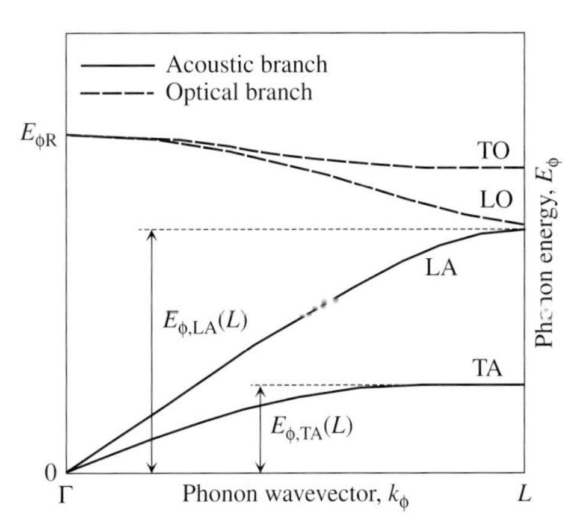

Figure 2.10 Schematic dispersion relation of phonons in a cubic crystal; only the section from the Γ to the L point is shown. The parameter $E_{\phi R}$ is the Raman phonon energy.

The question is, of course, whether the phonon can be a high-momentum particle while preserving a low enough energy to make it compatible with the total energy balance of the process, which is dominated by the large photon energy.

As already recalled, phonons are mechanical waves in the crystal. Taking into account that vibrational waves can be longitudinal (i.e., the matter displacement occurs parallel to the propagation direction) or transverse (the displacement occurs orthogonal to the propagation direction) and that two vibrational modes are possible according to whether nearby atoms oscillate in or out of phase (these are referred to as the *acoustical* [low energy] phonons and the *optical* [high energy] phonons), we have, in a cubic crystal, four possible vibrational modes, the longitudinal/transverse, optical/acoustic phonons (denoted as LA, TA, LO, TO, see Fig. 2.10). Since the crystal space supporting lattice vibrations is the same periodic space in which the electron motion takes place, the phonon dispersion relation is defined in the same reciprocal space as for the electrons, i.e., the FBZ, see Fig. 2.10. The maximum value for the phonon momentum within the

FBZ is therefore of the order of π/a. However, the phonon energy is small: while optical phonons have a maximum energy (called the Raman phonon energy, $E_{\phi\mathrm{R}}$) of the order of 60 meV in Si and 35 meV in GaAs, the maximum energies for the acoustic phonons are $E_{\phi,\mathrm{TA}}(L) \approx 20$ meV in Si and ≈ 10 meV in GaAs, $E_{\phi,\mathrm{LA}}(L) \approx 50$ meV in Si and ≈ 30 meV in GaAs; notice that the acoustical phonon energy vanishes at the Γ point. A phonon with $k = \pi/a$ will have propagation constant and wavelength

$$k_\phi = \frac{2\pi}{\Lambda} = \frac{\pi}{a} \rightarrow \Lambda = 2a.$$

However, the dispersion relation of acoustic phonons can be approximated, at low momentum, as

$$F = \frac{v_s}{\Lambda},$$

where $v_s \ll c_0$ is the velocity of sound in the medium. The maximum phonon energy will be

$$E_\phi = hF = \frac{hv_s}{\Lambda} \approx \frac{hv_s}{2a}.$$

Taking into account that the energy of a photon with wavelength λ is $E_{ph} = hf = hc_0/\lambda$, and that v_s is of the order of 10^4 m/s in a solid while $c_0 = 3 \times 10^8$ m/s, we have, assuming $2a \approx 1$ nm, $\lambda \approx 1$ μm, that

$$\frac{E_\phi}{E_{ph}} \approx \frac{v_s}{2a}\frac{\lambda}{c_0} \approx 10^3 \cdot \frac{10^4}{3 \times 10^8} \approx \frac{1}{30}.$$

For a photon energy of 1 eV this means a phonon energy near 33 meV. Note that the total phonon momentum and energy need not be necessarily carried by just a single phonon, but also by two or more phonons. It is clear, however, that a low number of phonons together can have at the same time a large momentum and a low enough energy with respect to the photon energy.

Example 2.2: Consider an indirect-bandgap semiconductor with $E_g \approx 1$ eV; suppose that a photon is absorbed and that an acoustic phonon (emitted or absorbed) provides the momentum difference. Assuming for the phonon energy $E_\phi = 30$ meV, evaluate the absorption shift in energy and wavelength.

We have that the minimum photon energy is $E_{ph} = 1000 \pm 30$ meV where the upper sign is for phonon emission, the lower sign for phonon absorption. Taking into account that $\lambda = 1240/E$ μm, where the energy is in meV, we have that, in terms of wavelength, the absorption edge for absorption is $\lambda_1 = 1240/1030 = 1.2038$ μm for phonon emission, and $\lambda_2 = 1240/970 = 1.2783$ μm for phonon absorption. The total shift is $\lambda_2 - \lambda_1 = 74.5$ nm.

Assuming a monotonic behavior of the dispersion relation of the valence and conduction bands (e.g., the parabolic band approximation in a direct-bandgap semiconductor),

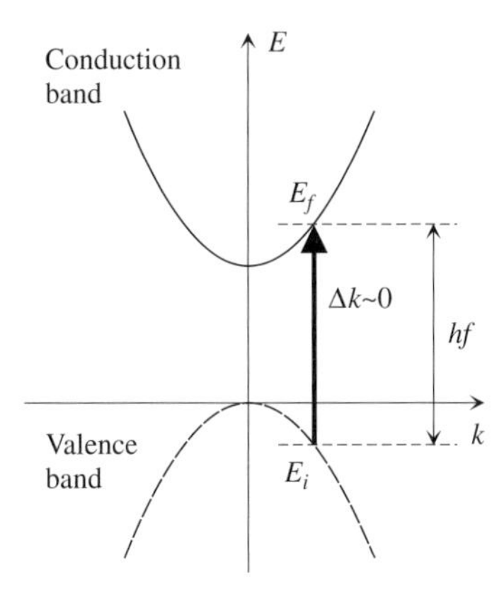

Figure 2.11 Energy conservation in a vertical transition (generation due to phonon absorption).

energy and momentum conservation lead to the following conclusion: having assigned the photon energy, the *initial* and *final* electron states (i.e., their energy and momentum) are uniquely determined. To show this, we use a simple parabolic model for the valence and conduction bands and assume vertical transitions (same momentum in the initial and final states, or negligible photon momentum). To fix the ideas, assume that the initial state is in the valence band, the final state in the conduction band (as in the case of absorption), see Fig. 2.11. The initial and final energies satisfy the dispersion relations for the valence and conduction bands, with $k_i = k_f = k$:

$$E_i = E_v - \frac{\hbar^2 k^2}{2m_h^*}, \quad E_f = E_c + \frac{\hbar^2 k^2}{2m_n^*}.$$

The absorbed photon energy is equal to the energy difference between final and initial states:

$$E_{ph} = \hbar\omega = E_f - E_i = E_c - E_v + \frac{\hbar^2 k^2}{2}\left(\frac{1}{m_h^*} + \frac{1}{m_n^*}\right) = E_g + \frac{\hbar^2 k^2}{2m_r^*},$$

where

$$m_r^* = \left[\left(m_h^*\right)^{-1} + \left(m_n^*\right)^{-1}\right]^{-1} \tag{2.2}$$

is the valence-conduction band *reduced mass* or *joint density of states (JDOS or JDS) reduced mass*. Solving for the initial or final momentum k, we obtain

$$\hbar^2 k^2 = 2m_r^*\left(\hbar\omega - E_g\right),$$

i.e.,

$$k = k_i = k_f = \frac{\sqrt{2m_r^*}}{\hbar}\sqrt{\hbar\omega - E_g}.$$

Substituting in the expression for the initial and final energies, we find that they are uniquely defined by the photon energy:

$$E_i = E_v - \frac{m_r^*}{m_h^*}\left(\hbar\omega - E_g\right) \equiv E_h(\hbar\omega)$$

$$E_f = E_c + \frac{m_r^*}{m_n^*}\left(\hbar\omega - E_g\right) \equiv E_e(\hbar\omega).$$

The minimum photon energy occurs when $k = 0$ and is equal to the energy gap:

$$E_{ph} = E_g \longrightarrow k = 0, \quad E_i = E_v, \quad E_f = E_c.$$

Therefore, E_g can be referred to as the *absorption edge*.

2.3.2 Perturbation theory and selection rules

According to quantum theory, the state of a system subject to time-independent forces is described by the solution of the *Schrödinger equation* with given Hamiltonian H_0 (related to total energy, e.g., of electrons and holes):

$$H_0\psi(\underline{r}, t) + \mathrm{j}\hbar\frac{\partial\psi(\underline{r}, t)}{\partial t} = 0,$$

where $\psi(\underline{r}, t)$ is the particle wavefunction and the Hamiltonian is related to the total particle energy as

$$H_0 = -\frac{\hbar^2}{2m}\nabla^2 + V(\underline{r}). \tag{2.3}$$

The first term in (2.3) is the quantum version of the kinetic energy, the second term the potential energy. A time-dependent (small) external force (here associated with an EM field, i.e., to a photon) can be described as *perturbation* of the Hamiltonian:

$$H_0 \rightarrow H_0 + H'.$$

The perturbed solution can be expressed (according to perturbation theory) as linear combination of unperturbed states; if the perturbing force is time-varying (in our case, harmonic) the probability of finding a particle belonging to the system (which was in the initial state i) in the final state f increases linearly with time. The per unit time (and sometimes per unit volume) transition probability from the initial to the final state is the *scattering rate* w_{fi}, see Fig. 2.12. If the *time duration of the interaction* and *size of the interaction region* are suitably large, quantum theory also suggests that the scattering rate is zero if energy and momentum are not conserved.

Given a certain amount of cause (e.g., optical power, photons, ...) the strength of the effect (e.g., the number of electrons promoted per unit time from the valence to the conduction band through photon absorption) depends on the magnitude of the

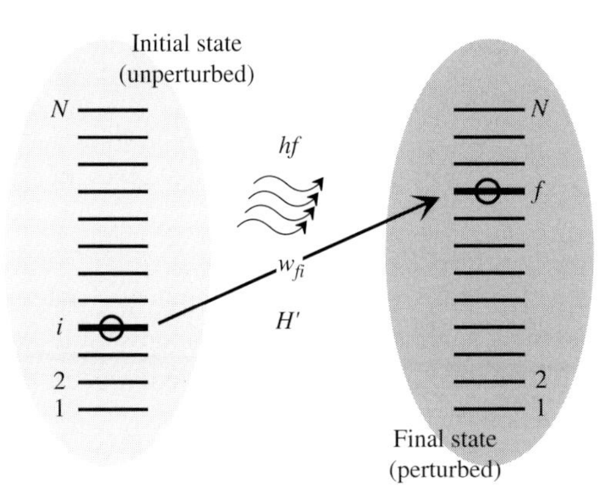

Figure 2.12 Perturbation theory picture of the transition between the initial and final state of a system.

scattering rate w_{fi}. Perturbation theory shows that the scattering rate is proportional to the magnitude square of the so-called *matrix element* M_{fi}:

$$M_{fi} = \int_{\underline{r}} \psi_f^*(\underline{r}) \cdot H'(\underline{r}, \nabla_{\underline{r}}) \cdot \psi_i(\underline{r}) \, d\underline{r}, \tag{2.4}$$

where $\psi_f(\underline{r})$ and $\psi_i(\underline{r})$ are the final and initial state wavefunctions, respectively, $H'(\underline{r}, \nabla_{\underline{r}})$ is the perturbation Hamiltonian, and $\nabla_{\underline{r}}$ is the gradient operator (operating on ψ_i). For an EM wave, the perturbation Hamiltonian is

$$H' \propto \underline{A} \cdot \nabla_{\underline{r}} \propto \frac{\partial}{\partial \xi},$$

where $\underline{A}$ is the EM wave *vector potential* (in a plane wave, $\underline{A}$ is parallel to the electric field and orthogonal to the propagation direction, see e.g. [10]) and ξ is the coordinate along the direction of the vector potential. In other words, the perturbation Hamiltonian operates on the initial state wavefunction as a *directional derivative operator*, with direction parallel to the EM field polarization. From the Bloch theorem (see Section 1.3.2), wavefunctions in a crystal are the product of a periodic function of space and of an atom-like wavefunction; near the Γ point of the FBZ, such wavefunctions can be approximated by their even or odd atom-like component (s-type or p-type for the conduction and valence band, respectively, see Fig. 1.11). Since a spatial derivative operator turns an even function of space into an odd function and vice versa, the integral kernel can, in some cases, be an odd function of space, leading to an identically zero matrix element in (2.4). For brevity, we will confine the discussion to some representative examples.

Consider, for example, an initial s-type state and a final p_x-state, which is odd along x, even along y and z. The two states can be synthetically described (assume functions are real for simplicity) as

$$s(x, y, z) = e(x)e(y)e(z), \quad p_x(x, y, z) = o(x)e(y)e(z),$$

where e is a (generic) even function, o an odd function of the argument. Assume, for instance, that the perturbation Hamiltonian is $\propto \partial/\partial x$. The application of H' on the initial state leads to an odd function of x; multiplying by the final p_x state we globally obtain an even function of x (and also of y, z):

$$M_{fi} \propto \int o(x)e(y)e(z)\frac{\partial}{\partial x}\left[e(x)e(y)e(z)\right]\mathrm{d}x\,\mathrm{d}y\,\mathrm{d}z = \int \left[o(x)e(y)e(z)\right]^2 \mathrm{d}x\,\mathrm{d}y\,\mathrm{d}z.$$

In this case the kernel is *even*, and $M_{fi} \neq 0$.

On the other hand, consider an initial s-type state and a final p_y-state, which is odd along y, even along x and z; assume again that the perturbation Hamiltonian is $\propto \partial/\partial x$. The application of H' to the initial state leads to an *odd* function of x; multiplying by the final p_y state, which is *even* along x, we have

$$\begin{aligned} M_{fi} &\propto \int e(x)o(y)e(z)\frac{\partial}{\partial x}\left[e(x)e(y)e(z)\right]\mathrm{d}x\,\mathrm{d}y\,\mathrm{d}z \\ &= \int e(x)o(y)e(z)o(x)e(y)e(z)\,\mathrm{d}x\,\mathrm{d}y\,\mathrm{d}z \\ &= \underbrace{\int e(x)o(x)\,\mathrm{d}x}_{\text{odd kernel}} \times \underbrace{\int o(y)e(y)\,\mathrm{d}y}_{\text{odd kernel}} \times \underbrace{\int e(z)e(z)\,\mathrm{d}z}_{\text{even kernel}} \quad = 0 \times 0 \times K = 0. \end{aligned}$$

Thus, the integral kernel is an *odd* function of x, an *odd* function of y, and an *even* function of z. In this case, therefore, the integral is zero and $M_{fi} = 0$. Generalizing the above result, it can be shown that, given an electric field directed along ξ, the *only interaction possible* is between p_ξ states and s states; for all other p states the matrix element is zero.

We can conclude that, while optical transitions must satisfy energy and momentum conservation, some transitions that are possible from this standpoint have in fact zero strength, i.e., zero matrix element. *Selection rules* state that the initial and final state pair must have an "even"–"odd" or "odd"–"even" nature to allow for a transition.

In bulk semiconductors, the matrix element depends on sets of states (initial or final) of the s and p type. It can be shown that, for *any field polarization*, there is always some interacting set of initial and final states (see Section 2.3.3); the matrix element is never zero and turns out to be polarization independent. This also occurs because light and heavy holes are *degenerate* and therefore contribute in a similar way to the transition.

The situation changes in quantum wells, since in this case the p-type or s-type wavefunctions are multiplied by the envelope wavefunction $\psi(z)$ (see (1.15) and suppose confinement takes place along z; if the QW is an epitaxial structure, z is orthogonal to the stratification), which can in turn be even or odd (typically even in the fundamental state, then alternately odd and even; in an asymmetric well the envelope functions will be only approximately even or odd). The presence of the envelope wavefunction alters the parity of the initial or final state wavefunctions, but only along the direction parallel to the QW stratification. Moreover, in a QW the degeneracy of light holes (LH) and heavy holes (HH) is removed and the different composition of the

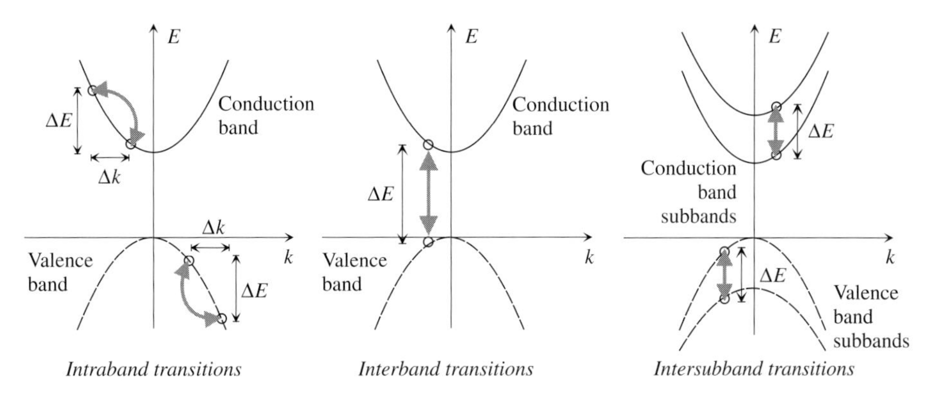

Figure 2.13 Examples of intraband (left), interband (center), and intersubband (right) transitions.

two wavefunctions, see (1.2) and (1.3), becomes effective in making the interaction polarization-dependent.

To introduce a specific example, let us recall some more vocabulary on radiative transitions. We talk about *intraband* transitions when the initial and final states are in the same band (e.g., conduction band or valence band); and of *interband* transitions if they are in different bands (e.g., initial conduction band, final valence band). In a quantum well we also have subbands, thus we have *intrasubband* transitions when the initial and final states are in the same subband, and *intersubband* transitions when they are in different subbands (e.g., initial and final in two different subbands of the conduction band, or one in a subband of the conduction band, the other in a subband of valence band); for some examples see Fig. 2.13. Note that intraband or intrasubband transitions are never direct, while intersubband transitions typically are, both if the two subbands belong to the same band (conduction or valence) and if they belong to two different bands (conduction and valence).

To discuss selection rules in a quantum well, consider a symmetrical QW where the envelope wavefunctions $\psi_i(z)$ are even or odd according to the index (the fundamental state $i = 1$ is even in z, the second state is odd, and so on). The envelope wavefunctions are slowly varying in z with respect to the conduction and valence subband s- and p-like wavefunctions (periodic in the crystal lattice). To obtain the total subband wavefunctions near the Γ point, valence (HH and LH) and conduction band wavefunctions are multiplied by $\psi_i(z)$, leading to the total wavefunctions:

$$\phi_{i,c}(x, y, z) = \psi_i^c(z)s(x, y, z), \quad i = 1, 2 \ldots$$
$$\phi_{i,v\xi}(x, y, z) = \psi_i^v(z)p_\xi(x, y, z), \quad i = 1, 2 \ldots, \quad \xi = x, y, z.$$

The matrix element of the transition between a valence band p state and a conduction band s state can therefore be expressed as

$$
\begin{aligned}
M_{fi} &\propto \int \phi_{i,c}^{*} \frac{\partial}{\partial \eta} \phi_{j,v\xi} \, \mathrm{d}\underline{r} = \int \psi_i^{*c}(z) s^*(\underline{r}) \frac{\partial}{\partial \eta} \left[\psi_j^v(z) p_\xi(\underline{r}) \right] \mathrm{d}\underline{r} \\
&= \int \left[\psi_i^{*c}(z) s^*(\underline{r}) p_\xi(\underline{r}) \frac{\partial \psi_j^v(z)}{\partial \eta} + \psi_i^{*c}(z) s^*(\underline{r}) \psi_j^v(z) \frac{\partial p_\xi(\underline{r})}{\partial \eta} \right] \mathrm{d}\underline{r} \\
&\approx \int_{-\infty}^{\infty} \psi_i^{*c}(z) \frac{\partial \psi_j^v(z)}{\partial \eta} \, \mathrm{d}z \times \underbrace{\frac{1}{\Omega} \int_\Omega s^*(\underline{r}) p_\xi(\underline{r}) \, \mathrm{d}\underline{r}}_{\text{always } 0} + \underbrace{\int_{-\infty}^{\infty} \psi_i^{*c}(z) \psi_j^v(z) \, \mathrm{d}z}_{\neq 0 \text{ if } i,j \text{ are both even/odd}} \\
&\times \underbrace{\frac{1}{\Omega} \int_\Omega s^*(\underline{r}) \frac{\partial p_\xi(\underline{r})}{\partial \eta} \, \mathrm{d}\underline{r}}_{\neq 0 \text{ if } \xi = \eta}, \quad i = 1, 2 \ldots, \quad \xi, \eta = x, y, z,
\end{aligned}
$$

where Ω is the elementary crystal cell volume. In the above development, we have made use of the 3D extension of the following result (where g is a periodic function with period L and f is a slowly varying function over the period L, like the envelope wavefunctions):

$$
\begin{aligned}
\int_{-\infty}^{\infty} f(z) g(z) \, \mathrm{d}z &= \sum_i \int_{iL}^{(i+1)L} f(z) g(z) \, \mathrm{d}z \\
&\approx \sum_i f(Li) \int_0^L g(z) \, \mathrm{d}z \approx \int_{-\infty}^{\infty} f(z) \, \mathrm{d}z \times \frac{1}{L} \int_0^L g(z) \, \mathrm{d}z.
\end{aligned}
$$

The integral

$$
I = \int_{-\infty}^{\infty} \psi_i^{*c}(z) \psi_j^v(z) \, \mathrm{d}z
$$

can in principle be different from zero even if the two envelope wavefunctions do not have the same index, because they refer to two different potential profiles (valence and conduction band wells); however, if the two wells are similar, or approximated, for example, by an infinite well, $I \approx \delta_{ij}$, i.e., M_{fi} is not negligible only if *the two indices are equal*.

We can conclude that the element matrix is zero unless two conditions are met:

- The indices of the envelope wavefunctions in the valence and conduction bands are both even or odd (1-1, 2-2, but also in general 1-3, etc.); however, while the 1-1, 2-2, etc. interaction is strong, the 1-3, etc. interaction is typically very weak.
- The interaction caused by an electric field polarized in the direction ξ involves the valence band state p_ξ.

In general, if the first condition is met, the interaction will have a *different weight* according to the *field polarization*. This happens because, while the LH states include all possible p_ξ wavefunctions, in the HH states one of these (p_z) does not appear, see (1.3) and (1.2). For HH states, therefore, the matrix element for a z-polarized field is zero, while LH states react (albeit with a different strength) to all polarizations. Since in a QW LH and HH states are not degenerate (and therefore the relevant interactions have

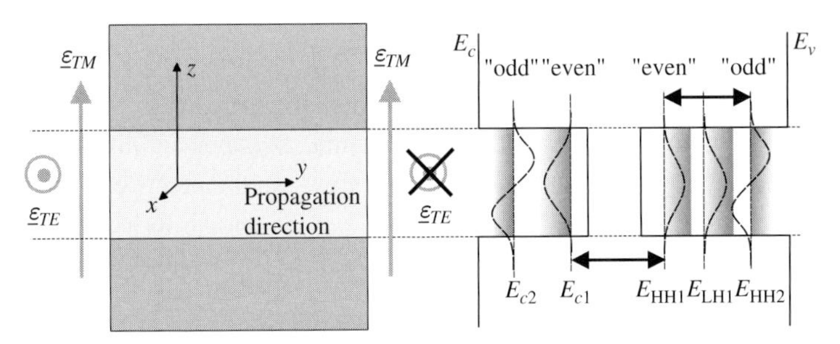

Figure 2.14 Examples of selection rules and polarization sensitivity in a QW with the stratification parallel to the z axis (left); the band structure is shown on the right. The thick arrows (right) define allowed interband or intersubband transitions. The horizontal (TE) field is absorbed by the well, the vertical (TM) does not interact.

a different minimum photon energy), a global *polarization sensitivity* of the allowed transitions results.

In summary, the *allowed transitions* between subbands in different bands are of the kind 1-1, 2-2, 3-1, etc. (i.e., the subband indices must be both even or odd), but strong transitions are typically with equal indices; the *forbidden transitions* are 1-2, 2-1, 1-4, etc. (i.e., the subband indices must be one even and the other odd); this is apparently the opposite of bulk rule (even–even, odd–odd rule), but is in fact the same rule, accounting for the envelope wavefunctions (see Fig. 2.14). On the other hand, intersubband transitions within the same band (conduction or valence) obey the even–odd rule as in bulk, thus 1-2 is allowed, 1-3 is forbidden, and so on. Allowed intersubband transitions (1-2, etc.) will, however, have low strength (at least in a symmetric well) due to the orthogonality of the envelope wavefunctions belonging to the initial and final states.

As an example of the *polarization sensitivity* arising in a QW from the removal of the degeneracy between HH and LH bands, consider a QW with wells orthogonal to the z axis. The TE polarization (electric field orthogonal to z) will cause both LH-C and HH-C transitions, while the TM polarization (electric field along z) will only cause LH-C transitions (C stands for conduction band). The relative transition strengths (also referred to as *oscillator strengths*) can be shown to be (see e.g. [11], Section 9.5):

- for TE polarization, 3/4 for HH-C, 1/4 for LH-C,
- for TM polarization, 0 for HH-C, 1 for the LH-C.

Consider for instance the QW in Fig. 2.14; suppose that the fundamental (lowest photon energy) transition is between the HH_1 band and the conduction band; HH-C transitions correspond to an absorption process. Assuming that the photon energy is larger than the HH-C transition threshold but lower than the LH-C transition threshold, only a TE field will interact (through the HH-C transition), while the TM field interaction strength will be zero. As a consequence, the TE field (parallel to the stratification) will be absorbed by the QW, while the TM field (orthogonal to the stratification) will be unaffected. A qualitative example of the resulting absorption and refractive index profiles is shown in Chapter 6, Fig. 6.34 (right).

2.3.3 Total scattering rates

Quantum mechanics enables as to evaluate the scattering rates (i.e., the probability of the transition, or the number of transitions per unit time, between an initial and a final state, see Fig. 2.12) for the three fundamental processes—spontaneous emission, stimulated emission, and absorption. The scattering rates can be shown to assume the following forms:[4]

$$\begin{aligned} w_{em}^{sp} &= w\delta(E_f - E_i + \hbar\omega) \\ w_{em}^{st} &= wn_{ph}\delta(E_f - E_i + \hbar\omega) \\ w_{abs} &= wn_{ph}\delta(E_f - E_i - \hbar\omega), \end{aligned}$$

where w (dimension J m^3 s^{-1}) is proportional to the matrix element M_{fi} describing the interaction strength, and is the same for all processes; for the expression in a bulk semiconductor, see (2.6). The δ function is a placeholder for energy conservation (it states that in a direct process the electron energy variation is equal to the photon energy), while n_{ph} is the *average photon number*, related to the photon density ρ_{ph} as

$$\rho_{ph} = \frac{n_{ph}}{V},$$

where V is the crystal volume considered. The photon density is in turn related to the EM wave power as

$$P_{op} = \underline{S} \cdot \hat{n} = \left(\hbar\omega\rho_{ph}\right)\frac{c_0}{n_r} \longrightarrow \rho_{ph} = n_r\frac{\underline{S} \cdot \hat{n}}{\hbar\omega c_0} = n_r\frac{P_{op}}{\hbar\omega c_0}, \tag{2.5}$$

where P_{op} is the optical power per unit surface, $\hbar\omega\rho_{ph} = E_{ph}\rho_{ph}$ is the photon energy density, and $\hat{n}$ is a unit vector parallel to the propagation direction.

From perturbation theory one has, for a bulk semiconductor,

$$w = \frac{2\pi}{\hbar}\frac{q^2}{m_0^2}\frac{\hbar}{2\omega\epsilon}p_{cv}^2, \tag{2.6}$$

where $\underline{p}_{cv}$ is the *momentum matrix element* in the so-called dipole approximation (or *dipole matrix element*):[5]

$$\underline{p}_{cv} = \int_\Omega u_{\underline{k}_c}^*(\underline{r})\left(-j\hbar\nabla_r\right)u_{\underline{k}_v}(\underline{r})\,d\underline{r} = \int_\Omega u_{\underline{k}_c}^*(\underline{r})\underline{p}u_{\underline{k}_v}(\underline{r})\,d\underline{r} \tag{2.7}$$

The functions $u_{\underline{k}_c}$ and $u_{\underline{k}_v}$ are the Bloch waves of the conduction and valence band states, respectively, see (1.1), having s-type or p-type behavior near the Γ point. In

[4] See [4], Section 4.3 and in particular (4.51), (4.54) and (4.55). The absorption scattering rate can be derived through the Fermi golden rule via a semiclassical approach; the stimulated emission follows from time reversal, but in order to derive spontaneous emission the thermodynamic equilibrium condition must be enforced. A direct evaluation of the scattering rates requires, on the other hand, the so-called *second quantization* process. The square magnitudes of the integrals appearing in (4.51), (4.52) of the quoted reference are the square magnitude of the dipole matrix element in (2.7) while the prefactor is $wn_{ph}\overline{p_{cv}^{2}}$ with w defined in (2.6).

[5] From quantum mechanics, $\underline{p} \rightarrow -j\hbar\nabla_{\underline{r}}$.

bulk materials, the dipole matrix element assumes the following values according to polarization (see e.g. [4], Section 4.4.1):

$$x\text{-polarized light} \rightarrow \begin{cases} \text{HH-C} \rightarrow p_{cv}^2 = \frac{1}{2}\mathfrak{p}_{cv}^2 \\ \text{LH-C} \rightarrow p_{cv}^2 = \frac{1}{6}\mathfrak{p}_{cv}^2 \end{cases}$$

$$y\text{-polarized light} \rightarrow \begin{cases} \text{HH-C} \rightarrow p_{cv}^2 = \frac{1}{2}\mathfrak{p}_{cv}^2 \\ \text{LH-C} \rightarrow p_{cv}^2 = \frac{1}{6}\mathfrak{p}_{cv}^2 \end{cases}$$

$$z\text{-polarized light} \rightarrow \begin{cases} \text{HH-C} \rightarrow p_{cv}^2 = 0 \\ \text{LH-C} \rightarrow p_{cv}^2 = \frac{2}{3}\mathfrak{p}_{cv}^2. \end{cases}$$

The parameter $\mathfrak{p}_{cv}$ (a momentum) is better expressed through the corresponding energy as

$$E_p = \frac{2\mathfrak{p}_{cv}^2}{m_0}, \tag{2.8}$$

where m_0 is the free electron mass and $E_p \approx 20\,\text{eV}$ for compound semiconductors.[6] Assuming unpolarized light, the average momentum matrix element is the same ($\mathfrak{p}_{cv}^2/3$) for HH and LH; thus, for unpolarized light and for an arbitrary mixture of light and heavy holes the average momentum matrix element will be[7]

$$\left\langle p_{cv}^2 \right\rangle = \frac{2}{3}\mathfrak{p}_{cv}^2.$$

Thus, for unpolarized light interacting with a bulk semiconductor, we have

$$w = \frac{2\pi}{\hbar}\frac{q^2}{m_0^2}\frac{\hbar}{2\omega\epsilon}\frac{2}{3}\mathfrak{p}_{cv}^2 \quad \text{J m}^3\,\text{s}^{-1}. \tag{2.10}$$

The scattering rates defined so far yield the transition probabilities between two existing states which are properly populated (the initial state is full, the final state is empty, according to the exclusion principle). Therefore, they do not take into account:

- The availability (i.e., the density) of initial and final states having the proper energy; it can be shown that this corresponds to a weight called the *joint density of states* (JDS) of the valence and conduction bands, $N_{cv}(\hbar\omega)$:[8]

$$N_{cv}(\hbar\omega) = \frac{(2m_r^*)^{3/2}}{2\pi^2\hbar^3}\sqrt{\hbar\omega - E_g} \quad \text{m}^{-3}\text{J}^{-1}. \tag{2.11}$$

[6] Such an energy is related to the effective mass, see, e.g., [11], Section 4.2.3 and 4.2.4:

$$\frac{2\mathfrak{p}_{cv}^2}{m_0\hbar\omega} \approx \frac{2\mathfrak{p}_{cv}^2}{m_0 E_g} \approx \frac{m_0}{m_n^*} - 1. \tag{2.9}$$

[7] Note that the same average value holds for light polarized along x, y, or z if we assume a 50% mixture of heavy and light holes.

[8] The joint density of states formally derives from the integration on all allowed k states, with the constraint of energy conservation. A variable change into energy introduces as the Jacobian the JDS.

- The state occupation probability (Fermi or quasi-Fermi statistics) f_n and f_h (electrons and holes):

$$f_n(E_n) = \frac{1}{1 + \exp\left(\dfrac{E_n - E_{Fn}}{k_B T}\right)} \tag{2.12}$$

$$f_h(E_h) = \frac{1}{1 + \exp\left(\dfrac{E_{Fh} - E_h}{k_B T}\right)}, \tag{2.13}$$

where E_{Fn} and E_{Fh} are the quasi-Fermi levels for electrons and holes.

Due to momentum conservation in direct transitions, the initial and final momenta coincide in magnitude; moreover, their value is related to the photon energy $E_{ph} = \hbar\omega$ as

$$k_n = \frac{\sqrt{2m_r^*}}{\hbar}\sqrt{\hbar\omega - E_g} = \frac{\sqrt{2m_r^*}}{\hbar}\sqrt{E_{ph} - E_g},$$

so that the initial and final energies are (for absorption; for emission, the two energies are interchanged)

$$E_i = E_v - \frac{\hbar^2 k_n^2}{2m_h^*} = E_v - \frac{m_r^*}{m_h^*}\left(E_{ph} - E_g\right) \tag{2.14}$$

$$E_f = E_c + \frac{\hbar^2 k_n^2}{2m_n^*} = E_c + \frac{m_r^*}{m_n^*}\left(E_{ph} - E_g\right). \tag{2.15}$$

Let us consider first the absorption process. The *total absorption scattering rate* including, together with the transition probability, also the information on the density of states and on their occupation probability, is obtained as

$$\mathcal{W}_{abs}(\hbar\omega) = w n_{ph} N_{cv}(\hbar\omega)\left[1 - f_h\left(E_i(\hbar\omega)\right)\right]\left[1 - f_n\left(E_f(\hbar\omega)\right)\right] \quad \mathrm{s}^{-1}. \tag{2.16}$$

Thus, to obtain the *total absorption scattering rate* $\mathcal{W}_{abs}$ we multiply w_{abs} by the joint density of states N_{cv}, and by the probability that the initial state (valence band) is filled by an electron (empty of a hole: $1 - f_h$) AND the final state (conduction band) is empty of an electron (probability $1 - f_n$); this, because of the exclusion principle.

Similarly, to obtain the *total emission scattering rate* $\mathcal{W}_{em} = \mathcal{W}_{em}^{sp} + \mathcal{W}_{em}^{st}$ we multiply $w_{em} = w_{em}^{sp} + w_{em}^{st}$ by the joint density of states N_{cv} and by the probability that the initial state (valence band) is filled by a hole, f_h, AND the final state (conduction band) is filled by an electron, f_n; we therefore obtain

$$\begin{aligned}\mathcal{W}_{em}(\hbar\omega) &= \mathcal{W}_{em}^{sp}(\hbar\omega) + \mathcal{W}_{em}^{st}(\hbar\omega) \\ &= w(1 + n_{ph})N_{cv}(\hbar\omega) f_n(E_n(\hbar\omega)) f_h(E_h(\hbar\omega)) \quad \mathrm{s}^{-1}.\end{aligned} \tag{2.17}$$

Note the following (quite obvious) result: the total scattering rates for stimulated and spontaneous emission are clearly related, since, from (2.17),

$$\mathcal{W}_{em}^{st} = n_{ph} w N_{cv} f_n f_h, \tag{2.18}$$
$$\mathcal{W}_{em}^{sp} = w N_{cv} f_n f_h. \tag{2.19}$$

Thus,

$$\mathcal{W}_{em}^{st} = n_{ph} \mathcal{W}_{em}^{sp}. \tag{2.20}$$

In what follows, we will also exploit the *scattering rates per unit volume*, (W, dimension $\mathrm{m}^{-3}\,\mathrm{s}^{-1}$) for the absorption and stimulated emission:

$$\begin{aligned} W_{abs}(\hbar\omega) &= \frac{1}{V}\mathcal{W}_{abs}(\hbar\omega) = \underbrace{\frac{2\pi}{\hbar}\frac{q^2}{m_0^2}\frac{\hbar\rho_{ph}}{2\omega\epsilon}\frac{2}{3}\mathrm{p}_{cv}^2}_{\frac{1}{V}wn_{ph}=w\rho_{ph}} N_{cv}(\hbar\omega) \\ &\quad \times \left[1 - f_h\left(E_i(\hbar\omega)\right)\right]\left[1 - f_n\left(E_f(\hbar\omega)\right)\right] \\ &= w\rho_{ph} N_{cv}(\hbar\omega)\left[1 - f_h\left(E_i(\hbar\omega)\right)\right]\left[1 - f_n\left(E_f(\hbar\omega)\right)\right] \end{aligned} \tag{2.21}$$

$$\begin{aligned} W_{em}^{st}(\hbar\omega) &= \frac{1}{V}\mathcal{W}_{em}^{st}(\hbar\omega) = \underbrace{\frac{2\pi}{\hbar}\frac{q^2}{m_0^2}\frac{\hbar\rho_{ph}}{2\omega\epsilon}\frac{2}{3}\mathrm{p}_{cv}^2}_{\frac{1}{V}wn_{ph}=w\rho_{ph}} N_{cv}(\hbar\omega) \\ &\quad \times f_h\left(E_i(\hbar\omega)\right) f_n\left(E_f(\hbar\omega)\right) \\ &= w\rho_{ph} N_{cv}(\hbar\omega) f_h\left(E_i(\hbar\omega)\right) f_n\left(E_f(\hbar\omega)\right). \end{aligned} \tag{2.22}$$

We can similarly define the scattering rates for spontaneous emission per unit volume $W_{em}^{sp} = \mathcal{W}_{em}^{sp}/V$ and the total scattering rate for emission per unit volume as

$$W_{em} = W_{em}^{st} + W_{em}^{sp} = \mathcal{W}_{em}/V, \tag{2.23}$$

where $\mathcal{W}_{em}$ is defined in (2.17).

Example 2.3: Check that at equilibrium (2.16) and (2.17) are compatible with the equilibrium Bose–Einstein photon statistics.

In fact, at thermodynamic equilibrium $\mathcal{W}_{abs} = \mathcal{W}_{em} = \mathcal{W}_{em}^{sp} + \mathcal{W}_{em}^{st}$, thus

$$\begin{aligned} &wn_{ph} \times N_{cv}(\hbar\omega) \times \left[1 - f_n(E_n(\hbar\omega))\right] \times \left[1 - f_h(E_h(\hbar\omega))\right] \\ &\quad = w(1 + n_{ph}) \times N_{cv}(\hbar\omega) \times f_n(E_n(\hbar\omega)) \times f_h(E_h(\hbar\omega)). \end{aligned}$$

Simplification of w and N_{cv}, taking into account that $E_{Fn} = E_{Fh} = E_F$, leads to

$$\exp\left(\frac{E_n - E_h}{k_B T}\right) = \exp\left(\frac{\hbar\omega}{k_B T}\right) = \frac{1}{n_{ph}} + 1.$$

The Bose–Einstein statistics for photons follows:

$$n_{ph} = n_{ph0} = \frac{1}{\exp\left(\dfrac{\hbar\omega}{k_B T}\right) - 1}. \tag{2.24}$$

2.4 The macroscopic view: the EM wave standpoint

The photon–semiconductor interaction is described, macroscopically, by parameters characterizing the behavior of an EM wave traveling through the semiconductor. These are the complex dielectric constant $\epsilon(hf) = \epsilon' - \mathrm{j}\epsilon''$ and, equivalently, the complex propagation constant $\gamma = \omega\sqrt{\epsilon\mu_0} = \bar{\alpha} + \mathrm{j}\beta$. The absorption coefficient can, in the presence of stimulated emission, become negative, corresponding to *gain* $\overline{g}$ and to *net gain* $g = \overline{g} - \alpha$. In modeling the interaction between the EM wave and the semiconductor, this can be defined as the *EM wave standpoint*. From the *semiconductor standpoint*, on the other hand, the interaction with the EM wave can be characterized by the radiative recombination rate R_o, the radiative generation rate G_o or the net radiative recombination rate $U_o = R_o - G_o$.

We will now relate the microscopic behavior (scattering rates) to the macroscopic parameters of the EM wave (absorption, gain, net gain). Let us assume that the electric field in the medium propagates as a (monochromatic) plane wave

$$\underline{E} = \underline{E}_0 \exp(-\gamma z) = \underline{E}_0 \left[\exp(-\bar{\alpha} z)\exp(-\mathrm{j}\beta z)\right],$$

with optical power (power density)

$$P_{op} = P_{op}(0)\exp(-2\bar{\alpha}) \equiv P_{op}(0)\exp(-\alpha z),$$

where we have denoted by $\bar{\alpha}$ the field attenuation, while $\alpha = 2\bar{\alpha}$ is the absorption coefficient. Describe the EM wave as a set of traveling photons (photon gas); the photon density ρ_{ph} must satisfy, in steady state, the continuity equation:

$$\left.\frac{\mathrm{d}\rho_{ph}}{\mathrm{d}t}\right|_{em,abs} = W_{em} - W_{abs} = \frac{\mathrm{d}\Phi_{ph}}{\mathrm{d}x} - \frac{\mathrm{d}}{\mathrm{d}x}\left(\rho_{ph}\frac{c_0}{n_r}\right), \tag{2.25}$$

where Φ_{ph} is the photon flux (photon density by photon velocity). Expressing the scattering rates from (2.21) and (2.23), one has

$$\frac{\mathrm{d}}{\mathrm{d}x}\left(\rho_{ph}\frac{c_0}{n_r}\right) = \rho_{ph} w N_{cv}\left[f_n f_h - (1 - f_n)(1 - f_h)\right] + \frac{1}{V} w N_{cv} f_n f_h. \tag{2.26}$$

The first term in the r.h.s. (including the stimulated emission and absorption rates) is proportional to ρ_{ph}, while the second term is an incoherent source, associated with spontaneous emission. While the first term will lead to gain or loss of radiation, the second term does not contribute to increasing or decreasing the power of an EM wave propagating in the medium, but rather generates photons propagating in random directions. For this reason, we do not consider the second term when evaluating the material absorption or gain.

In the presence of absorption or gain, the photon density increases or decreases exponentially in the propagation direction:

$$\rho_{ph}(x) = \rho_{ph}(0)\exp(-\alpha x)\exp(\overline{g}x),$$

where α is the absorption, $\overline{g}$ is the gain due to stimulated emission. Substituting in (2.26) we readily identify

$$\alpha = \frac{n_r}{c_0} w N_{cv}(1 - f_n)(1 - f_h) \tag{2.27}$$

$$\overline{g} = \frac{n_r}{c_0} w N_{cv} f_n f_h. \tag{2.28}$$

Finally, the net gain $g = \overline{g} - \alpha$ can be expressed as

$$g = \frac{n_r}{c_0} w N_{cv} f_n f_h - \frac{n_r}{c_0} w N_{cv}(1 - f_n)(1 - f_h) = \frac{n_r}{c_0} w N_{cv}(f_n + f_h - 1).$$

To understand when the net gain is positive and when, in contrast, absorption prevails leading to negative net gain, we have to consider the occupation probabilities f_n and f_h.

In "ordinary" conditions, the occupation probability of the conduction and valence band states is described by the Boltzmann tail of the Fermi–Dirac statistics, i.e., both occupations are ≈ 0. In such conditions, the net gain is negative and absorption prevails.

On the other hand, the occupation probability increases in the presence of large injection of electrons and holes, until all valence and conduction band states are full, so that f_n and f_h may be close to 1 (at least, in a certain carrier energy range). Now, radiative recombination is fostered and stimulated emission prevails, leading to positive gain. Due to the effect of stimulated emission and gain, a semiconductor can amplify the light traveling through it, as in optical amplifiers and lasers.

The condition for positive net gain can be expressed in terms of the electron and hole quasi-Fermi levels as follows. Taking into account that

$$g = \frac{n_r}{c_0} w N_{cv}(f_n + f_h - 1), \tag{2.29}$$

where f_n and f_h are given by (2.12) and (2.13), and $g > 0$ if $f_n + f_h - 1 > 0$. The condition

$$f_n + f_h - 1 > 0 \longrightarrow \frac{1}{1 + \exp\left(\dfrac{E_n - E_{Fn}}{k_B T}\right)} + \frac{1}{1 + \exp\left(\dfrac{E_{Fh} - E_h}{k_B T}\right)} > 1,$$

directly implies

$$\exp\left(\frac{E_n - E_h - E_{Fn} + E_{Fh}}{k_B T}\right) < 1 \longrightarrow E_{Fn} - E_{Fh} > E_n - E_h.$$

Since the minimum allowed electron–hole energy difference is $E_n - E_h = E_g$, the positive gain condition (called the *population inversion condition*: notice that this condition has nothing to do with the inversion condition in a MOS system) is

$$E_{Fn} - E_{Fh} > E_g.$$

In inversion conditions $np \gg n_i^2$; this only happens in a forward biased pn junction or heterojunction. Conversely, in a depleted region $np \ll n_i^2$ and therefore absorption prevails.

2.4.1 The semiconductor gain energy profile

The qualitative behavior of $g(E_{ph})$, taking as a parameter the concentration of the injected carrier populations (for simplicity assume quasi-neutrality, $n \approx p$), can be obtained as follows. We start from (2.29) and approximate the Fermi distribution at 0 K as a step function. For $E_{ph} < E_g$, the joint density of states is zero and $g = 0$. If the population inversion condition is not met ($n < n_{inv}$), on the other hand, $f_n(E_{ph}) \approx f_h(E_{ph}) \approx 0$ for all photon energies, and

$$g(E_{ph}) \approx -\frac{n_r}{c_0} w N_{cv}(E_{ph}) = \alpha(E_{ph}) \quad \forall E_{ph}.$$

Let us assume now that, for a given carrier injection corresponding to population inversion ($n > n_{inv}$), $f_n(E_{ph}) \approx f_h(E_{ph}) \approx 1$ for $E_{ph} < E_0$ (where $E_0 = E_{Fh} + E_{Fn} + E_g$), while $f_n(E_{ph}) \approx f_h(E_{ph}) \approx 0$ for $E_{ph} > E_0$; then

$$g(E_{ph}) \approx \frac{n_r}{c_0} w N_{cv}(E_{ph}), \quad E_{ph} < E_0$$

$$g(E_{ph}) \approx -\frac{n_r}{c_0} w N_{cv}(E_{ph}) = \alpha(E_{ph}), \quad E_{ph} > E_0.$$

Thus, the *gain* is positive and proportional to the joint density of states up to E_0, while the *absorption* is proportional to the joint density of states for $E > E_0$. For increasing population injection, E_0 increases, leading to the qualitative behavior shown in Fig. 2.15 for a bulk material (above) and a quantum well (below), with staircase density of states (for simplicity, exciton resonances, see Section 2.4.3, were not considered). In the energy range considered we have also assumed that w is weakly dependent on the photon energy.

For temperatures above 0 K the gain profile vs. the photon energy can be evaluated numerically. As already stated, the inversion population condition is implemented in a direct-bias pn junction or double heterojunction where quasi-neutrality holds ($p \approx n$). From $n = p$ and the photon energy $\hbar\omega$, we derive the initial and final electron energy and the electron and hole quasi-Fermi levels. More explicitly, in (2.29) we express the Fermi distributions (2.12) and (2.13) by setting $E_n - E_{Fn} = (E_n - E_c) - (E_{Fn} - E_c)$, $E_{Fh} - E_h = (E_{Fh} - E_v) - (E_h - E_v)$. The terms $E_{Fn} - E_c$ and $E_{Fh} - E_v$ can be approximated as a function of the population density $n \approx p$ through the Joyce–Dixon formulae (1.12) and (1.13), while $E_n - E_c$ and $E_h - E_v$ are derived from the photon

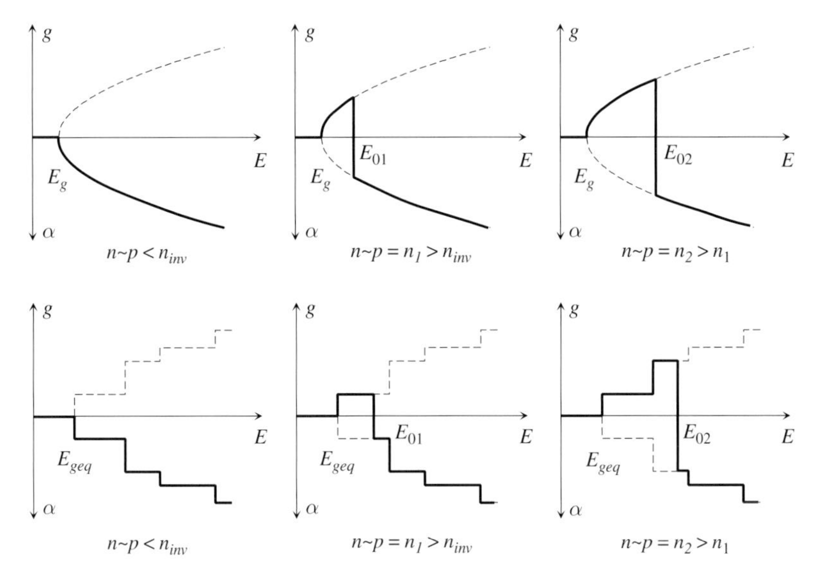

Figure 2.15 Qualitative behavior of gain at $T = 0$ K as a function of the photon energy from below inversion (left) and for increasing injected carrier density above inversion (center and right). Above: bulk (3D); below: QW (2D) with staircase joint density of states.

energy $\hbar\omega$ according to (2.14) and (2.15), i.e., using energy conservation in vertical transitions, as

$$E_n - E_c = \frac{m_r^*}{m_n^*}\left(\hbar\omega - E_g\right), \quad E_h - E_v = -\frac{m_r}{m_h^*}\left(\hbar\omega - E_g\right).$$

As an example, the resulting gain profile $g(\hbar\omega, n = p)$ of an AlGaAs/GaAs double heterostructure is shown in Fig. 2.16. The net gain is zero for photon energies below E_g because of the zero joint density of states; it increases for photon energies above E_g and then decreases again as a function of $\hbar\omega$, since at high energy fewer and fewer states are available. A larger $n = p$ corresponds to larger gain, for small $n = p$ the gain is negative, i.e., absorption prevails.

Gain in QW structures follows a similar behavior but, as already remarked, the shape of gain vs. the photon energy is different due to the presence of the staircase density of states. A qualitative picture at 300 K is shown in Fig. 2.17.

The gain profile $g = n_r c_0^{-1} w N_{cv} f_h f_n$ derived so far depends on the photon energy and on the carrier population and is based on the assumption that the electron and hole distributions f_n and f_h can be approximated through the *quasi-Fermi distributions*. In laser diodes above threshold, however, electrons and holes of a specific energy E_n and E_h selectively recombine to emit (through stimulated emission) photons with energy $\hbar\omega = E_n - E_h$. In high injection conditions (corresponding to high photon concentration and emitted optical power) the carrier lifetime is in the picosecond range and the restoring mechanisms of the quasi-Fermi distribution (typically, phonon scattering), whose characteristic time constants are around 1 ps, are unable to

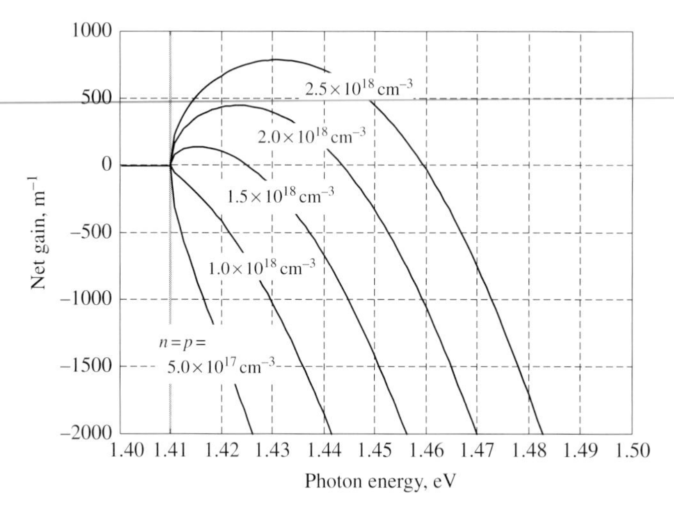

Figure 2.16 Gain profile in a AlGaAs/GaAs double heterojunction as a function of the electron and hole injected populations.

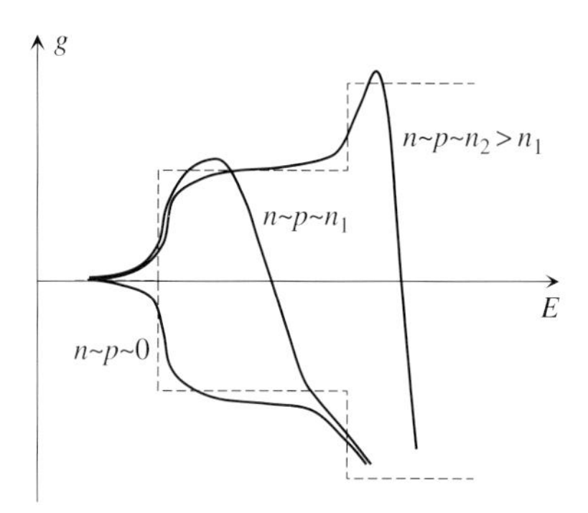

Figure 2.17 Qualitative behavior of the gain of a QW for different injected population density, starting from depletion conditions.

immediately compensate for the carrier depletion arising in narrow energy intervals around E_n and E_h (the effect is also called *spectral hole burning*). As a consequence, the occupation probabilities f_n and f_h decrease and so does the gain. Let ρ_{ph} be the density of photons emitted through stimulated emission; gain compression can be approximated as

$$g(E_{ph}) = \frac{g_F(E_{ph})}{1 + \epsilon_c \rho_{ph}} \approx g_F(E_{ph}) \left(1 - \epsilon_c \rho_{ph}\right), \tag{2.30}$$

where g_F is the unperturbed (quasi-Fermi distributions) gain profile, and ϵ_c is the nonlinear gain compression (or suppression) coefficient, with values of the order of $10^{-16} - 10^{-17}$ cm^3, see e.g. [11], Section 11.2.1. Gain compression ultimately limits the stimulated carrier lifetime and therefore the dynamic of lasers; see Section 2.5.1.

2.4.2 The semiconductor absorption energy profile

In a depleted region (e.g., a *pn* junction in reverse bias) $f_n \approx f_h \approx 0$ and absorption prevails; by expanding w from (2.10), one has from (2.29) for *direct-bandgap bulk semiconductors* (we assume bulk material, unpolarized light):

$$\alpha = \frac{n_r}{c_0} w N_{cv}(\hbar\omega) = \frac{\pi q^2 \hbar n_r}{2\epsilon m_0 c_0} \frac{2}{3} \left(\frac{2\mathfrak{p}_{cv}^2}{m_0} \right) \frac{1}{\hbar\omega} N_{cv}(\hbar\omega), \tag{2.31}$$

where N_{cv} is the JDOS defined in (2.11) and $\mathfrak{p}_{cv}$ is the dipole matrix element, related to the energy $E_p \approx 20$ eV, see (2.8). For GaAs, $E_p \approx 22.71$ eV, for InP $E_p \approx 17$ eV, see [12] for example. The overall behavior of the absorption coefficient is dominated by the JDOS,[9] since the $1/\hbar\omega$ term becomes significant for energies at which the effective mass approximation breaks down. Making explicit JDOS in (2.31), one has

$$\alpha = \frac{q^2 n_r (2m_r^*)^{3/2}}{2\pi\epsilon\hbar^2 m_0 c_0} \frac{2}{3} \left(\frac{2\mathfrak{p}_{cv}^2}{m_0} \right) \frac{1}{\hbar\omega} \sqrt{\hbar\omega - E_g} \approx \frac{K}{\hbar\omega} \sqrt{\hbar\omega - E_g},$$

where $K \approx 5.6 \times 10^4$ for GaAs if α is in cm^{-1} and all energies are in eV.[10] In direct-bandgap semiconductors, the absorption profile increases sharply for $hf > E_g$, due to the $\sqrt{\hbar\omega - E_g}$ term, see Fig. 2.18. Such a behavior dominates the absorption profile close to the absorption edge in compound semiconductors like GaAs, InP, CdTe, and GaN, see Fig. 2.20.

In *indirect-bandgap semiconductors*, the photon absorption requires the emission or absorption of one or more phonons. This weakens the interaction strength (due to the multibody interaction), and changes the absorption threshold. Denoting by E_ϕ the average phonon energy (meV at ambient temperature), the total absorption is $\alpha = \alpha_e + \alpha_a$,

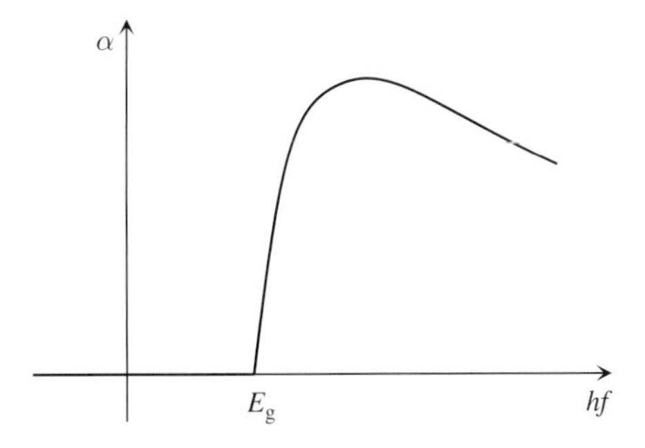

Figure 2.18 Theoretical behavior of the absorption coefficient in a direct-bandgap semiconductor.

[9] A subtle point concerns what kind of hole effective mass has to be exploited in the definition of the reduced mass: m_{lh}, m_{hh} or some average. A possibility is to separately consider the HH and LH contributions, see e.g. [13], Section 4.2. Alternatively, an average hole mass can be assumed, which is often taken as m_{hh} (for $k \neq 0$ the HH-C transition dominates).

[10] This value results from a reduced GaAs mass $m_r^* = 0.065 m_0$.

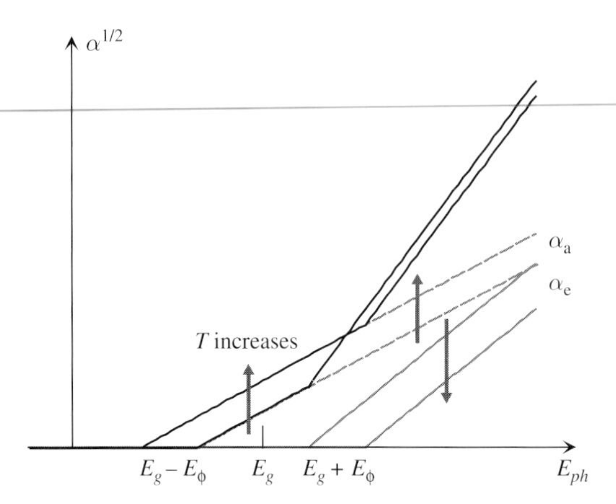

Figure 2.19 Absorption coefficient of an indirect-bandgap semiconductor at different temperature.

where α_e is the absorption with phonon emission, α_a that with phonon absorption. Second-order perturbation theory yields (see e.g. [4], Section 4.5)

$$\alpha_e \propto (\hbar\omega - E_g - E_\phi)^2, \quad \hbar\omega > E_g + E_\phi$$
$$\alpha_a \propto (\hbar\omega - E_g + E_\phi)^2, \quad \hbar\omega > E_g - E_\phi.$$

Since E_ϕ increases with T, the absorption cutoff is also temperature dependent. The overall behavior of the absorption coefficient vs. the photon energy is shown in Fig. 2.19; the phonon contribution shifts the absorption edge to lower energies. The theoretical behavior is confirmed by experimental data (for Si see, e.g., [14]); above threshold, the absorption coefficient has a temperature sensitivity of the order of $0.2\,\text{cm}^{-1}/\text{K}$. In direct-bandgap semiconductors, on the other hand, the absorption edge depends only on temperature through the temperature variation of the energy gap (E_g increases with decreasing T, leading to a blue shift in the absorption edge).

The indirect bandgap behavior can be affected by the presence of secondary direct minima, as in Ge, where the main minimum is indirect, but a direct (Γ point) minimum exists, see Fig. 1.12(c), so that for $hf < 0.66\,\text{eV}$, absorption is negligible; for $0.66 < hf < 0.9\,\text{eV}$, the material behaves as an indirect-bandgap material, absorption is low and increases slowly vs. the photon energy; for $hf > 0.9\,\text{eV}$, finally, the dominant process is *direct*, and absorption increases sharply. Such a composite behavior is shown in Fig. 2.20.

2.4.3 The QW absorption profile

In a QW, conduction and valence subbands exist. Conduction band subbands can be labeled, according to increasing minimum energy, as C_1, C_2, etc. Valence band subbands include both heavy and light hole subbands, which generally are nondegenerate. In an unstrained QW, typically the first valence band sublevel is HH_1, the second is LH_1 and then HH_2, LH_2, etc. follow. The resulting absorption profile inherits the staircase behavior of the QW JDOS, but only the steps relative to allowed transitions having

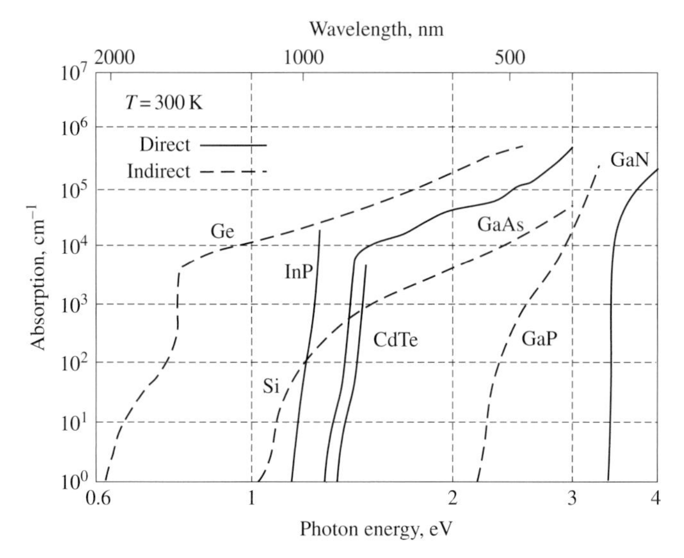

Figure 2.20 Absorption profile of some direct- and indirect-bandgap semiconductors (adapted from [1]).

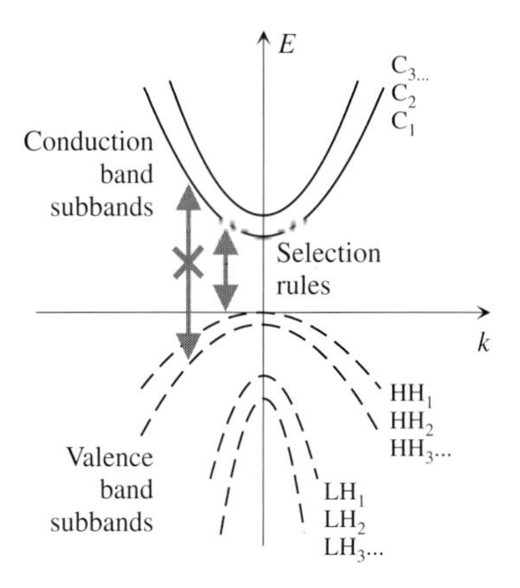

Figure 2.21 Dispersion relation for a QW with a qualitative picture of the HH and LH bands.

nonnegligible strength are present; see the qualitative picture in Fig. 2.21 in which the ordering of subbands is not represented exactly. The staircase profile of the absorption is shown in Fig. 2.22; the absorption edge is shifted with respect to the bulk material value. However, the measured absorption profile of a QW differs from the ideal staircase profile because sharp transitions are enhanced and overlap with a new effect, the *exciton resonance peaks.* Such peaks can also be detected in bulk materials, but only at low temperature, while in QW they appear at ambient temperature and can be strongly modulated by an applied electric field.[11]

[11] Excitons also change the theoretical absorption profile near the absorption edge, leading to absorption enhancement for $E_{ph} > E_g$. For a more detailed treatment on excitons, see, for example, [11], Section 13.3 and references therein.

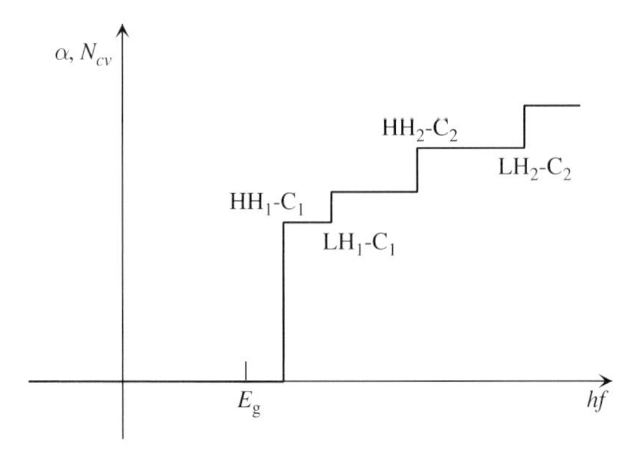

Figure 2.22 Absorption profile (neglecting exciton peaks) and joint density of states for a QW: staircase behavior with steps corresponding to the allowed transitions.

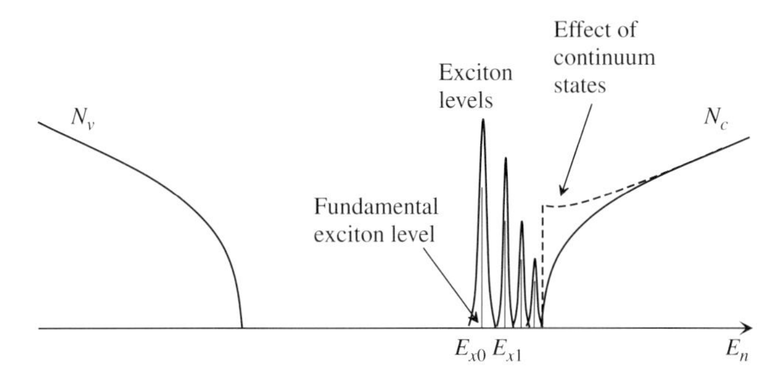

Figure 2.23 Qualitative picture of exciton levels in the semiconductor density of states.

Excitons can be introduced in a simplified way as follows. The electron and hole charges are $-q$ and q, respectively. An electron and a hole therefore experience Coulomb attraction in real space, leading to a weak form of bond similar, in principle, to the (strong) bond existing between the hydrogen nucleus and the hydrogen atom electron. While in the hydrogen atom the binding energy is large (more than 10 eV), the weak (typically less than 10 meV in bulk semiconductors) electron–hole bond is broken at ambient temperature, where the average energy of particles is of the order of 40 meV. A weakly bound e-h pair is called an *exciton*. Excitons can be shown to introduce a set of closely packed discrete levels in the forbidden band of the semiconductor, see Fig. 2.23, plus an additional continuum set of states that merge with the conduction band state density.

The position of the exciton levels can be explained from the fact that a small amount of energy breaks the exciton, bringing the bound electron into the conduction band. The exciton level can therefore be understood as an energy level immediately below the conduction band, in much the same way as a shallow donor level is located immediately below the conduction band edge. In such conditions, an ionization energy of the order of 10 meV is enough to promote an electron trapped in the shallow level into the conduction band. It can be shown, in fact, that the exciton dispersion relation (in the effective mass approximation) can be expressed as

$$E_{nx}(\underline{k}_x) = E_g - \frac{R^*_\infty}{n^2} + \frac{\hbar^2 k_x^2}{2m_x^*}, \quad n = 1, 2, \ldots \infty,$$

where $m_x^* = m_h^* + m_n^*$ is the exciton effective mass, and R^*_∞ is the effective Rydberg constant,

$$R^*_\infty = \frac{m_r^*}{m_0} \frac{1}{n_r^4} R_\infty,$$

while $R_\infty = 13.605$ eV is the Rydberg constant. Due to the effect of the reduced mass and of the refractive index, $R^*_\infty \ll R_\infty$.

Example 2.4: Evaluate the fundamental exciton energy in GaAs and GaN.

For GaAs assume $m_n^* = 0.067m_0$, $m_h^* = 0.51m_0$, $\epsilon_r = 13$. The reduced mass is therefore $m_r^* = (1/m_n^* + 1/m_h^*)^{-1} = (1/0.067 + 1/0.51)^{-1} m_0 \approx 0.06m_0$. The binding energy for the fundamental state will be

$$E_x = E_g - E_{1x}(0) = R^*_\infty = \frac{m_r^*}{m_0} \frac{1}{n_r^4} R_\infty = \frac{0.06}{13^2} \times 13.605 = 4.8 \text{ meV}$$

which is much lower than the average thermal energy at ambient temperature. On the other hand, in GaN $m_n^* = 0.22m_0$, $m_h^* \approx m_0$, $m_r^* = (1/m_n^* + 1/m_h^*)^{-1} = (1/0.22 + 1)^{-1} m_0 \approx 0.18m_0$, $\epsilon_r \approx 10$. With respect to GaAs, the effective mass is much larger (consistent with the larger energy gap, 3.4 eV). As a result we have

$$E_x = \frac{m_r^*}{m_0} \frac{1}{n_r^4} R_\infty = \frac{0.18}{10^2} \times 13.605 \approx 25 \text{ meV},$$

so that at room temperature $E_x \approx k_B T$ and the fraction of nonionized excitons is significant.

Excitons can interact with photons having energies immediately below the absorption edge E_g, thus creating strong absorption peaks and an increase of the absorption profile for $E_{ph} \approx E_g$. Such peaks appear close to the absorption edge energy in bulk semiconductors or close to each staircase step in the QW response, and merge with the step due to the effect of finite linewidth.

In a bulk material like GaAs, exciton peaks disappear at ambient temperature and are only visible at low T, see Fig. 2.24 [15]. In a quantum well, however, the binding energy of excitons is substantially larger; in fact, it can be shown that, in this case, the binding energy of the fundamental exciton state is

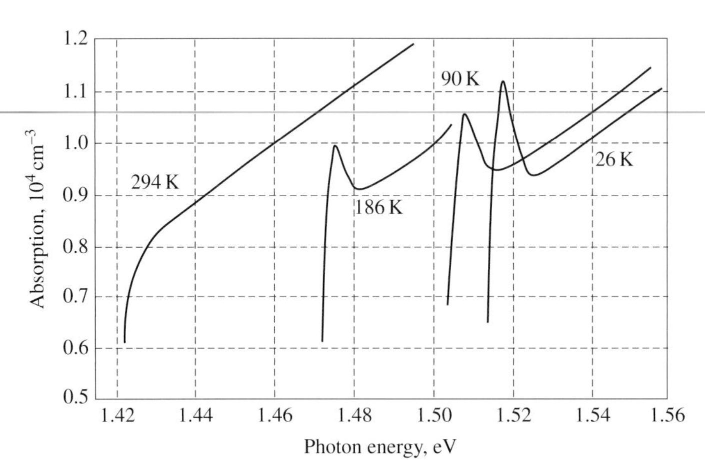

Figure 2.24 Absorption profiles in bulk GaAs vs. temperature, showing the effect of excitons. From [15], Fig. 3, (©1962 American Physical Society).

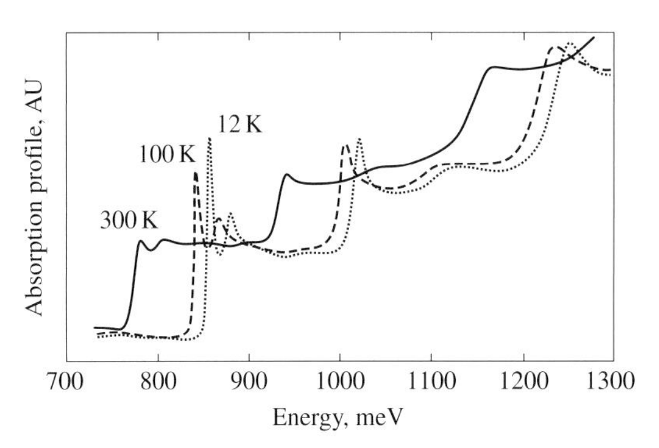

Figure 2.25 Measured absorption profile of InGaAs/InAlAs 10nm/200m undoped MQW. The exciton peaks are clearly visible also at ambient temperature. From [16] Fig. 2 (a) (©1988 IEEE).

$$E_x = E_{g,\text{eff}} - E_{1x}(0) = \frac{4m_r^*}{m_0}\frac{1}{n_r^4}R_\infty,$$

where $E_{g,\text{eff}}$ is the effective gap (i.e., the distance between the fundamental level in the conduction band and the fundamental level in the valence band). The energy E_x turns out to be four times larger than in the bulk semiconductor, thus in GaAs, for example, making the exciton effect significant at room temperature also. In fact, in the QW the exciton resonant peaks also do not disappear at $T = 300$ K [16] [17]; see Fig. 2.25 for the profile at ambient and low temperature. Exciton peaks are particularly important in the design of QW electroabsorption modulators, since they are strongly modulated in energy and amplitude by the application of an electric field. Due to the different oscillator strengths relative to the HH and LH interactions with conduction band states, the QW absorption profile is also polarization dependent; a qualitative behavior for TE and TM polarizations is shown in Fig. 6.34.

2.4.4 Spontaneous emission spectrum

The *spontaneous emission spectrum* describes the spectral distribution of the EM energy radiated by the semiconductor (at or outside equilibrium) due to photon spontaneous emission. It can be derived by summing the spontaneous emission scattering rate over *all possible photon states* satisfying energy conservation. To this end, we select as a volume a cube with side L and apply to the photons the so-called *pseudo-quantization*, i.e., we require that the photon wavenumber satisfy the resonance condition along x, y, z:

$$\underline{k}_{ph} = l\frac{2\pi}{L}\widehat{x} + m\frac{2\pi}{L}\widehat{y} + n\frac{2\pi}{L}\widehat{z},$$

where l, m, n are integers; the space allocated to each state is $(2\pi)^3/2L^3 = 4\pi^3/V$, where the factor 2 accounts for the presence of two polarization states (TE and TM).

Summing over all photon wavevectors, and taking into account that $\mathcal{W}_{em}^{sp}(\underline{k}_{ph}) = \mathcal{W}_{em}^{sp}(k_{ph})$, and that $k_{ph} = \omega/c$, we express the total spontaneous emission rate per unit volume, R_o^{sp}, in the form[12]

$$\begin{aligned}
R_o^{sp} &= \frac{1}{V}\sum_{\underline{k}_{ph}} \mathcal{W}_{em}^{sp}(\underline{k}_{ph}) \approx \frac{1}{4\pi^3}\int \mathcal{W}_{em}^{sp}(k_{ph})\,\mathrm{d}\underline{k}_{ph} \\
&= \frac{1}{4\pi^3}\int_0^\infty \mathcal{W}_{em}^{sp}(k_{ph})k_{ph}^2\,\mathrm{d}k_{ph}\int_0^\pi \sin\theta\,\mathrm{d}\theta\int_0^{2\pi}\mathrm{d}\phi \\
&= \frac{1}{\pi^2}\int_0^\infty \mathcal{W}_{em}^{sp}(k_{ph})k_{ph}^2\,\mathrm{d}k_{ph} = \frac{1}{\pi^2\hbar^3c^3}\int_{E_g}^\infty \mathcal{W}_{em}^{sp}(\hbar\omega)(\hbar\omega)^2\,\mathrm{d}\hbar\omega \\
&= \int_{E_g}^\infty \mathcal{W}_{em}^{sp}(\hbar\omega)g_{ph}(\hbar\omega)\,\mathrm{d}\hbar\omega,
\end{aligned}$$

where

$$g_{ph}(\hbar\omega) = \frac{2(\hbar\omega)^2}{2\pi^2\hbar^3c^3} = \frac{2\omega^2 n_r^3}{2\pi^2\hbar c_0^3}\ \mathrm{J}^{-1}\,\mathrm{m}^{-3} \tag{2.32}$$

is the *photon density of states* per unit energy and volume. Substituting for the scattering rate from (2.19), we obtain

$$\begin{aligned}
R_o^{sp} &= \frac{1}{\pi^2\hbar^3c^3}\int_{E_g}^\infty w N_{cv}(\hbar\omega) f_h\left(E_i(\hbar\omega)\right) f_n\left(E_f(\hbar\omega)\right)(\hbar\omega)^2\,\mathrm{d}\hbar\omega \\
&= \int_{E_g}^\infty r_o^{sp}(\hbar\omega)\,\mathrm{d}\hbar\omega,
\end{aligned} \tag{2.33}$$

[12] Since each spontaneous emission corresponds to a recombination event, we exploit for the emission and recombination rates the same symbol R_o^{sp} or r_o^{sp}, where the "r" suggests "recombination."

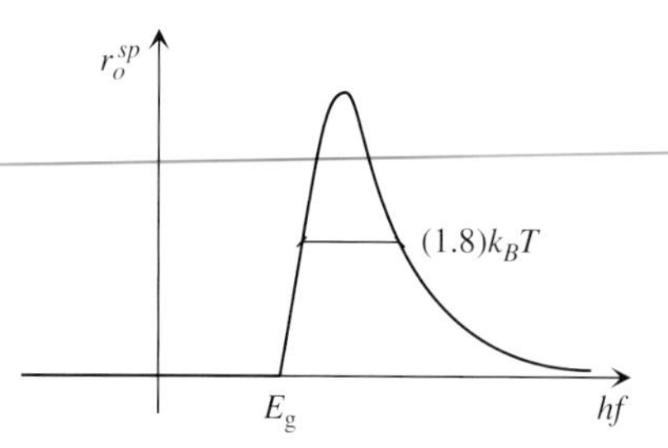

Figure 2.26 Energy behavior of the spontaneous emission spectrum (low injection or nondegenerate semiconductor).

where r_o^{sp} (dimension $\mathrm{cm^{-3}\,s^{-1}\,J^{-1}}$) is the *spontaneous emission spectrum*. Taking into account (2.10) and (2.11), we have

$$r_o^{sp}(\hbar\omega) = \frac{q^2(2m_r^*)^{3/2}}{3\pi^3\hbar^5 m_0^2 c^3 \epsilon} \mathrm{p}_{cv}^2 \sqrt{\hbar\omega - E_g}\, f_h\left(E_i(\hbar\omega)\right) f_n\left(E_f(\hbar\omega)\right)\hbar\omega. \tag{2.34}$$

A useful alternative form is

$$r_o^{sp}(\hbar\omega) = \frac{1}{\tau_0}\frac{\hbar\omega}{E_g} N_{cv}(\hbar\omega)\, f_e\left(E_e, E_{Fn}\right) f_h\left(E_h, E_{Fh}\right), \tag{2.35}$$

where the *spontaneous radiative lifetime* τ_0 is defined in (2.47). The spontaneous emission spectrum has a maximum immediately above E_g and a (half-power) linewidth of the order of $1.8k_BT$ in energy (47 meV at 300 K) in low-injection conditions; see Fig. 2.26. In high-injection conditions the spectrum broadens; see Example 2.5. For further details on the spontaneous emission linewidth in low- and high-injection (nondegenerate and degenerate) conditions, see Section 5.2.5.

Example 2.5: Evaluate the spontaneous emission spectrum in GaAs, for different carrier densities, assuming quasi-neutrality. Assume $\tau_0 \approx 0.5$ ns.

We exploit for $r_o^{sp}(\hbar\omega)$ the expression (2.35), including as limiting cases the degenerate and nondegenerate semiconductor, and we evaluate the joint density of states from (2.11). For GaAs, assume $m_n^* = 0.067m_0$, $m_h = 0.4m_0$, from which

$$m_r^* = \left(1/m_h + 1/m_n^*\right)^{-1} = 0.057m_0$$

($m_0 = 9.11 \times 10^{-31}$ kg), $\hbar\omega = E_g = 1.42$ eV, $N_c = 4.45 \times 10^{17}\,\mathrm{cm^{-3}}$, $N_v = 7.72 \times 10^{18}\,\mathrm{cm^{-3}}$. From $n \approx p$, N_c, and N_v, we derive $E_{Fn} - E_c$ and $E_v - E_{Fh}$ through the Joyce–Dixon approximation (1.12) and (1.13). The resulting behavior is shown in Fig. 2.27; for each density the spectrum is normalized with respect to the maximum. Notice that for low carrier density the spectral width is of the order of $1.8k_BT$ but increases in high injection.

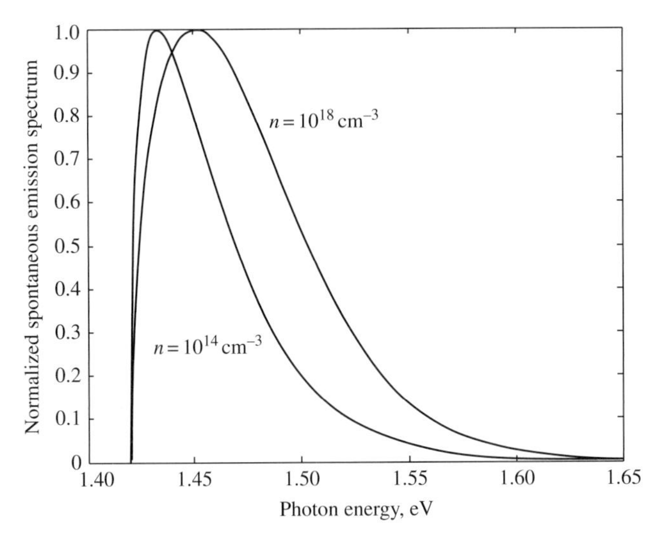

Figure 2.27 Normalized spontaneous emission spectrum for GaAs as a function of the photon energy and of the electron population (quasi-neutrality assumed).

2.4.5 Spontaneous emission, gain, and absorption spectra

In the analysis of some devices we conveniently exploit an alternative expression for the spontaneous emission rate *for a specific photon wavevector* $\overline{r}_o^{sp} \equiv \mathcal{W}_{em}^{sp} = wN_{cv} f_n f_h$ (s^{-1}); see (2.19).[13] Such an expression relates the spontaneous emission spectrum to the gain spectrum, or, at or near equilibrium, to the absorption spectrum.

From the definition of the net gain (2.29) and introducing the quasi-Fermi statistics (2.12) and (2.13), we have

$$\frac{c_0}{n_r} g = wN_{cv}(f_n + f_h - 1)$$
$$= wN_{cv} \frac{1 - \exp\left(\dfrac{hf - E_{Fn} + E_{Fh}}{k_B T}\right)}{\left[1 + \exp\left(\dfrac{E_n - E_{Fn}}{k_B T}\right)\right]\left[1 + \exp\left(\dfrac{E_{Fh} - E_h}{k_B T}\right)\right]}, \tag{2.36}$$

where $E_n - E_h = hf$. On the other hand, from (2.19), (2.12), and (2.13) we have

$$\overline{r}_o^{sp} = wN_{cv} f_n f_h = \frac{wN_{cv}}{\left[1 + \exp\left(\dfrac{E_n - E_{Fn}}{k_B T}\right)\right]\left[1 + \exp\left(\dfrac{E_{Fh} - E_h}{k_B T}\right)\right]}. \tag{2.37}$$

[13] In evaluating $\overline{r}_o^{sp}$ we do not integrate the spontaneous rate over all possible wavevectors, but select photons with a specific E and $\underline{k}$ state. The concept is useful, for example, in the analysis of lasers, where we are interested in the amount of spontaneous emission falling in the laser cavity photon states, rather than in the whole spontaneous emission spectrum.

Comparing (2.36) and (2.37), we obtain

$$\overline{r}_{o}^{sp}(h\omega) = \frac{\dfrac{c_0}{n_r} g(h\omega)}{1 - \exp\left(\dfrac{hf - E_{Fn} + E_{Fh}}{k_B T}\right)} = \frac{c_0}{n_r} g(h\omega) n_{sp}(h\omega), \tag{2.38}$$

where we have introduced the nondimensional *spontaneous emission factor* n_{sp}:[14]

$$n_{sp} = \frac{1}{1 - \exp\left(\dfrac{\hbar\omega - E_{Fn} + E_{Fh}}{k_B T}\right)}. \tag{2.39}$$

Equation (2.38) relates the *spontaneous emission spectrum* (with reference to a specific photon state with energy $\hbar\omega$ and wavenumber $\underline{k}_{ph}$) to the *net gain spectrum*. Assuming, (as in laser diodes, where population inversion occurs) that the separation of the quasi-Fermi levels is larger than the gap ($E_{Fn} - E_{Fh} > E_g$), while $\hbar\omega \approx E_g$, the exponential turns out to have a negative argument, possibly large in absolute value, due to the small value of $k_B T$; we therefore have $n_{sp} > 1$ but typically $n_{sp} \approx 2 - 3$.

At or near *thermodynamic equilibrium* the quasi-Fermi levels coincide and $g \approx -\alpha$; (2.38) therefore yields

$$\overline{r}_{o0}^{sp}(h\omega) = \frac{\dfrac{c_0}{n_r} \alpha(h\omega)}{\exp\left(\dfrac{h\omega}{k_B T}\right) - 1} = \frac{c_0}{n_r} \alpha(h\omega) n_{ph0}(\hbar\omega), \tag{2.40}$$

where $n_{ph0}(\hbar\omega)$ is the equilibrium Bose–Einstein photon distribution, see (2.24). Equation (2.40) holds for a specific photon state; we can extend the result to consider *all photon states having a certain energy* by summing over all possible photon states and exploiting the photon state density weight (2.32), as outlined in Section 2.4.4. We obtain

$$R_{o0}^{sp} \approx \frac{1}{V} \sum_{\underline{k}_{ph}} W_{em}^{sp}(\underline{k}_{ph}) \approx \int_{E_g}^{\infty} \frac{\dfrac{c_0}{n_r} \alpha(\hbar\omega)}{\exp\left(\dfrac{\hbar\omega}{k_B T}\right) - 1} g_{ph}(\hbar\omega)\, d\hbar\omega = \int_{E_g}^{\infty} r_{o0}^{sp}(\hbar\omega)\, d\hbar\omega,$$

[14] The parameter n_{sp} has, unfortunately, the same name as β_k, the fraction of spontaneous emission into mode k of a laser. An alternative name for n_{sp} is *population inversion factor* (see [18], Section 11.2.3 and 11.2.16).

where[15]

$$r_{o0}^{sp}(\hbar\omega) = \frac{n_r^2(\hbar\omega)^2}{\pi^2\hbar^3 c_0^2} \frac{\alpha(\hbar\omega)}{\exp\left(\dfrac{h\omega}{k_B T}\right) - 1} \tag{2.41}$$

$$r_{o0}^{sp}(f) = \frac{8\pi n_r^2 f^2}{c_0^2} \frac{\alpha(hf)}{\exp\left(\dfrac{hf}{k_B T}\right) - 1}. \tag{2.42}$$

Equations (2.41) or (2.42), called the *van Roosbroeck–Shockley relation* [19], connect the *equilibrium spontaneous emission spectrum* r_{o0}^{sp} ($\mathrm{J^{-1}\,m^{-1}\,s^{-1}}$ or $\mathrm{Hz^{-1}\,m^{-1}\,s^{-1}}$) with the *absorption spectrum*. From a physical standpoint, (2.41) or (2.42) imply that, to satisfy the equilibrium condition, the number of photons absorbed by the material and the number of photons generated by spontaneous emission must be (energy by energy) related.

2.5 The macroscopic view: the semiconductor standpoint

From *the EM wave standpoint*, absorption and stimulated emission cause wave attenuation or amplification, respectively, while spontaneous emission can be interpreted as an incoherent photon source term not leading to the amplification of an incident EM wave.

From the *semiconductor standpoint*, absorption and emission (stimulated and spontaneous) correspond to generation and recombination, respectively, of e-h pairs. Deriving the semiconductor *optical* or *radiative* generation and recombination rates (i.e., the number of electrons and holes generated or recombined per unit time and volume) corresponding to the interaction with a monochromatic EM wave with specific propagation constant $\underline{k}$ is indeed straightforward, since each photon absorption event corresponds to the generation of an e-h pair, and each photon emission corresponds to the recombination of an e-h pair. It follows that:

- The generation rate G_o equals the absorption rate (per unit volume) W_{abs}.
- The recombination rate R_o equals the emission rate (per unit volume) W_{em}.
- The net recombination rate U_o equals the rate difference $W_{em} - W_{abs}$.

[15] Remember the variable change

$$\int F(\hbar\omega)\,\mathrm{d}\hbar\omega = \int 2\pi\hbar F(2\pi\hbar f)\,\mathrm{d}f = \int \Phi(f)\,\mathrm{d}f,$$

with $\Phi(f) = 2\pi\hbar F(2\pi\hbar f)$; for brevity we often identify (somewhat ambiguously) $\Phi \equiv F$.

We thus have, expressing the scattering rates per unit volume from (2.21) and (2.23):

$$U_o \equiv \left.\frac{\mathrm{d}n}{\mathrm{d}t}\right|_{RG} \equiv \left.\frac{\mathrm{d}p}{\mathrm{d}t}\right|_{RG} \equiv \left.\frac{\mathrm{d}\rho_{ph}}{\mathrm{d}t}\right|_{em,abs} = W_{em}(\hbar\omega) - W_{abs}(\hbar\omega)$$

$$= \rho_{ph} w N_{cv}\left[f_n f_h - (1-f_n)(1-f_h)\right] + \frac{1}{V} w N_{cv} f_n f_h .$$

We can conveniently relate the total recombination rate to the overall parameters of the EM wave, such as absorption and gain, see (2.27) and (2.28):

$$\begin{aligned} U_o(\hbar\omega, n, p) &= \rho_{ph} w N_{cv} f_n f_h - \rho_{ph} w N_{cv}(1-f_n)(1-f_h) + \frac{1}{V} w N_{cv} f_n f_h \\ &= \overline{g}\rho_{ph}\frac{c_0}{n_r} - \alpha\rho_{ph}\frac{c_0}{n_r} + \frac{1}{V} w N_{cv} f_n f_h \\ &= \underbrace{\overline{g}\frac{P_{op}}{\hbar\omega} + \frac{1}{V} w N_{cv} f_n f_h}_{R_o^{st}(\hbar\omega)+\overline{r}_o^{sp}(\hbar\omega)\cdot V^{-1}} - \underbrace{\alpha\frac{P_{op}}{\hbar\omega}}_{G_o(\hbar\omega)} \end{aligned} \tag{2.43}$$

where P_{op} is the power flux (W/cm^2), R_o^{st} is the stimulated recombination rate, $\overline{r}_o^{sp} V^{-1}$ the spontaneous recombination rate (per unit volume) for a specific k-state, G_o is the generation rate.

For $R_o^{st}(\hbar\omega)$ and $G_o(\hbar\omega)$, we can assume that photons are single-energy or narrow-band (as in a laser, where photons may belong to a specific k-state, or in a photodetector illuminated by a single-frequency optical beam); if the material is illuminated by an optical signal with assigned spectral content, the generation (or stimulated recombination) rates can be integrated over the photon energy profile, i.e., over the spectral distribution of P_{op}. Spontaneous generation, on the other hand, is intrinsically broadband, as discussed in Section 2.4.4. The total associated spontaneous recombination rate should, therefore, account for the (broad) spontaneous emission spectrum, which also includes the photon density of states (2.32).

To further clarify this point, we note that recombination rates can refer to a single specific photon k-states, or to all photons having a certain energy, or to all possible photon states, integrated over the energy spectrum. In discussing *stimulated* recombination, we typically refer to a *specific photon state,* as in a laser cavity, and denote the relevant recombination rate as R_o^{st}. Concerning *spontaneous* recombination, we must distinguish several relevant cases according to the following notation:

- R_o^{sp} is the total spontaneous recombination rate per unit volume, integrated over all possible photon states and energies; it coincides with the total spontaneous emission rate per unit volume. R_o^{sp} is obtained by integrating, over all photon energies, r_o^{sp}.
- r_o^{sp} is the spontaneous recombination rate per unit volume accounting for all photon states of *given energy*; it coincides with the spontaneous emission spectrum (2.34).
- $\overline{r}_o^{sp} \equiv \mathcal{W}_{em}^{sp}$ is the spontaneous recombination rate for a *specific photon state*; it coincides with the scattering rate for spontaneous emission.

- $\overline{r}_o^{sp} V^{-1} \equiv \overline{R}_o^{sp}$ is the spontaneous recombination rate per unit volume for a *specific photon state*.[16]

To express r_o^{sp} we directly exploit (2.34) and we assume, for the moment, that the semiconductor is *nondegenerate*, e.g., is in low injection conditions. Taking into account (1.9), (1.10), (1.11) and expressing $E_e - E_h$ in terms of the photon energy $\hbar\omega$ we have, in the Boltzmann approximation:

$$f_n(E_e, E_{Fn}) f_h(E_h, E_{Fh}) = \mathrm{e}^{\frac{E_{Fn}-E_e}{k_B T}} \mathrm{e}^{\frac{E_h - E_{Fh}}{k_B T}} = \frac{np}{n_i^2} \mathrm{e}^{-\frac{\hbar\omega}{k_B T}}.$$

Thus, the spontaneous recombination rate per unit volume (the spontaneous recombination spectrum) has the following energy distribution:

$$r_{o,\mathrm{ND}}^{sp}(\hbar\omega) = K' \hbar\omega \sqrt{\hbar\omega - E_g}\, \mathrm{e}^{-\frac{\hbar\omega}{k_B T}}, \quad K' = \frac{q^2 (2m_r^*)^{3/2}}{3\pi^3 \hbar^5 m_0^2 c^3 \epsilon} \mathrm{p}_{cv}^2 \frac{np}{n_i^2}. \tag{2.44}$$

The energy behavior of the low-injection spontaneous recombination spectrum coincides with the spontaneous emission spectrum and is therefore shown in Fig. 2.26. The *total* spontaneous recombination rate R_o^{sp} defined above can be recovered by integrating over all photon energies; note that, in the nondegenerate case, the total rate will be proportional to pn, i.e., it follows the elementary law $R_{o,\mathrm{ND}}^{sp} \propto pn$.

2.5.1 Carrier radiative lifetimes

In the lifetime approximation the recombination rates are expressed as

$$R_n = \frac{n}{\tau_n}, \quad R_h = \frac{p}{\tau_h},$$

where $R_n = R_h$ for band-to-band processes. In general, however, lifetimes will depend on the carrier population and a constant lifetime can be only defined in *small-signal conditions* and (often) in *low-injection conditions*.[17] However, many optoelectronic devices (e.g., semiconductor sources) operate in high injection, quasi-neutrality conditions ($n \approx p$); in this case, the lifetime concept can be still applied, but τ will depend on the carrier concentration.

In fact, in nondegenerate conditions we have

$$R_n = r_n pn \approx r_n n^2 = \frac{n}{\tau_n}$$

[16] The symbol $\overline{R}_o^{sp}$ will be exploited in the analysis of lasers; we will approximate $\overline{R}_o^{sp} \approx \beta_k R_o^{sp}$, where $\beta_k \ll 1$ is the *spontaneous emission factor*, see (5.60) and (5.61).

[17] Low injection occurs in a semiconductor when the excess minority carrier concentration is much larger than the equilibrium concentration, while the majority carrier concentration is practically at the equilibrium condition, e.g., in an n-type material $n \approx N_D$, $p \approx p'$, p' excess hole concentration.

where r_n is a proper recombination coefficient and quasi-neutrality has been assumed; in this case,

$$\tau_n = \frac{1}{r_n n},$$

i.e., the lifetime depends *inversely* on the carrier concentration.

Let us now discuss the lifetime associated with recombination due to *spontaneous* and *stimulated emission* of photons.

We start by considering the emission of photons having a specific energy and wavenumber, and extend the treatment to the case of spontaneous emission, where a broad emission spectrum is involved. The discussion is done in two steps: *nondegenerate semiconductors* (i.e., the Fermi levels are within the forbidden gap, and the Boltzmann statistics hold); and *degenerate semiconductors* (the Fermi levels are within the conduction and valence bands, the injection level is high, and the Fermi statistics must be used).

Using the Boltzmann approximation for densities and substituting for the JDS and for n and p, the following expression is obtained for the net radiative recombination rate of a *nondegenerate semiconductor*, in the presence of a photon wave with assigned energy and momentum (single-photon state), see (2.43):

$$U_{o,\mathrm{ND}} = \left[R^{st}_{o,\mathrm{ND}}(\hbar\omega) - G_o(\hbar\omega)\right] + \overline{R}^{sp}_{o,\mathrm{ND}},$$

where

$$R^{st}_{o,\mathrm{ND}}(\hbar\omega) - G_o(\hbar\omega) = \frac{n_r P_{op}}{\hbar\omega c_0} w \frac{\sqrt{2} m_r^{*3/2}}{\pi^2 \hbar^3 n_i^2} \sqrt{\hbar\omega - E_g} \left[\exp\left(-\frac{\hbar\omega}{k_B T}\right) np - n_i^2\right]$$

$$\overline{R}^{sp}_{o,\mathrm{ND}} \equiv \frac{1}{V} \overline{r}^{sp}_{o,\mathrm{ND}} = \frac{1}{V} w \frac{\sqrt{2} m_r^{*3/2}}{\pi^2 \hbar^3 n_i^2} \sqrt{\hbar\omega - E_g} \exp\left(-\frac{\hbar\omega}{k_B T}\right) np.$$

In thermodynamic equilibrium, $U_{o,\mathrm{ND}}$ vanishes if two conditions are met: the photon number follows the Bose–Einstein statistics (2.24), and the carrier equilibrium condition $np = n_i^2$ holds.[18] Out of equilibrium, and assuming quasi-neutrality ($n \approx p$), we can separate the recombination rates due to stimulated and spontaneous emission as[19]

$$\overline{R}^{sp}_{o,\mathrm{ND}} = \frac{n}{\overline{\tau}^{sp}_{n,\mathrm{ND}}(\hbar\omega)} \approx \frac{w}{V} \frac{\sqrt{2} m_r^{*3/2}}{\pi^2 \hbar^3 n_i^2} \sqrt{\hbar\omega - E_g} e^{-\frac{\hbar\omega}{k_B T}} n^2 = H n^2$$

$$R^{st}_{o,\mathrm{ND}} = \frac{n}{\tau^{st}_{n,\mathrm{ND}}(\hbar\omega)} \approx \frac{n_r w}{\hbar\omega c_0} \frac{\sqrt{2} m_r^{*3/2}}{\pi^2 \hbar^3 n_i^2} \sqrt{\hbar\omega - E_g} e^{-\frac{\hbar\omega}{k_B T}} n^2 P_{op} = K n^2 P_{op},$$

[18] Note that the net recombination rate does not follow the elementary law $U_0 = r_o(np - n_i^2)$, unless the photons are near equilibrium (i.e., the net recombination rate is zero only when equilibrium holds for photons *and* carriers).

[19] The notation for the recombination rates was also formally extended to lifetimes.

where K and H do not depend on the carrier population. The two lifetimes $\overline{\tau}^{sp}_{n,\mathrm{ND}}$ and $\tau^{st}_{n,\mathrm{ND}}$ therefore exhibit the following dependence on the carrier population:

$$\overline{\tau}^{sp}_{n,\mathrm{ND}} = \frac{1}{Hn} \tag{2.45}$$

$$\tau^{st}_{n,\mathrm{ND}} = \frac{1}{K P_{op} n}. \tag{2.46}$$

As already recalled, the spontaneous lifetime defined here ($\overline{\tau}^{sp}_{n}$) is for *single-frequency and propagation vector* photons; to obtain the total recombination rate R^{sp}_{o} (and the total spontaneous lifetime τ^{sp}_{n}) we integrate the recombination (emission) spectrum (2.34) over all photon energies. In the nondegenerate case $r^{sp}_{o}(\hbar\omega)$ is provided by (2.44). Exploiting the integral

$$I = \int_{E_g}^{\infty} \xi\sqrt{\xi - E_g}\,\mathrm{e}^{-\frac{\xi}{k_B T}}\,\mathrm{d}\xi = \mathrm{e}^{-\frac{E_g}{k_B T}}(k_B T)^{3/2}\frac{\sqrt{\pi}}{2}\left[E_g + \frac{3k_B T}{2}\right]$$
$$\approx \mathrm{e}^{-\frac{E_g}{k_B T}}(k_B T)^{3/2}\frac{\sqrt{\pi}E_g}{2},$$

and expressing the intrinsic concentration from (1.7) and the effective densities of states of the valence and conduction bands from (1.6), we obtain

$$R^{sp}_{o,\mathrm{ND}} = \int_{E_g}^{\infty} r^{sp}_{o,\mathrm{ND}}(\hbar\omega)\,\mathrm{d}\hbar\omega = \frac{1}{2\tau_0}\left(\frac{2\pi\hbar^2 m^*_r}{k_B T m^*_h m^*_n}\right)^{3/2} np = \frac{1}{2\tau_0}\frac{N_{cv}}{N_c N_v}np,$$

where τ_0 is the *spontaneous radiative lifetime*:[20]

$$\frac{1}{\tau_0} = \frac{2q^2 n_r E_g}{3\pi\hbar^2 c_0^3 m_0^2 \epsilon_0}\mathrm{p}^2_{cv}, \tag{2.47}$$

with typical values around 0.1–0.5 ns (τ_0 decreases with increasing energy gap and is equal to 0.44 ns for $E_g = 1$ eV), and the joint effective density of states N_{cv} is

$$N_{cv} = 2\frac{(2\pi m^*_r k_B T)^{3/2}}{h^3}.$$

Example 2.6: Evaluate the spontaneous radiative lifetime in GaAs. For GaAs we assume $n_r = 3.4$, $E_g = 1.42$ eV, and $2\mathrm{p}^2_{cv}/m_0 = 23$ eV.

From (2.47) we obtain

$$\frac{1}{\tau_0} = \frac{q^2 n_r E_g}{3\pi\epsilon_0 m_0 c_0^3\hbar^2}\left(\frac{2\mathrm{p}^2_{cv}}{m_0}\right)$$
$$= \frac{(1.6\times10^{-19})^2\cdot3.4\cdot1.42\times1.6\times10^{-19}\cdot23\times1.6\times10^{-19}}{3\pi\cdot8.86\times10^{-12}\cdot9.11\times10^{-31}\cdot(3\times10^8)^3\cdot(1.055\times10^{-34})^2} = \frac{1}{0.32\,\mathrm{ns}}.$$

---

[20] The expression used is in agreement with the parameter τ_R in [12], Eq. (7.37bis).

Assuming quasi-neutrality ($p \approx n$) we have $R_o^{sp} \equiv R_{o,\mathrm{ND}}^{sp} = n/\tau_{n,\mathrm{ND}}^{sp}$ where the result for the total spontaneous lifetime for the *nondegenerate case*, $\tau_{n,\mathrm{ND}}^{sp}$, is

$$\frac{1}{\tau_{n,\mathrm{ND}}^{sp}} = \frac{2q^2 n_r h E_g}{3\sqrt{2\pi}\epsilon_0 m_0^2 c_0^3} \mathfrak{p}_{cv}^2 \left(\frac{m_r^*}{k_B T m_h^* m_n^*}\right)^{3/2} n \approx \frac{1}{2\tau_0} \frac{N_{cv}}{N_c N_v} n$$
$$= \frac{2(\pi)^{3/2}\hbar^3}{\sqrt{2}(k_B T)^{3/2}} \frac{(m_r^*)^{3/2}}{(m_n^* m_h^*)^{3/2}} \frac{n}{\tau_0}. \tag{2.48}$$

Apparently, from (2.48) we have that $\tau_{n,\mathrm{ND}}^{sp} \to 0$ for $n \to \infty$. The paradox of zero lifetime for extremely large carrier densities can be removed by accounting for *degeneracy*. In the degenerate case, we can obtain an approximate expression of the lifetime asymptotic value by assuming the quasi-Fermi distributions at 0 K, and quasi-neutrality; we obtain (see Example 2.7) for the recombination rate $R_o^{sp} \equiv R_{o,\mathrm{D}}^{sp}$:

$$R_{o,\mathrm{D}}^{sp} = \frac{n}{\tau_{n,\mathrm{D}}^{sp}} \approx \frac{2q^2 n_r E_g \mathfrak{p}_{cv}^2}{3\pi\hbar^2 c_0^3 m_0^2 \epsilon_0} \left[1 + \frac{n^{2/3} 3^{5/3} \pi^{4/3} \hbar^2}{10 E_g m_r^*}\right] n \approx \frac{n}{\tau_0}, \tag{2.49}$$

since the second term in square brackets in (2.49) is typically $\ll 1$.[21] With this approximation, the spontaneous radiative lifetime under degenerate conditions tends to the constant value $\tau_{n,\mathrm{D}}^{sp} = \tau_0$. The spontaneous radiative lifetime τ_0 can therefore be assumed as the *limiting value* of the lifetime due to spontaneous emission for very large population densities.

Example 2.7: Derive the asymptotic value for the spontaneous lifetime (2.49).

Integrating (2.35), we have

$$R_o^{sp} = \frac{1}{\tau_0} \int_{E_g}^{\infty} \frac{\hbar\omega}{E_g} N_{cv}(\hbar\omega) f_n(E_n(\hbar\omega)) f_h(E_h(\hbar\omega))\, \mathrm{d}\hbar\omega. \tag{2.50}$$

The integral can be approximated by assuming the quasi-Fermi statistics at 0 K and quasi-neutrality ($n \approx p$), i.e.,

$$f_n(E_e, E_{Fn}) \approx 1 \quad \text{if} \quad E_n < E_{Fn}, \ 0 \text{ otherwise}$$
$$f_h(E_h, E_{Fh}) \approx 1 \quad \text{if} \quad E_h > E_{Fh}, \ 0 \text{ otherwise.}$$

[21] In fact, assuming, e.g., for GaAs $m_n^* = 0.067 m_0$, $m_h^* = 0.4 m_0$, and therefore

$$m_r^* = \left(1/m_h^* + 1/m_n^*\right)^{-1} = 0.057 m_0,$$

with $m_0 = 9.11 \times 10^{-31}$ kg, $E_g = 1.42$ eV, one has

$$n^{2/3} \frac{3^{5/3}\pi^{4/3}\hbar^2}{10 E_g m_r^*} \approx n^{2/3} \frac{3^{5/3}\pi^{4/3} \cdot \left(10^{-34}\right)^2}{10 \cdot 1.42 \cdot 1.6 \times 10^{-19} \cdot 0.057 \cdot 9.11 \times 10^{-31}} = \left(\frac{n}{n_0}\right)^{2/3},$$

where $n_0 \approx 3 \times 10^{20}$ cm^{-3}. Also for large electron densities, the term will be $\ll 1$.

The conditions $E_n < E_{Fn}$ and $E_h > E_{Fh}$ imply in turn, from energy conservation, see (2.14) and (2.15):

$$E_n = E_c + \frac{m_r^*}{m_n^*}\left(\hbar\omega - E_g\right) < E_{Fn} \to \hbar\omega < E_g + \frac{m_n^*}{m_r^*}\left(E_{Fn} - E_c\right) \tag{2.51}$$

$$E_h = E_v - \frac{m_r^*}{m_h^*}\left(\hbar\omega - E_g\right) > E_{Fh} \to \hbar\omega < E_g + \frac{m_h^*}{m_r^*}\left(E_v - E_{Fh}\right). \tag{2.52}$$

With the same approximation, let us evaluate n and p; we have

$$\begin{aligned} n &\approx \int_{E_c}^{E_{Fn}} N_c\left(E_n\right) \mathrm{d}E_n = \frac{\sqrt{2} m_n^{*3/2}}{\pi^2\hbar^3} \int_{E_c}^{E_{Fn}} \sqrt{E_n - E_c}\, \mathrm{d}E_n \\ &= \frac{2\sqrt{2} m_n^{*3/2}}{3\pi^2\hbar^3}\left(E_{Fn} - E_c\right)^{3/2} \\ p &\approx \int_{E_{Fh}}^{E_v} N_v\left(E_h\right) \mathrm{d}E_h = \frac{\sqrt{2} m_h^{*3/2}}{\pi^2\hbar^3} \int_{E_{Fh}}^{E_v} \sqrt{E_v - E_h}\, \mathrm{d}E_h \\ &= \frac{2\sqrt{2} m_h^{*3/2}}{3\pi^2\hbar^3}\left(E_v - E_{Fh}\right)^{3/2}. \end{aligned} \tag{2.53}$$

Quasi-neutrality now implies

$$m_n^*\left(E_{Fn} - E_c\right) = m_h^*\left(E_v - E_{Fh}\right).$$

It follows that the limitations in $\hbar\omega$ given by (2.51) and (2.52) become symmetrical, i.e., $\hbar\omega < E_{\max}$, where

$$E_{\max} = E_g + \frac{m_n^*\left(E_{Fn} - E_c\right)}{m_r^*} = E_g + \frac{m_h^*\left(E_v - E_{Fh}\right)}{m_r^*}. \tag{2.54}$$

Using the above approximations we obtain

$$\begin{aligned} R_o^{sp} &\approx \frac{1}{2\pi^2\hbar^3 c^3} \frac{2\pi}{\hbar} \frac{q^2}{m_0^2} \frac{\hbar^2}{\epsilon} \frac{2}{3} \mathrm{p}_{cv}^2 \int_{E_g}^{E_{\max}} N_{cv}(\hbar\omega)\hbar\omega\, \mathrm{d}\hbar\omega \\ &= \frac{1}{2\pi^2\hbar^3 c^3} \frac{2\pi}{\hbar} \frac{q^2}{m_0^2} \frac{\hbar^2}{\epsilon} \frac{2}{3} \mathrm{p}_{cv}^2 \frac{\sqrt{2} m_r^{*3/2}}{\pi^2\hbar^3} \int_{E_g}^{E_{\max}} \sqrt{\hbar\omega - E_g}\,\hbar\omega\, \mathrm{d}\hbar\omega. \end{aligned} \tag{2.55}$$

From the integral,

$$I = \int_{E_g}^{E_{\max}} \sqrt{\xi - E_g}\,\xi\, \mathrm{d}\xi = \frac{2}{3} E_g \left[\frac{m_n^*}{m_r^*}\left(E_{Fn} - E_c\right)\right]^{\frac{3}{2}} \left[1 + \frac{3}{5}\frac{m_n^*}{m_r^*}\left(\frac{E_{Fn} - E_c}{E_g}\right)\right],$$

expressing $E_{Fn} - E_c$ as a function of n from (2.53),

$$\left(E_{Fn} - E_c\right)^{3/2} = n\frac{3\pi^2\hbar^3}{2\sqrt{2} m_n^{*3/2}}, \tag{2.56}$$

and using (2.55), we finally obtain (2.49).

For *stimulated lifetime*, a similar result is obtained. In fact, from (2.43) one has

$$R_o^{st} = \frac{g P_{op}}{\hbar\omega}$$

but the (maximum) gain is approximately proportional to the carrier density as $g \approx an$, where a is the *differential gain*, so that the stimulated lifetime becomes

$$\frac{1}{\tau_n^{st}} = \frac{a P_{op}}{\hbar\omega}, \tag{2.57}$$

i.e., for large carrier injection, the stimulated lifetime is constant with respect to the charge density but decreases with increasing optical power. This implies that τ_n^{st} can be typically decreased below the value of spontaneous lifetime, with consequences for the relative speed of devices operating on stimulated emission (lasers) rather than on spontaneous emission (LEDs).

For increasing optical power, however, gain compression has to be taken into account. Since the optical power is proportional to the photon density ρ_{ph}, see (2.5), as $P_{op} = \hbar\omega(c_0/n_r)\rho_{ph}$, we can from (2.30) approximate the gain for large carrier densities as

$$g \approx \frac{an}{1 + \epsilon_c \rho_{ph}};$$

we therefore obtain

$$R_o^{st} = \frac{g P_{op}}{\hbar\omega} \approx \frac{c_0}{n_r}\frac{an\rho_{ph}}{1 + \epsilon_c \rho_{ph}},$$

with carrier lifetime

$$\frac{1}{\tau_n^{st}} = \frac{c_0}{n_r}\frac{a\rho_{ph}}{1 + \epsilon_c \rho_{ph}} \underset{\rho_{ph}\to\infty}{\to} \frac{c_0}{n_r}\frac{a}{\epsilon_c}. \tag{2.58}$$

Assuming $a = 10^{-15}\,\mathrm{cm}^2$ for the differential gain, $\epsilon_c = 10^{-17}\,\mathrm{cm}^3$ for the gain compression coefficient, $n_r \approx 3$, we obtain for the limiting value of stimulated lifetime:

$$\frac{1}{\tau_n^{st}} = \frac{c_0}{n_r}\frac{a}{\epsilon_c} = \frac{3 \times 10^{10}}{3}\frac{10^{-15}}{10^{-17}} = \frac{1}{10^{-12}}\,\mathrm{s}^{-1},$$

i.e., τ_n^{st} is of the order of 1 ps, corresponding to a cutoff frequency $1/(2\pi\tau_n^{st}) \approx 160\,\mathrm{GHz}$. This purely indicative value at least suggests that the dynamics related with stimulated emission is much faster than that associated with spontaneous emission.

Example 2.8: Evaluate the GaAs spontaneous lifetime as a function of the population density, assuming quasi-neutrality.

We can evaluate R_o^{sp} from (2.50) and then compute the lifetime as $\tau_n(\hbar\omega) = n/R_o^{sp}$. The integral in (2.50) has to be performed through numerical quadrature. To relate the population density to the quasi-Fermi levels we can exploit the Joyce–Dixon formulae which are, however, slightly inaccurate for very high densities; a more accurate approach is the Nilsson approximation [20]. For GaAs, we assume $m_n^* = 0.067m_0$,

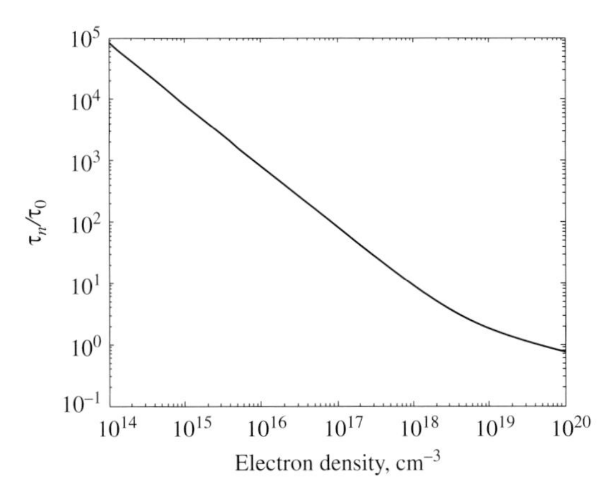

Figure 2.28 Normalized electron lifetime in GaAs as a function of the carrier population.

$m_h = 0.4m_0$, from which $m_r^* = \left(1/m_h + 1/m_n^*\right)^{-1} = 0.057m_0$ ($m_0 = 9.11 \times 10^{-31}$ kg), $\hbar\omega = E_g = 1.42\,\text{eV}$, $N_c = 4.45 \times 10^{17}\,\text{cm}^{-3}$, $N_v = 7.72 \times 10^{18}\,\text{cm}^{-3}$. The result is shown in Fig. 2.28; for large carrier density the lifetime is asymptotically close to τ_0.

2.6 Questions and problems

2.6.1 Questions

1. Explain why the dielectric properties and losses of a material (as a function of the photon energy) are not independent.
2. What is the qualitative behavior of the absorption of a direct-bandgap semiconductor as a function of the photon energy?
3. Explain the effect of excitons on the low-temperature absorption profile vs. the photon energy for a direct-bandgap semiconductor.
4. Define *interband*, *intraband*, *intersubband* transitions in a bulk and in a QW semiconductor.
5. Define *indirect* and *direct* transitions in a semiconductor.
6. Explain why some semiconductors are called *direct bandgap* and others *indirect bandgap*.
7. Explain the role of photons in direct transitions.
8. Explain the role of phonons in indirect transitions.
9. Classify as direct/indirect bandgap the following semiconductors: GaAs, InP, InGaAsP, AlGaAs, AlAs, SiC, Ge, Si, SiGe, HgCdTe, GaSb, GaN.
10. What is a *phonon*?
11. Qualitatively explain the role of selection rules in bulk semiconductors and quantum wells.

12. Justify the *even–odd* selection rule in bulk interband transitions. How does the even–odd rule transform into a even–even, odd–odd rule for intersubband transitions in a quantum well (subbands in the conduction and valence bands)?
13. Consider an intersubband direct transition between two subbands of a symmetric QW. What selection rule has to be applied in this case?
14. Explain why photons cannot effectively assist alone an indirect transition, and why phonons are needed to make the transition possible.
15. Qualitatively draw the ideal behavior of the *absorption* in a bulk indirect-bandgap semiconductor as a function of the photon energy. What happens if the temperature is changed?
16. Sketch the behavior of the *absorption* in a quantum well as a function of the photon energy.
17. Given the absorption coefficient α and the photon energy $\hbar\omega$, how is the optical generation rate G_o related to the optical power per unit area?
18. Select a material well suited for making a detector in FIR, UV, NIR, and visible range among HgCdTe, GaN, GaAs, InGaAsP.
19. Explain why the absorption profile of Ge becomes sharper (as a function of the photon energy) at energies of the order of 0.9 eV.
20. Explain why the density of states enters into the absorption coefficient, while the occupation probability (Fermi function) does not.
21. What is the spontaneous radiative lifetime τ_o? What is the order of magnitude of τ_o?
22. Explain why the stimulated radiative lifetime can be made smaller than the spontaneous lifetime.
23. Explain the reason why gain decreases for very high optical power (gain compression).
24. Discuss the effect of gain compression on the stimulated lifetime.
25. Explain why a quantum well is polarization sensitive, while an isotropic bulk material is not.
26. Sketch the absorption profile of a quantum well for TE and TM polarization.

2.6.2 Problems

1. A semiconductor has a direct bandgap of 2.5 eV. What is the maximum wavelength absorbed?
2. A photon impinges on a material with $E_g = 1.41$ eV. What is the minimum photon energy $E_{ph,\min}$ needed to generate an electron–hole pair? Assuming that the photon energy is $hf = 2$ eV, what is the destination of the extra energy $hf - E_{ph,\min}$? What is the photon momentum? What is the kinetic energy of an electron (assume GaAs) with the same momentum?
3. A photon with energy $E_{ph} = 1.6$ eV is absorbed by an InP sample. Evaluate the energy of the electron and heavy hole after the generation, assuming $m_n^* = 0.082 m_0$, $m_{hh}^* = 0.85 m_0$, and $E_g = 1.35$ eV.

4. Light is incident at one end of a semiconductor sample of thickness $d = 10\,\mu m$. If 15% of the light is absorbed per μm, calculate the absorption coefficient and the fraction of light that is transmitted.
5. For GaAs, compare the theoretical absorption spectrum with experimental values.
6. Evaluate the spontaneous radiative lifetime in an HgCdTe alloy with a bandgap of 0.1 eV. Assume for the momentum matrix element $E_P = 20\,eV$ and for the refractive index $n_r = 3.4$.
7. Evaluate the absorption in a HgCdTe alloy (bandgap of 0.1 eV) as a function of the energy and compare it with the GaAs absorption. Assume for the momentum matrix element $E_P = 20\,eV$. To estimate the electron effective mass m_n^* suppose that this varies with the gap according to (2.9); for the hole effective mass assume $m_h^* = 0.443m_0$ and for the refractive index $n_r = 3.4$.
8. In an InP sample the minority carrier nonradiative lifetime is $\tau_{nr} = 1\,\mu s$. Estimate the carrier concentration needed to have a radiative lifetime $\tau_r = \tau_{nr}/10$. Assume quasi-neutrality and $m_n^* = 0.082m_0$, $m_h^* = 0.85m_0$; for the InP limit radiative lifetime assume $\tau_0 = 0.5\,ns$.
9. Calculate the electron diffusion length in GaAs at $T = 300\,K$ assuming a radiative lifetime of $\tau_n^r = 1\,ns$. Repeat the computation in Si ($\tau_n^r = 100\,ns$). The carrier mobilities are $\mu_n = 5000\,cm^2\,V^{-1}\,s^{-1}$ (GaAs) and $\mu_n = 1000\,cm^2\,V^{-1}\,s^{-1}$ (Si); the nonradiative lifetime is $\tau_n^{nr} = 10\,ns$ in both materials.

3 High-speed semiconductor devices and circuits

3.1 Electronic circuits in optical communication systems

As well as optoelectronic devices such as optical sources, modulators (electrooptic or electroabsorption), optical amplifiers and detectors, high-speed optical communication systems also include dedicated electronic circuits and subsystems. Although most of these are in the low-speed digital sections of the system, and can therefore be implemented with conventional Si-based technologies, some strategic components and subsystems operate at the maximum system speed, e.g., at 10 Gbps or 40 Gbps, often with rather demanding requirements in terms of noise or output voltage (current driving) capabilities. Since high-speed digital data streams are ultimately transmitted and received in baseband, high-speed (high-frequency) subsystems must also possess *ultrawide bandwidth*. Relevant examples are the *driver amplifiers* of lasers or modulators (in direct or indirect modulation systems, respectively), and the detector *front-end amplifiers*.

The enabling technologies in high-speed circuits for optoelectronic systems are the same as found in RF, microwave and millimeter-wave analog integrated circuits. In these domains, silicon-based (CMOS or bipolar) electronics with conventional integrated circuit (IC) approaches are replaced, at increasing frequency, by ICs based on SiGe or III-V compound semiconductors (GaAs or InP). Such circuits exploit, as active devices, advanced bipolar transistors (heterojunction bipolar transistors, HBTs) or heterostructure-based field-effect transistors (such as the high electron mobility transistors, HEMTs).

Besides active devices, high-speed circuits also include passive (distributed or concentrated) elements. Examples of distributed components amenable to monolithic integration are planar transmission lines such as the microstrip and the coplanar lines on semiconductor substrates. The transmission line theory also enables one to readily introduce, as a typical high-frequency modeling and characterization tool for linear multiports, the scattering parameters.

3.2 Transmission lines

Transmission lines (TXLs) are a convenient model for 1D wave propagation; besides their direct importance as high-speed circuit components, the analysis of

TXLs is also useful because they are also a basic building block in electro-optic distributed components, such as traveling wave photodiodes or distributed modulators.

Two parallel ideal conductors (one is the active or signal conductor, the other the return or ground conductor) surrounded by a homogeneous, lossless medium, see Fig. 3.1(a), support a transverse electromagnetic (TEM) propagation mode in which both the electric and the magnetic fields lie in the line cross section and are orthogonal to the line axis and wave propagation direction. In such a TEM TXL, the electric field can be rigorously derived from a potential function that satisfies the Laplace equation in the line cross section. The transverse electric potential is uniquely determined by the conductor potentials, or, assuming the ground conductor as the reference, by the signal line potential $v(z, t)$, where z is parallel to the line axis and propagation direction. Under the same conditions, the transverse magnetic field is related to the total current $i(z, t)$ flowing in the signal conductor. From the Maxwell equations, v and i can be shown to satisfy the partial differential equation system (called the *telegraphers' equations*):

$$\frac{\partial}{\partial z} i(z, t) = -\mathcal{C} \frac{\partial}{\partial t} v(z, t) \tag{3.1}$$

$$\frac{\partial}{\partial z} v(z, t) = -\mathcal{L} \frac{\partial}{\partial t} i(z, t), \tag{3.2}$$

where $\mathcal{L}$ is the per-unit-length (p.u.l.) line inductance and $\mathcal{C}$ is the p.u.l. line capacitance. The telegraphers' equations are compatible with the voltage and current Kirchhoff equations applied to the lumped equivalent circuit of a line cell of infinitesimal length, see Fig. 3.1(b). The p.u.l. parameters have a straightforward meaning – they correspond to the total series inductance of the unit-length cell and to the total capacitance between the two conductors in the unit-length cell.

The circuit model can be extended to account for losses by introducing a series p.u.l. resistance $\mathcal{R}$ (associated with ohmic losses in the conductors) and a parallel p.u.l. conductance $\mathcal{G}$ (associated with the dielectric losses in the surrounding medium), see Fig. 3.1(c). In fact, series losses cause small longitudinal field components, thus making the field distributions slightly different from the ideal TEM pattern; however, the TXL model can be heuristically extended to cases in which the propagating mode is not exactly TEM due to the presence of metal losses, or because the cross section is inhomogeneous (i.e., it includes materials with different dielectric permittivities). In the latter case, we say that the structure supports a *quasi-TEM* mode. Quasi-TEM propagation can be approximately modeled as a TXL with frequency-dependent propagation parameters. In both the TEM and quasi-TEM cases the operating bandwidth is wide, ranging from DC to an upper frequency limit associated with the onset of high-order modes or sometimes to limitations related to line losses, and the frequency dispersion of the propagation parameters (due to modal dispersion in the quasi-TEM case but also to ohmic losses) is low, at least in the high-frequency range.

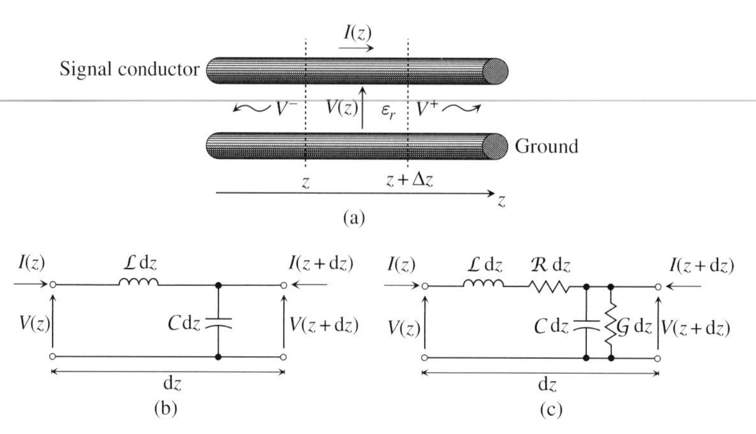

Figure 3.1 Example of a TEM transmission line (a) and equivalent circuit of a line cell of length dz in the lossless (b) and lossy (c) cases.

The telegraphers' equations admit, in the lossless cases, a general solution in terms of forward (V_+, I_+) or backward (V_-, I_-) propagating waves:

$$v(z,t) = V_\pm(z \mp v_f t) \tag{3.3}$$

$$i(z,t) = I_\pm(z \mp v_f t). \tag{3.4}$$

Substitution into system (3.1), (3.2) yields

$$\frac{\partial}{\partial z} I_\pm(z \mp v_f t) = -\mathcal{C}\frac{\partial}{\partial t} V_\pm(z \mp v_f t) \rightarrow I'_\pm = \pm \mathcal{C} v_f V'_\pm$$

$$\frac{\partial}{\partial z} V_\pm(z \mp v_f t) = -\mathcal{L}\frac{\partial}{\partial t} I_\pm(z \mp v_f t) \rightarrow V'_\pm = \pm \mathcal{L} v_f I'_\pm,$$

where the derivative with respect to the argument is denoted by the primed function symbol. Elimination of the voltage unknown leads to

$$I'_\pm = \mathcal{L}\mathcal{C} v_f^2 I'_\pm,$$

from which the line phase velocity results as

$$v_f = \frac{1}{\sqrt{\mathcal{L}\mathcal{C}}}.$$

The voltage and current waveforms are related by the *characteristic impedance* Z_0:

$$V'_\pm = \pm\sqrt{\frac{\mathcal{L}}{\mathcal{C}}} I'_\pm \rightarrow V_\pm = \pm Z_0 I_\pm, \quad Z_0 = \sqrt{\frac{\mathcal{L}}{\mathcal{C}}}. \tag{3.5}$$

From (3.3) and (3.4), we conclude that a lossless TXL supports *undistorted* wave propagation. Time-harmonic voltages or currents of frequency f and angular frequency

$\omega = 2\pi f$ yield propagating waves of the form:

$$\begin{aligned}
v_{\pm}(z,t) &= V_{\pm}(z \mp v_f t) = \sqrt{2}\,\mathrm{Re}\left[V^{\pm}\exp\left(\mathrm{j}\omega t \mp \mathrm{j}\frac{\omega}{v_f}z\right)\right] \\
&= \sqrt{2}\,\mathrm{Re}\left[V^{\pm}\exp\left(\mathrm{j}\omega t \mp \mathrm{j}\beta z\right)\right] \\
i_{\pm}(z,t) &= I_{\pm}(z \mp v_f t) = \sqrt{2}\,\mathrm{Re}\left[I^{\pm}\exp\left(\mathrm{j}\omega t \mp \mathrm{j}\beta z\right)\right] \\
&= \frac{1}{Z_0}\sqrt{2}\,\mathrm{Re}\left[V^{\pm}\exp\left(\mathrm{j}\omega t \mp \mathrm{j}\beta z\right)\right],
\end{aligned}$$

where $V^{\pm}$ is a complex proportionality constant to be determined through the initial and boundary conditions,[1] while

$$\beta = \frac{\omega}{v_f} = \omega\sqrt{\mathcal{L}\mathcal{C}}$$

is the propagation constant of the line. The time-periodic waveform with period $T = 1/f$ is also periodic in space with spatial periodicity corresponding to the *guided wavelength* λ_g as

$$\beta = \frac{2\pi}{\lambda_g} \rightarrow \lambda_g = \frac{v_f}{f} = \frac{\lambda_0}{n_{\mathrm{eff}}},$$

where $n_{\mathrm{eff}} = \sqrt{\epsilon_{\mathrm{eff}}}$ is the line effective refractive index and ϵ_{eff} is the line effective (relative) permittivity.

Undistorted propagation is typical of lossless TXLs, where the signal phase velocity is frequency independent. For lossy lines the telegraphers' equations can be modified, by inspection of the related equivalent circuit, as

$$\frac{\partial}{\partial z}i(z,t) = -\mathcal{C}\frac{\partial}{\partial t}v(z,t) - \mathcal{G}v(z,t) \tag{3.6}$$

$$\frac{\partial}{\partial z}v(z,t) = -\mathcal{L}\frac{\partial}{\partial t}i(z,t) - \mathcal{R}i(z,t). \tag{3.7}$$

In this case, propagation is no longer undistorted and the simple solution outlined so far is not generally valid. The lossy case can be conveniently addressed in the frequency domain, i.e., for time-harmonic v and i. We assume now that the time-domain solution has the form:

$$\begin{aligned}
v_{\pm}(z,t) &= \sqrt{2}\,\mathrm{Re}\left[V^{\pm}(z,\omega)\exp\left(\mathrm{j}\omega t\right)\right] \\
i_{\pm}(z,t) &= \sqrt{2}\,\mathrm{Re}\left[I^{\pm}(z,\omega)\exp\left(\mathrm{j}\omega t\right)\right],
\end{aligned}$$

where $V^{\pm}(z,\omega)$ and $I^{\pm}(z,\omega)$ are the space-dependent *phasors* associated with $v_{\pm}$ and $i_{\pm}$, such as

$$V^{\pm}(z,\omega) = V^{\pm}\exp\left(\mp\alpha z \mp \mathrm{j}\beta z\right) = V^{\pm}\exp\left(\mp\gamma z\right) \tag{3.8}$$

$$I^{\pm}(z,\omega) = I^{\pm}\exp\left(\mp\alpha z \mp \mathrm{j}\beta z\right) = \frac{V^{\pm}(z,\omega)}{Z_0} = \frac{V^{\pm}}{Z_0}\exp\left(\mp\gamma z\right), \tag{3.9}$$

[1] The $\sqrt{2}$ factor is introduced to normalize $V^{\pm}$ to the effective value rather than the peak value of the voltage.

where α is the line attenuation,[2] $\gamma = \alpha + j\beta$ is the *complex propagation constant*, Z_0 is the (now possibly complex) characteristic impedance, and $V^{\pm}$ is a constant to be determined from initial and boundary conditions. In general, an arbitrary solution $v(z, t)$ $(i(z, t))$ can be associated with a phasor $V(z, \omega)$ $(I(z, \omega))$, resulting from the superposition of forward and backward propagating waves $V^{\pm}(z, \omega)$ $(I^{\pm}(z, \omega))$.

For time-harmonic signals, system (3.6), (3.7) becomes

$$\frac{d}{dz}V(z, \omega) = -(j\omega\mathcal{L} + \mathcal{R})I(z, \omega) \tag{3.10}$$

$$\frac{d}{dz}I(z, \omega) = -(j\omega\mathcal{C} + \mathcal{G})V(z, \omega). \tag{3.11}$$

Substituting $V^{\pm}$ and $I^{\pm}$ from (3.8) and (3.9) we obtain, for the complex propagation constant γ and the complex characteristic impedance Z_0:

$$\gamma = \alpha + j\beta = \sqrt{(j\omega\mathcal{L} + \mathcal{R})(j\omega\mathcal{C} + \mathcal{G})} \tag{3.12}$$

$$\frac{V^{\pm}(z, \omega)}{I^{\pm}(z, \omega)} = \pm\sqrt{\frac{j\omega\mathcal{L} + \mathcal{R}}{j\omega\mathcal{C} + \mathcal{G}}} \equiv \pm Z_0. \tag{3.13}$$

The dispersive behavior of the line is thus apparent already when the p.u.l. parameters are frequency independent; in fact, $\mathcal{R}$, $\mathcal{L}$, and $\mathcal{G}$ are themselves frequency dependent because of the presence of metal and dielectric losses. Equations (3.12) and (3.13) can be written in a more compact and general form by introducing the series p.u.l. impedance $\mathcal{Z} = \mathcal{R} + j\omega\mathcal{L}$ and p.u.l. admittance $\mathcal{Y} = \mathcal{G} + j\omega\mathcal{C}$ of the line as

$$\gamma = \sqrt{\mathcal{Z}(\omega)\mathcal{Y}(\omega)}, \quad Z_0 = \sqrt{\frac{\mathcal{Z}(\omega)}{\mathcal{Y}(\omega)}}. \tag{3.14}$$

In a lossless or lossy line, power is transmitted by forward and backward waves in the positive and negative z direction, respectively. The net power transmitted in the positive z direction is in fact (assume for simplicity that the characteristic impedance is approximately real, as in the high-frequency approximation):

$$\begin{aligned} P(z) = \mathrm{Re}(VI^*) &= \frac{1}{Z_0}\mathrm{Re}\left\{\left[V^+(z) + V^-(z)\right]\left[V^+(z) - V^-(z)\right]^*\right\} \\ &= \frac{1}{Z_0}\mathrm{Re}\left[\left|V^+(z)\right|^2 - \left|V^-(z)\right|^2 + V^{+*}(z)V^-(z) - V^+(z)V^{-*}(z)\right] \\ &= \frac{1}{Z_0}\left|V^+(z)\right|^2 - \frac{1}{Z_0}\left|V^-(z)\right|^2 = \frac{1}{Z_0}\left|V^+\right|^2 e^{-2\alpha z} - \frac{1}{Z_0}\left|V^-\right|^2 e^{2\alpha z} \\ &= P^+(z) - P^-(z) = P^+(0)e^{-2\alpha z} - P^-(0)e^{2\alpha z}, \end{aligned}$$

[2] From now on we denote by α both the attenuation and the absorption, the difference being clear from the context.

i.e., the net power can be decomposed into a forward and backward (or incident and reflected) power. Forward power is absorbed in the positive z direction with an *absorption* equal to 2α, while backward power is absorbed in the negative z direction with the same characteristic length (the line absorption length) $L_\alpha = (2\alpha)^{-1}$.

3.2.1 *RG*, *RC*, and high-frequency regimes

The line parameters have been assumed so far to be frequency independent. In fact, the presence of lossy metals and dielectrics implies that $\mathcal{L}$, $\mathcal{R}$, and $\mathcal{G}$ depend on the operating frequency. Due to the finite metal conductivity, electric and magnetic fields penetrate the line conductors down to an average thickness called the *skin penetration depth* δ:

$$\delta = \sqrt{\frac{2}{\mu\sigma\omega}} = \sqrt{\frac{1}{\pi\mu\sigma f}}, \tag{3.15}$$

where $\mu \approx \mu_0 = 4\pi \times 10^{-7}$ H/m is the metal magnetic permeability (we assume conductors to be nonmagnetic). If δ is much smaller than the conductor thickness, the current flow is limited to a thin surface layer having sheet impedance[3]

$$Z_s(\omega) = R_s + \mathrm{j}X_s = \frac{1+\mathrm{j}}{\sigma\delta} = (1+\mathrm{j})\sqrt{\frac{\omega\mu}{2\sigma}}. \tag{3.16}$$

Thus, the high-frequency p.u.l. resistance follows the law

$$\mathcal{R}(f) \approx \mathcal{R}(f_0)\sqrt{\frac{f}{f_0}},$$

while the high-frequency p.u.l. inductance can be split into two contributions: the external inductance $\mathcal{L}_{ex}$ (related to the magnetic energy stored in the dielectric surrounding the line), and the frequency-dependent internal inductance $\mathcal{L}_{in}$ (related to the magnetic energy stored within the conductors). Since the corresponding reactance $X_{in}(f)$ behaves as $X_{in}(f) \approx X_{in}(f_0)\sqrt{f/f_0}$, one has

$$\mathcal{L}(f) = \mathcal{L}_{ex} + \mathcal{L}_{in}(f) \approx \mathcal{L}_{ex} + \mathcal{L}_{in}(f_0)\sqrt{\frac{f_0}{f}} \underset{f\to\infty}{\approx} \mathcal{L}_{ex}.$$

At high frequency, therefore, the total inductance can be approximated by the external contribution. For different reasons, the p.u.l. conductance will also be frequency dependent; in fact, this is associated with the complex permittivity of the surrounding dielectrics $\epsilon = \epsilon' - \mathrm{j}\epsilon'' = \epsilon'\left(1 - \mathrm{j}\tan\bar{\delta}\right)$, where $\tan\bar{\delta} \approx \bar{\delta}$ is the (typically small, $10^{-2} - 10^{-4}$) dielectric *loss tangent*.[4] Consider, in fact, a parallel-plate capacitor of

[3] The sheet impedance is the impedance of a square piece of conductor; it is often expressed in ohm per square ($\Omega/\square$).

[4] The loss angle $\bar{\delta}$ is usually denoted as δ; we use a different notation to avoid confusion with the skin penetration depth.

area A and electrode spacing h; the capacitor admittance will be

$$Y = j\omega\epsilon\frac{A}{h} = j\omega\epsilon'\frac{A}{h} + \omega\epsilon''\frac{A}{h} = j\omega C + \omega C\tan\delta = j\omega C + G(\omega),$$

where the conductance scales linearly with frequency, i.e., $G(\omega) = (\omega/\omega_0)G(\omega_0)$. (We assume that the loss angle is weakly dependent on frequency.) The result can be generalized to a transmission line with transversally homogeneous (or inhomogeneous) lossy dielectrics, where in general

$$\mathcal{G}(f) \approx \frac{f}{f_0}\mathcal{G}(f_0),$$

i.e., the line conductance increases linearly with frequency. Materials characterized by heavy conductor losses (such as doped semiconductors), on the other hand, have frequency-independent conductivity.

In a lossy line, the propagation parameters γ and Z_0 are real at DC and very low frequency:

$$\alpha + j\beta \approx \sqrt{\mathcal{R}\mathcal{G}}, \quad Z_0 \approx \sqrt{\frac{\mathcal{R}}{\mathcal{G}}},$$

since in this case the line works as a resistive distributed attenuator. In an intermediate frequency range $j\omega\mathcal{C} + \mathcal{G} \approx j\omega\mathcal{C}$, while $j\omega\mathcal{L} + \mathcal{R} \approx \mathcal{R}$ in most lines because typically series losses prevail over parallel losses. The line performances are therefore dominated by the p.u.l. resistance and capacitance (*RC regime*), with parameters

$$\alpha + j\beta \approx \frac{1+j}{\sqrt{2}}\sqrt{\omega\mathcal{C}\mathcal{R}}, \quad Z_0 \approx \frac{1-j}{\sqrt{2}}\sqrt{\frac{\mathcal{R}}{\omega\mathcal{C}}}.$$

In the RC regime the line is strongly dispersive and the characteristic impedance is complex. Finally, in the *high-frequency regime* $j\omega\mathcal{C} \gg \mathcal{G}$ and $j\omega\mathcal{L} \gg \mathcal{R}$; the imaginary part of Z_0 can be neglected and the complex propagation constant can be approximated as

$$Z_0 \approx Z_{0l} = \sqrt{\frac{\mathcal{L}}{\mathcal{C}}} \tag{3.17}$$

$$\gamma = \alpha + j\beta \approx \frac{\mathcal{R}(f)}{2Z_0} + \frac{\mathcal{G}(f)\,Z_0}{2} + j\omega\sqrt{\mathcal{L}\mathcal{C}} = \alpha_c(f) + \alpha_d(f) + j\beta_l, \tag{3.18}$$

where Z_{0l} is the impedance of the lossless line, $\alpha_c \propto \sqrt{f}$ and $\alpha_d \propto f$ are the conductor and dielectric attenuation, respectively (usually $\alpha_c \gg \alpha_d$ in the RF and microwave range), and β_l is the propagation constant of the lossless line. Therefore, in the high-frequency regime a wideband signal with little or no low-frequency content propagates undistorted, apart from the signal attenuation. The onset of the high-frequency regime depends on line parameters; integrated structures with micrometer-scale dimensions can operate in the RC range for frequencies as high as a few gigahertz. Moreover, the impact of losses is related to the length of the TXL; in short structures (like those arising in some distributed optoelectronic devices) signal distortion can be modest even though the line operates under very broadband excitation.

3.2.2 The reflection coefficient and the loaded line

The voltage and current phasors in a TXL are, in general, a superposition of forward and backward waves:

$$V(z) = V^+(z) + V^-(z) = V_0^+ e^{-\gamma z} + V_0^- e^{\gamma z} \tag{3.19}$$

$$I(z) = I^+(z) + I^-(z) = \frac{V^+(z)}{Z_0} - \frac{V^-(z)}{Z_0} = \frac{V_0^+}{Z_0} e^{-\gamma z} - \frac{V_0^-}{Z_0} e^{\gamma z}. \tag{3.20}$$

An alternative formulation results from introducing the (voltage) *reflection coefficient* of the line at section z, $\Gamma(z)$:

$$\Gamma(z) \equiv \frac{V^-(z)}{V^+(z)} = \frac{V_0^-}{V_0^+} \exp(2\gamma z) = \Gamma_0 \exp(2\gamma z), \tag{3.21}$$

where $\Gamma_0 = \Gamma(0)$. In terms of the reflection coefficient (3.19), (3.20) become

$$V(z) = V^+(z)\left[1 + \Gamma(z)\right] = V_0^+ e^{-\gamma z}\left[1 + \Gamma_0 e^{2\gamma z}\right] \tag{3.22}$$

$$I(z) = \frac{V^+(z)}{Z_0}\left[1 - \Gamma(z)\right] = \frac{V_0^+}{Z_0} e^{-\gamma z}\left[1 - \Gamma_0 e^{2\gamma z}\right]. \tag{3.23}$$

At a section z the V/I ratio, which can be interpreted as the line input impedance $Z(z)$, can be now expressed as

$$Z(z) = \frac{V(z)}{I(z)} = Z_0 \frac{1 + \Gamma(z)}{1 - \Gamma(z)} = Z_0 \frac{1 + \Gamma_0 e^{2\gamma z}}{1 - \Gamma_0 e^{2\gamma z}}; \tag{3.24}$$

conversely, Γ can be uniquely derived from the *normalized impedance* $\zeta(z) = Z(z)/Z_0$ as

$$\Gamma(z) = \frac{Z(z) - Z_0}{Z(z) + Z_0} = \frac{\zeta(z) - 1}{\zeta(z) + 1}. \tag{3.25}$$

Using the reflection coefficient, we can readily decompose power into an incident and reflected component as

$$P(z) = P^+(z) - P^-(z) = P^+(z)\left[1 - |\Gamma(z)|^2\right].$$

Consider now a line closed on a load Z_L at $z = z_L$ (Fig. 3.2) and assume we want to evaluate the input impedance in section z, $z < z_L$. The load imposes the boundary condition $V(z_L)/I(z_L) \equiv Z(z_L) = Z_L$, which becomes

$$\Gamma(z_L) \equiv \Gamma_L = \frac{Z_L - Z_0}{Z_L + Z_0} = \frac{\zeta_L - 1}{\zeta_L + 1}, \tag{3.26}$$

with $\zeta_L = Z_L/Z_0$. From (3.21) we have

$$\Gamma_0 = \Gamma_L e^{-2\gamma z_L} \rightarrow \Gamma(z) = \Gamma_L e^{2\gamma(z - z_L)}.$$

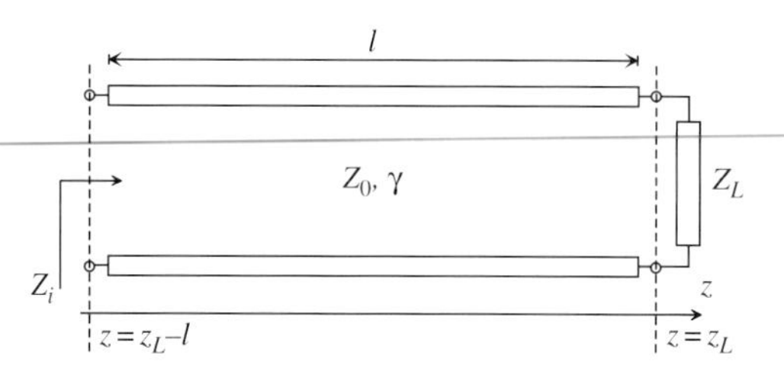

Figure 3.2 Input impedance of a loaded transmission line.

The input impedance at section z (such as $z_L - z = l$, l = line length) can therefore be expressed as

$$Z_i = Z(z) = Z_0 \frac{1 + \Gamma(z)}{1 - \Gamma(z)} = Z_0 \frac{1 + \Gamma_L e^{-2\gamma l}}{1 - \Gamma_L e^{-2\gamma l}}.$$

Expanding the load reflection coefficient and expressing the exponential in terms of hyperbolic functions we obtain

$$Z_i = Z_0 \frac{Z_L \cosh(\gamma l) + Z_0 \sinh(\gamma l)}{Z_L \sinh(\gamma l) + Z_0 \cosh(\gamma l)}.$$

Notice that for $l \to \infty$, $Z_i \to Z_0$ independent of the load. For a lossless line, however, the input impedance is periodic vs. the line length, with periodicity $\lambda_g/2$ (due to the tan function)

$$Z_i = Z_0 \frac{Z_L + jZ_0 \tan(\beta l)}{Z_0 + jZ_L \tan(\beta l)}.$$

In particular, if the load is a short ($Z_L = 0$) or an open ($Y_L = 0$) we have

$$Z_i(Z_L = 0) = jZ_0 \tan(\beta l)$$
$$Z_i(Y_L = 0) = -jZ_0 \cot(\beta l);$$

a reactive load is therefore obtained, alternatively inductive and capacitive according to the value of the *line electrical angle* $\phi = \beta l = 2\pi l/\lambda_g$. Finally, it can readily be shown by inspection than the input impedance of a shorted lossy line for $l \to 0$ is $Z_i \approx j\omega\mathcal{L}l + \mathcal{R}l$, while the input impedance of a short line in open circuit is $Z_i \approx (j\omega\mathcal{C}l + \mathcal{G}l)^{-1}$.

Equations (3.24) and (3.25) establish a biunivocal correspondence between (normalized) impedances and reflection coefficients. Impedances with positive real part map into the unit circle of the Γ plane $|\Gamma| = 1$; in fact, denoting $\zeta = r + jx$, for reactive impedances ($\zeta = jx$) one has

$$\Gamma(z) = \frac{jx - 1}{jx + 1} \to |\Gamma| = 1.$$

On the other hand, real impedances map onto the real Γ axis, $\Gamma = 0$ corresponding to $Z = Z_0$ (impedance matching condition), while $Z = 0$ corresponds to $\Gamma = -1$ (short circuit) and $Y = 0$ corresponds to $\Gamma = 1$ (open circuit). The image of the r = const. and x = const. lines within the Γ plane unit circle is a set of circles and circular segments, often referred to as the *Smith chart*; see Fig. 3.3.

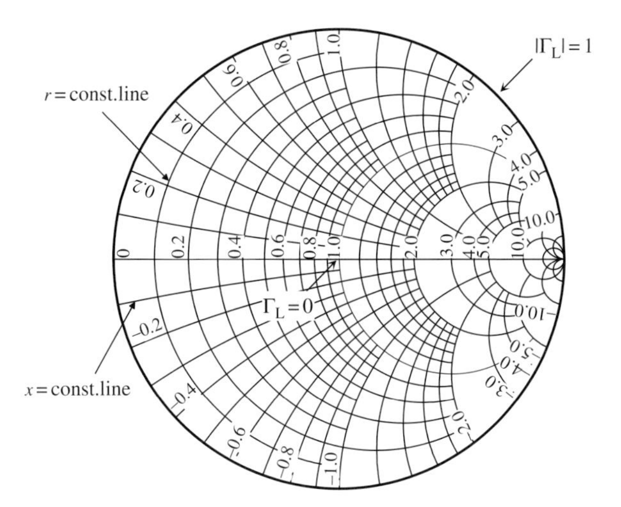

Figure 3.3 Constant normalized resistance and reactance lines in the Γ_L plane: the Smith chart.

3.2.3 Planar integrated quasi-TEM transmission lines

The practical implementation of hybrid or monolithic integrated circuits on a dielectric or semiconductor substrate typically requires that the TXL have an inhomogeneous cross section (partly air, partly dielectric) rather than a uniform dielectric as in coaxial cables and striplines, see Fig. 3.4. Due to the nonuniform cross section of the line, the propagation mode is quasi-TEM rather than TEM. This implies that the phase velocity is weakly dependent on frequency, at least for frequencies well below the onset of the first higher-order propagation mode. In quasi-TEM lines an effective permittivity ϵ_{eff} can be introduced, such as

$$\lambda_g = \frac{\lambda_0}{\sqrt{\epsilon_{\text{eff}}}} = \frac{\lambda_0}{n_{\text{eff}}}, \quad v_f = \frac{c_0}{\sqrt{\epsilon_{\text{eff}}}} = \frac{c_0}{n_{\text{eff}}},$$

where $n_{\text{eff}} = \sqrt{\epsilon_{\text{eff}}}$ is the effective refractive index and c_0 the velocity of light *in vacuo*. The quasi-TEM line propagation parameters can be expressed in a more convenient way by introducing the *in-air* (or *in-vacuo*) p.u.l. capacitance and inductance, $\mathcal{L}_a$ and $\mathcal{C}_a$, defined as the parameters of a (TEM) line with homogeneous cross section and $\epsilon = \epsilon_0$. For such a line we have

$$v_f = \frac{1}{\sqrt{\mathcal{L}_a \mathcal{C}_a}} = c_0 \rightarrow \mathcal{L}_a = \frac{1}{c_0^2 \mathcal{C}_a}, \tag{3.27}$$

since the phase velocity of a TEM mode equals the velocity of light in the dielectric medium. However, assuming that the line cross section does not include magnetic media, the p.u.l. line inductance does not depend on the dielectric properties of the line cross section, i.e., $\mathcal{L} \equiv \mathcal{L}_a$. In a lossless or low-loss (high-frequency approximation) line we can therefore write

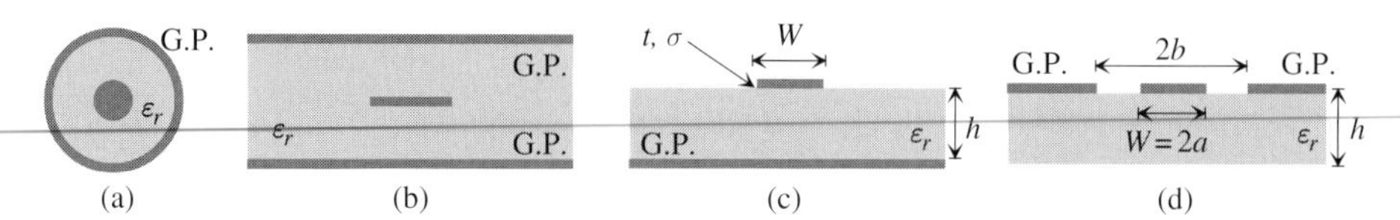

Figure 3.4 TEM lines: (a) coaxial cable, (b) stripline. Quasi-TEM lines: (c) microstrip, (d) coplanar waveguide (CPW). G.P. stands for ground plane.

$$Z_0 = \sqrt{\frac{\mathcal{L}}{\mathcal{C}}} = \sqrt{\frac{\mathcal{L}_a}{\mathcal{C}}} = \frac{1}{c_0}\sqrt{\frac{1}{\mathcal{C}\mathcal{C}_a}} \tag{3.28}$$

$$v_f = \frac{1}{\sqrt{\mathcal{L}\mathcal{C}}} = c_0\sqrt{\frac{\mathcal{C}_a}{\mathcal{C}}} \to \mathcal{C} = \epsilon_{\text{eff}}\mathcal{C}_a = \frac{\sqrt{\epsilon_{\text{eff}}}}{c_0 Z_0}. \tag{3.29}$$

The effective permittivity therefore is a proper "average" value of the cross section permittivity. Moreover, (3.27) suggests that a high-inductance line is a low-capacitance line and vice versa, while from (3.28) a high-capacitance line has low impedance and vice versa. Low- and high-impedance lines can be immediately recognized from their high- and low-capacitance characteristics, e.g., a microstrip with a wide signal conductor ($W/h \gg 1$) has high $\mathcal{C}$ and $\mathcal{C}_a$, low Z_0, while a narrow microstrip ($W/h \ll 1$) has low $\mathcal{C}$ and $\mathcal{C}_a$, high Z_0.

3.2.4 Microstrip lines

In the most common planar transmission line, the microstrip, see Fig. 3.4(c), the signal conductor lies on a dielectric substrate backed by a ground plane. Propagation is quasi-TEM due to the inhomogeneous cross section (partly air, partly dielectric). Microstrip lines on composite semiconductor substrates can be exploited in distributed optoelectronic components, such as traveling-wave photodiodes or electroabsorption modulators; in such cases, the analysis becomes complex due to the inhomogeneous, lossy substrate. Also in the simplest case of a low-loss dielectric substrate, no closed-form expressions exist for the characteristic parameters (impedance, effective permittivity and losses); however, several approximations are available in the literature; see, e.g., [21], [22].

An example of the behavior of the characteristic impedance and effective refractive index vs. the normalized strip width W/h is shown in Fig. 3.5. As anticipated, the impedance decreases with increasing W/h; in hybrid circuit substrates (e.g., ceramic, such as alumina, $\epsilon_r \approx 10$) or semiconductor substrates ($\epsilon_r \approx 13$), Z_0 ranges from 120 to 25 Ω. Typical semiconductor substrate thicknesses are between 500 and 100 μm; the minimum strip width is limited by the resolution of lithographic processes and (above all) by the increase of ohmic losses to ≈ 30 μm, while for $W/h \gg 1$ spurious (lateral) resonances arise, thus limiting the maximum width. The frequency dispersion of the effective permittivity can be approximated by fitting formulae, such as

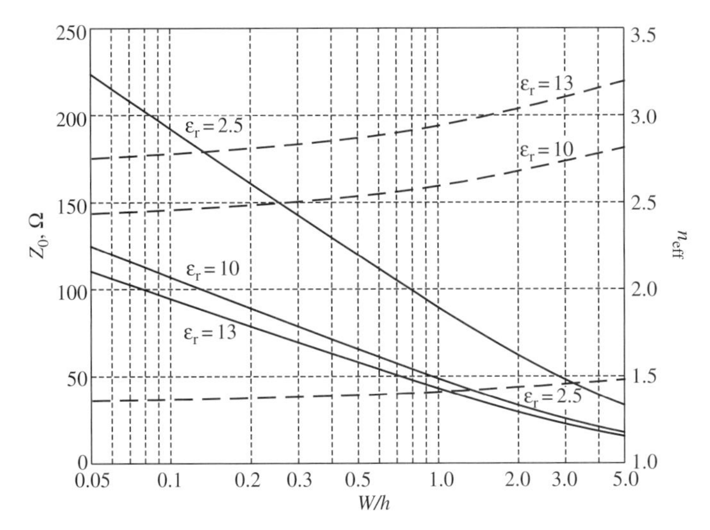

Figure 3.5 Microstrip impedance (continuous line, left-hand axis) and effective refractive index (dashed line, right-hand axis) vs. aspect ratio W/h for several values of the substrate permittivity.

$$\epsilon_{\text{eff}}(f) = \left[\sqrt{\epsilon_{\text{eff}}(0)} + \frac{\psi^{1.5}}{1+\psi^{1.5}}\left(\sqrt{\epsilon_r} - \sqrt{\epsilon_{\text{eff}}(0)}\right)\right]^2, \tag{3.30}$$

where $\psi \propto hf\sqrt{\epsilon_r - 1}$. Equation (3.30) yields increasing effective permittivity with increasing frequency, with asymptotic value $\epsilon_{\text{eff}} \to \epsilon_r$ for $f \to \infty$. Since the onset of the first higher-order mode occurs for frequencies such as $\psi \approx 1$, a small substrate thickness h is required for high-frequency operation.

An example of conductor and dielectric attenuation behavior vs. the line aspect ratio is shown in Fig. 3.6; the conductor attenuation α_c is higher for narrow strips, while it decreases for wider strips. The dielectric attenuation increases for increasing strip width, since in this case an increasing fraction of the electric field energy is in the substrate.

3.2.5 Coplanar lines

While in microstrip lines the impedance depends on the substrate thickness, thus requiring careful technological control of this parameter, in coplanar waveguides (CPWs) on a thick substrate the line impedance only depends on the strip width $W = 2a$ and on the lateral ground plane spacing $W + 2G = 2b$, where G is the gap width, and can therefore be accurately controlled by lithographic processes. Moreover, under such conditions the high-frequency effective permittivity is independent of the line impedance. According to a rather conservative estimate, such favorable properties occur if the substrate thickness h is of the order of the ground plane spacing $2b$.

CPWs are at present quite popular in millimeter-wave monolithic integrated circuits; moreover, they are exploited as a building block in a number of distributed

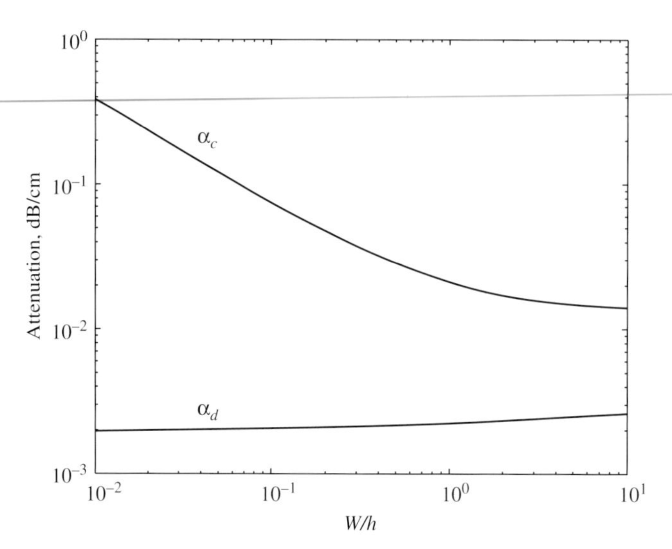

Figure 3.6 Example of microstrip attenuation (metal and dielectric) vs. aspect ratio W/h. The frequency is $f = 1$ GHz, the strip thickness is $t = 5$ μm with conductivity $\sigma = 4.1 \times 10^7$ S/m, the substrate thickness is $h = 500$ μm with permittivity $\epsilon_r = 10$ and loss tangent $\tan\bar{\delta} = 10^{-3}$.

optoelectronic devices such as distributed electrooptic modulators on ferroelectric substrates.

For $h \to \infty$, exact closed-form expressions are available for the line parameters (see, e.g., [23], also for approximations holding if h is finite or in asymmetric CPWs):

$$Z_0 = \frac{30\pi}{\sqrt{\epsilon_{\text{eff}}}} \frac{K(k')}{K(k)} \tag{3.31}$$

$$\epsilon_{\text{eff}} = \frac{\epsilon_r + 1}{2}, \tag{3.32}$$

where $k = a/b = W/(W + 2G)$, $k' = \sqrt{1 - k^2}$ and $K(k)$ is the complete elliptic integral of the first kind. The ratio $K(k)/K(k')$ can be approximated as [24]:

$$\frac{K(k)}{K(k')} \approx \begin{cases} \dfrac{1}{\pi} \log\left(2\dfrac{1 + \sqrt{k}}{1 - \sqrt{k}}\right), & 0.5 \le k^2 < 1 \\ \left[\dfrac{1}{\pi} \log\left(2\dfrac{1 + \sqrt{k'}}{1 - \sqrt{k'}}\right)\right]^{-1}, & 0 < k^2 \le 0.5. \end{cases} \tag{3.33}$$

The frequency dispersion of the effective permittivity leads to an increasing behavior with increasing frequency, as in the microstrip; the characteristic frequency corresponding to the upper limit of the line operating range is given by the onset of the first TE surface mode $f_{TE} = c_0/(4h\sqrt{\epsilon_r - 1})$. Finally, conductor and dielectric losses can be expressed as

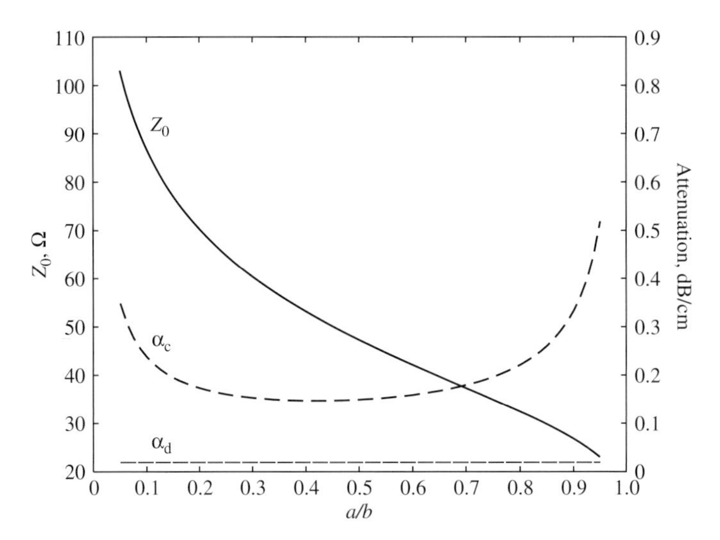

Figure 3.7 Coplanar waveguide impedance (continuous line, left-hand axis) and dielectric and conductor attenuation (dashed line, right-hand axis) vs. aspect ratio a/b on an infinitely thick substrate ($\epsilon_r = 10$, loss angle 0.001). The ground plane spacing is $2b = 600$ μm, the strip thickness is $t = 5$ μm with conductivity $\sigma = 4.1 \times 10^7$ S/m.

$$\alpha_c = \frac{8.68 R_s \sqrt{\epsilon_{\text{eff}}}}{480\pi K(k) K(k')(1-k^2)} \times \left\{ \frac{1}{a}\left[\pi + \log\left(\frac{8\pi a(1-k)}{t(1+k)}\right)\right] + \frac{1}{b}\left[\pi + \log\left(\frac{8\pi b(1-k)}{t(1+k)}\right)\right]\right\} \quad \text{dB/m}$$

$$\alpha_d = 27.83 \frac{\tan\bar{\delta}}{\lambda_0} \frac{\epsilon_r}{2\sqrt{\epsilon_{\text{eff}}}} \quad \text{dB/m},$$

where the parameter $R_s = \sqrt{\omega\mu/2\sigma}$ is the skin-effect surface resistance (σ is the metallization conductivity, μ is the metal magnetic permeability). The impedance range of a CPW on ceramic or semiconductor substrates is similar to the microstrip case; see Fig. 3.7. However, both very narrow and very wide lines exhibit high conductor losses. In fact, for $a \to 0$ the line resistance diverges, and so does the conductor attenuation; for $a \to b$ the line impedance $Z_0 \to 0$, and therefore $\alpha_c \approx \mathcal{R}/2Z_0 \to \infty$. Conductor losses obviously depend on the CPW size, which is reduced down to $a = 5$ μm in some electrooptic components (e.g., lithium niobate modulators). In such cases, the line thickness can be increased to reduce ohmic losses; however, in the high-frequency (high-speed) regime the line resistance depends on the conductor periphery rather than on the conductor area, due to the skin effect, thus making conductor thickening less effective than in low-speed operation.

3.3 The scattering parameters

Consider a linear, nonautonomous (i.e., without independent sources) two-port network in the frequency domain. The port voltage $\underline{V} = (V_1, V_2)^T$ and current

$\underline{I} = (I_1, I_2)^T$ phasors are related as

$$\underline{V} - \mathbf{Z}\underline{I}, \tag{3.34}$$

where $\mathbf{Z}$ is a 2×2 complex matrix called the *impedance matrix*. Similarly, we can define $\mathbf{Y} = \mathbf{Z}^{-1}$ as the two-port *admittance matrix*. In the same way as we characterize a one-port through its reflection coefficient rather than its impedance, we can describe a two-port (or, in general, an N-port) through the *scattering matrix*. In fact, consider that any voltage and current can be uniquely decomposed (also in a lumped-parameter circuit) into a forward and backward component, which are conveniently normalized as

$$V_i = \sqrt{R_{0i}}a_i + \sqrt{R_{0i}}b_i, \quad I_i = \frac{a_i}{\sqrt{R_{0i}}} - \frac{b_i}{\sqrt{R_{0i}}}.$$

The parameters (dimension $\mathrm{W}^{1/2}$) a_i and b_i are the forward and backward *power waves*, while R_{0i} is the *normalization resistance* (or impedance) of port i. The two relations can be inverted to provide the power waves from the total voltages and currents as

$$a_i = \frac{1}{2\sqrt{R_{0i}}}V_i + \frac{\sqrt{R_{0i}}}{2}I_i, \quad b_i = \frac{1}{2\sqrt{R_{0i}}}V_i - \frac{\sqrt{R_{0i}}}{2}I_i.$$

If the voltages and currents are replaced in (3.34) by the power waves, on defining $\underline{a} = (a_1, a_2)$ and $\underline{b} = (b_1, b_2)$ we derive the representation:

$$\underline{b} = \mathbf{S}\underline{a}, \tag{3.35}$$

where $\mathbf{S}$ is the scattering matrix of the two-port. Although the normalization impedances can be different for each port, typically they are chosen all equal ($R_{0i} = R_0$), often with the default value $R_0 = 50\ \Omega$. In this case, the scattering matrix is related to the impedance matrix as

$$\mathbf{S} = (\mathbf{Z} - R_0\mathbf{I})(\mathbf{Z} + R_0\mathbf{I})^{-1}, \tag{3.36}$$

where $\mathbf{I}$ is the identity matrix. For a one-port, (3.36) clearly reduces to (3.25), i.e., $S \equiv \Gamma$.

To appreciate the advantages of the scattering matrix representation, we can conveniently expand (3.35) in scalar form as

$$\begin{aligned} b_1 &= S_{11}a_1 + S_{12}a_2 \\ b_2 &= S_{21}a_1 + S_{22}a_2; \end{aligned}$$

thus, the scattering matrix elements are, by definition,

$$S_{11} = \left.\frac{b_1}{a_1}\right|_{a_2=0}, \quad S_{21} = \left.\frac{b_2}{a_1}\right|_{a_2=0}$$

$$S_{12} = \left.\frac{b_1}{a_2}\right|_{a_1=0}, \quad S_{22} = \left.\frac{b_2}{a_2}\right|_{a_1=0}.$$

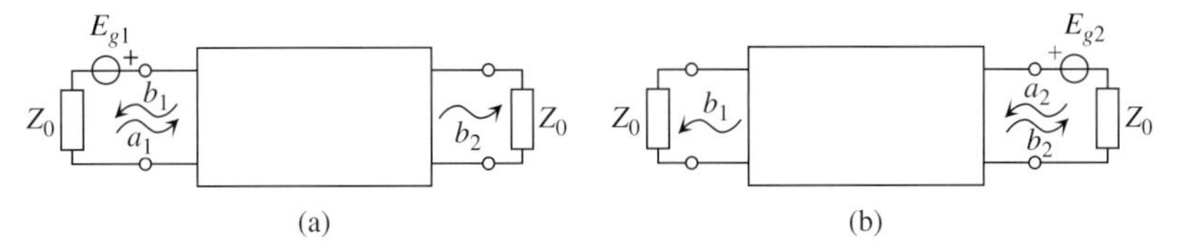

Figure 3.8 Loading conditions for measuring (a) S_{11} and S_{21} and (b) S_{12} and S_{22}.

The condition $a_i = 0$ can be achieved by properly loading port i; see Fig. 3.8. In fact, loading port i with a resistance R_{0i} implies $V_i = -R_{0i} I_i$ (the current positively enters the port), i.e.,

$$a_i = \frac{1}{2\sqrt{R_{0i}}} V_i + \frac{\sqrt{R_{0i}}}{2} I_i = -\frac{R_{0i} I_i}{2\sqrt{R_{0i}}} + \frac{\sqrt{R_{0i}}}{2} I_i = 0.$$

Thus, S_{11} is the reflection coefficient seen at port 1 when port 2 is closed on the normalization resistance, S_{21} is the transmission coefficient between ports 1 and 2 under the same conditions, and similarly for S_{22} and S_{12}. From a practical standpoint, measuring the scattering matrix elements is easier and more convenient than measuring the impedance or admittance matrix elements. Indeed, the scattering matrix and power wave description also applies for structures in which voltages and currents do not, strictly speaking, exist (such as a non-TEM metal or dielectric waveguide, as an optical fiber). Moreover, ideal wideband matched loads can be far more easily implemented than short or open circuits (required for the Y or Z parameter characterization, respectively). Finally, most high-speed transistors happen to be *unstable* when closed on reactive terminations, thus making the direct measurement of Z or Y parameters virtually impossible.

3.3.1 Power and impedance matching

The term *power waves* applied to the a and b parameters naturally follows from the definition of the total power entering port i:

$$P_i = \mathrm{Re}\left(V_i I_i^*\right) = \mathrm{Re}\left[(a_i + b_i)\left(a_i^* - b_i^*\right)\right] = |a_i|^2 - |b_i|^2 = |a_i|^2\left(1 - |\Gamma|^2\right),$$

where $|a_i|^2$ is the incident power into port i, and $|b_i|^2$ the reflected power from port i. If a two-port is lossless, the total power entering the two-port is zero, i.e., $P_1 + P_2 = 0$; this implies

$$P_1 + P_2 = \underline{a} \cdot \underline{a}^{T*} - \underline{b} \cdot \underline{b}^{T*} = \underline{a} \cdot \left(\mathbf{I} - \mathbf{S} \cdot \mathbf{S}^{*T}\right) \cdot \underline{a}^{T*} = 0,$$

where $\mathbf{I}$ is the identity matrix. Setting $\mathbf{S}^{T*} = \mathbf{S}^{\dagger}$ we obtain

$$\mathbf{S}^{\dagger}\mathbf{S} = \mathbf{I},$$

i.e., the scattering matrix is Hermitian. Finally, the scattering matrix of a reciprocal two-port (N-port) is symmetric, $S_{ij} = S_{ji}$.

Assume now that a two-port is connected to a generator with internal impedance Z_g at port 1 and to a load with impedance Z_L at port 2.[5] From circuit theory one immediately obtains that, if the generator and load are directly connected, the maximum power transfer from the generator to the load occurs when $Z_L = Z_g^*$ (*conjugate matching* or *power matching* conditions). In this case the power delivered to the load is the *generator available power* P_{av}:

$$P_{av} = \frac{|E_g|^2}{4R_g},$$

where E_g is the phasor associated with the generator open-circuit voltage (normalized with respect to the effective value). A dual expression holds for a generator with short-circuit current A_g and internal admittance $Y_g = G_g + \mathrm{j}B_g$:

$$P_{av} = \frac{|A_g|^2}{4G_g}. \tag{3.37}$$

If we assume now that the generator and load are connected through the two-port (which could be the model of a linear amplifier), maximum power transfer occurs when port 1 *and* port 2 are conjugately matched:

$$Z_{in} = Z_g^*, \quad Z_L = Z_{out}^*, \tag{3.38}$$

where Z_{in} and Z_{out} are the input and output impedance of the two-port when loaded at port 1 with Z_g and at port 2 with Z_L, respectively. Conditions (3.38) can be also expressed in terms of reflection coefficients as

$$\Gamma_g = \Gamma_{in}^*(\mathbf{S}, \Gamma_L), \quad \Gamma_L = \Gamma_{out}^*(\mathbf{S}, \Gamma_g), \tag{3.39}$$

where $\mathbf{S}$ is the two-port scattering matrix. The load and generator reflection coefficients that simultaneously meet (3.39) yield maximum power transfer between the generator and the load. However, conjugate matching is not always possible, but only if the two-port is *unconditionally stable* (i.e. if $|\Gamma_{in,out}| < 1$ for any $|\Gamma_{L,g}| < 1$).[6]

Power matching corresponds to the maximum power transfer between a load and a generator connected by a two-port (e.g., an amplifier). This is an optimum condition in narrowband systems; however, in wideband design and in the presence of distributed components (such as transmission lines) another matching approach (often called *impedance matching*) is preferred, since it allows minimization of reflections and the related signal distortion. For the sake of definiteness, consider a lossless transmission line with real characteristic impedance Z_0, and load and generator resistances R_g and R_L; the transfer function between the load voltage V_L and the generator voltage E_g can be evaluated from (6.24) and (6.25) taking into account that

[5] In the present chapter we use Z_g to avoid confusion with the symbol G as the initial of "gate"; elsewhere the generator impedance will be denoted by Z_G.

[6] In what follows, we will assume that the active components considered (in particular, the amplifiers) are unconditionally stable; this is also a convenient choice in wideband design. For a discussion on stability issues see e.g. [25].

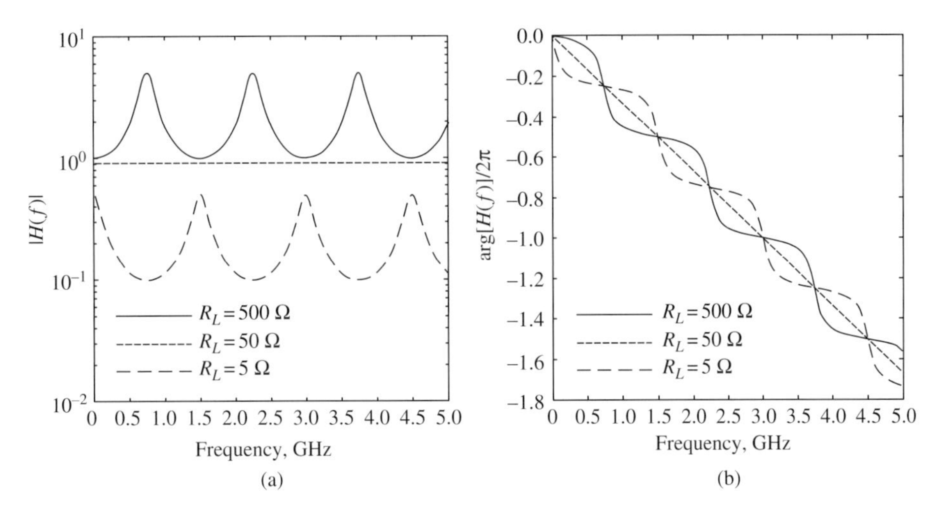

Figure 3.9 Frequency behavior of the magnitude (a) and phase (b) of $H = V_L/E_g$ for a 50 Ω line loaded with R_L. The generator resistance is 5 Ω.

$V(L) = V^+ e^{-j\beta L} + V^- e^{j\beta L}$ $(\beta = \omega/v_f)$:

$$H(\omega) = \frac{V_L}{E_g} = \frac{Z_0}{Z_0 + Z_g} \frac{1 + \Gamma_L}{1 - \Gamma_L \Gamma_g \exp(-2j\beta L)} \exp(-j\beta L).$$

If the load and generator are impedance matched to the line ($R_L = R_g = Z_0$), one has

$$H(\omega) = \frac{Z_0}{Z_0 + Z_g} \exp(-j\beta L),$$

i.e., the magnitude of H is constant and the phase is linear, corresponding to constant group delay and no linear distortion. The same situation in fact arises also if the line is mismatched at one end only ($R_L = Z_0$ or $R_g = Z_0$) since in both cases $\Gamma_L \Gamma_g = 0$ and no *multiple reflections* take place between the generator and the load. However, if both the generator and the load are mismatched the response shows magnitude and phase ripples, as shown in Fig. 3.9; those lead to signal distortion and should therefore be reduced or eliminated by properly terminating the distributed element. Note that generator and load impedance matching with a line having real Z_0 also corresponds to the power-matching condition.

3.4 Passive concentrated components

Passive components for high-speed hybrid and monolithic integrated circuits can be either *distributed* or *concentrated* (or *lumped*). Distributed components are based on single or multiconductor (coupled) transmission lines; TXLs can be exploited to design impedance-matching sections, filters, reactive equalizers, directional couplers, and other low-loss components. The typical size of distributed components at centerband is a fraction of the guided wavelength (e.g., $\lambda_g/4$); thus, such components have a comparatively

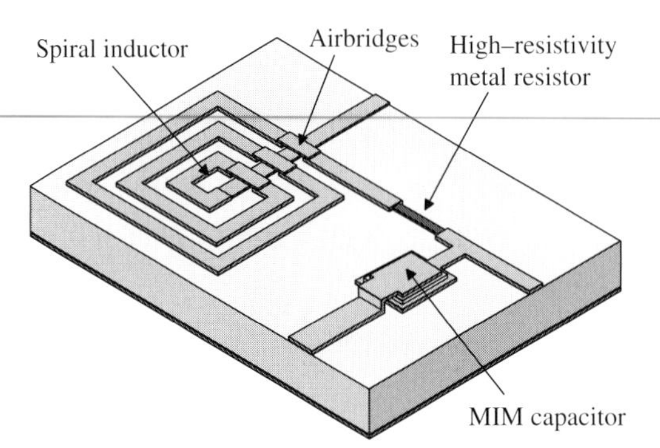

Figure 3.10 Examples of integrated RF passive lumped components.

large size. Moreover, the design based on distributed components is often narrowband.[7] Finally, low-loss TXL components are reactive and cannot therefore be exploited in resistors, attenuators, or other dissipative components.

Lumped capacitors, resistors, or inductors are implemented in hybrid and integrated circuits according to several approaches. Such components have *small size* with respect to the operating wavelength and can be potentially wideband. However, losses typically limit the operating bandwidth of reactive components (capacitors and, above all, inductors).

Many lumped components can be integrated monolithically on semiconductor substrates. *Integrated inductors* can be realized through short, high-impedance transmission lines or, more conveniently, as spiral inductors; see Fig. 3.10. Since no magnetic cores are available for frequencies above a few hundred megahertz, all RF inductors are in air, with rather low inductance values.[8] The maximum inductance of planar integrated spiral inductors is of the order of 100 nH; however, the larger the inductance, the larger is the parasitic capacitance and therefore the smaller is the resonance frequency. Spiral inductors can be exploited for bias T design or for low-Q filters. *Integrated capacitors* usually follow the so-called MIM (metal insulator metal) approach, where a first-level metal is exploited as a ground plane, a suitable insulator (silicon dioxide, silicon nitride, polymide) acts as a dielectric, and the second-level metal is the signal conductor. Finally, *integrated resistors* can be obtained through high-resistivity metals or low-doping layers. Examples of integrated resistors and capacitors are shown in Fig. 3.10. Integrated lumped elements allow one to obtain concentrated forms of components such as directional couplers, power dividers, matching sections; such lumped versions typically have a reduced size with respect to the distributed ones, but also higher losses and a lower-frequency operation range.

[7] An important exception is given by the so-called *traveling-wave design* where impedance matched TXLs are exploited. Relevant examples are distributed amplifiers and traveling-wave optoelectronic components such as distributed photodiodes and modulators.

[8] Some attempt has been made recently to increase the inductance through the use of magnetic conductors, e.g., permalloy. The approach has some advantages in the RF range, but is complex from the technological standpoint.

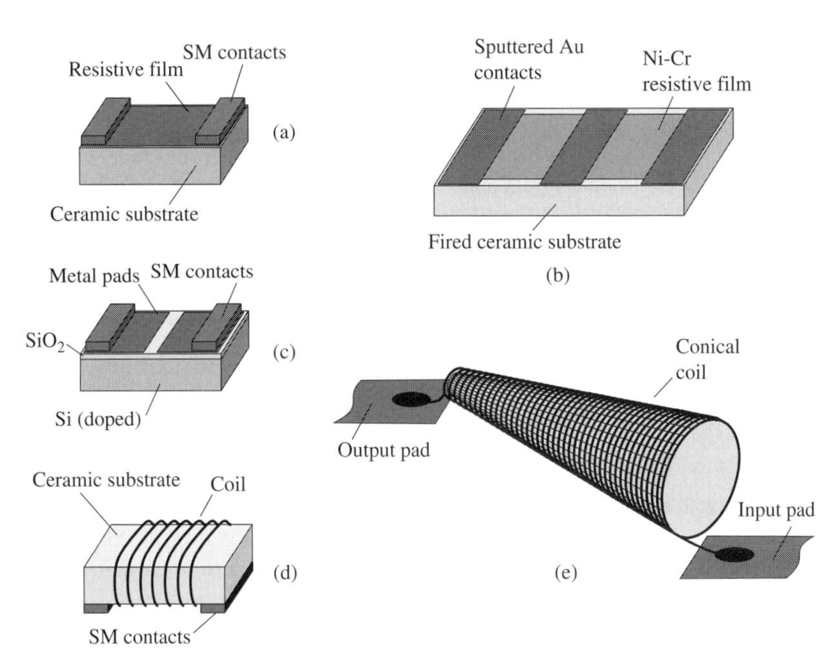

Figure 3.11 Examples of discrete RF lumped components: (a) thin-film chip resistor for surface mount; (b) coplanar broadband chip resistor up to 40 GHz on fired alumina substrate; (c) chip capacitor; (d) chip inductor; (e) ultrabroadband conical inductor.

Discrete lumped components can be externally inserted into *hybrid integrated circuits* (i.e., circuits where the substrate is dielectric, and the active semiconductor devices are not monolithically integrated), usually as surface-mount chip resistors, inductors, or capacitors. Chip resistors are commonly exploited as external loads in a number of integrated optoelectronic components; see Fig. 3.11(b) where a coplanar resistor made of two resistors in parallel (the lateral pads are connected to the ground plane, the central pad to the active conductor) is realized by depositing a NiCr high-resistivity alloy on a fired alumina substrate

Chip resistors are obtained by deposing a resistive thin film over a dielectric (e.g., ceramic) chip. Wrap-around or flip-chip contacts are then added, allowing for surface mounting (SM) on a microstrip or coplanar circuit. An example of such structures (shown bottom up) can be found in Fig. 3.11(a); the side size of the component is often well below 1 mm. Chip capacitors can be obtained by depositing a dielectric layer (e.g., SiO_2) on a conductor or semiconductor (e.g., Si); the dielectric layer is then coated with metal so as to define the external contacts, which can be surface mounted through flip-chip (i.e., by connecting the component upside down); see Fig. 3.11(c).

While chip resistors can be properly manufactured so as to achieve spectacular bandwidths (e.g., from DC to millimeter waves), thus making it possible to provide ultrabroadband matched terminations in traveling-wave components (such as 40 Gbps distributed electrooptic modulators), broadband inductors are difficult to obtain because of the increase of losses with frequency and the upper limitation related to the *LC* resonant frequency. The quality factor of RF and microwave inductors typically peaks in a very narrow band, with maximum values well below 10^2. An example of an RF and microwave chip inductor is shown in Fig. 3.11(d); achievable inductance values typically decrease with increasing operating frequency.

However, ultrabroadband bias Ts (see Section 3.4.1) for optoelectronic devices, and the related electronic circuitry and instrumentation, require broadband inductors as RF blocks. Conical inductors (Fig. 3.11(e)) are a particular technology allowing for very broadband behavior, due to a strong reduction of the parasitic capacitance and to the scaling invariance of the design (see, e.g., [26]).

3.4.1 Bias Ts

Most electronic or optoelectronic devices have to be properly DC biased at the input and output. DC bias is typically imposed under the form of a DC voltage applied through an (almost) ideal DC voltage source. The total applied voltage is therefore $v(t) = V_{DC} + v_{RF}(t)$, where the second contribution is the (zero-average) signal voltage. However, $v(t)$ cannot be applied by a single generator including V_{DC} in series with $v_{RF}(t)$, because the DC and RF generator impedances should be different (e.g., 50 Ω for the RF, ≈ 0 for the DC), and the impedance of a DC source is beyond control at RF. The easiest way to implement such a separation between the DC and RF parts is through a 3-pole component called the *bias T* because of its shape; see Fig. 3.12(a). The operation of the bias T is trivial if the capacitors and inductor values are assumed to be very large; in this case, we can analyze the structure in two limiting cases: DC ($f \to 0$) and RF ($f \to \infty$). In DC, C_1 and C_2 are open and L is a short; thus, the RF signal is blocked by the open C_1 while the DC bias is connected to the output through L (acting as a DC short). C_2 is open and does not interfere with the DC path. At RF, C_1 and C_2 are shorted while L is an open circuit, thus blocking the DC signal from reaching the output. At the same time, the RF signal is shorted to the input, but cannot reach the DC bias node (DC IN) because of the RF blocking inductor. C_2 adds an additional RF block, shorting the RF signal to ground; it could be omitted in an ideal design with arbitrarily large component values. In Fig. 3.12(b) two bias Ts are exploited to bias the input and output of a field-effect transistor.

In practice, however, the bias T has to be realized with finite-valued components. This leads to two operating bands, the "DC" in the interval $0 < f < f_1$ and the "RF" in the interval $f_2 < f < \infty$. The bias T design becomes difficult if f_1 is very close to f_2 and f_2 is low, as in many wideband amplifiers for optical communication systems or instrumentation; see Example 3.1.

Example 3.1: Suppose a bias T is connected to the input of a FET as in Fig. 3.12(b); for simplicity, the FET input is considered as an open circuit, while the DC bias generator (V_G) internal impedance is negligible. Using a simple structure with $C_2 = 0$, find the $C_1 \equiv C$ and L values such as $f_1 = 100$ kHz and $f_2 = 10$ MHz. As a condition, impose that in the "DC" range the input impedance seen from RF IN is $|Z_{in}| > 1$ kΩ (so that the RF generator is effectively decoupled from the device input in the "DC" range),

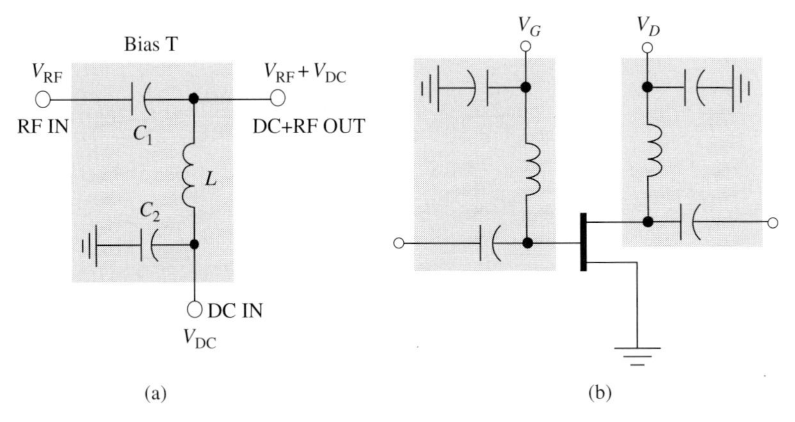

Figure 3.12 Applying a DC bias through a bias T: (a) possible structure of a lumped bias T; (b) transistor bias through an input and output bias T.

while in the "RF" range the input impedance seen from RF IN is again $|Z_{in}| > 1\ \text{k}\Omega$ (implying that the path through the blocking inductor L to the DC generator is an open circuit in the "RF" range). Repeat for (b) $f_1 = 100$ kHz, $f_2 = 1$ GHz and for (c) $f_1 = 100$ kHz, $f_2 = 200$ kHz.

Taking into account that the FET input is an open circuit, we can select the LC circuit so that the series resonance is between f_1 and f_2. At low frequency, the impedance seen between the RF input and the DC input (which is shorted to ground; remember that the internal impedance of the DC source is assumed as negligible) will be large and capacitive (the inductor is almost a short in this range), and at high frequency will be large and inductive (the capacitor is almost a short in this range). This yields the two conditions

$$|Z_{in}(f_1)| = \frac{1 - \omega_1^2 LC}{\omega_1 C}, \quad |Z_{in}(f_2)| = \frac{\omega_2^2 LC - 1}{\omega_2 C},$$

and, conveniently selecting the resonance frequency as

$$\frac{1}{\sqrt{LC}} = 2\pi \sqrt{f_1 f_2},$$

we have

$$|Z_{in}(f_1)| = \frac{1}{2\pi C}\left(\frac{1}{f_1} - \frac{1}{f_2}\right) = \frac{1}{2\pi C}\frac{\Delta f}{f_1 f_2} = |Z_{in}(f_2)|.$$

Thus,

$$C = \frac{1}{2\pi\, |Z_{in}|}\frac{\Delta f}{f_1 f_2}, \quad L = \frac{1}{(2\pi)^2 f_1 f_2 C} = \frac{|Z_{in}|}{2\pi \Delta f}.$$

For $f_1 = 100$ kHz, $f_2 = 10$ MHz we have

$$C = \frac{1}{2\pi \cdot 1 \times 10^3} \frac{10 \times 10^6 - 100 \times 10^3}{10 \times 10^6 \cdot 100 \times 10^3} = 0.157 \text{ nF}$$

$$L = \frac{1 \times 10^3}{2\pi \left(10 \times 10^6 - 100 \times 10^3\right)} = 16 \text{ μH};$$

while for $f_1 = 100$ kHz, $f_2 = 1$ GHz, $C = 0.159$ nF, $L = 15.9$ nH; for $f_1 = 100$ kHz, $f_2 = 200$ kHz, $C = 0.79$ nF, $L = 1.59$ mH. While the capacitance value is almost constant (f_1 is in fact constant), the inductance becomes extremely large if the transition between "DC" and "RF" is narrow. In such cases, more complex circuit schemes have to be exploited to allow for reasonably valued components.

3.5 Active components

High-speed circuits exploit as active components both field-effect (FET) and bipolar junction transistors (BJT), implemented in several conventional and compound semiconductor technologies. Si-based MOSFETs with nanometer gate length have demonstrated microwave and even millimeter-wave operation; such devices are not well suited to high-power (high breakdown voltage) applications, but can be conveniently exploited, for example, in receiver stages with some noise penalty in comparison with compound semiconductor technologies. Compound semiconductor FETs based on GaAs and InP substrates are currently implemented as high electron mobility transistors (HEMTs) in lattice-matched (LMHEMT) or pseudomorphic (PHEMT) form. Although conventional, Si-based bipolar transistors are limited to RF operation, heterojunction bipolars (HBTs) in compound semiconductor technologies (GaAs or InP) or based on the SiGe material system have shown good performances up to millimeter waves. HBTs are characterized by features similar to the conventional bipolars, i.e., larger current density and current driving capability with respect to FETs.

3.5.1 Field-effect transistors (FETs)

High-speed field-effect transistors (FETs) are based on a conducting channel whose current (driven by the potential difference between the drain and source electrodes) is modulated by the potential applied to the control electrode (the gate), which is isolated from the channel (by a metal-oxide-semiconductor junction in MOSFETs or by a reverse-biased Schottky junction in compound semiconductor FETs); see Fig. 3.13.

The first high-speed field-effect transistor able to operate at microwave frequencies was the GaAs-based MESFET (metal-semiconductor FET), in which the channel is a layer (implanted or epitaxial) of highly doped bulk semiconductor. A simplified cross section of a high-breakdown-voltage MESFET is shown in Fig. 3.14; in such devices,

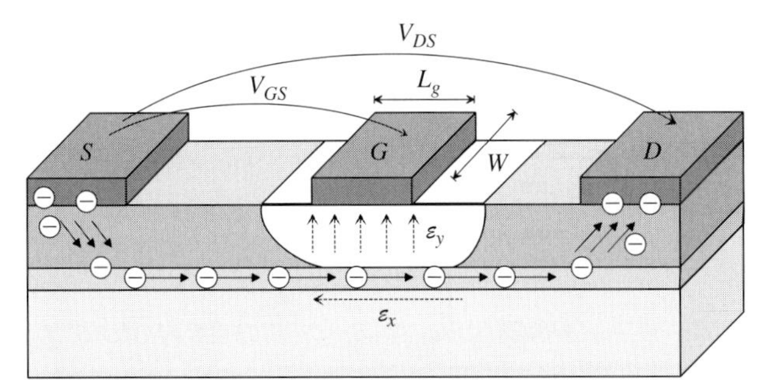

Figure 3.13 Schematic operation principle of an n-channel FET: electrons are driven by the channel electric field induced by the V_{DS} potential difference between drain and source, and the resulting channel is modulated by the control electrode (the gate, here a reverse-biased Schottky junction) through the application of a vertical control electric field. L_g is the gate length, W the gate periphery.

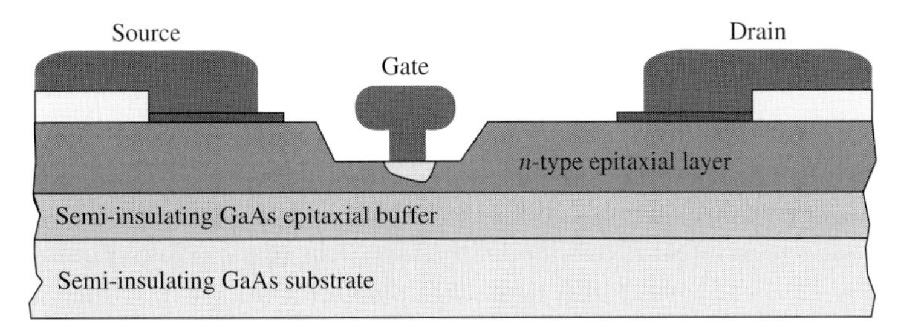

Figure 3.14 Cross section of an epitaxial recessed-gate power MESFET.

the control action of the gate occurs by depleting the active layer underneath the Schottky contact, thus reducing its cross section and, as a consequence, the total device current. Recessed-gate power MESFETs exhibit drain breakdown voltages in excess of 20 V and can still be exploited for frequencies up to 10 GHz; in high-performance applications they have generally been replaced by heterojunction-based FETs (HEMTs and PHEMTs).

The qualitative behavior of the DC characteristics of the MESFET can be derived by inspection of Fig. 3.15. Suppose that most of the V_{DS} potential difference falls in the active channel located underneath the Schottky gate; for low V_{DS}, the potential difference between the gate and each point in the conducting channel is almost constant; as a consequence, the *channel potential* ϕ_{ch} increases linearly from the grounded source to the drain, but the depletion region is uniform along the gate, and the channel cross section is constant. In such a condition (Fig. 3.15(a)) the device behaves as a variable resistor; the resistance depends on the applied V_{GS}. For $V_{GS} = V_{TH}$ (the threshold voltage) the conducting layer is completely depleted and the channel conductance is zero. Increasing V_{DS}, the local potential difference between the gate and the channel becomes more and more nonuniform (Fig. 3.15(b)) and, as a consequence, the channel cross section narrows from the source to the gate, giving rise to a nonlinear channel potential distribution. For large enough V_{DS} (Fig. 3.15(c)) the electric field at the drain edge of the gate increases so as to lead to velocity saturation of

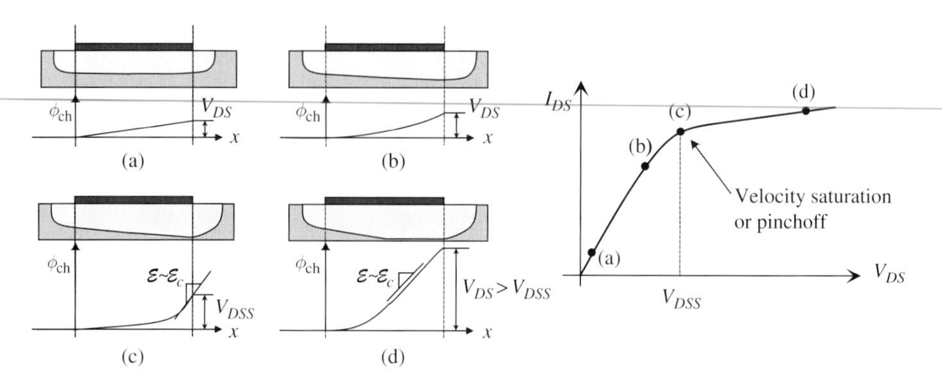

Figure 3.15 Qualitative explanation of the DC n-channel MESFET characteristics for constant V_{GS} and increasing V_{DS}.

the carriers (this happens in GaAs for electric fields ε much larger than a critical field $\varepsilon_c \approx 3.2\,\text{kV/cm}$). Beyond this point, the channel current approximately saturates, and for larger V_{DS} the saturation point is displaced toward the source, leading to a small decrease in the length of the nonsaturated (ohmic) part of the channel (Fig. 3.15(d)); this causes a small increase of the drain current I_D, which turns out to be more evident in short-gate devices (e.g., with gate length $L_g < 0.5\ \mu\text{m}$). In materials where the velocity saturation effect is less abrupt than in GaAs (e.g., in Si), current saturation is due to the channel pinchoff, that is, it takes place at a voltage V_{DS} such that the cross section of the conducting channel vanishes at the drain edge of the gate.

In conclusion, the FET output characteristics (at the input no DC current is absorbed since the gate is isolated) have a *linear* region for low V_{DS}, a *triode region* where the drain conductance begins to decrease, and finally, beyond a certain knee voltage, the *saturation region*, where the drain current is approximately independent of V_{DS}. In an n-channel FET, decreasing V_{GS} increases the amount of channel depletion and therefore leads to a lower current. For very large V_{DS} and/or large currents, breakdown occurs in the form of a sometimes catastrophic increase of the drain current. In high-speed Schottky gate FETs the breakdown voltage is low in the *on-state* (high current) and high in the *off-state* (low current, device at threshold). The resulting DC characteristics are qualitatively shown in Fig. 3.16; note that the I_D corresponding to $V_{GS} = 0$ is often referred to as I_{DSS}; since the gate is a Schottky barrier, the maximum V_{GS} is limited by the Schottky barrier built-in voltage; indicative values are 0.6–0.7 V for GaAs and 0.9–1 V for AlGaAs. The soft breakdown effect shown in Fig. 3.16 is a peculiarity of HEMTs and PHEMTs, due to a substrate charge injection mechanism; it is considerably alleviated by technology optimization.

3.5.2 FET DC model

A simplified analytical model for the DC current of any FET can easily be developed according to the channel pinchoff saturation mechanism; current saturation due to carrier velocity saturation will be discussed in Section 3.5.4.

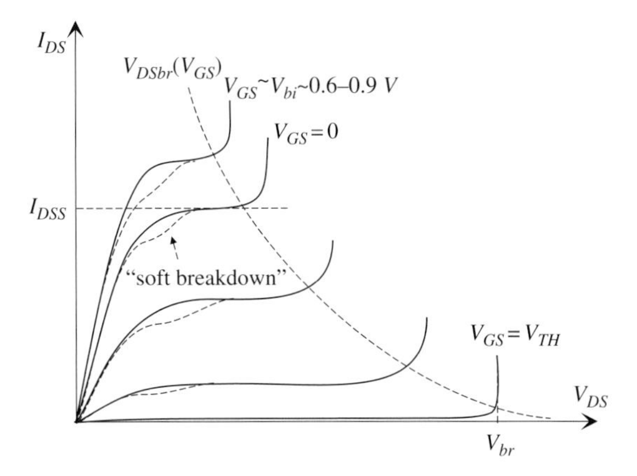

Figure 3.16 DC drain characteristics for Schottky gate FET. The kinks are typical of HEMT devices and are referred to as the kink or soft breakdown effect.

Consider an n-type FET with channel along x; let us denote by $-Q_{\text{ch}}(x)$ the total mobile charge (per unit surface) in the channel; in a MESFET we simply have

$$-Q_{\text{ch}}(x) = -q \int_{\text{channel}} n(x, y)\, \mathrm{d}y,$$

where $n(x, y)$ is the channel electron density ($n \approx N_D$ in a MESFET). The channel drift current is

$$I_D = W \mu_{n0} Q_{\text{ch}}(x) \frac{\mathrm{d}\phi_{\text{ch}}}{\mathrm{d}x}, \tag{3.40}$$

in which W is the gate periphery, μ_{n0} is the low-field electron mobility, while the channel electric field (directed along x) is

$$\mathcal{E}_{\text{ch}} = -\frac{\mathrm{d}\phi_{\text{ch}}(x)}{\mathrm{d}x};$$

ϕ_{ch} is the *channel potential*. In general, $Q_{\text{ch}}(x)$ depends on the potential difference between the gate and section x of the channel:

$$Q_{\text{ch}}(x) = Q_{\text{ch}}\left(V_{GS} - \phi_{\text{ch}}(x)\right).$$

Integrating (3.40) from $x = 0$ to $x = L_g$ (corresponding to $\phi_{\text{ch}} = V_S = 0$ and $\phi_{\text{ch}} = V_D - V_S = V_{DS}$), we obtain

$$\begin{aligned}\int_0^{L_g} I_D\, \mathrm{d}x &= W \mu_{n0} \int_0^{L_g} Q_{\text{ch}}\left(V_{GS} - \phi_{\text{ch}}\right) \frac{\mathrm{d}\phi_{\text{ch}}}{\mathrm{d}x}\, \mathrm{d}x \\ &= W \mu_{n0} \int_0^{V_{DS}} Q_{\text{ch}}\left(V_{GS} - \phi_{\text{ch}}\right) \mathrm{d}\phi_{\text{ch}},\end{aligned}$$

i.e.,

$$I_D = \frac{W \mu_{n0}}{L_g} \int_0^{V_{DS}} Q_{\text{ch}}\left(V_{GS} - \phi_{\text{ch}}\right) \mathrm{d}\phi_{\text{ch}}. \tag{3.41}$$

We now approximate the function $Q_{\text{ch}}(V_{GS} - \phi_{\text{ch}})$ in a linear way (this is indeed almost exact in some FETs, such as the MOSFET), as follows:

$$|Q_{\text{ch}}| = \begin{cases} C_{\text{ch}} \left(V_{GS} - \phi_{\text{ch}} - V_{TH}\right), & V_{GS} - \phi_{\text{ch}} - V_{TH} \geq 0 \\ 0, & V_{GS} - \phi_{\text{ch}} - V_{TH} < 0 \end{cases}$$

where C_{ch} has the dimension of a capacitance per unit surface. Substituting in (3.41) and integrating, we obtain

$$\begin{aligned} I_D &= \frac{W\mu_{n0}}{L_g} \int_0^{V_{DS}} C_{\text{ch}} \left(V_{GS} - \phi_{\text{ch}} - V_{TH}\right) \mathrm{d}\phi_{\text{ch}} \\ &= \frac{W\mu_{n0}C_{\text{ch}}}{L_g} \left[\left(V_{GS} - V_{TH}\right) V_{DS} - \frac{1}{2} V_{DS}^2 \right]. \end{aligned} \quad (3.42)$$

For increasing V_{DS}, (3.42) yields a current that first increases linearly (linear region), then begins to saturate (triode region) to finally reach a maximum I_{Ds} for $V_{DS} = V_{DSS}$, corresponding to channel pinchoff at $x = L_g$, and therefore to current saturation. From (3.42) the saturation voltage V_{DSS} reads

$$V_{DSS} = V_{GS} - V_{TH};$$

for $V_{DS} > V_{DSS}$ (3.42) no longer holds and the current is approximately constant with value $I_{\text{Ds}} = I_D(V_{DSS})$:

$$I_{\text{Ds}} = I_D(V_{DSS}) = \frac{W\mu_{n0}C_{\text{ch}}}{2L_g} \left(V_{GS} - V_{TH}\right)^2. \quad (3.43)$$

Although saturation in compound semiconductor FETs is typically due to velocity saturation, the above model can serve as a simple analytical tool to explore some of the main features of different classes of high-speed FETs.

3.5.3 FET small-signal model and equivalent circuit

The analog operation of transistors often corresponds to small-signal conditions, i.e., a small-amplitude signal source is superimposed on the DC bias. The small-signal device response can be approximated by a linear, frequency-dependent model, e.g., under the form of the admittance or scattering matrix. (FETs often operate in common source configuration, i.e., the source is signal grounded, the gate acts as the input and the drain as the output.) It is, however, useful to associate with the small-signal behavior a circuit model (the *small-signal equivalent circuit*), whose elements can be physically mapped into the FET structure as shown Fig. 3.17 (for simplicity, we refer to a MESFET, but the approach is readily extended to other FET families). The external parasitic resistances R_S, R_D, R_G appear here as access resistances of the source, drain, and gate terminals, respectively, while the intrinsic resistance R_I is the resistance of the ohmic part of the channel. The capacitances C_{GS} and C_{GD} derive from partitioning the total capacitance associated with the gate depletion region, while C_{DS} is mostly a geometrical capacitance between drain and source. The other elements of the equivalent

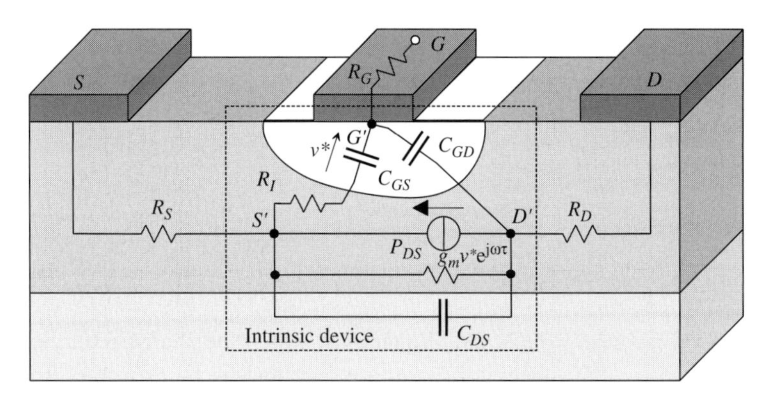

Figure 3.17 Physical mapping of a small-signal circuit into the MESFET cross section.

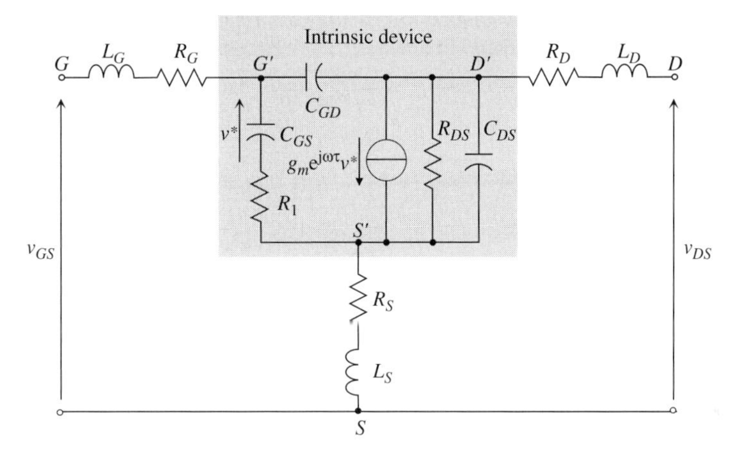

Figure 3.18 High-speed FET equivalent circuit; the intrinsic part is enclosed in the box.

circuit of the intrinsic device (describing its specific transistor action) can be derived by differentiating the $I_D(V_{GS}, V_{DS})$ relation as

$$g_m = \frac{\partial I_D}{\partial V_{GS}}, \quad R_{DS}^{-1} = \frac{\partial I_D}{\partial V_{DS}}.$$

The *transconductance* g_m is the device amplification between the input driving voltage (the voltage v^* applied between the intrinsic gate and the channel, across the depletion region) and the output (drain) current. The parameter τ describes an additional phase delay between the driving voltage v^* and the generated current $g_m v^*$; it is often associated with the delay that carriers experience drifting below the gate contact, thus justifying the name *transit time*. A more readable form of the equivalent circuit, in which also some external inductive parasitics have been added, is shown in Fig. 3.18.

The parameters of the equivalent circuit can be exploited to derive two figures of merit related to the device speed or, equivalently, to the maximum operating bandwidth: the *cutoff frequency* f_T and the *maximum oscillation frequency* $f_{\max}$.

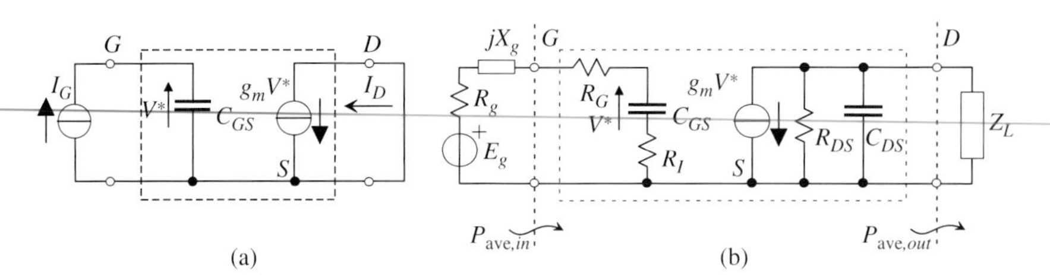

Figure 3.19 Evaluating the cutoff frequency (a) and the maximum oscillation frequency (b) of a FET.

The *cutoff frequency* f_T is the frequency at which the short-circuit current gain of the transistor has unit magnitude. Such a parameter can be estimated from measurements, either directly or from extrapolation; however, a straightforward connection to the equivalent circuit parameters is possible if a simplified intrinsic equivalent circuit is exploited in the analysis. By inspection of Fig. 3.19(a) one has (we assume harmonic generators and phasor notation)

$$I_D = \frac{g_m}{\mathrm{j}\omega C_{GS}} I_G,$$

i.e., imposing $|I_D/I_G| = 1$ at $\omega_T = 2\pi f_T$,

$$f_T = \frac{g_m}{2\pi C_{GS}}.$$

The *maximum oscillation frequency* is the frequency at which the maximum available power gain (MAG, the ratio between the available power at the device output and the generator available power) is unity. Since the MAG decreases with frequency, for $f > f_{\max}$ the device becomes passive and cannot provide gain even if it is power-matched at the input and output (i.e., if $Z_g = Z_{in}^*$, $Z_L = Z_{out}^*$, thus leading to maximum power transfer). Again, $f_{\max}$ can be related to the parameters of a simplified equivalent circuit; some extra elements have to be added to avoid singular values in the output available power and in the input power. Assuming that the output is power matched, the output available power will be (see (3.37)):

$$P_{av,out} = \frac{g_m^2 |V^*|^2}{4R_{DS}^{-1}}.$$

However, if the input is power matched, the generator reactance compensates for the C_{GS} reactance, $R_g = R_G + R_I$ and the gate current is

$$I_G = \frac{E_g}{R_G + R_I} \rightarrow V^* = \frac{E_g}{\mathrm{j}\omega C_{GS}(R_G + R_I)}.$$

Thus,

$$P_{av,out} = \frac{g_m^2 |E_g|^2 R_{DS}}{4\omega^2 C_{GS}^2 (R_G + R_I)^2} = \frac{g_m^2 R_{DS}}{\omega^2 C_{GS}^2 (R_G + R_I)} P_{av,in},$$

since $P_{av,in} = |E_g|^2/(4R_g)$ and $R_g = R_G + R_I$.

The available power gain $G_{av} = P_{av,out}/P_{av,in}$ corresponds to the MAG, because power matching is implemented at the input; moreover, it coincides (since the output is also matched) with the operational power gain $G_{op} = P_{out}/P_{in}$. The condition to be imposed to evaluate $f_{\max}$ is

$$\text{MAG} = \frac{g_m^2 R_{DS}}{(2\pi f_{\max})^2 C_{GS}^2 (R_G + R_I)} = 1,$$

leading to

$$f_{\max} = f_T \sqrt{\frac{R_{DS}}{R_G + R_I}}. \tag{3.44}$$

Since $R_{DS} \gg R_G + R_I$ (typically by one order of magnitude), $f_{\max} > f_T$ unless the input resistance is high. Equation (3.44) also holds for bipolar transistors in common emitter configuration with the substitutions $R_G \to R_B$ (the base resistance), $R_{DS} \to R_{CE}$ (the collector–emitter output resistance).

3.5.4 High-speed FETs: the HEMT family

High-speed field-effect transistors (FETs) today exist in a number of technologies based both on Si and on compound semiconductors. Deep scaling down to nanometer gate lengths of Si-based MOSFETs has made available transistors with cutoff frequencies in the millimeter wave range; due to the comparatively high gate input resistance, the maximum oscillation frequencies are in fact less outstanding. Heterostructure FETs are currently manufactured on several compound semiconductor substrates: GaAs, InP and also GaN. Devices of choice for high-speed electronic and optoelectronic applications are the GaAs- or InP-based pseudomorphic high electron mobility transistor (PHEMT) and, in the future, the GaN HEMT (mainly for high-power or high-voltage applications).

The basic building block of heterojunction FETs is the so-called *modulation doped heterojunction*, consisting of a highly doped (n-type) widegap layer on top of an intrinsic narrowgap layer; see Fig. 3.20. In a modulation doped heterostructure in equilibrium, the doped layer (*supply layer*) donors are ionized, and therefore transfer their electrons into the conduction band potential well, originated by the bandgap discontinuity. A thin conducting channel is generated and confined in the interface well, in a way that is not dissimilar to the MOS surface inversion channel. Moreover, the electrons in the potential well suffer little or no impurity scattering, since the narrowgap material is undoped; this leads to an improvement in the channel electron mobility with respect to FETs (like the MESFET) exploiting a doped channel.[9] A last important advantage of the modulation doped approach is the possibility of making the doped widegap layer (supply layer) thin, but highly doped, thus reducing the distance between the controlling gate and

[9] Huge mobility increases that led to the (historical) name of the HEMT can, however, be detected only at cryogenic temperature, where phonon scattering becomes ineffective and the mobility is dominated by impurity scattering; at ambient temperature, phonon scattering prevails and the mobility advantage is less dramatic.

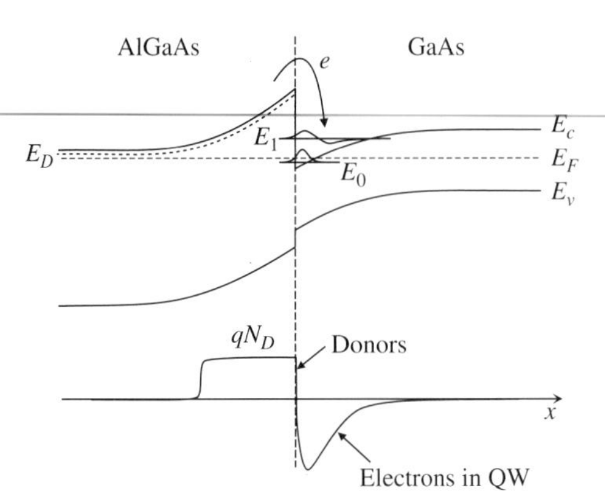

Figure 3.20 Example of the band diagram for modulation-doped structure with triangular potential profile QW; the charge distribution (ionized donors in the supply layer, free electrons in the QW) is shown below.

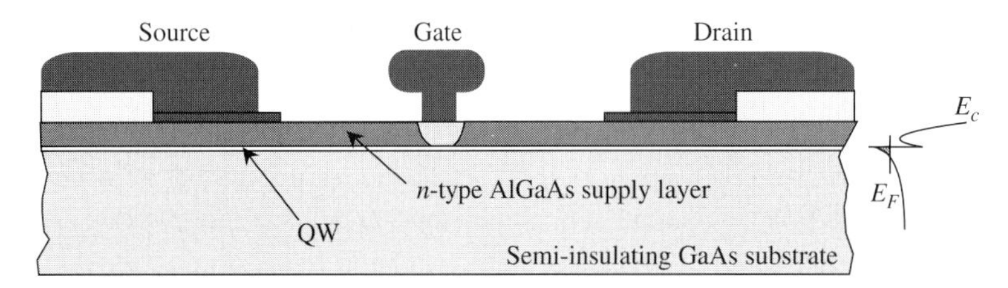

Figure 3.21 Cross section of a conventional AlGaAs/GaAs HEMT.

the channel. Since electrons in the channel are confined in the vertical direction, the modulation doping heterojunction typically is a QW for the channel electrons, which form a charge sheet often referred to as a *two-dimensional electron gas* (2DEG).

The (conventional) HEMT, shown in Fig 3.21, is a Schottky gate FET where a reverse bias applied to the gate electrostatically modulates the 2DEG. The supply layer is typically AlGaAs (widegap), while the substrate is semi-insulating GaAs (narrowgap). The $Al_xGa_{1-x}As$ layer has an Al fraction x between 23% and 30%; larger Al fractions would be desirable to increase the conduction band discontinuity ΔE_c and the channel confinement; unfortunately, for $x > 30\%$ the AlGaAs supply layer develops a trap level (the so-called *DX centers*) which actually limits the effectiveness of the doping. In order to decrease the surface impurity scattering from the donors in the supply layer, a thin undoped AlGaAs *spacer layer* is epitaxially grown between the GaAs substrate and the doped supply layer.

State-of-the art HEMTs are based on a double heterostructure, wherein a narrowgap material is sandwiched between the widegap substrate and the widegap supply layer. The resulting rectangular (rather than triangular) conduction band QW has superior confinement properties and may also have better transport performance than in the conventional HEMT. According to whether the narrowgap layer is lattice-matched or in a strained (tensile or compressive) condition, we have the LMHEMT (lattice-matched HEMT) or the PHEMT (pseudomorphic HEMT).

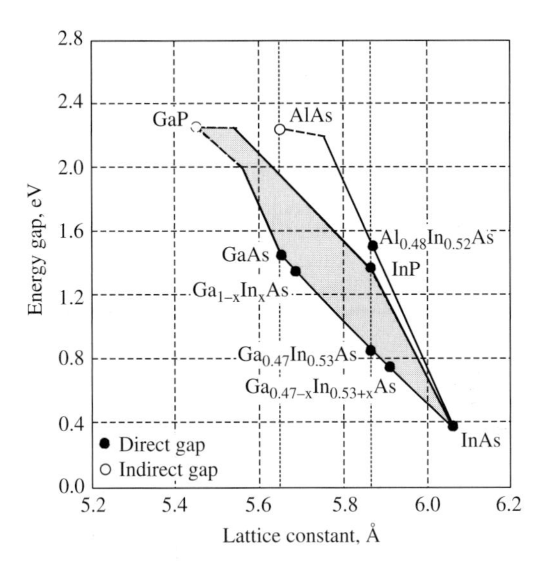

Figure 3.22 Energy gap–lattice constant diagram for the AlGaAs, the InGaAsP and the InAlAs alloys.

Several material systems and modulation doping techniques in the supply layer (including δ doping instead of uniform doping) can be exploited in PHEMTs. Representative examples are:

- The GaAs-based PHEMT, where the supply layer is doped AlGaAs, the channel is $In_{1-x}Ga_xAs$ with low indium content (see the black dot close to GaAs in the energy gap–lattice constant diagram in Fig. 3.22), and the substrate is GaAs. Due to the strained (pseudomorphic) channel, the conduction band discontinuity is increased, thus improving the confinement and sheet carrier density in the conducting channel. The potential barrier between the channel and the substrate is low, however.
- The InP-based PHEMT, for which several choices exist. Taking into account that $Al_{0.48}In_{0.52}As$ and $Ga_{0.47}In_{0.53}As$ are lattice matched to InP, a possible lattice-matched structure with an $Al_{0.48}In_{0.52}As$ supply layer (doped), a $Ga_{0.47}In_{0.53}As$ channel (undoped), and an InP substrate (undoped) is shown in Fig. 3.23. Implementing the channel through a pseudomorphic layer of $Ga_{0.47-x}In_{0.53+x}As$, x small, we increase the bandgap discontinuity toward both the supply layer and the substrate, thus obtaining a PHEMT with improved performances.

The GaN-based HEMT has a conventional HEMT structure made of a widegap supply layer of AlGaN grown on an undoped GaN substrate. A peculiar feature of the AlGaN/GaN system is the presence of a piezoelectrically induced electron charge in the interface QW, even without any supply layer doping. The AlGaN-GaN HEMT therefore has excellent properties in terms of channel charge, which can be further improved by intentional doping.

InP-based PHEMTs are probably the devices of this class offering the best performance in terms of high cutoff and maximum oscillation frequencies, with record

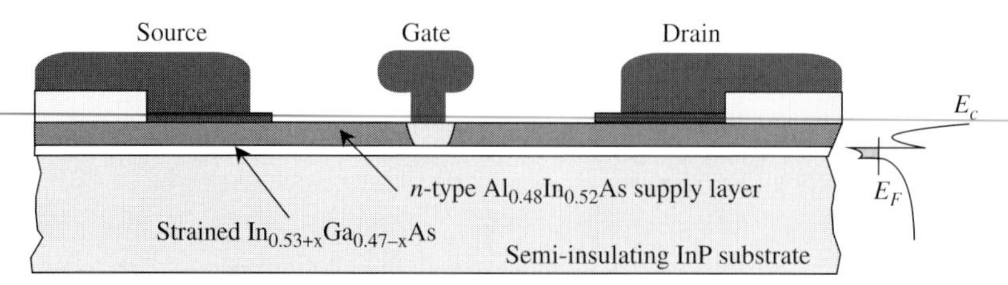

Figure 3.23 Cross section of InP-based PHEMT.

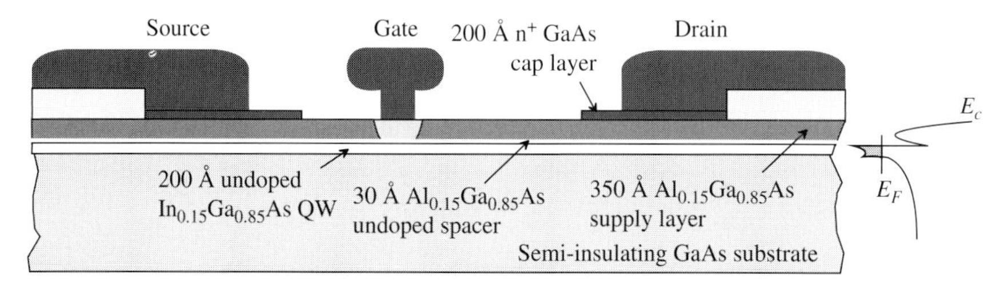

Figure 3.24 Cross section of GaAs-based power PHEMT.

values well into the millimeter-wave range. GaAs-based PHEMTs offer some advantages in terms of breakdown voltage (which can be pushed into the 10–20 V range) but are confined to applications below 50–60 GHz. However, the superior development of the GaAs technology and lower cost with respect to InP make this the device of choice for many applications, e.g., in 40 Gbps systems. An example of a GaAs-based PHEMT structure is shown in Fig. 3.24; note the asymmetric placement of the gate, typical of power devices and aimed at decreasing the maximum field in the drain–gate region, i.e., at increasing the breakdown voltage.

A comparison between a number of competing Si-based and III-V based FET technologies is shown in Fig. 3.25 [6], [27], [28]. Experimental data suggest that InP-based pseudomorphic or lattice-matched HEMTs have record cutoff frequency; GaAs PHEMTs, while slightly inferior, perform better than GaAs MESFETs. Si-based NMOS exhibit very high cutoff frequencies for nanometer gate lengths, but the corresponding $f_{\max}$ is often compromised by the large input resistance. One may note that, for decreasing gate length, the Si penalty related to the lower initial mobility is somewhat mitigated by the fact that nanometer-scale MOSFETs operate in velocity saturation conditions. In this respect, the large threshold field for velocity saturation in InP is a significant advantage over GaAs or Si. An alternative approach to direct growth on InP substrates, which would considerably decrease the device cost, has been proposed under the name of the *metamorphic* approach. In general, metamorphic devices are grown on substrates mismatched with respect to the active region (GaAs or even Si; see e.g., [29]); properly designed buffer layers have to be interposed in order to avoid defects. As a simple example, an InP-based HEMT can be grown on a GaAs substrate topped by a graded InGaAs epitaxial buffer leading from the GaAs to the InP lattice constant. Recently, metamorphic HEMTs (MHEMTs) have shown good reliability and performances comparable to lattice-matched or pseudomorphic HEMTs.

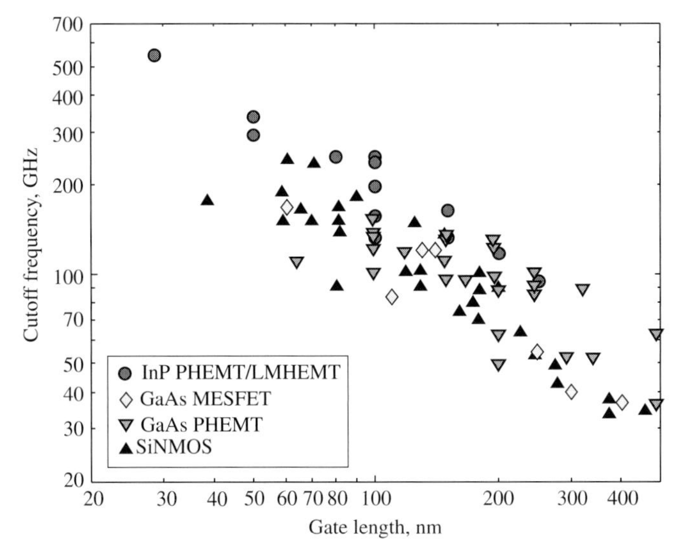

Figure 3.25 Gate length vs. cutoff frequency for a number of FET technologies (InP, GaAs, Si-based NMOS) with submicrometer gate length (0.04–0.5 μm). Data from [27] (Fig. 5) and [6].

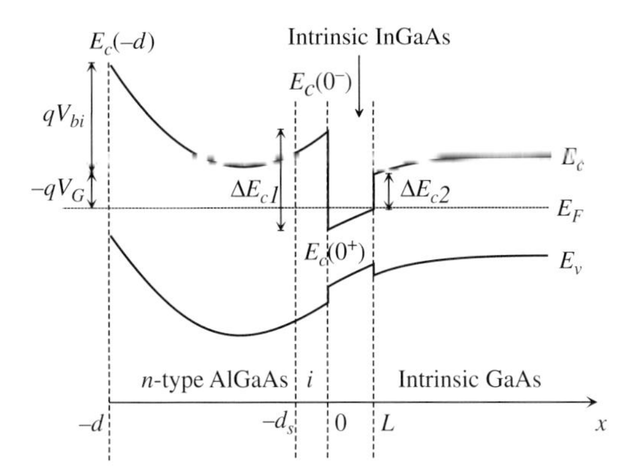

Figure 3.26 GaAs-based PHEMT band structure in equilibrium.

The carrier density hosted by the interface potential well of a modulation doped structure can be evaluated taking into account the modified density of states of the QW and the equilibrium band structure shown in Fig. 3.26. The complete analysis will be omitted here; we only remark that increasing the Fermi level E_F with respect to $E_c(0)$ causes an increase of the E_c slope in 0^+ and therefore an increase of the QW surface electric field in 0^+, $\mathcal{E}_s$. However, from the Gauss law, this is related to the total surface charge density n_s in the potential well:

$$\mathcal{E}_s = \frac{q}{\epsilon} n_s. \tag{3.45}$$

In general, n_s depends on $E_F - E_c(0)$ according to an implicit law that has to be made explicit through numerical techniques. However, in a suitable range of energies and

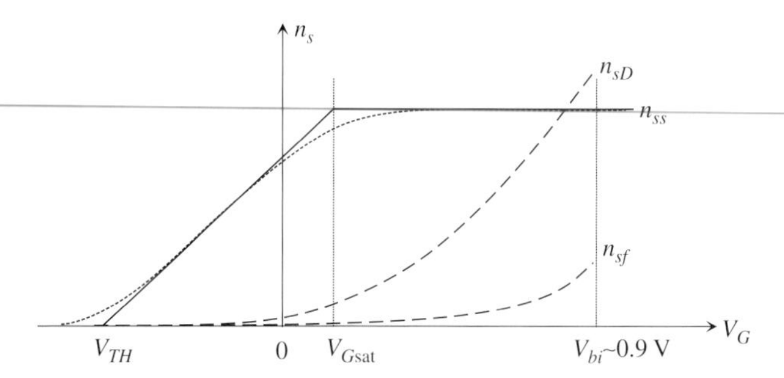

Figure 3.27 Behavior of sheet carrier density n_s for a PHEMT as a function of the gate bias. V_{TH} is the threshold voltage, V_{Gsat} is the gate voltage at which the sheet carrier density saturates at n_{ss}, V_{bi} is the Schottky gate barrier built-in voltage. The sheet carrier densities n_{sD} and n_{sf} refer to the donor trapped carriers and free carriers in the supply layer, respectively.

surface charge densities the relation can be linearly approximated as

$$E_F - E_c(0) \approx a n_s, \tag{3.46}$$

where a is a fitting parameter of the order of 10^{-17} eV m^2; in MKS units the parameter has to be expressed as qa, where q is the electron charge.

The sheet carrier density n_s can be modulated by acting on the external bias of a reverse-biased Schottky junction connected to the supply layer. Denote by d the thickness of the supply layer, by d_s the thickness of the spacer layer (meant to screen the channel carriers from surface donor scattering); if the Schottky barrier reverse bias is increased the $E_c(0^+)$ slope decreases, leading to a decrease of n_s; conversely, if the Schottky negative bias is reduced, the slope increases, thus increasing the sheet carrier density n_s. The threshold voltage V_{TH} corresponds to the (flatband) condition in which $E_c(0^+)$, and therefore the channel charge, vanishes; the increase in n_s is, however, limited by the fact that, for positive Schottky contact (gate) bias, the supply layer ultimately ceases to be depleted, thus decoupling the population n_s from the gate control. Thus, n_s vanishes at threshold, then increases to saturate at a value n_{ss}; such behavior is shown in Fig. 3.27.

In order to relate the gate bias V_G to the sheet electron concentration n_s, the Poisson equation can be suitably solved in the depleted supply layer plus spacer. Taking into account the jump condition in 0,

$$E_c(0^-) = E_c(0^+) + \Delta E_c,$$

and the boundary condition in $-d$,

$$E_c(-d) = qV_{bi} - qV_G + E_F,$$

we finally have, from the solution of Poisson's equation,

$$qV_{bi} - qV_G + E_F = \frac{q^2 N_D}{2\epsilon}(d - d_s)^2 - q\mathcal{E}_s d + E_c(0^+) + \Delta E_c,$$

where N_D is the supply layer doping, and ϵ is the supply layer permittivity. Solving by the surface electric field we obtain

$$\mathcal{E}_s = \frac{qN_D}{2d\epsilon}(d - d_s)^2 - \frac{V_{bi}}{d} + \frac{V_G}{d} + \frac{E_c(0^+) - E_F + \Delta E_c}{qd}. \tag{3.47}$$

Taking into account that, for $V_G = V_{TH}$, $n_s \propto \mathcal{E}_s \propto E_F - E_c(0^+) \approx 0$ by definition, we can define the threshold voltage V_{TH} as

$$V_{TH} = -\frac{qN_D}{2\epsilon}(d - d_s)^2 + V_{bi} - \frac{\Delta E_c}{q}. \tag{3.48}$$

Finally, since from (3.45) $\epsilon\mathcal{E}_s = qn_s$ and

$$n_s \approx \frac{E_F - E_c(0^+)}{aq},$$

we obtain from (3.47) and (3.48) the result

$$\frac{1}{\epsilon}\left(d + \frac{\epsilon a}{q}\right) qn_s = (V_G - V_{TH}).$$

Defining the equivalent thickness of the 2DEG, Δd, as

$$\Delta d = \frac{\epsilon a}{q} \approx \frac{13 \cdot 8.86 \times 10^{-12} \cdot 10^{-17}}{1.69 \times 10^{-19}} \approx 7 \text{ nm},$$

we evaluate the surface mobile charge associated with the 2DEG, $Q_s = qn_s$, as

$$Q_s = \frac{\epsilon}{d + \Delta d}(V_G - V_{TH}) = C_{eq}(V_G - V_{TH}), \tag{3.49}$$

where C_{eq} is the equivalent 2DEG capacitance. Equation (3.49) provides a simple tool with which to analyze the PHEMT through the linear charge control approximation. Of course (3.49) ceases to be valid when the 2DEG saturates to n_{ss}; a detailed analysis, based again on the solution of the Poisson equation, leads to the result

$$n_{ss} \approx \sqrt{\frac{\Delta E_c}{q}\frac{2N_D\epsilon}{q}},$$

which shows that the saturation density increases with increasing supply layer doping, but also with the conduction band discontinuity between the supply layer and the narrowgap channel. The saturation gate voltage can be derived, assuming $Q_s = qn_{ss}$ and $d + \Delta d \approx d$ in (3.49), as

$$V_{G\text{sat}} \approx V_{TH} + \frac{dq}{\epsilon}n_{ss}. \tag{3.50}$$

Starting from the charge control relation (3.49) and assuming, more realistically, that the drain current saturation is due to *velocity saturation* rather than to channel pinchoff, we can easily develop an approximate model for the drain current. To this end, assume a simplified piecewise velocity–field curve for the electrons, in which abrupt saturation occurs for the threshold field $\mathcal{E}_{th} \approx v_{n,\text{sat}}/\mu_{n0}$, where μ_{n0} is the low-field mobility.

The drain current can be obtained by assuming that velocity saturation takes place at the drain edge of the gate ($x = L_g$) for $V_{DS} = V_{DSSv}$. In the velocity saturated region the current is

$$I_{Ds} = W v_{n,\text{sat}} q n_s = W v_{n,\text{sat}} \frac{\epsilon}{d + \Delta d} [V_{GS} - V_{DSSv} - V_{TH}],$$

where W is the gate periphery, since $V_{GS} - V_{DSSv}$ is the potential difference between gate and channel at $x = L_g$ in velocity saturation conditions. Since the current must be continuous from the ohmic to the velocity-saturated part of the channel, we should have

$$W v_{n,\text{sat}} \frac{\epsilon}{d + \Delta d} [V_{GS} - V_{DSSv} - V_{TH}] = \frac{W \mu_{n0}}{L_g} \frac{\epsilon}{d + \Delta d} \left[(V_{GS} - V_{TH}) V_{DSSv} - \frac{1}{2} V_{DSSv}^2 \right],$$

where we have exploited (3.42). We therefore obtain

$$V_{DSSv} = (V_{GS} - V_{TH} + L_g \mathcal{E}_s) - \sqrt{(V_{GS} - V_{TH})^2 + (L_g \mathcal{E}_s)^2},$$

where $\mathcal{E}_s = v_{n,\text{sat}} / \mu_{n0}$, and the saturation current I_{Ds} due to velocity saturation is

$$I_{Ds} = W v_{n,\text{sat}} \frac{\epsilon}{d + \Delta d} \left[\sqrt{(V_{GS} - V_{TH})^2 + (L_g \mathcal{E}_s)^2} - L_g \mathcal{E}_s \right]. \qquad (3.51)$$

Equation (3.51) defines the saturation transcharacteristics of the PHEMT; by differentiation, we obtain the device transconductance:

$$g_m = \frac{W \mu_{n0}}{L_g} \frac{\epsilon}{d + \Delta d} \frac{V_{GS} - V_{TH}}{\sqrt{\left(\frac{V_{GS} - V_{TH}}{L_g \mathcal{E}_s} \right)^2 + 1}}. \qquad (3.52)$$

For short gate devices we can approximate the maximum transconductance (at $V_{GS} = 0$) by letting $L_g \to 0$ in (3.52), as

$$g_m \approx W v_{n,\text{sat}} \frac{\epsilon}{d + \Delta d}, \qquad (3.53)$$

and the maximum cutoff frequency approximately is

$$2\pi f_T = \frac{g_m}{C_{GS}} \approx W v_{n,\text{sat}} \frac{\epsilon}{d + \Delta d} \times \frac{d + \Delta d}{W L_g \epsilon} = \frac{v_{n,\text{sat}}}{L_g} = \frac{1}{\tau_t}.$$

The expression for g_m in (3.53) suggests a major advantage of PHEMTs over MESFETs and other FETs: the fact that g_m can be increased by reducing the supply layer thickness d (and increasing its doping) without compromising the current (as would happen in a MESFET due to the thinner active layer) or introducing extra impurity scattering due to the increase in doping. The cutoff frequency can be readily interpreted in terms of the transit time of electrons below the gate, τ_t.

3.5.5 High-speed heterojunction bipolar transistors

Conventional bipolar transistors are based on two *pn* junctions with a common side, leading to the so-called *pnp* and *npn* dual structures. In the *npn* device (preferred for analog applications due to the superior transport properties of electrons with respect to holes) the *np* emitter–base junction is in forward bias and injects electrons into the thin *p*-type base layer. A small fraction of such electrons recombine in the base, attracting holes from the base contact to sustain e-h pair recombination and thus giving rise to a small base current I_B. Most of the electrons injected from the emitter are swept by the electric field in the reverse-bias base–collector junction, so as to be ultimately collected by the collector contact as the collector current $I_C = -I_B - I_E = -\alpha I_E$ (we assume all currents are positive entering). The parameter $\alpha < 1$ (but close to 1) is denoted as the common base current gain. We thus have, solving for I_C:

$$I_C = \frac{\alpha}{1-\alpha} I_B = \beta I_B, \quad \beta = \frac{\alpha}{1-\alpha} \gg 1,$$

where β is the common emitter current gain. The common base current gain α can be shown to be expressed as the product of two factors:

$$\alpha = \gamma b,$$

where $\gamma < 1$ is the *emitter efficiency*, $b < 1$ is the *base transport factor*. The emitter efficiency γ accounts for the fact that, in an *npn* transistor, the emitter current has two carrier components – electrons injected from the emitter into the base and holes injected from the base into the emitter. The latter component is useless, since it is not finally collected by the collector, and does not contribute to current gain. The parameter $\gamma < 1$ is the ratio between the useful component of I_E and the total emitter current and should therefore be made as close as possible to unity. On the other hand, the base transport factor b is the fraction of injected electrons that successfully travel through the base to reach the collector. To achieve a large current gain β, α (and therefore *both* γ and b) should be almost unity.

The base transport factor b can be optimized by making the base thickness (length) as small as possible (in conventional transistors, with respect to the carrier diffusion length in the base), or, equivalently, by making the transit time of minority carriers in the base (electrons in an *npn*) much smaller than the minority carriers' lifetime in the base. Epitaxial growth allows reduction of the base thickness to the nanometer scale (although this of course increases the base distributed resistance, unless the emitter width is reduced through self-aligning techniques similar to those exploited in submicrometer CMOS gate technology).

Optimization of the emitter efficiency γ traditionally required the emitter–base junction to be strongly asymmetrical, with $N_{DE} \gg N_{AB}$. Assume for completeness that a bandgap difference $\Delta E_g = E_{gE} - E_{gB}$ may exist between the emitter and the base; it can be shown that

$$\gamma = \frac{1}{1 + \dfrac{N_{AB} D_{hE} W_B N_{vE} N_{cE}}{N_{DE} D_{nB} W_E N_{vB} N_{cB}} \exp\left(-\dfrac{\Delta E_g}{k_B T}\right)}, \tag{3.54}$$

where D_{hE} is the hole diffusivity in the emitter, D_{nB} is the electron diffusivity in the base, W_B and W_E are the base and emitter thickness, respectively, and $N_{v\alpha}$ ($N_{c\alpha}$) are the effective valence (conduction) band state densities in the emitter or base ($\alpha = E, B$). In a conventional, homojunction bipolar $\Delta E_g = 0$ and $N_{\alpha E} = N_{\alpha B}$, $\alpha = c, v$; taking into account that $D_{hE} \approx D_{nB}$ (the hole diffusivity is in fact lower, but of the same order of magnitude as that of the electrons) and that $W_B < W_E$ (but again with similar orders of magnitude), the only way to have γ close to unity is to set

$$\frac{N_{AB}}{N_{DE}} \ll 1,$$

i.e., to make the emitter doping much larger (e.g., one order of magnitude) than the base doping. Since the base doping should be larger than the collector doping,[10] we finally have the conventional bipolar design rule $N_{DE} \gg N_{AB} > N_{DC}$. Assuming $b \approx 1$ and using the Einstein relation $D = (k_B T/q)\,\mu$, we have from (3.54)

$$\beta \approx \left(\frac{\mu_{nB} W_E N_{vB} N_{cB}}{\mu_{hE} W_B N_{vE} N_{cE}}\right) \frac{N_{DE}}{N_{AB}} \exp\left(\frac{\Delta E_g}{k_B T}\right), \tag{3.55}$$

which will be exploited in the rest of the discussion.

As in field-effect transistors, two figures of merit can be introduced to characterize the bipolar transistor speed: the cutoff frequency f_T and the maximum oscillation frequency $f_{\max}$. Faster transistors can generally be obtained by scaling down the device geometry and scaling up the doping level. Increasing the doping level is mandatory in order to properly scale the junction depletion region sizes (roughly, dopings scale as l^{-2}, where l is some characteristic dimension, so that a size scaling down of 100 corresponds approximately to an increase in the doping level of a factor of 100). Conventional, homojunction bipolars are, unfortunately, affected by basic limitations if the cutoff frequency has to be pushed beyond a few gigahertz. In fact, the increase in the doping level in the emitter (the region with the highest doping in the whole device) leads to the so-called *bandgap narrowing* effect, whereby the material gap decreases slightly for high doping according, or example, to the Lanyon–Tuft model [30], [31]:

$$\Delta E_{gE} = \frac{3q^3}{16\pi \epsilon_s^{3/2}} \sqrt{\frac{N_{\alpha E}}{k_B T}},$$

where ϵ_s is the semiconductor permittivity, $N_{\alpha E}$ is the donor or acceptor emitter doping, T is the absolute temperature, and ΔE_{gE} is the emitter gap decrease. For Si at ambient temperature one has

$$\Delta E_{gE} \approx 22.5\sqrt{\frac{N_{\alpha E}}{1 \times 10^{18}}} \quad \text{meV}$$

[10] This is typically needed for two purposes: decreasing the width of the base depletion layer in the base–collector junction and its sensitivity to the applied V_{CE}: this minimizes the so-called Early effect and as a consequence the small-signal output resistance of the transistor, increasing the breakdown voltage of the base–collector junction and therefore the transistor's maximum output power.

with the doping in cm^{-3} units. At $N_{DE} \approx 1 \times 10^{19}$ cm^{-3} the bandgap narrowing is of the order of 100 meV; since, from (3.55),

$$\beta(N_{DE}) \approx \beta(0) \exp\left(-\frac{\Delta E_{gE}}{k_B T}\right),$$

where $\beta(0)$ is the low-doping current gain, we immediately see that even a small bandgap narrowing leads to a decrease in the common emitter current gain. Another important point to consider is the fact that the low base doping required by the condition $N_{DE} \gg N_{AB}$ implies a high value for the input base distributed resistance, which in turn compromises the transistor f_{max}. In conclusion, homojunction bipolars have limited space for optimization for operation above a few gigahertz because this would need to increase the base doping to reduce the input resistance, which would imply extremely large emitter dopings, leading in turn to emitter bandgap narrowing.

A possible way out of this stalemate is obtained by a bipolar design where the bandgap of the base is different (in particular, smaller) than the emitter bandgap. In this case we can satisfy the condition

$$\frac{N_{AB}}{N_{DE}} \exp\left(-\frac{\Delta E_g}{k_B T}\right) \ll 1$$

by exploiting a suitably large ΔE_g, even if $N_{AB} \approx N_{DE}$. This allows a decrease in the base resistance and, as a consequence, an increase in f_{max} (of course, the emitter width also has to be scaled down to submicrometer size to reduce the base resistance). The band diagram of a heterojunction bipolar transistor in the *forward active region* (i.e., when the emitter–base junction is in forward bias and the base–collector junction is in reverse bias) is shown in Fig. 3.28. The increase in the emitter efficiency can also readily be interpreted in terms of a potential barrier opposing the hole back-diffusion into the emitter; in fact, the global effect of the heterojunction is related to the entire bandgap difference ΔE_g, which also has an influence on the relative intrinsic concentrations. In the example shown, the base and collector are narrowgap and the base–emitter heterojunction is abrupt; alternative designs can be obtained through a graded base–emitter heterojunction and by using a widegap collector.

3.5.6 HBT equivalent circuit

Bipolar transistors (both conventional and heterojunction) are junction-based devices for which a few fundamental theoretical relations hold, at least approximately. In the direct active region (we refer to a *npn* device) the emitter current follows the junction law vs. the driving voltage V_{BE}:

$$I_E \approx I_{E0}\left(e^{V_{BE}/V_T} - 1\right),$$

where $V_T = k_B T/q$ (26 mV at 300 K). Since $I_E \approx I_C$ we also have

$$I_C \approx I_{C0}\left(e^{V_{BE}/V_T} - 1\right) \approx I_{C0} e^{V_{BE}/V_T},$$

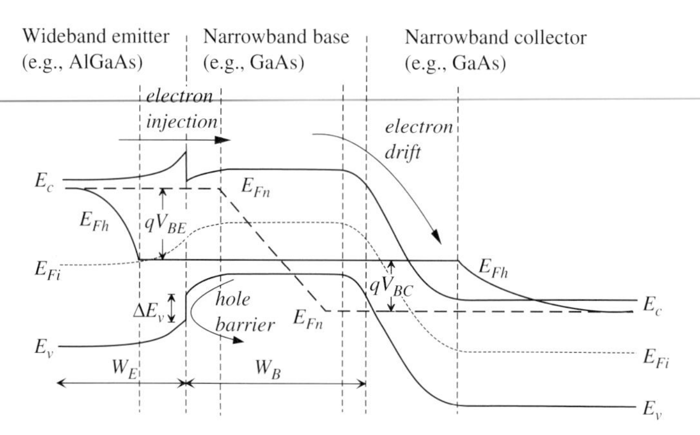

Figure 3.28 Bandstructure of a *npn* heterojunction bipolar in the direct active region. In the example shown, the emitter is widegap, the base and collector are narrowgap.

so that the intrinsic bipolar transconductance g_m can be expressed as

$$g_m = \frac{\partial I_C}{\partial V_{BE}} = \frac{I_C}{V_T}.$$

On the other hand, $I_B = I_C/\beta$; thus, the input differential conductance of the bipolar can be expressed as

$$G_{B'E} = \frac{\partial I_B}{\partial V_{BE}} \approx \frac{I_B}{V_T} = \frac{I_C}{\beta V_T} = \frac{g_m}{\beta}.$$

The intrinsic base has been denoted as B', as opposed to the external (extrinsic) base contact. Due to the input junction structure, the input capacitance of the bipolar, C_{BE}, is the capacitance of a forward-biased *pn* junction, and therefore appears in the equivalent small-signal circuit in parallel to $G_{B'E} = R_{B'E}^{-1}$. Addition of other parasitic capacitances and inductances, together with the output resistance R_{CE} arising from the weak dependence of I_C on V_{CE} (the so-called Early effect) finally leads to the equivalent circuit in Fig. 3.29. The distributed base resistance $R_{BB'}$ models the resistive path between the transistor input and the intrinsic base (the base current flows in the narrow base layer orthogonal to the collector and emitter current densities); due to the very small base thickness such a resistance tends to be large (therefore negatively affecting the device f_{max}) unless the base is suitably doped and the emitter is very narrow. The bipolar cutoff frequency

$$f_T = \frac{g_m}{2\pi C_{BE}} = \frac{I_C}{2\pi C_{BE} V_T}$$

increases with the collector bias current, but for high values of I_C saturation occurs due to *high-injection effects*, which causes a drop of the transistor current gain β. In conclusion, the cutoff frequency typically exhibits a maximum vs. I_C, and decreases for low and high I_C values. Plotting I_C and I_B as a function of V_{BE} in semilog scale we obtain the result shown in Fig. 3.30: for low currents I_B is mainly due to leakage effects and the base current may be larger than the collector current. In an intermediate range

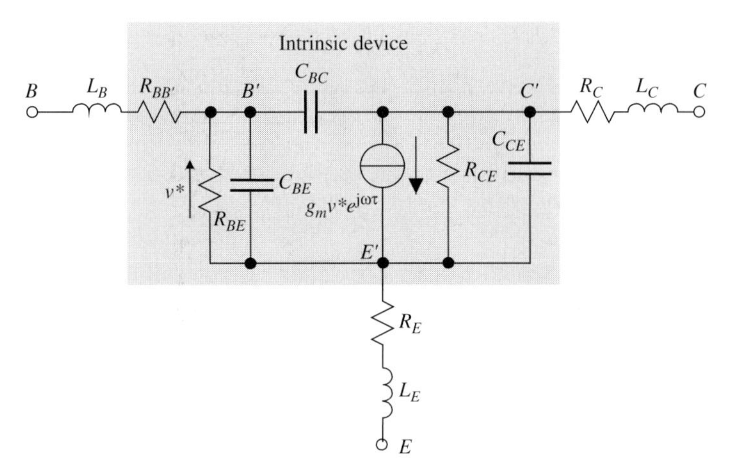

Figure 3.29 Small-signal equivalent circuit of a bipolar transistor. The gray box is the intrinsic device.

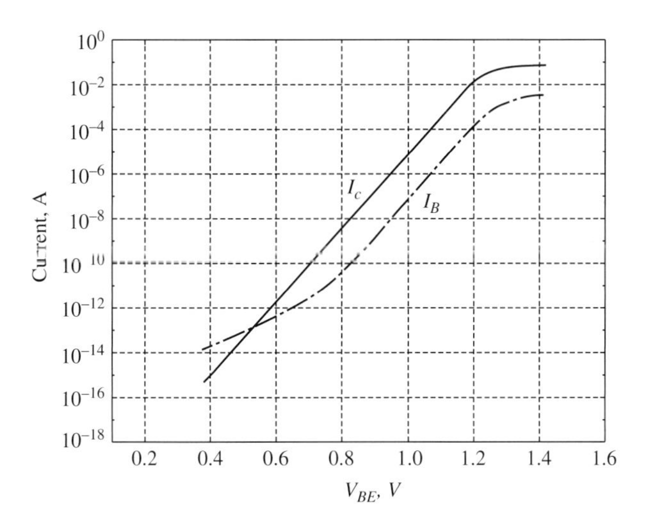

Figure 3.30 Qualitative Gummel plot of heterojunction bipolar transistor.

$I_C = \beta I_B$, as foreseen by ideal transistor operation; finally, for large V_{BE} both currents saturate and β decreases. The diagram in Fig. 3.30 is referred to as the *Gummel plot* of the transistor. InP-based HBTs currently reach cutoff frequencies of the order of several hundreds of gigahertz and maximum oscillation frequencies of the order of 1 THz.

3.5.7 HBT choices and material systems

Heterojunction bipolar transistors can be implemented in a variety of material systems. III-V-based HBTs exploit either GaAs or InP substrates. For the GaAs-based device, the widegap emitter is obtained through lattice-matched AlGaAs (another choice is to exploit lattice-matched InGaP); in AlGaAs the Al content must be kept below 30% since above this value a trap level (the *DX centers*) appears in the widegap material, leading

Table 3.1 Epitaxial structure of InP-based HBT; t is the layer thickness. After [32], Table 1. (©1995 Elsevier)

Layer	Material	t (nm)	Doping (cm^{-3})
Cap 1	InAs	15	$N_D = 3 \times 10^{19}$
Cap 2	$In_{0.53}Ga_{0.47}As$	100	$N_D = 3 \times 10^{19}$
Cap 3	$In_{0.52}Al_{0.48}As$	50	$N_D = 3 \times 10^{19}$
Emitter	$In_{0.52}Al_{0.48}As$	150	$N_D = 1 \times 10^{17}$
Spacer	$In_{0.53}Ga_{0.47}As$	30	undoped
Base	$In_{0.53}Ga_{0.47}As$	150	$N_A = 5 \times 10^{18} - 1.3 \times 10^{19}$
Spacer	$In_{0.53}Ga_{0.47}As$	30	undoped
Collector	$In_{0.53}Al_xGa_{0.47-x}As$	400	$N_D = 7 \times 10^{16}$
Buffer	$In_{0.53}Ga_{0.47}As$	200	$N_D = 3 \times 10^{19}$
Substrate	InP	350 μm	$N_D = 1 \times 10^{18}$

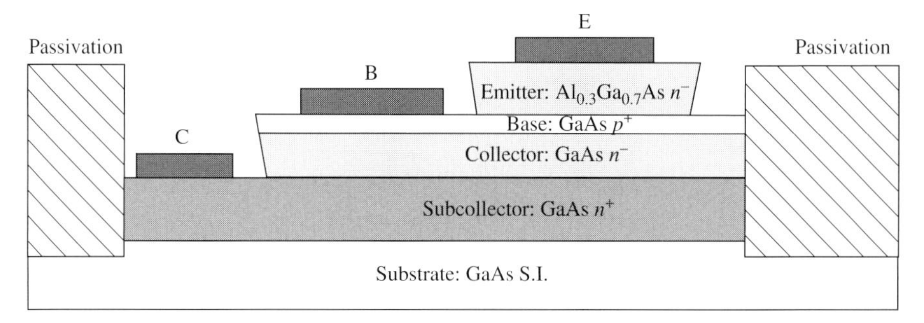

Figure 3.31 Cross section of epitaxial GaAs-based HBT.

to an increase of the leakage currents. The base can be either GaAs (lattice matched) or InGaAs (pseudomorphic), while the collector, subcollector, and substrate are GaAs. A simplified example of GaAs-based epitaxial HBT structure is shown in Fig. 3.31.

Although the GaAs-based technology is well consolidated, still better performance can be obtained with InP-based devices, thanks to the larger bandgap difference between the emitter and the base. Typical devices have an InAlAs emitter, a lattice-matched or pseudomorphic InGaAs base, and an InP collector, subcollector, and substrate. An example of epitaxial structure is reported in Table 3.1 [32].

Of particular interest are HBTs exploiting Si substrates. The SiGe alloy is a narrow-band material, which can be epitaxially grown over a Si substrate in pseudomorphic form. In fact, the $Si_{1-x}Ge_x$ alloy exhibits a lattice mismatch with respect to Si equal to $\approx 4x\%$, where x is the Ge fraction. An example of advanced SiGe HBT is shown in Fig. 3.32; the only SiGe layer is the narrowgap base; the device is made with a self-centered polysilicon emitter process and deep trench isolation between neighboring devices. SiGe HBTs are an interesting alternative to other high-speed transistors for low-power analog circuits and digital circuits, although the process is more complex than for CMOS-based circuits. SiGe bipolars can reach millimeter wave operation,

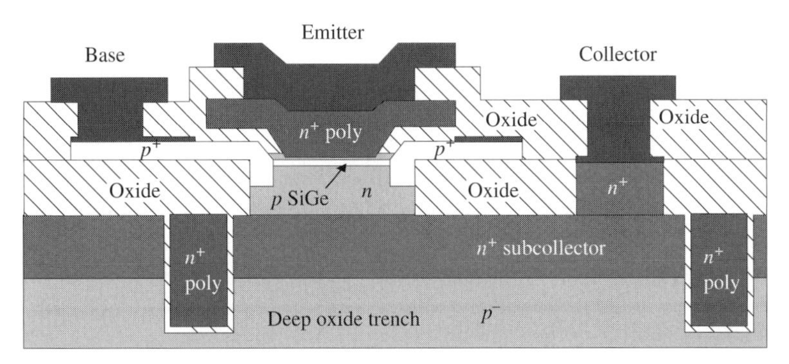

Figure 3.32 Cross section of a SiGe HBT.

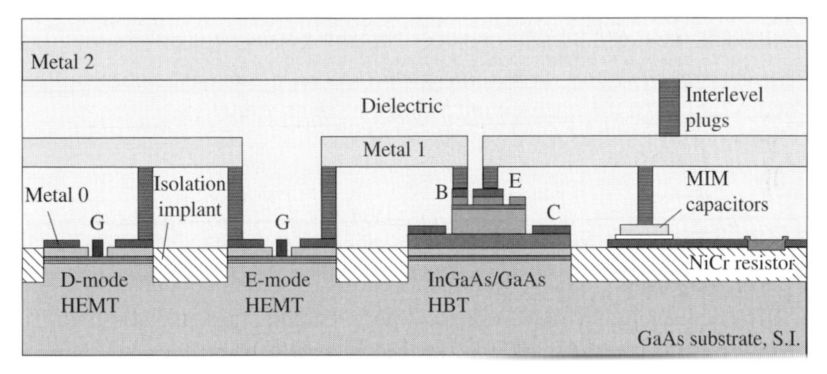

Figure 3.33 Example of mixed bipolar-FET integration on GaAs substrate (not to scale): the cross section shows a depletion-mode HEMT, an enhancement-mode HEMT, a GaAs-based HBT, a NiCr resistor and a MIM capacitor. Adapted from [33], Fig. 1.

but with low breakdown voltages (below 3 V; the breakdown voltage decreases with increasing cutoff frequency, see Ch. 6, Fig. 6.61), while GaAs-based (and, to a lesser extent, InP-based) transistors exhibit larger breakdown voltages and power densities.

Finally, the advent of HBTs has made it possible to extend mixed bipolar-FET process such as the BICMOS (where bipolars are SiGe HBTs) but also HBT-HEMT integration, as shown in Fig. 3.33 [33]. Depletion-mode HEMTs (D-HEMTs) are in the on state for $V_{GS} = 0$, while enhancement-mode HEMTs (E-HEMTs) are in the off state for $V_{GS} = 0$ and begin to conduct for $V_{GS} > 0$. The integration of D-HEMTs and E-HEMTs is important in applications such as fast logical circuits (e.g., multiplexers and demultiplexers for 40 Gbps operation).

3.6 Noise in electron devices

In real circuits, all electrical signals are affected by small-amplitude random fluctuations (the noise); noisy signals can be modeled as the superposition of an ideal, noiseless signal, and of a zero-average stochastic process. Since noise is a signal with nonzero root mean square (r.m.s.) value, it is associated, at a circuit level, with power exchange. Noise power depends on the second-order statistical properties of the signal fluctuations, such as the autocorrelation function, the power spectrum, the r.m.s. value, or the quadratic mean. For simplicity, we will assume all devices considered to be linear or in small-signal operation. In large-signal conditions the treatment becomes more involved, since noise sources are modulated by the instantaneous working point.

3.6.1 Equivalent circuit of noisy *N*-ports

A noisy one-port can be modeled by adding to a noiseless equivalent circuit a random voltage or current source, according to a series (Thévenin) or a parallel (Norton) approach; see Fig. 3.34. The one-port available noise power can be obtained by integrating the available noise power spectrum $p_n(f)$ (W/Hz) over the operating bandwidth, and is related to the open-circuit voltage or short-circuit current power spectra $S_{e_n}(f)$ (V^2/Hz) or $S_{i_n}(f)$ (A^2/Hz) as follows:

$$P_{n,av} = \frac{\overline{v_n^2}}{4R} = \int_0^\infty \frac{S_{e_n}(f)}{4R}\,\mathrm{d}f = \frac{\overline{i_n^2}}{4G} = \int_0^\infty \frac{S_{i_n}(f)}{4G}\,\mathrm{d}f = \int_0^\infty p_n(f)\,\mathrm{d}f,$$

where R and G are the one-port internal resistance or conductance.

The equivalent circuit of noisy two-ports, see Fig. 3.35, includes two correlated random voltage or current sources; the noise model therefore consists of the two-port small-signal parameters (e.g., the impedance, admittance or scattering matrix), and of the generator's power (real) and correlation (complex) spectra. For the series and parallel cases:

$$S_{i_{n1}}(\omega), \quad S_{i_{n1}i_{n2}}(\omega), \quad S_{i_{n2}}(\omega),$$
$$S_{e_{n1}}(\omega), \quad S_{e_{n1}e_{n2}}(\omega), \quad S_{e_{n2}}(\omega).$$

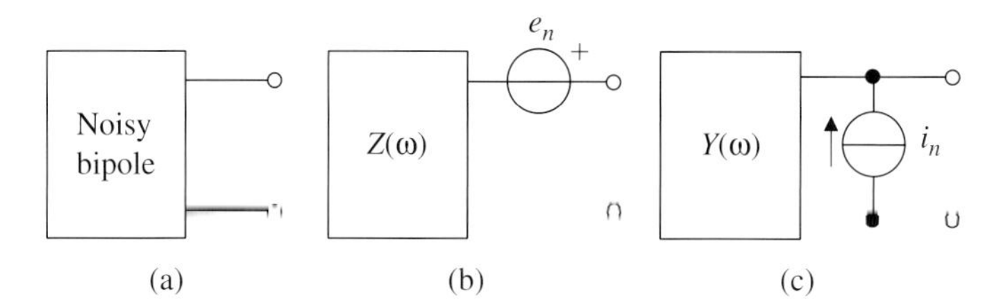

Figure 3.34 Noisy one-port (a); series (b) and parallel (c) equivalent circuit with noise generators.

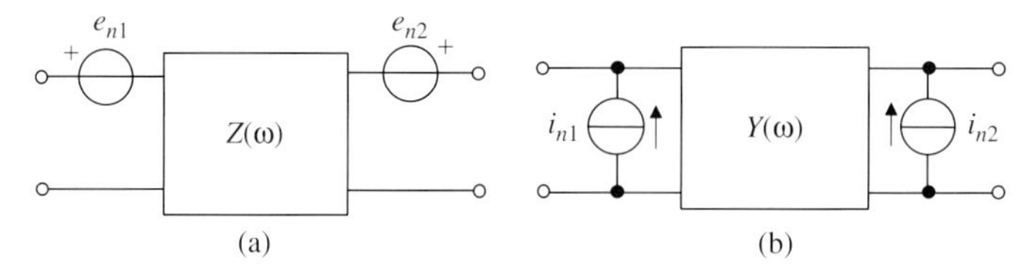

Figure 3.35 Series (a) and parallel (b) equivalent circuit of a noisy two-port with correlated generators.

3.6.2 Noise models of active and passive devices

Semiconductor noise is associated with fluctuations of the carrier velocity and population. *Diffusion noise* is a general term addressing fluctuations associated with transport (reducing to *thermal noise* at or near thermodynamic equilibrium), while *generation–recombination (GR) noise* refers to population fluctuations associated with optical, thermal, or avalanche (Auger) GR mechanisms. In some cases, noise models exist relating the noise behavior to the DC or small-signal parameters of the device. Examples are passive circuits, diodes, and, with some added complexity, bipolar and field-effect transistors.

According to the Nyquist law, the (white, i.e., frequency-independent) power spectra of the open-circuit noise voltage or short-circuit noise current of a *resistor* are

$$S_{v_n}(f) = 4k_B T R \ \text{V}^2/\text{Hz}, \quad S_{i_n}(f) = 4k_B T G \ \text{A}^2/\text{Hz},$$

with available power spectral density

$$p_n(f) = \frac{S_{v_n}(f)}{4R} = \frac{S_{i_n}(f)}{4G} = k_B T \quad \text{W/Hz}.$$

The r.m.s. values of the open-circuit voltage and short-circuit current are, respectively,

$$v_{n,\text{rms}} = \sqrt{\overline{v_n^2}} = \sqrt{4k_B T R B} \ \text{V}, \quad i_{n,\text{rms}} = \sqrt{\overline{i_n^2}} = \sqrt{4k_B T G B} \ \text{A},$$

where B is the system bandwidth ($B \approx B_r$, the bit rate). For a passive RLC one-port, the generalized Nyquist law holds:

$$S_{v_n}(\omega) = 4k_B T \,\text{Re}\,[Z(\omega)] \ \text{V}^2/\text{Hz}, \quad S_{i_n}(\omega) = 4k_B T \,\text{Re}\,[Y(\omega)] \ \text{A}^2/\text{Hz},$$

where Z and Y are the one-port impedance and admittance, respectively.

Diffusion and GR noise from junction devices (like diodes and bipolar transistors) can be modeled through a *shot noise model* according to which noise is described by a Poissonian process, whose power spectrum is proportional to the process mean (DC) value according to Campbell's theorem [34]. For a diode, the power spectrum of the short-circuit noise current is

$$S_{i_n}(\omega) = 2q\,(I_D + 2I_0) \approx 2qI_D,$$

where I_D is the total diode current and I_0 the reverse saturation current. Note that in reverse bias $I_D = -I_0$ and $S_{i_n}(\omega) = 2qI_0$. This model also holds in photodiodes, where the total reverse current is dominated by the photocurrent I_L, while $I_0 \equiv I_d$ is the dark current. The bipolar transistor noise is also described by a shot noise model.

For field-effect transistors (FETs), compact noise models exist relating the short-circuit gate and drain current fluctuation power and correlation spectra to a set of small-signal parameters. An example is the so-called *PRC* or Cappy high-frequency noise model for the intrinsic FET noise [35], yielding

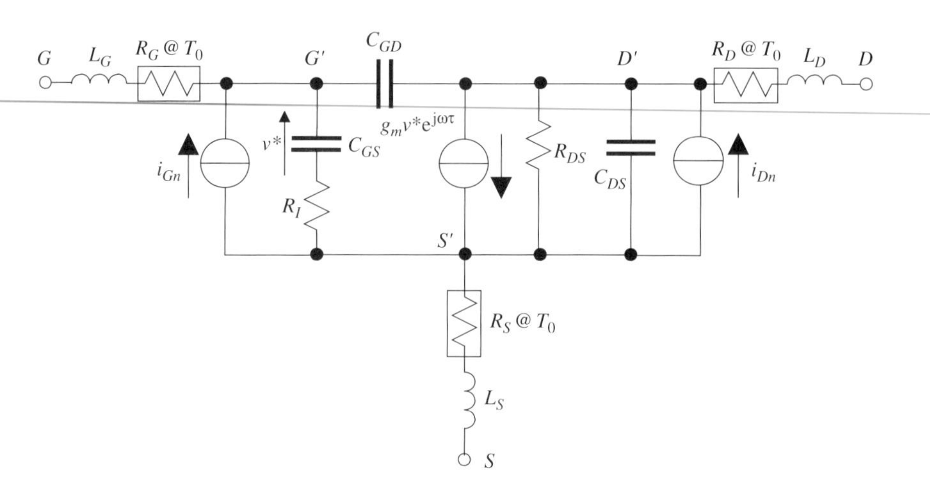

Figure 3.36 Noise model for a high-frequency field-effect transistor.

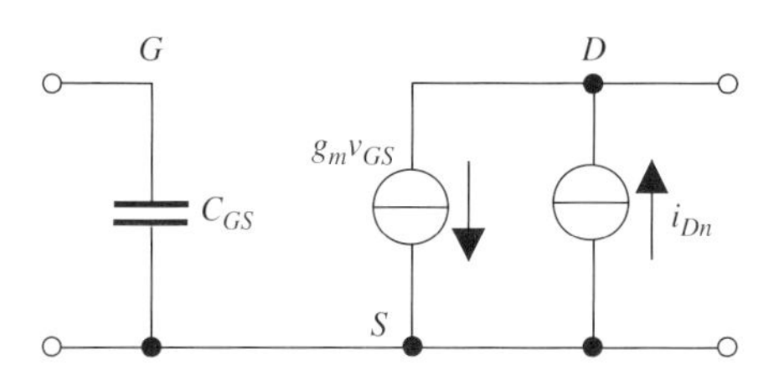

Figure 3.37 Simplified FET intrinsic noise equivalent circuit.

$$S_{i_{Dn}} \approx 4k_B T_0 g_m P \tag{3.56}$$

$$S_{i_{Gn}} \approx 4k_B T_0 \frac{\omega^2 C_{GS}^2}{g_m} R \tag{3.57}$$

$$S_{i_{Gn} i_{Dn}} \approx \mathrm{j} C \sqrt{S_{i_{Dn}} S_{i_{Gn}}}, \tag{3.58}$$

where g_m is the device transconductance, C_{GS} is the input capacitance, according to the equivalent circuit shown in Fig. 3.36, and T_0 is a reference temperature at which the parameters P, R, and C are measured (the default value is the ambient temperature or, more exactly, $T_0 = 290$ K). The circuit also includes thermal noise models for the parasitic resistances. The parameters P, R, and C can be considered as fitting factors; P (also called β in MOSFET modeling) has an ideal value $P = 2/3$ in long-gate devices; for R the ideal value is $R = 5/4$, while C expresses the correlation magnitude (typically $C \approx 0.6 - 0.7$).[11] The imaginary correlation spectrum implies that short-circuit gate and drain current fluctuations are in quadrature due to the capacitive coupling of channel current fluctuations to the gate. At low frequency, the gate noise spectrum $S_{i_{Gn}}$ and the correlation spectrum $S_{i_{Gn} i_{Dn}}$ become negligible, and only the output drain noise (white) is significant. A simplified intrinsic circuit, neglecting parasitics, is shown in Fig. 3.37.

[11] For a more detailed treatment, see, e.g., [36].

3.7 Monolithic and hybrid microwave integrated circuits and optoelectronic integrated circuits

High-speed electronic integrated circuits (ICs) can be implemented through two complementary approaches, the hybrid IC and the monolithic IC. Integrated circuits operating in the microwave range (i.e., up to 30–40 GHz or 40 Gbps) are often denoted as (monolithic) microwave integrated circuits, (M)MICs.

In the hybrid approach, the circuit is realized on a dielectric substrate, integrating all distributed components and, possibly, some lumped components (which may, however, also be inserted as discrete lumped elements through wire bonding or surface mount techniques). In the hybrid approach, the semiconductor active elements are inserted as lumped components and connected again through wirebonding or surface mount.

On the other hand, monolithic circuits integrate all active and passive elements on a semiconductor substrate. While hybrid circuits often exploit distributed components based on transmission line approaches, at least for narrowband applications, in monolithic circuits the lumped approach is preferred, owing to the possibility of reducing the circuit size (lumped components are much smaller than the guided wavelength, while distributed elements, as already recalled, have characteristic sizes of the order of $\lambda_g/4$ at centerband). Monolithic integrated circuits can be based on GaAs, InP, or Si substrates and as active elements may exploit FETs or bipolars (typically HBTs).

According to the transmission medium used, we may have microstrip or coplanar integrated circuits. Microstrip circuits are more compact in size due to the lower ground plane, but require a precise control of the dielectric thickness, while coplanar circuits are preferred at very high frequency (mm waves) or, as in mixed optoelectronic circuits, because the coplanar waveguide is immediately compatible with the layout of some optoelectronic devices (e.g., electrooptic modulators). Microstrip circuits in fact only allow for straightforward connection of series elements (parallel elements are required to reach the lower ground plane, often by etching a hole in the substrate, the so-called *via hole*), while a coplanar circuit allows for the connection of both series and parallel elements. Finally, the on-wafer high-frequency characterization requires connection of the integrated circuit to the measurement setup through *coplanar probes*; to this purpose, coplanar ground planes must be made available (e.g., through via holes) at the circuit input and output.

A final step toward monolithic integration is the integration of electronic circuits with optoelectronic devices, leading to the so-called optoelectronic integrated circuits (OEICs). In practice, the development of OEICs is confined to a few specific components, such as detectors and the related front-end amplifier stages. In such cases, the monolithic integration drastically reduces the interconnection parasitics; however, the implementation of OEIC integration is fraught with problems associated with the difficulty of realizing both transistors and optoelectronic components on the same semiconductor substrate with possibly different epitaxial layer structures. A few techniques for realization of OEICs are reported in Fig. 3.38; the regrowth option allows us to achieve the maximum degree of freedom in terms of the E and OE epitaxial structure, but can create reliability problems, while in cases (a) and (b) etching is required to

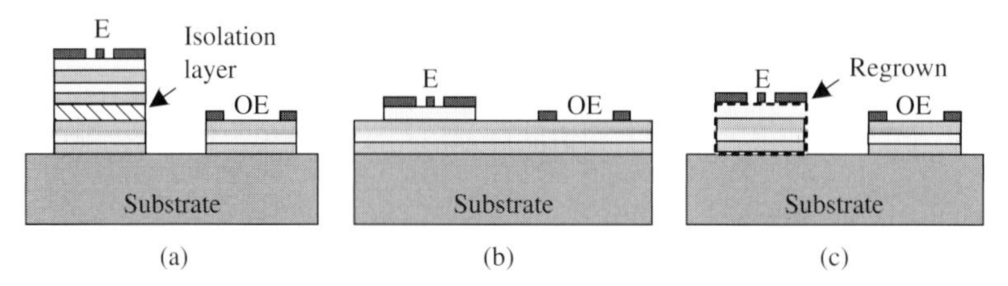

Figure 3.38 Techniques for optoelectronic integration (OEICs): (a) growth of the electronic part (E) through a unique epitaxial structure in which the optoelectronic (OE) and E part have been separated through isolation; (b) growth of the E and the OE part using the same layers; (c) epitaxial regrowth of the E part after the OE has been realized. Adapted from [37], Fig. 12.2.

remove the extra layers in the OE part. Note that most OE devices are junction devices requiring doped layers, while E devices should be grown on a semi-insulating substrate. In some cases, the same epitaxial structure can be used for the E and OE parts (see Fig. 3.38(b)), with some compromise on layer optimization. Note that optoelectronic integration with long-wavelength detectors or sources also requires the use of InP-based electronics rather than of the more convenient GaAs-based electronics.

A qualitative example of the layout of a hybrid or monolithic integrated circuit in microstrip or coplanar technology can be introduced as a simple, single-stage open-loop amplifier with two lumped bias Ts and input and output matching section. Figure 3.39 presents two possible circuit implementations, with distributed matching sections (a), typically (but not necessarily) hybrid, or with lumped matching sections (b), usually monolithic. Figure 3.39 (a) also shows the equivalent circuit of two microstrip to coaxial connectors, modeled through a low-pass filter.

Figure 3.40 shows a simplified hybrid microstrip implementation of the single-stage amplifier, in which the input and output matching sections have been separately realized on two different ceramic substrates. The active device is introduced in packaged form and exploits as the ground plane (and also as the heat sink) a ridge in the metal package. Bias Ts are implemented using as series inductors the parasitic wire bonding inductance; chip capacitors connected to the package as the ground are also part of the bias T. The same amplifier can be implemented in monolithic form, as shown in Fig. 3.41, through lumped input and output matching sections. Via holes are used quite liberally to provide local grounding, in addition to the ground pads needed for the input and output coplanar connectors. The circuit shown is unpackaged. Finally, Fig. 3.42 shows a coplanar waveguide monolithic implementation exploiting distributed matching sections. Due to the typically small size of MMICs, such a solution is realistic only if the frequency is high enough to make distributed elements compact, e.g., for millimeter wave operation. While coplanar waveguides easily allow for open- and short-circuit line stubs (i.e., short pieces of transmission lines for the implementation of the distributed matching sections), the layout is globally less compact, and ground planes have to be connected together by air bridges to suppress spurious modes where the two ground planes are at different potentials. Both in the microstrip and in the coplanar layout, a source air bridge is used in the active component. The active device

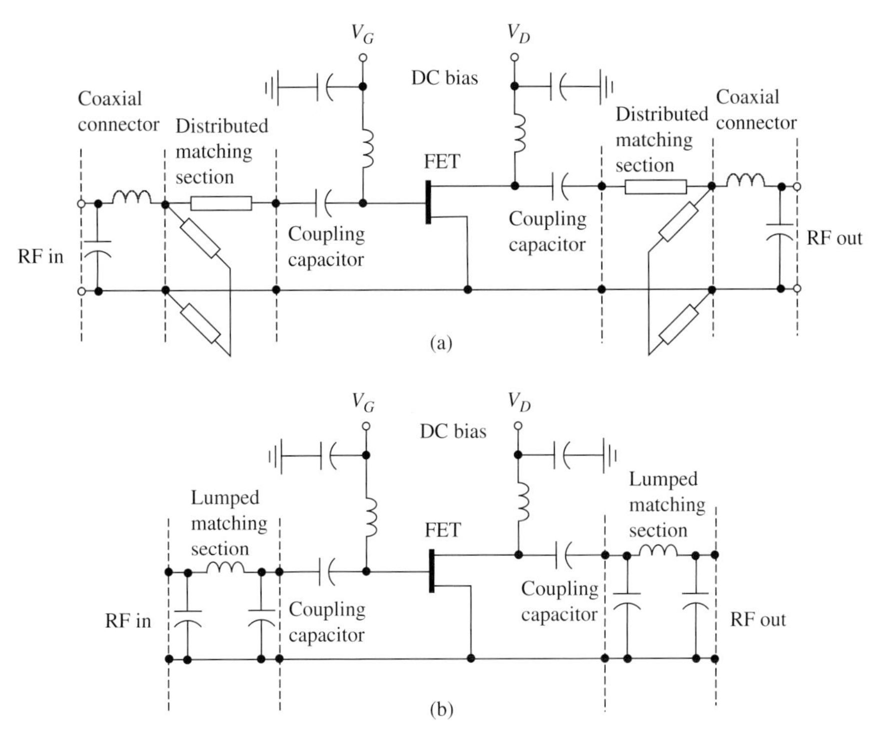

Figure 3.39 Circuit implementation of a simple one-stage amplifier with input and output matching sections: (a) with distributed elements, mainly hybrid; (b) with lumped elements, mainly integrated.

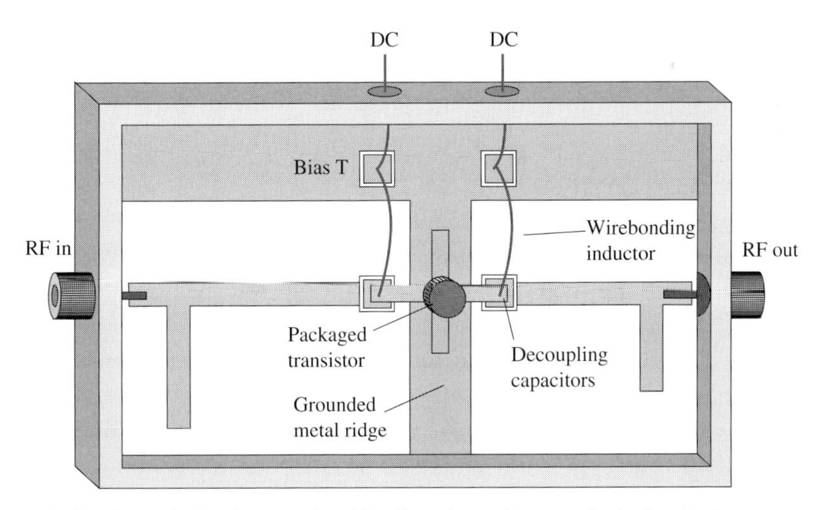

Figure 3.40 Hybrid microstrip implementation (distributed matching section) of a single-stage amplifier.

layout has been kept the same in the microstrip and coplanar versions, although the difference in operation frequency (microwave vs. millimeter-wave) also has an impact on the FET layout (e.g., on the length of the gate fingers, which is decreasing with increasing frequency).

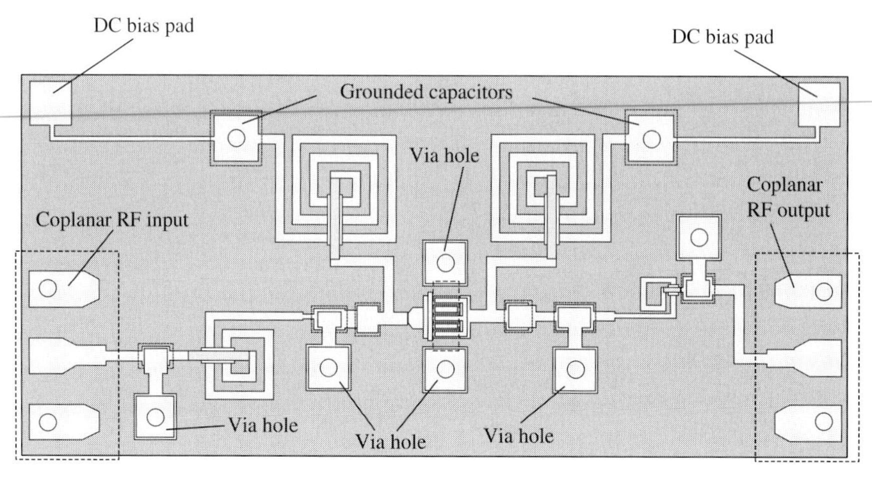

Figure 3.41 Example of monolithic microstrip implementation of single-stage microwave amplifier (lumped matching sections). The layout is qualitative, with elements only approximately to scale. The chip size is $\approx$ 0.5 mm $\times$ 1 mm. The source air bridge area is shown with a dashed line.

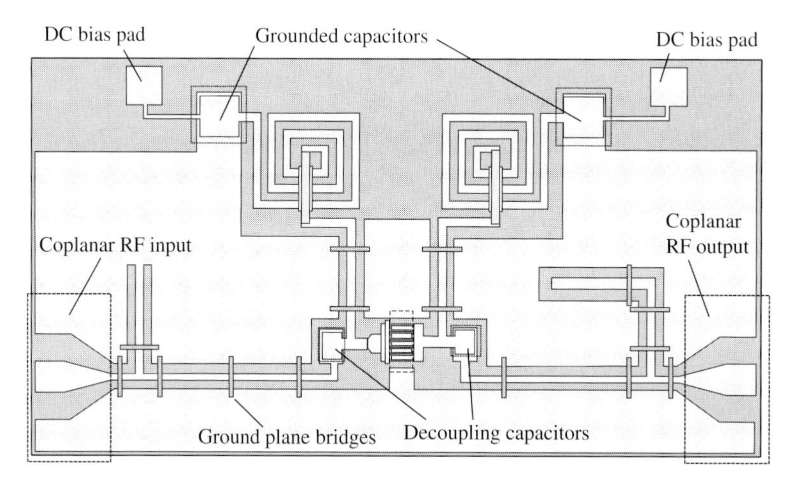

Figure 3.42 Example of monolithic coplanar implementation of millimeter-wave single-stage amplifier (distributed matching sections). The layout is qualitative, with elements only approximately to scale. The chip size is $\approx$ 1 mm $\times$ 2 mm. The source air bridge area is shown with a dashed line.

Both hybrid and monolithic integrated circuits are typically packaged (in a metal or dielectric enclosure) and connected to other subsystems through electrical connectors. RF connectors are often coaxial, with diameter decreasing with increasing frequency, and they correspond to microstrip–coaxial transitions. Sometimes, unpackaged MMICs are directly connected to a hybrid circuit that is in turn packaged, as shown in Fig. 3.43.

A few coaxial to microstrip transitions ordered by increasing frequency operating range are shown in Fig. 3.44. Soldered connectors can be used up to a few gigahertz, while at higher frequency wire bonding or ribbon bonding (having lower parasitic p.u.l.

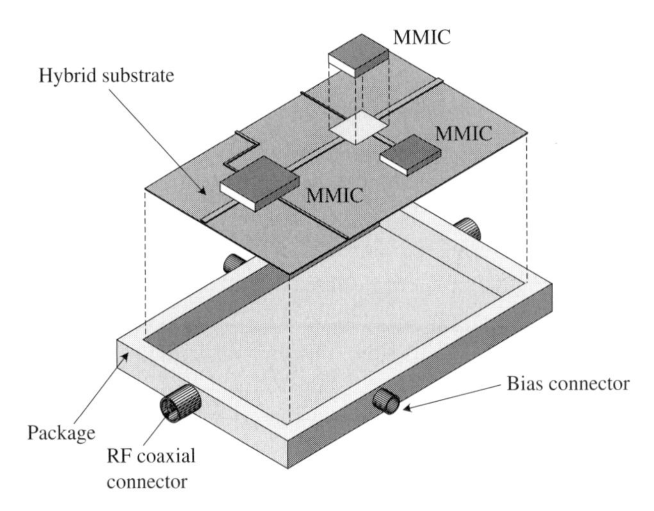

Figure 3.43 MMIC on-chip connection into a hybrid circuit, which in turn is packaged and connected to other circuits through coaxial connectors.

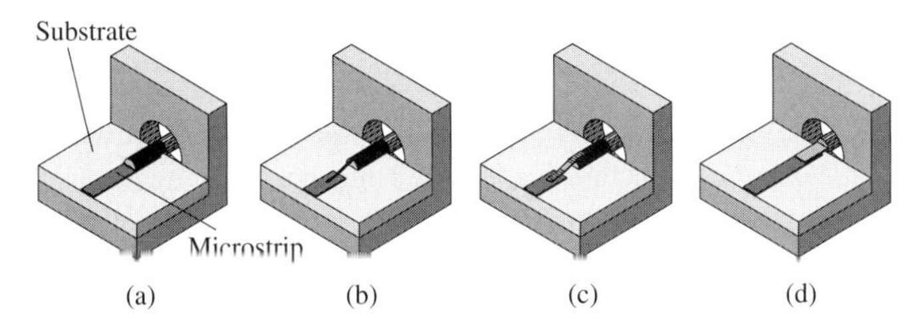

Figure 3.44 Microstrip to coaxial connectors: (a) soldered; (b) wire bonding; (c) ribbon bonding; (d) unsoldered connector.

inductance) are exploited. High-frequency connectors operating, e.g., at 40 GHz and beyond are not based on wire or ribbon bonding, but rather on contact connectors.[12]

3.8 Questions and problems

3.8.1 Questions

1. Identify the digital and analog electronic subsystems in a high-speed optical transceiver (transmitter and receiver). What components are more critical?
2. Define the frequency ranges corresponding to RF and microwaves. In which frequency range does the electronic part of a 10 Gbps system operate?

[12] High-frequency coaxial connectors are denoted by conventional names, some of them referring to the frequency band they were initially meant to cover. Thus we have the K connectors (up to 40 GHz), the V connectors (up to 60 GHz) and the so-called W1 connectors (Anritsu name, 1 mm radius) up to 110 GHz, which is currently the highest frequency exploited in standard instrumentation. The connector size decreases with increasing frequency.

3. Explain the difference between a lumped-parameter and a distributed-parameter high-speed circuit.
4. Explain why impedance or admittance parameters are difficult to measure at microwave frequencies.
5. An impedance has scattering parameter 0.5 with respect to 50 Ω. What is the impedance value?
6. An impedance has scattering parameter j with respect to 50 Ω. What is the impedance value? What component (lumped) could implement it?
7. Explain the meaning of conjugate matching.
8. Can a non-TEM waveguide be exploited to transmit a digital signal in baseband? Justify the answer.
9. Explain the difference between a TEM and a quasi-TEM waveguide.
10. Sketch a microstrip and a coplanar line.
11. Explain why in a coplanar waveguide losses are not monotonic vs. the characteristic impedance, whereas they are in a microstrip.
12. What is the effective permittivity in a quasi-TEM line?
13. A quasi-TEM line is on a dielectric with permittivity 2. The effective permittivity is 3.5. Explain why something is wrong with the information provided.
14. Sketch the behavior of the line attenuation vs. frequency for a quasi-TEM line.
15. Explain the skin effect in a quasi-TEM transmission line.
16. Explain why lumped inductors for MMICs are not implemented with magnetic materials.
17. Justify the fact that the frequency range of a lumped inductor is narrower when the inductance is higher.
18. Justify the shape of the velocity–field curve in GaAs and InP. Why can the maximum carrier velocity be higher than the saturation carrier velocity?
19. Explain why a heterostructure can be profitably used in a field-effect transistor.
20. Explain the motivation behind the evolution from HEMTs to PHEMTs.
21. Suppose you want to exploit MOSFETs for the electronic part of a 40 Gbps system, and assume to need a cutoff frequency at least 3 times the maximum operating frequency. What should the approximate gate length be?
22. Describe the structure of a InP-based PHEMT.
23. Explain the operation of the HBT.
24. Justify the interest in SiGe HBTs for high-speed optical communication systems.
25. In a SiGe HBT-based circuit, where do you find Ge?
26. What is the purpose of the bias T in a high-speed circuit?
27. Explain why the conjugate matching approach for an amplifier design cannot be applied as such to a wideband amplifier. What is the purpose of the input and output matching network in this case? (Hint: a wideband amplifier must have flat gain, but the MAG decreases with frequency.)
28. What are the main differences between a coplanar and a microstrip circuit layout?

3.8.2 Problems

1. The conductivity of a 2 μm thick conductor is $\sigma = 1 \times 10^5$ S/m. Evaluate the frequency at which the skin-effect penetration depth is equal to the conductor thickness. Repeat the problem using a metal conductivity ($\sigma = 1 \times 10^7$ S/m).
2. A transmission line has (high-frequency regime) characteristic impedance $Z_0 = 30\ \Omega$, effective refractive index $n_{\mathrm{eff}} = 2$, loss equal to 0.2 dB/mm at 1 GHz. Supposing that the loss is due to metal losses, evaluate the per-unit-length parameters of the line. What is the p.u.l. resistance at 100 GHz?
3. What is (approximately) the impedance Z_0 of a microstrip with substrate permittivity $\epsilon_r = 10$, substrate thickness 0.5 mm, strip width 0.5 mm?
4. Evaluate the input impedance of a transmission line of length l, propagation constant β, attenuation α, closed by a load Z_L and with characteristic impedance Z_0. What happens if $Z_L = Z_0$ or if Z_L is a short or an open? Is it always true that the input impedance of an infinitely long line is Z_0?
5. Evaluate the transmission line parameter expression (3.18) in the high-frequency approximation. (Hint: assume that the characteristic impedance is real and equal to the limit for $f \to \infty$ and expand the complex propagation constant.)
6. A PHEMT with gate periphery $W = 0.5$ mm has a process transconductance of 200 mS/mm. The gate length is $L_g = 150$ nm, the AlGaAs thickness is $d = 100$ nm and the relative permittivity is $\epsilon_r = 13$. Estimate the cutoff frequency f_T and, assuming a total input resistance $R_{in} = 10\ \Omega$ and an output resistance $R_{DS} = 100\ \Omega$, the maximum oscillation frequency.
7. A PHEMT with $g_m = 200$ mS, $f_T = 200$ GHz, $P = 0.9$ is connected at the input with a generator with internal resistance $R_g = 50\ \Omega$. Evaluate the ratio between the total output short-circuit drain noise current due to the cumulative effect of the noisy input generator and of the PHEMT noise and the same quantity due to the noisy input generator only. This parameter is called the device noise figure (NF). How does the NF behave as a function of frequency? Suppose that the system bandwidth is $B = 1$ Hz and compute the NF at low frequency and at $f_T/2$.

4 Detectors

4.1 Photodetector basics

Photodetectors (PDs) are the first block in the system receiver chain.[1] Their purpose is to convert an optical (analog or digital) signal into an electrical signal, typically a current (the *photocurrent*, i_L). The physical mechanism at the basis of semiconductor detectors is the *optical generation* of electron–hole (e-h) pairs through the *absorption* of incident photons. Photogenerated e-h pairs are then separated and collected to the external circuit by an electric field. Such a collecting field can be induced by an external voltage bias in a reverse-biased junction (as in *pn*, *pin*, and Schottky detectors), or in bulk (as in photoconductors). In some cases, a third step (after photogeneration and collection) is present: the photocurrent is *amplified* through external or built-in gain processes.

In the absence of illumination, detectors still have an output current, the *dark current* i_d. The photocurrent is often linearly related to the input optical power $p_{in}(t)$ as $i_L(t) = \Re p_{in}(t)$, where $\Re$ is the detector *responsivity* (A/W); however, for large input optical power the generated photocarriers ultimately screen the collecting electrical field, leading to current saturation. Since the photocurrent depends on the amount of photogenerated carriers and, therefore, on the material absorption profile vs. the wavelength λ of the modulated optical carrier, the responsivity will depend on λ, $\Re = \Re(\lambda)$, with bandpass behavior. However, the optical bandwidth of most detectors is wide.

Finally, the quasi-static, memoriless relation $i_L = \Re p_{in}$ holds only when the optical power varies slowly with time, or, for time-harmonic input optical power, with a modulation frequency lower than the device cutoff frequency. When the input power varies too rapidly with time, the output current does not follow its instantaneous value, because of low-pass or delay mechanisms, such as the effect of the device capacitance and the effect of the *transit time* that photocarriers experience before being collected. We can therefore define, in small-signal conditions and for a harmonic input optical power, a frequency domain (complex) responsivity $\Re(\omega)$ relating the amplitude and phase of the small-signal photocurrent component at ω to the amplitude and phase of the harmonic input optical power. The function $\Re(\omega)$ is typically low-pass.

[1] The input optical signal may be optically amplified before detection by a semiconductor optical amplifier (SOA); this closely corresponds to the architecture of RF receivers, where the first function is low-noise amplification, followed by signal demodulation through a mixer or (in the simplest case) an envelope detector. Photodetectors in direct detection receivers operate, in fact, as envelope detectors.

The output of the PD is a current, i.e., from the electrical standpoint the PD is a high-output impedance device. In order to drive the receiver circuits connected to the detector, the PD current has to be converted into a voltage through a load resistance or, better, a transimpedance amplifier (TIA) providing power amplification. Besides the photocurrent and dark current, the PD also generates an output *noise current*, that can be modeled, in junction devices, according to a shot noise approach. PD noise can be negligible composed with the noise from the front-end amplifier, or dominant; in the former case the receiver operates in the *thermal noise limit*, in the latter it operates in the *shot noise limit*.

4.2 Photodetector structures

Photogenerated carriers are collected by an electric field $\mathcal{E}$, induced by an applied voltage in a bulk semiconductor or in a reverse-bias junction. Possible semiconductor-based detector structures are:

- Bulk: photoresistors (photoconductors);
- *pn* junction-based: *pn* photodiodes, *pin* photodiodes, avalanche photodiodes (APD), phototransistors;
- Metal–semiconductor junction based: Schottky barrier photodiodes, metal–semiconductor–metal (MSM) photodiodes.

Among nonsemiconductor device choices, we have *vacuum detectors* and *organic detectors* (somewhat similar to semiconductor detectors), which will not be discussed here.

In photoconductors (Fig. 4.1(a)), photocarrier generation takes place in a neutral or lightly doped bulk semiconductor, i.e., in a resistive region. The photogenerated excess

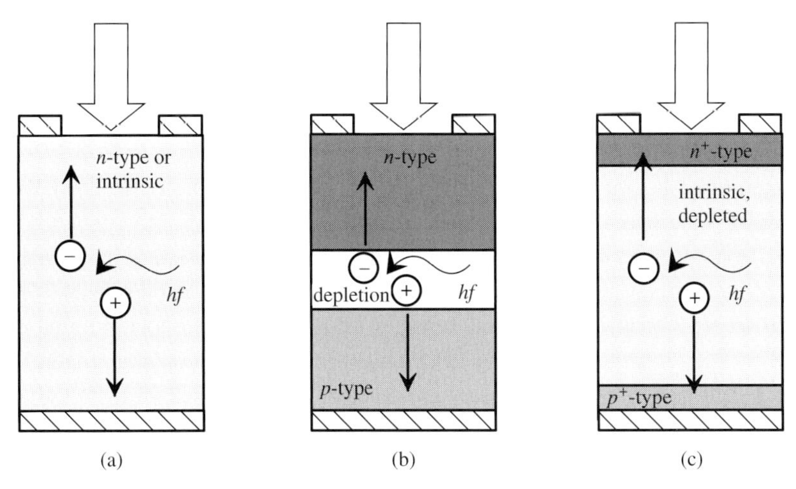

Figure 4.1 Simplified scheme of (a) photoresistor; (b) *pn* photodiode; (c) *pin* photodiode. The bias circuit is not shown.

carrier density modulates the conductivity, thus leading to current modulation under constant bias voltage. Carriers are removed to the external circuit by the almost uniform electric field induced by the voltage bias. The device turns out to have high gain, but the bandwidth is limited by the photocarrier lifetime. The dark current is very high (but can be capacitively decoupled from the load) and (thermal) noise is typically large; notice that the photoresistor is always working in the thermal noise limit. Photoconductors are simple, low-speed devices, not very well suited for high-performance telecom applications.

Junction-based devices exploit the photocarrier generation in a reverse-bias *pn*, *pin*, or Schottky junction. Photocarriers are removed to the external circuit by the junction reverse electric field, thus increasing the diode reverse saturation current (which, in the absence of illumination, is the dark current). A somewhat naive implementation of the concept is the *pn* photodiode, see Fig. 4.1(b); owing to the very small width of the depletion region, photons are also absorbed in the adjacent diffusion regions, leading to poor frequency response, limited by transit time and lifetime (besides RC capacitive effects). As for other junction photodetectors, the dark current is small and noise is mainly shot noise.

Optimization of the photodiode structure leads to *pin* photodiodes (Fig. 4.1(c)), in which e-h generation occurs in a large intrinsic region sandwiched between high-doping layers. The width of the intrinsic layer can be made large enough with respect to the absorption length to render the related photocurrent contribution dominant over the photocurrent originating from the diffusion regions; moreover, in PiN heterostructure devices the doped layers are widegap and do not absorb light altogether at the operating wavelength. Photocarriers are removed to the external circuit by the almost uniform electric field induced in the intrinsic region by the applied reverse bias. The frequency response is limited by transit time and RC effects; noise is shot noise and the dark current is low. The *pin* photodiode is a high-performance device, with achievable bandwidths in excess of 40 GHz; due to the unit gain, the sensitivity is not outstanding.

Figure 4.2(a) shows a simplified scheme of the avalanche photodiode (or APD). In the version shown, the structure is a *pin* diode to which an additional *pn* junction has been added with a highly doped n side. Photogenerated electrons in the intrinsic layer are removed to the high-field region associated with the n^+p junction depletion layer, and undergo avalanche multiplication. From the standpoint of responsivity, the APD can be interpreted as a *pin* device to which a photocarrier gain (multiplication) block has been added. The increase in responsivity is, however, obtained at the expense of increased noise (multiplied shot noise and excess noise) and a reduction of bandwidth (due to the additional delay introduced by the avalanche buildup). Despite the increase in noise, the larger responsivity typically leads, at least in a thermal noise-limited receiver, to a better (i.e., smaller) sensitivity. APDs were traditionally implemented in Si or Ge, but in recent years more advanced heterostructure devices based on InGaAs have appeared as competitors of *pin* photodiodes, at least up to 10 Gbps.

In phototransistors (Fig. 4.2(b)) photocarriers are generated in the base of a bipolar or heterojunction bipolar transistor. The resulting base current is amplified by the transistor common emitter current gain β, thus leading to an output collector current $I_C \approx \beta \Re P_{in}$

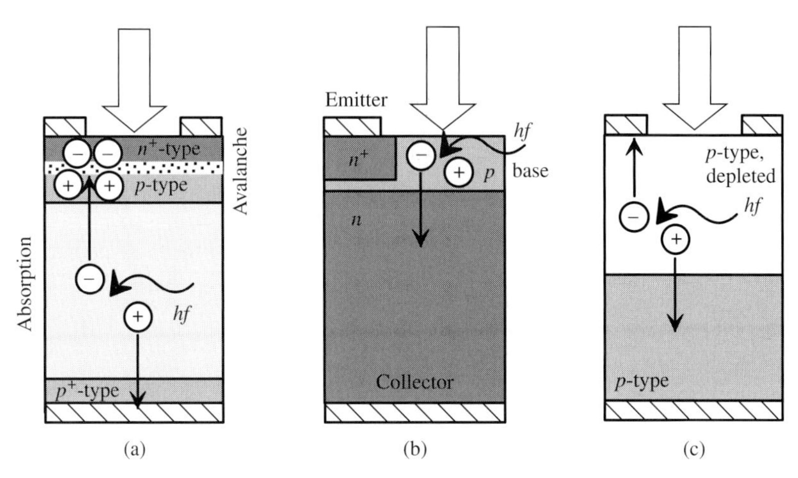

Figure 4.2 Simplified scheme of (a) avalanche photodetector; (b) phototransistor; (c) Schottky photodiode. The bias circuit is not shown.

where $\Re = I_B/P_{in}$ is the responsivity, referred to the base current I_B. The phototransistor has high gain, shot noise of the amplified collector current (but no excess noise as in APDs); the bandwidth used to be limited by the low cutoff frequency of conventional bipolar transistors but, at least in the laboratories, high-speed heterojunction phototransistors reach speeds in excess of 10 Gbps.

Schottky or MSM detectors are based on Schottky (metal–semiconductor) junctions in reverse bias (Fig. 4.2(c)); the operation is somewhat similar to that of *pn* or *pin* diodes (although an additional photogeneration mechanism is introduced by carriers photoexcited from the metal into the semiconductor) and the device structure is simpler; however, illumination of the device area is an issue, due to metal absorption, thus requiring interdigitated electrode structures. Moreover, the frequency response is often affected by slow tails, which make the device less appealing for high-speed applications.

4.3 Photodetector materials

While the collection of photogenerated e-h pairs requires an electric field to be applied to photocarriers – a step whose implementation depends on the detailed device structure – the very possibility of carrier generation is related to the material absorption profile and, in particular, to the *absorption threshold*, i.e. the minimum energy photons must have to be absorbed. From (2.1), absorbed photons must satisfy the condition:

$$E_{ph} = \hbar\omega \geq E_g \quad \longrightarrow \quad \lambda\,[\mu\text{m}] \leq \frac{1.24}{E_g\,[\text{eV}]}.$$

Both direct-bandgap (GaAs, InGaAs, InP,...) and indirect-bandgap (Si, Ge) semiconductors can be exploited in PDs; direct-bandgap materials typically have higher absorption, which results in smaller absorption volumes and higher speed.

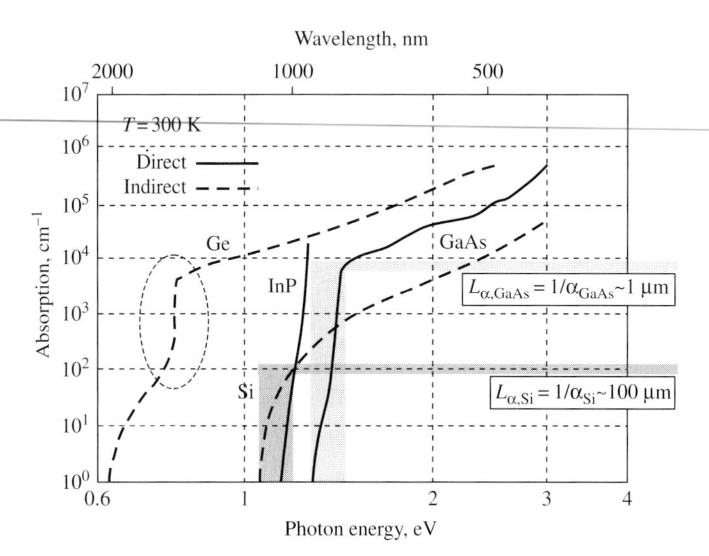

Figure 4.3 Absorption coefficients of some direct-bandgap (InP, GaAs) and indirect-bandgap (Ge, Si) semiconductors and comparison between the absorption lengths of Si and GaAs at the same distance from the absorption edge; note the Ge composite behavior.

In fact, the exponential decrease of the optical power in the absorption region can be expressed as

$$P_{in}(x) = P_{in}(0)\exp(-\alpha x) = P_{in}(0)\exp(-x/L_\alpha), \tag{4.1}$$

where $L_\alpha = 1/\alpha$ is the *absorption length*, i.e., the distance wherein light attenuates by a factor $\exp(-1) = 0.37$ or $-10\log_{10}(0.37) = 4.31$ dB. To absorb light almost entirely, the thickness d of the detector's active (or absorption) region must be suitably larger than L_α, or at least $d \approx L_\alpha$. Considering the absorption length of typical direct-bandgap and indirect-bandgap detectors (see Fig. 4.3), we see that in a direct-bandgap material the thickness of the absorption region should be $d \approx 1\,\mu$m, while in indirect-bandgap materials (e.g., in Si) $d \approx 100\,\mu$m. A large d corresponds to a high *transit time*, the delay with which photocarriers are collected; this is a major limitation to the detector speed.

In junction-based detectors (pn, pin and Schottky), however, the (depleted) absorption region acts as a capacitor, whose capacitance $C \propto d^{-1}$ limits the detector speed, due to RC cutoff. Increasing d, the transit-time-limited speed *decreases*, while the RC-limited speed *increases*, thus leading to the need for a design trade-off, see Section 4.9.5. Si-based detectors are typically limited by transit time, while in direct-bandgap detectors a compromise is sought between transit time and RC cutoff.

Example 4.1: Compare the transit times for a direct-bandgap detector (assume $\alpha_1 \approx 10^4\,\text{cm}^{-1}$) and an indirect-bandgap detector ($\alpha_2 \approx 10^2\,\text{cm}^{-1}$). Suppose that photocarriers travel at saturation speed ($v \approx 10^7$ cm/s).

We assume $\alpha L \approx 1$ or $L = L_\alpha = 1/\alpha$, with a transit time $\tau = L/v = 1/\alpha v$. In the two cases considered, we obtain $\tau_1 = (\alpha_1 v)^{-1} = (10^4 \cdot 10^7)^{-1} = 10\,$ps;

$\tau_2 = (\alpha_2 v)^{-1} = (10^2 \cdot 10^7)^{-1} = 1$ ns. The transit-time-limited bandwidth will be of the order of $f_T = (2\pi\tau)^{-1}$, i.e., $f_{T1} = (2\pi\tau_1)^{-1} = (2\pi \cdot 10^{-11})^{-1} = 16$ GHz and $f_{T2} = (2\pi\tau_2)^{-1} = (2\pi \cdot 10^{-9})^{-1} = 0.16$ GHz, respectively.

To introduce an overview on the practically important photodetector materials, let us consider the relation between the energy gaps of two semiconductors operating as the *emitter* and the *receiver* in the same optical link. Since the emitted photon energy is of the order of the emitting material energy gap $E_{g,e}$, this has to be larger than the receiver material energy gap $E_{g,r}$, i.e., $E_{g,e} \geq E_{g,r}$. In principle, detectors have a wide optical bandwidth, but this is limited by the behavior of the responsivity versus the photon energy; such behavior suggests that the photon energy should be suitably larger than the receiver gap, but not too large. Figure 4.4 shows the energy gap of many relevant semiconductors and semiconductor alloys as a function of the lattice constant, and includes three horizontal lines, one corresponding to AlGaAs/GaAs emission (around 0.8 μm) and the others to InGaAsP emission (1.3 and 1.55 μm, exploited in long-haul optical systems). AlGaAs/GaAs sources can exploit, as a receiver, materials such as Si, Ge, and GaAs- or InP-based alloys. For long-wavelength sources (1.3 and 1.55 μm), possible receiver materials are Ge (or SiGe with a small Si fraction), InGaAs, InGaAsP, InGaAsSb, and CdHgTe. Due to the relative immaturity of antimonides and of CdHgTe, the receiver materials of choice are, in this case, InGaAs and (to a certain extent) Ge. Si-based receivers cannot be exploited at all for long-wavelength detection.

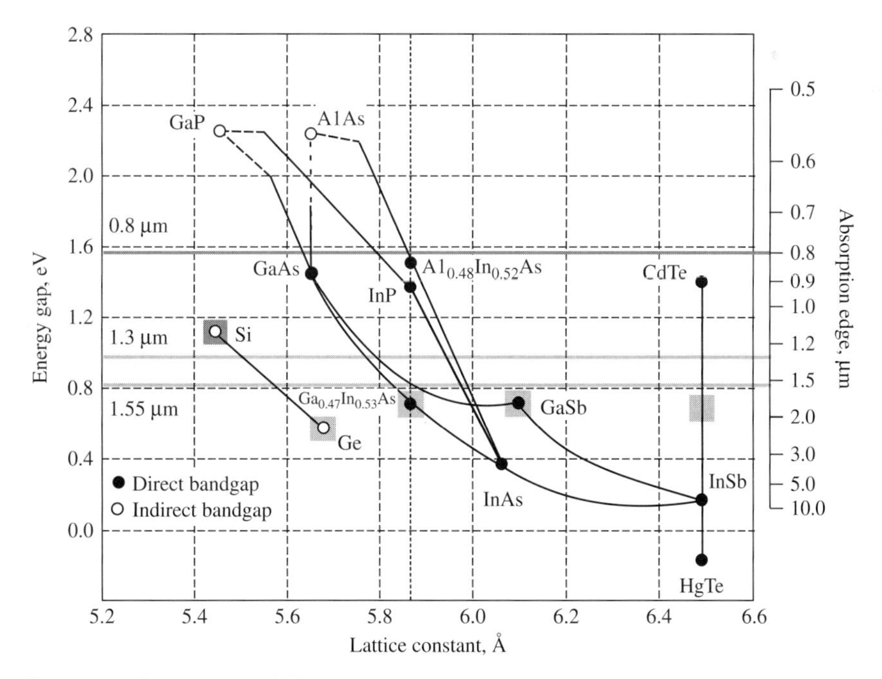

Figure 4.4 Detector and source materials.

The most popular material for 1.55 and 1.3 μm high-performance detectors is probably InGaAs. This ternary alloy is a direct-bandgap material, with tunable E_g; it can be grown through lattice-matched heteroepitaxy on InP substrates, with composition $In_{0.53}Ga_{0.47}As$; in this case, of course, the bandgap is uniquely defined. The corresponding direct-bandgap quaternary alloy (InGaAsP) can be grown lattice-matched to InP substrates, with variable gap. Although InGaAsP is the material of choice for long-wavelength sources, in PDs the ternary alloy is usually preferred. Antimonide-based alloys, such as the direct-bandgap ternary AlGaSb, can be grown on GaSb; however, the technology of antimonides (including the quaternary InGaAsSb) is not competitive at present with the InP technology.

Detectors for short-haul (e.g., local area network, LAN) applications (0.8 μm wavelength) can be based on AlGaAs/GaAs. This direct-bandgap (up to 45% Al mole fraction) ternary alloy can be grown lattice-matched on GaAs substrates and has a mature technology and a comparatively low cost. Due to the very similar avalanche ionization coefficients of electrons and holes, however, this material is not well suited to avalanche detectors; see Section 4.12. Silicon used to be a material of choice for 0.8 μm detectors, and obviously has the advantage of mature technology and low cost, besides being well suited to avalanche detector design due to the very different hole and electron avalanche ionization coefficients. However, the comparatively low absorption confines silicon to low-speed applications, at least if conventional (vertical) photodetector structures are used.

Germanium is an indirect bandgap material ($E_{g,i} \approx 0.66\,\text{eV}$) with a larger direct bandgap ($E_{g,d} \approx 0.8\,\text{eV}$). The corresponding absorption edge wavelengths are

$$\lambda_i \approx \frac{1.24}{0.66} = 1.87\,\mu\text{m}, \quad \lambda_d \approx \frac{1.24}{0.8} = 1.55\,\mu\text{m}.$$

The secondary, direct absorption edge therefore allows for the detection of long-wavelength radiation (in particular 1.3 μm). The substrate technology is mature enough and the avalanche coefficients of electrons and holes are quite different (so that good quality, low noise avalanche detectors can be developed). Emerging SiGe technologies have given this material interesting perspectives for integration on a Si substrate; unfortunately, Si-rich alloys, with low mismatch with respect to the Si substrate, have optical properties close to those of Si. Ge-rich alloys, on the other hand, exhibit large mismatch with the Si substrate, and the related growth technology on Si substrates still poses problems.

HgCdTe (cadmium mercury telluride, also called MERCAD or MERCATEL) is a ternary alloy whose components are a semiconductor (CdTe) and a compound with metal bandstructure (HgTe). Assuming that a metal has negative energy gap (i.e., the valence band edge is above the conduction band edge, and the two bands overlap), the MERCAD gap can be tuned down to almost zero, making this material a suitable alloy for FIR detection. The substrate of choice is CdTe. Despite the complex technology, this material still is very popular for FIR detectors (much less for long-wavelength communication detectors), although the small gap and resulting large intrinsic population require low-temperature operation.

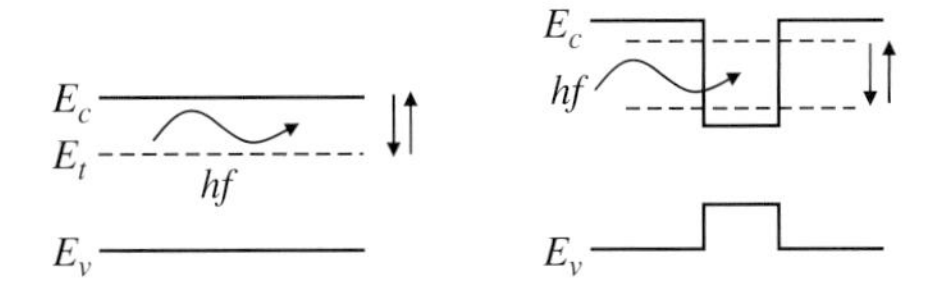

Figure 4.5 Extrinsic absorption in bulk, involving a trap level (left), and in a quantum well (right), as an intersubband transition.

4.3.1 Extrinsic and QW detectors

The detectors discussed so far, called *intrinsic detectors*, are based on band-to-band absorption processes, and their responsivity vanishes for photon energies below the material absorption edge E_g. This limitation can be somewhat overcome in the so-called *extrinsic detectors*. Bulk extrinsic detectors make use of low-energy transitions between trap levels and the conduction or valence band, and can therefore detect FIR radiation, albeit with low responsivity (Fig. 4.5, left). The same result can be achieved by exploiting low-energy intersubband transitions, as in quantum well (QW) detectors; see Fig. 4.5, right. Compared with low-gap HgCdTe FIR detectors, the QW solution allows for a larger energy gap, and therefore is able to relax constraints on low-temperature operation related to the need to lower the intrinsic carrier concentration (however, low-T operation is also needed to suppress background noise). Another promising area for QW intersubband detectors based on widegap semiconductors (e.g., GaN) may be the development of long-wavelength receivers.

4.4 Photodetector parameters

4.4.1 PD constitutive relation

From the electrical standpoint, photodetectors are *one-ports* with a second (optical) input port. Assume that the (modulated) input optical power around wavelength λ is $p_{in}(t)$ and that the output current is i_{PD} (including the photocurrent i_L and the dark current i_d); the PD is generally characterized by the constitutive relation

$$i_{PD}(t) = f\left(p_{in}(t), v_{PD}(t); \frac{\mathrm{d}}{\mathrm{d}t}, \lambda\right), \tag{4.2}$$

where v_{PD} is the detector voltage, and λ is the wavelength of the optical carrier (influencing the material absorption and therefore the amount of generated photocarriers). The time derivative implies that the relation between p_{in}, v_{PD} and the output current i_{PD} will not, in general, be memoriless. In fact, as already mentioned, the photocurrent response vs. p_{in} is lowpass and, moreover, detectors typically include reactive electrical elements (such as junction capacitances and stray inductances). The output current (taken as positive entering, see Fig. 4.6, left) can be decomposed as

$$i_{PD} = i_L + i_d,$$

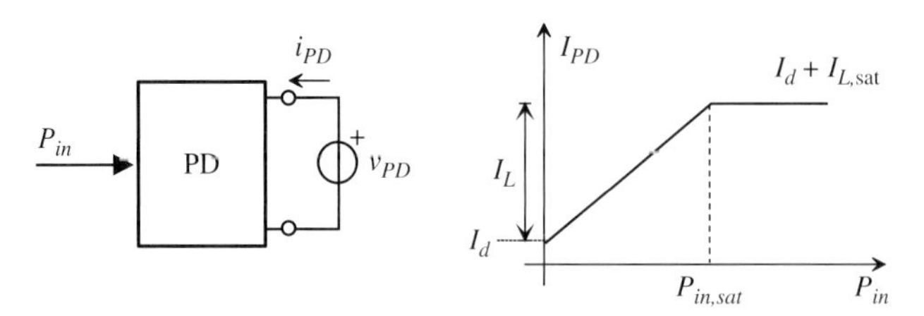

Figure 4.6 Photodetector input–output diagram (left); photodetector current as a function of the optical power in stationary (DC) or quasi-stationary conditions (right).

where the dark current i_d (the current in the absence of optical power) and the photocurrent i_L (the current contribution due to incident light) can be defined as

$$i_d = f\left(0, v_{PD}(t); \frac{\mathrm{d}}{\mathrm{d}t}, \lambda\right) \tag{4.3}$$

$$i_L = f\left(p_{in}(t), v_{PD}(t); \frac{\mathrm{d}}{\mathrm{d}t}, \lambda\right) - i_d. \tag{4.4}$$

In DC stationary conditions we have

$$I_{PD} = f(P_{in}, V_{PD}; 0, \lambda) = I_L + I_d,$$

where $I_d = f(0, V_{PD}; 0, \lambda)$ is the DC dark current, $I_L = f(P_{in}, V_{PD}; 0, \lambda) - I_d$ is the DC photocurrent.

Equation (4.2) implies that, in general, the relation between the photodiode current and the optical power is *nonlinear* (i.e., the photocurrent is not simply proportional to the input optical power) and *dispersive* or with memory (according to the already-mentioned low-pass behavior). However, for a slowly varying $p_{in}(t)$, (4.2) reduces to a memoriless (quasi-static) relation that can be often linearly approximated as

$$i_{PD}(t) = i_L + i_d \approx \mathfrak{R}(\lambda, v_{PD})p_{in}(t) + i_d(v_{PD}), \tag{4.5}$$

where $\mathfrak{R}$ is the photodetector *responsivity*, in general (as already mentioned) a function of the optical carrier wavelength, but also, in some devices (e.g., in avalanche detectors), of the photodiode output voltage. In many detectors (e.g., photodiodes in reverse bias), both the dark current and the responsivity are, however, virtually independent of v_{PD}, and the dark current (the diode reverse saturation current) is small, so that

$$i_{PD}(t) = \mathfrak{R}(\lambda)p_{in}(t) + I_d \approx \mathfrak{R}(\lambda)p_{in}(t).$$

The linear dependence for the photocurrent i_L, see Fig. 4.6, right, typically holds for input optical powers $P_{in} < P_{in,\text{sat}}$ where $P_{in,\text{sat}}$ is the saturation optical power at which the photocurrent saturates at $I_{L,\text{sat}}$ (the total photodetector current saturates at $I_{PD,\text{sat}} = I_{L,\text{sat}} + I_d \approx I_{L,\text{sat}}$). The decrease in the responsivity at high optical power and the ultimate saturation (Fig. 4.6, right) is due to device-specific intrinsic effects (e.g., space-charge screening of the electric field collecting photocarriers); however, the total photodetector current can also saturate due to the circuit loading conditions.

4.4.2 Responsivity and quantum efficiency

The detector photocurrent (and therefore the responsivity) can in principle be derived by integrating the optical generation rate G_o, see (2.43), over the device active volume:

$$I_L = q \int_V G_o(\underline{r}, P_{in}) \, \mathrm{d}\underline{r}.$$

From I_L the device responsivity can in turn be obtained as

$$\mathfrak{R} = \frac{I_L}{P_{in}} \text{ or } \mathfrak{R}_{\text{diff}} = \frac{\mathrm{d}I_L}{\mathrm{d}P_{in}}.$$

The first definition is called the *incremental* responsivity, the second the *differential* responsivity; they coincide if the current–power characteristic is linear, see (4.5).

A more elementary derivation of G_o can, however, be useful to derive an ideal, best-case limit to the detector sensitivity. As a first step, we directly relate G_o to the optical power as follows. Differentiating (4.1) with respect to x and defining $\widetilde{P}_{in} = P_{in}/A$ as the optical power density (W/m^2), with A the detection area, we have

$$\frac{\mathrm{d}\widetilde{P}_{in}(x)}{\mathrm{d}x} = -\alpha \widetilde{P}_{in}(x) \rightarrow \frac{\text{Energy lost due to absorption}}{t \cdot V} = -\frac{\Delta \widetilde{P}_{in}}{\Delta x} = \alpha \widetilde{P}_{in}.$$

Dividing by the photon energy $E_{ph} = \hbar\omega$, we obtain

$$\begin{aligned} \frac{(\text{Energy lost}) \,/\, (t \cdot V)}{\text{Photon energy } \hbar\omega} &= \frac{\alpha \widetilde{P}_{in}}{\hbar\omega} = \frac{\text{Number of photons absorbed}}{t \cdot V} \\ &= \frac{\text{Number of e-h pairs generated}}{t \cdot V} = G_o. \end{aligned}$$

That is,

$$G_o = \frac{\alpha \widetilde{P}_{in}}{\hbar\omega}, \tag{4.6}$$

where G_o is the optical generation rate associated with the external photon flux, i.e., the number of e-h pairs generated per unit time and volume. Equation (4.6) coincides (with $P_{op} \equiv \widetilde{P}_{in}$) with the result in (2.43), derived from perturbation theory. Since the optical power density decreases exponentially with x, the same behavior is followed by the optical generation rate:

$$G_o(x) = \frac{\alpha \widetilde{P}_{in}(x)}{\hbar\omega} = \frac{\alpha \widetilde{P}_{in}(0)}{\hbar\omega} \exp(-x/L_\alpha) = G_o(0) \exp(-x/L_\alpha).$$

Assume now that all the incoming optical power is absorbed, and all of the generated e-h pairs are collected as a current flowing in the external circuit (each photogenerated e-h pair, once collected in the external circuit, causes *one* electron to flow through it).

We have

$$\frac{\text{Number of electrons in the external circuit}}{t} = \frac{I_L}{q}$$

$$= V \cdot \frac{\text{Number of e-h pairs generated}}{t \cdot V} = \frac{\text{Number of photons absorbed}}{t \cdot V}$$

$$= A \int_0^\infty G_o(x)\,\mathrm{d}x = A \int_0^\infty \frac{\alpha \widetilde{P}_{in}(x)}{\hbar\omega}\,\mathrm{d}x = -\frac{A}{\hbar\omega}\int_0^\infty \frac{\mathrm{d}\widetilde{P}_{in}(x)}{\mathrm{d}x}\,\mathrm{d}x \approx \frac{P_{in}(0)}{\hbar\omega},$$

i.e.,

$$\frac{I_L}{q} = \frac{P_{in}(0)}{\hbar\omega},$$

where $P_{in}(0)$ is the incident power. From this simplified model, it follows that the photocurrent indeed depends linearly on $P_{in}(0)$ through the responsivity $\mathfrak{R}$:

$$I_L = \frac{q}{\hbar\omega} P_{in}(0) = \mathfrak{R} P_{in}(0). \tag{4.7}$$

Using power and current densities, we similarly have $J_L = \mathfrak{R}\widetilde{P}_{in}(0)$. The responsivity is thus measured in A/W (as the current–power ratio, or, equivalently, the current density–power density ratio).

The above analysis is based on the assumption that each incident photon generates an electron in the external circuit, and leads to an ideal, best-case value for the responsivity. From (4.7) one has

$$\mathfrak{R} = \frac{q}{\hbar\omega} = \frac{q}{E_{ph}}, \tag{4.8}$$

holding when all of the incident photons are absorbed and converted into the external short-circuit current. In such best-case conditions, the responsivity has a maximum $\mathfrak{R}_{\text{max}}$ as a function of the photon energy that can be derived as follows: for photon energies below the absorption threshold, the responsivity is zero; assuming a sharp increase of α above threshold, from (4.8) one has that $\mathfrak{R}$ is maximum for $E_{ph} \approx E_g$, i.e.,

$$\mathfrak{R}_{\text{max}} \approx \frac{q}{E_g} = \frac{1}{E_g\ [\text{eV}]} \approx \frac{\lambda[\mu\text{m}]}{1.24}. \tag{4.9}$$

For $E_{ph} > E_g$, the responsivity ideally decreases with increasing E_{ph} as

$$\mathfrak{R}(E_{ph}) \approx \mathfrak{R}_{\text{max}} \frac{E_g}{E_{ph}},$$

according to the behavior in Fig. 4.7; for energies close to the threshold, the responsivity approximately follows the absorption coefficient, while for higher energies it decreases as the inverse of the photon energy.[2]

[2] From a physical standpoint, the decrease of responsivity with increasing photon energy is related to the fact that an energetic photon having an excess energy with respect to the threshold $\approx E_g$ still generates just one electron–hole pair, the extra energy being dissipated through phonon emission, i.e., heat. For example, a photon with $E_{ph} = 2E_g$ still generates the same current as a photon with $E_{ph} \approx E_g$, at the expense of twice the optical power; as a consequence, the responsivity is one half of the maximum value at $E_{ph} \approx E_g$.

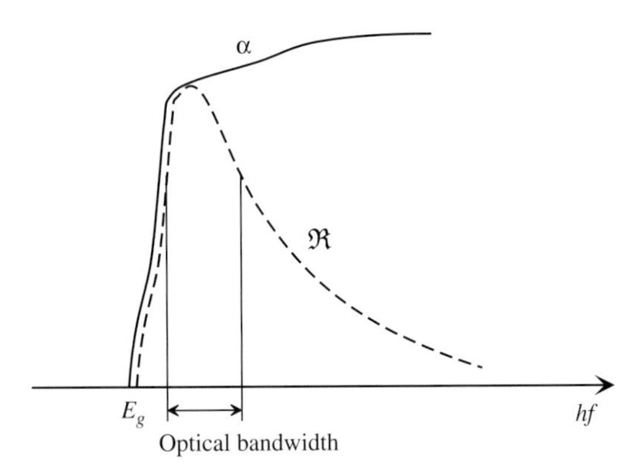

Figure 4.7 Behavior of absorption α and responsivity $\mathfrak{R}$ (in arbitrary units, normalized so that $\mathfrak{R} \approx \alpha$ near E_g) vs. the photon energy.

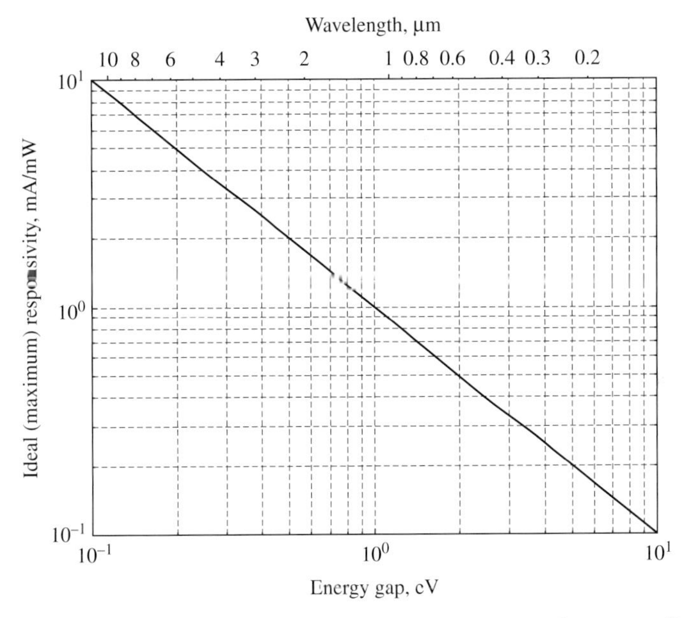

Figure 4.8 Ideal maximum responsivity $1/E_g[\mathrm{eV}]$ vs. energy gap and corresponding wavelength.

Due to the inverse dependence of $\mathfrak{R}_{\mathrm{max}}$ on the energy gap, see Fig. 4.8, very large maximum responsivities are achieved in far infrared detectors; for long-wavelength infrared detectors for optical communications the maximum responsivity has an order of magnitude of 1 A/W. Responsivities in excess of the maximum ideal value can be achieved in devices with internal gain.

In real devices, the number of electrons flowing in the external circuit can be substantially lower than the number of incident photons, leading to a responsivity smaller than the ideal value given by (4.8). This happens because the incident light has to undergo a number of steps before being converted into a current, see Fig. 4.9:

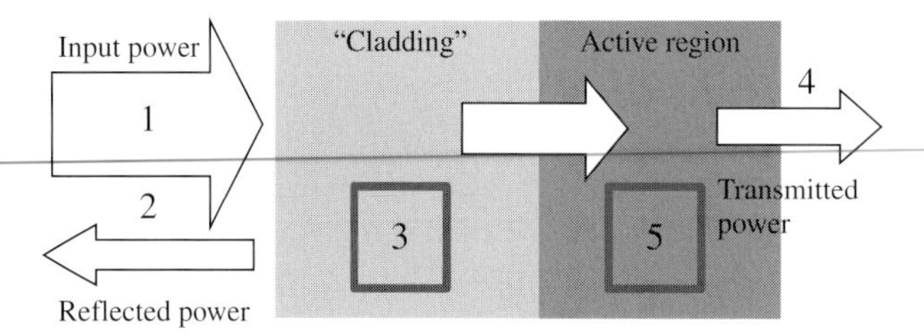

Figure 4.9 Nonideality mechanisms affecting the detector response.

1. The optical power $P_{in}(0)$ is incident on the photodetector.
2. Part of the power is reflected at the PD interface due to dielectric mismatch.
3. Part of the power is absorbed in regions of where it does not contribute to useful output current.
4. Part of the power is transmitted through the PD without being absorbed.
5. Finally, part of the power is absorbed and yields a useful current component.

Additional detector figures of merit are the internal quantum efficiency η_Q and the external (or device) quantum efficiency η_x. The internal quantum efficiency is defined as

$$\eta_Q = \frac{\text{generated pairs}}{\text{photons reaching the active region}};$$

typically $\eta_Q \approx 1$. On the other hand, η_x is directly related to the responsivity, since

$$\eta_x = \frac{\text{collected pairs}}{\text{incident photons}} = \frac{I_L/q}{P_{in}/\hbar\omega} = \frac{\hbar\omega}{q}\Re < \eta_Q. \tag{4.10}$$

In general, $\eta_x \leq 1$ in the absence of gain. If we assume ideal operation $\eta_x = 1 \equiv \eta_Q$ and we obtain again (4.8) and (4.9).

The responsivity of real devices is smaller than the ideal value, but clearly exhibits the theoretical bandpass behavior versus the operating wavelength. An example is reported in Fig. 4.10, referring to a Si *pin* photodiode and a *PiN* InGaAs photodiode.[3] The measured responsivity shows an abrupt increase corresponding to the absorption edge, and then decreases with increasing energy. For the InGaAs *PiN* diode, the abrupt fall of the responsivity at short wavelength is due to the fact that, for such energies, absorption also takes place in the surrounding cladding widegap layers, in particular in the upper cladding layer (producing no useful output current). Both devices exhibit a very large optical bandwidth, well in excess of 200 nm. The same remark applies to a Ge photodiode realized with an innovative Ge on Si process [38]; Fig. 4.10 also shows, for comparison, the maximum responsivity $\Re_{\max}$ of an ideal detector with unit external efficiency, see (4.9); the ideal value is remarkably close to the peak responsivity of real devices. From the material standpoint, InGaAs offers excellent properties at long wavelength (1.55 and 1.3 μm), and so does Ge; Si, as already remarked, is only effective at short wavelength, e.g., around 0.8 μm.

[3] Data from Hamamatsu Photonics web site [39]; the S5971 Si diode has a cutoff frequency of 1 GHz, the G8196 InGaAs diode has a cutoff frequency of 3 GHz.

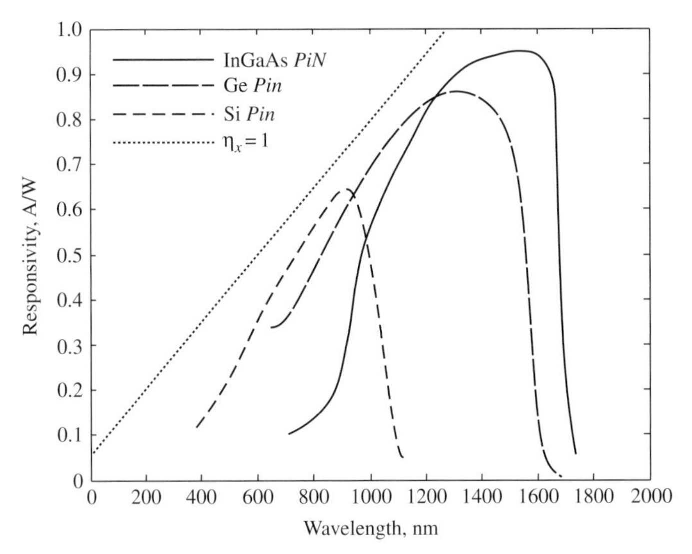

Figure 4.10 Responsivity versus wavelength for a Si homojunction, an InGaAs heterojunction high-speed *pin* photodiode (Hamamatsu Photonics, [39]), and a Ge-on-Si photodiode (data from [38], Fig. 1). The dotted curve is the ideal case with external quantum efficiency $\eta_x = 1$ and responsivity $\Re = q/(\hbar\omega)$; see (4.10).

4.4.3 PD electrical bandwidth and equivalent circuit

The responsivity concept can readily be extended to describe the detector frequency response, assuming that the device operates linearly (or is linearized, e.g., around a DC condition). We start from the constitutive relation in (4.2), for brevity dropping the wavelength dependence:

$$i_{PD}(t) = f\left(p_{in}(t), v_{PD}(t), \frac{\mathrm{d}}{\mathrm{d}t}\right).$$

We separate the DC and signal component (the subscript 0 denotes the DC working point):

$$P_{in} = P_{in,0} + \hat{p}_{in}(t), \quad V_{PD} = V_{PD,0} + \hat{v}_{PD}(t), \quad I_{PD} = I_{PD,0} + \hat{\imath}_{PD}(t),$$

and, assuming sinusoidal modulation of light, we associate phasors with the signal components as follows:

$$\hat{p}_{in}(t) = \mathrm{Re}\left(\hat{P}_{in}\mathrm{e}^{\mathrm{j}\omega t}\right), \quad \hat{v}_{PD}(t) = \mathrm{Re}\left(\hat{V}_{PD}\mathrm{e}^{\mathrm{j}\omega t}\right), \quad \hat{\imath}_{PD}(t) = \mathrm{Re}\left(\hat{I}_{PD}\mathrm{e}^{\mathrm{j}\omega_m t}\right),$$

where ω is the light angular modulation frequency. Linearizing around a DC working point we obtain

$$I_{PD,0} + \hat{\imath}_{PD}(t) = \underbrace{f\left(P_{in,0}, V_{PD,0}, 0\right)}_{I_{PD,0}} + \left.\frac{\partial f(\mathrm{d}/\mathrm{d}t)}{\partial p_{in}}\right|_0 \hat{p}_{in}(t) + \left.\frac{\partial f(\mathrm{d}/\mathrm{d}t)}{\partial v_{PD}}\right|_0 \hat{v}_{PD}(t),$$

where the second and third terms are the small-signal photocurrent $\hat{\imath}_L$ and dark current $\hat{\imath}_d$, respectively. Exploiting the phasor notation, we can express the small-signal detector current $\hat{\imath}_{PD}$ as

$$\hat{\imath}_{PD}(t) = \hat{\imath}_L(t) + \hat{\imath}_d(t) = \mathrm{Re}\left(\mathfrak{R}(\omega)\hat{P}_{in}\mathrm{e}^{\mathrm{j}\omega t}\right) + \mathrm{Re}\left(Y_{PD}(\omega)\hat{V}_{PD}\mathrm{e}^{\mathrm{j}\omega t}\right),$$

where $\mathfrak{R}(\omega)$ is the (complex) small-signal responsivity, $Y_{PD}(\omega)$ the detector small-signal admittance. The phasor associated with $\hat{\imath}_{PD}(t)$ is therefore

$$\hat{I}_{PD}(\omega) = Y_{PD}(\omega)\hat{V}_{PD}(\omega) + \hat{I}_L(\omega),$$

where the signal photocurrent phasor $\hat{I}_L(\omega)$ is linearly related to the signal optical power phasor as

$$\hat{I}_L(\omega) = \mathfrak{R}(\omega)\hat{P}_{in}(\omega).$$

In the linearized model, $\hat{I}_{PD} = \hat{I}_L$ for $\hat{V}_{PD} = 0$; thus, $\hat{I}_L$ is often referred to as the *short-circuit photocurrent* (i.e., the photocurrent of the detector whose small-signal load is a short).

The *complex responsivity* $\mathfrak{R}(\omega)$, describing the detector small-signal frequency response, is typically a low-pass function of the modulation frequency; see Fig. 4.11. Narrowband photodetectors can, however, be designed for analog applications (i.e., for detecting analog narrowband signals modulating an optical carrier). A normalized responsivity $\mathfrak{r}(\omega)$ can be defined from

$$\frac{\hat{I}_L(\omega)}{\hat{I}_L(0)} = \frac{\mathfrak{R}(\omega)}{\mathfrak{R}(0)}\frac{\hat{P}_{in}(\omega)}{\hat{P}_{in}(0)} = \mathfrak{r}(\omega)\frac{\hat{P}_{in}(\omega)}{\hat{P}_{in}(0)}.$$

Assuming constant $\hat{P}_{in}(\omega)$ we have

$$\mathfrak{r}(\omega) = \frac{\hat{I}_L(\omega)}{\hat{I}_L(0)} = \frac{\mathfrak{R}(\omega)}{\mathfrak{R}(0)} \to |\mathfrak{r}(\omega)|_{\mathrm{dB}} = 20\log_{10}\left|\frac{\mathfrak{R}(\omega)}{\mathfrak{R}(0)}\right|.$$

For low-pass detectors, the 3 dB bandwidth is defined as the frequency $f_{3\mathrm{dB}}$ (or f_T) at which the normalized responsivity drops by 3 dB with respect to the DC value, i.e.,

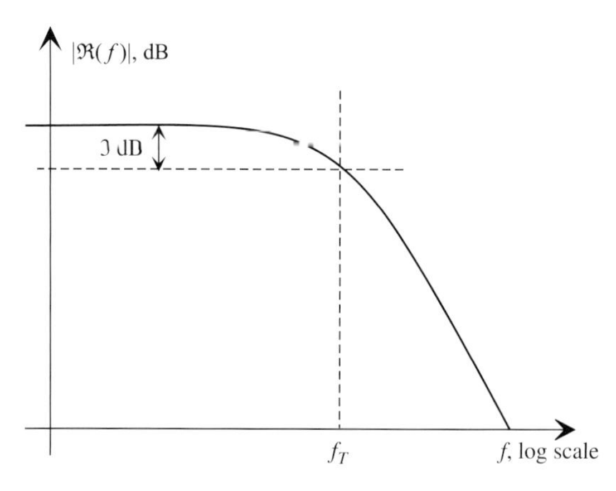

Figure 4.11 Typical low-pass behavior of the photodetector frequency response under sinusoidally modulated light.

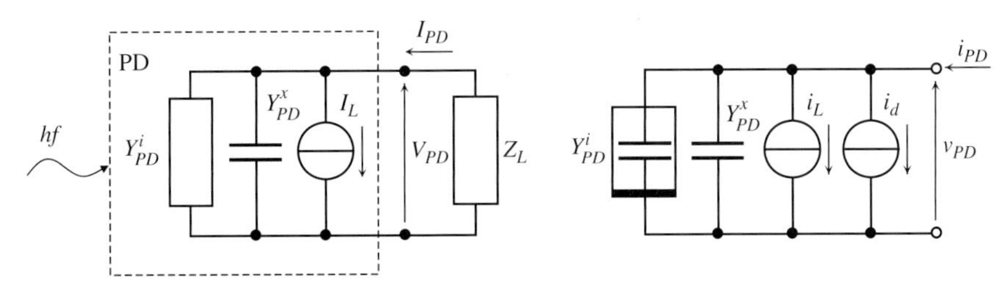

Figure 4.12 Small-signal equivalent circuit of loaded photodetector (left) and large-signal simplified model for a photodiode-like detector (right). The boxed capacitor symbol denotes a nonlinear capacitor.

$$|\mathfrak{r}(\omega_{3\text{dB}})|_{\text{dB}} = -3 \to 20\log_{10}\left|\frac{\Re(\omega_{3\text{dB}})}{\Re(0)}\right| = -3 \to \Re(f_{3\text{dB}}) = \frac{1}{\sqrt{2}}\Re(0).$$

The above (intrinsic) cutoff frequency refers to the short-circuit photocurrent and is therefore independent of the detector loading. Transit time, avalanche buildup delay, phototransistor current gain high-frequency cutoff are typically accounted for in the intrinsic cutoff frequency. On the other hand, the overall detector response is also affected by the load impedance and by parasitic (extrinsic) elements. The main load-related cutoff mechanism is the RC cutoff, caused by the combined effect of the device internal and extrinsic capacitance with the load resistance.[4]

A quantitative evaluation of the total cutoff frequency can be based on the simplified equivalent circuit of the photodetector in Fig. 4.12. In the frequency domain, the photodetector can be modeled by the current–voltage phasor relation:

$$I_{PD}(\omega) = \left[Y^i_{PD}(\omega) + Y^x_{PD}(\omega)\right] V_{PD}(\omega) + I_L(\omega),$$

where $I_L = \Re(\omega) P_{in}$ is the short-circuit photocurrent component at ω (modeled as a current source), Y^i_{PD} is the detector intrinsic admittance, and Y^x_{PD} is the detector parasitic (usually capacitive) admittance. The above representation leads directly to the circuit in Fig. 4.12, left. The load impedance Z_L (modeling, for example, the input capacitance or resistance of a front-end amplifier) clearly adds to the detector capacitive and resistive loading, and therefore influences the bandwidth of the loaded detector. As already mentioned, we include in the frequency-dependent responsivity $\Re(\omega)$ intrinsic cutoff effects only, while the RC cutoff is handled at a circuit level. Assuming $Z_L = R_L$ and a total detector capacitance C_{PD}, the current on the load $I_{R_L} = -I_{PD}$ will be

$$I_{R_L}(\omega) = -\frac{I_L(\omega)}{1 + \mathrm{j}\omega R_L C_{PD}} \to \left|I_{R_L}(\omega)\right| = \frac{|I_L(\omega)|}{\sqrt{1 + \omega^2 R_L^2 C_{PD}^2}}.$$

Therefore, even if $I_L(\omega) = \Re P_{in}$, $\Re$ frequency independent, the responsivity of the loaded detector,

$$|\Re_l(\omega)| = \frac{\Re}{\sqrt{1 + \omega^2 R_L^2 C_{PD}^2}},$$

[4] As discussed later, the RC bandwidth limitation can be overcome through distributed (traveling-wave) photodetectors, in which the bandwidth is limited by velocity mismatch with the optical signal and/or losses.

will be frequency dependent, with RC-limited cutoff frequency

$$f_{3\text{dB}} = \frac{1}{2\pi R_L C_{PD}}.$$

Since in junction-based detectors the DC current is small (dark current) and often bias-independent, in the first approximation the detector *large-signal model* is simply the parallel of a capacitive (sometimes nonlinear) admittance with two current generators modeling the photocurrent i_L (linearly dependent on the optical power) and the dark current i_d (often negligible); see Fig. 4.12, right. Other effects, such as a voltage-dependent photocurrent (due, e.g., to a voltage-dependent internal gain) or a nonlinear detector input admittance (as in *pn* photodiodes), can be readily implemented at a circuit level, as well as more complex parasitic networks including series connector resistances, wire inductances and distributed transmission line elements.

4.4.4 Photodetector gain

In some detectors (e.g., *pn* and *pin* photodiodes), the number of collected pairs is approximately equal to the number of generated pairs. However, certain detectors have internal gain mechanisms whereby the number of collected pairs may be much larger than the number of generated pairs. The internal gain implies the amplification A_I of the photocurrent that would have been (in theory) generated if no gain were present (called the *primary* photocurrent). If we define a comparison ideal device with primary photocurrent only and responsivity $\mathfrak{R}|_{A_I=1}$, we have

$$\mathfrak{R} = A_I\, \mathfrak{R}|_{A_I=1}\,.$$

Examples of photodetectors with gain are the avalanche photodiode (due to the avalanche multiplication of carriers), the phototransistor (due to the transistor current gain), and also the photoresistor. Apart from intrinsic mechanisms, photodetector gain can also be obtained by cascading the detector with active blocks:

- in front of the detector, as an optical amplifier (e.g., a semiconductor optical amplifier, SOA);
- after the detector, as an electronic front-end amplifier, possibly integrated with the detector into an integrated receiver.

4.5 Photodetector noise

The short-circuit photocurrent is affected by random fluctuations, i.e., by noise. Photodetector noise is the result of noise in the input light (e.g., laser noise, see Section 5.13.4, plus additional noise from optical amplifiers) converted into current, plus the contribution of the detector; see Fig. 4.13. We neglect for the moment the

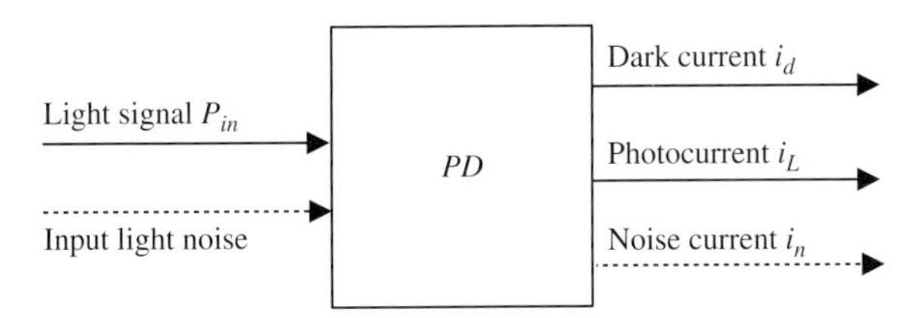

Figure 4.13 Generation of noise from the photodetector.

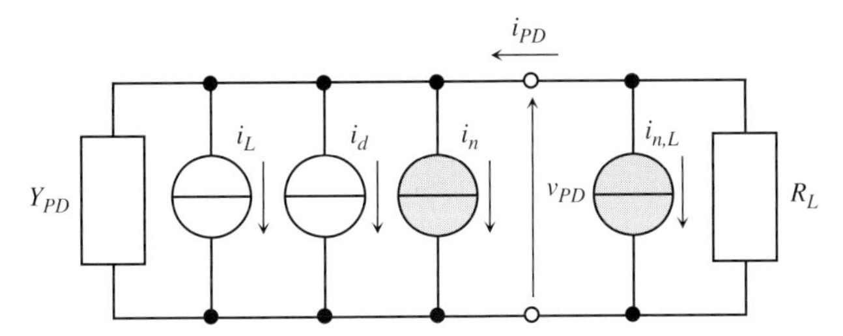

Figure 4.14 Simplified equivalent circuit of a detector connected to a load, including noise generators (in gray). For simplicity the load is resistive and the detector is modeled through a linear admittance.

converted input noise,[5] and focus on the detector noise contribution. This can be represented, from a circuit standpoint, as a zero-average noise current i_n superimposed on the detector current as

$$i_{PD} = \mathcal{R} p_{in} + i_d + i_n,$$

where i_n is a random process with power spectrum $S_{i_n}(\omega)$ and square mean value ($i_{n,\mathrm{rms}}$ is the root mean square, r.m.s., value):

$$i^2_{n,\mathrm{rms}} = \overline{i^2_n} = \int_0^\infty S_{i_n}(f)\,\mathrm{d}f.$$

A simplified equivalent circuit of the detector including noise generators is shown in Fig. 4.14. In many cases, the power spectrum is white (i.e., constant with frequency) and

$$\overline{i^2_n} = S_{i_n} B,$$

where B is the system bandwidth ($B \approx B_r$, the bit rate).

Detector noise can be modeled as thermal noise (in photoresistors), shot noise (in photodiodes and phototransistors), or multiplied shot noise (in avalanche detectors). For photoresistors we have, from the Nyquist law, $S_{i_n} = 4k_B T G$, where k_B is the Boltzmann constant, T the operating temperature, and G the photoresistor conductance; the r.m.s. value of the noise current is

$$i_{n,\mathrm{rms}} = \sqrt{\overline{i^2_n}} = \sqrt{4k_B T G B}.$$

[5] Converted input noise is light noise transformed into electrical (current) noise by detection. In the ideal case, light noise is only affected by the shot noise deriving from the corpuscular nature of light; conversion of such a light quantum noise leads to the so-called quantum noise limit of the receiver.

In photodiodes operating with constant current I_{PD} the shot noise model yields the power spectrum:

$$S_{i_n} = 2qI_{PD} = 2qI_L + 2qI_d = 2q\Re P_{in} + 2qI_d \approx 2q\Re P_{in}. \quad (4.11)$$

If the photodiode operates in small-signal conditions (as in an analog link), the shot noise process is second-order stationary and (4.11) applies, where I_{PD} is the photodiode bias current. In digital operation, the photodiode current ideally assumes two values corresponding to the one and zero levels (e.g., 0 for the low level and I_L for the high level), and the shot noise process becomes nonstationary. However, for realistic bit rates a quasi-stationary approximation can be made, and a slowly varying power spectrum with average value between $2qI_L + 2qI_d$ and $2qI_d \approx 0$ can be assumed as the noise spectrum. The r.m.s. noise current for the high level is therefore

$$\overline{i_n^2} = 2qI_L B + 2qI_d B \approx 2q\Re P_{in} B,$$

neglecting the shot noise contribution of the dark current, and the averaged value (over high and low states, assuming 50% probability) is $\overline{i_n^2}/2$.

Noise is obviously a major concern in detectors, owing to their position as the first block in the receiver chain, and affects the receiver sensitivity S, i.e., the *minimum input power* needed to achieve a desired SNR (signal to noise ratio) at the receiver output. Neglecting input laser noise, the detector output SNR can be expressed as the ratio of the signal and noise available powers. We refer here to a digital link, with a simple intensity modulation (IM) scheme. In the high state, the input power is P_{in} and the detector current $I_{PD}^2 \approx I_L^2$; in the low state, the input power is zero and $I_{PD}^2 = I_d^2 \approx 0$. We can now compute the photodiode output SNR, averaging both the output available detector signal and noise powers and assuming a detector output conductance G_{PD} (which actually simplifies out) as follows:[6]

$$\text{SNR} = \frac{P_{s,av}}{P_{n,av}} = \frac{\frac{1}{2}\left(I_L^2/4G_{PD}\right)}{\frac{1}{2}\left(\overline{i_n^2}/4G_{PD}\right)} = \frac{I_L^2}{\overline{i_n^2}} = \frac{I_L^2}{2qI_L B} = \frac{I_L}{2qB} = \frac{\Re P_{in}}{2qB}, \quad (4.12)$$

where we have neglected the dark-current shot noise. The averaged SNR is therefore simply *the SNR in the high level*; we will exploit this result in what follows without further discussion.

Since in a loaded detector the SNR is also influenced by the noise (typically thermal) introduced by the front-end amplifier stage, (4.12), corresponds to an ideal case (called the *shot noise limit*) in which the detector load is noiseless and the SNR increases as the photocurrent, i.e., as the optical power. Suppose now that we consider the load thermal noise contribution as a noise current $i_{n,L}$ in parallel to the photodiode noise current i_n, such as

$$\sqrt{\overline{i_{n,L}^2}} = \sqrt{4k_B T G_n B},$$

[6] The discussion holds for a digital modulation scheme, or for an analog, large-signal, modulation scheme where the signal amplitude is comparable to the bias; in (strictly) small-signal analog modulation, noise is established by the DC working point and the r.m.s. noise current is independent of the signal current.

where G_n is the load noise conductance (coinciding with the load conductance for a simple resistive load); we have for the SNR:

$$\text{SNR} = \frac{I_L^2}{\overline{i_n^2} + \overline{i_{n,L}^2}} = \frac{I_L^2}{2qI_L B + 4k_B T G_n B}.$$

If the load noise is dominant with respect to the detector shot noise we have

$$\text{SNR} \approx \frac{I_L^2}{4k_B T G_n B} = \frac{(\mathfrak{R} P_{in})^2}{4k_B T G_n B},$$

where the SNR increases as the square of the optical power. This condition is the *thermal noise limit* of the detector. From the desired SNR, the sensitivity S can be obtained directly; in the shot noise limit one has

$$S = \frac{2qB}{\mathfrak{R}}\text{SNR},$$

i.e., the sensitivity increases as the bandwidth or the bit rate.

The SNR and the sensitivity can be considered as system-level figures of merit for the receiver. More specific device-level noise parameters are:

- The noise equivalent power, NEP or P_{NEP}, defined as the optical input power yielding a short-circuit SNR equal to one. The NEP defines the detector *noise floor*, since in such conditions the detector output SNR is unity. In the short-circuit case, the load is $R_L = 0$ and therefore $i_{L,n} = 0$; thus, the device operates in the shot noise limit, and we must impose

  $$\text{SNR} = \frac{I_L^2}{2qI_L B + 2qI_d B} = \frac{\mathfrak{R}^2 P_{NEP}^2}{2q\mathfrak{R}P_{NEP}B + 2qI_d B} = 1,$$

  from which P_{NEP} can be derived; we also have two limiting cases according to whether the photocurrent is larger or smaller than the dark current (remember that in these conditions the input optical power and therefore the photocurrent is very small):
  - Shot-noise-limited NEP: $P_{NEP} = 2qB/\mathfrak{R}$;
  - Dark-current-limited NEP: $P_{NEP} = \sqrt{2qI_d B}/\mathfrak{R}$.
- The photodetector *detectivity*, defined as the inverse of the NEP:

  $$D = \frac{1}{\text{NEP}}.$$

  Moreover, the *specific detectivity* (D^*) is defined, in the dark current limit, by normalizing with respect to the device area A and the bandwidth:

  $$D^* = \sqrt{AB}D = \frac{\sqrt{AB}}{\text{NEP}}.$$

  Finally, an *optics-independent specific detectivity* (D^{**}) can be defined with reference to the numerical aperture of the optical system NA:

  $$D^{**} = D^* \cdot \text{NA}.$$

Further details on photodetector noise will be provided when discussing specific devices and front-end amplifier choices.

4.6 Photodiodes

In general, the DC photodiode current can be modeled, taking into account the Shockley junction diode law (yielding the dark current) and the photocurrent $I_L = \Re P_{in}$, as

$$I_{PD} = \Re P_{in} - I_0 \left[\exp\left(-\frac{V_{PD}}{\eta V_T} \right) - 1 \right],$$

where the current positive sign is taken in the reverse direction (i.e., the forward current is negative, the photocurrent is positive). V_{PD} is the reverse bias voltage, I_0 is the reverse saturation current; see Fig. 4.15. The thermal voltage is $V_T = k_B T/q = 26$ mV at $T = 300$ K, and η is the ideality factor ($1 \leq \eta \leq 2$). For $V_{PD}/\eta V_T \gg 1$ (i.e., in reverse bias) one has

$$I_{PD} \approx \Re P_{in} + I_d.$$

That is, the photodiode current is proportional to the input power, and $I_d = I_0$ is the dark current. The region in which $I_{PD} > 0$, $V_{PD} > 0$ is the *photodiode region*; see Fig. 4.15. If the device is under weak direct bias ($V_{PD} < 0$), the photocurrent may dominate $I_{PD} > 0$; this is the *photovoltaic region*, in which the device acts as a power source, i.e., converts the input optical power into electrical power. In this region, the applied bias is negative but the current is positive, leading to a net power flow from the photodiode to the external circuit. The photovoltaic region is the operation mode of the photodiode as a solar cell. However, the signal properties of the photodiode degrade, since the photocurrent is increasingly masked by the direct current; moreover, the responsivity will rapidly saturate for increasing input optical power. Finally, in the *direct region* the

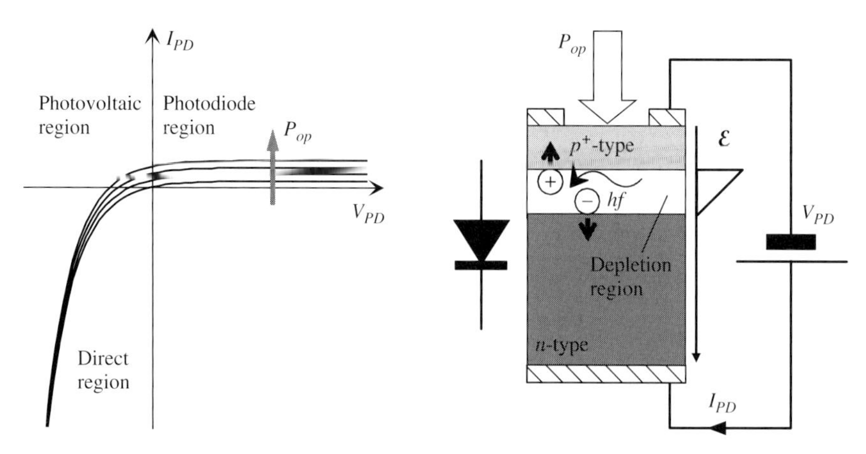

Figure 4.15 V–I characteristics of a photodiode (left) biased by an ideal voltage source and under illumination (right).

photocurrent is masked by the large direct current, and net electrical power is absorbed – an operation mode unsuited to both telecom and energy applications.

In practical cases, of course, the photodiode has to be connected to a load, rather than to an ideal voltage source; linearity of the current–power response of the loaded device is preserved until the instantaneous device working point is in the photodiode region, but (extrinsic) current compression and saturation arise as soon as the input optical power increases so as to bring the instantaneous working point into the photovoltaic region.

4.7 The *pn* photodiode

A pn photodiode usually has a thin p-type, highly doped layer lying on the device surface, on top of a n-type, less doped substrate; see Fig. 4.16. The total photocurrent results from e-h pairs generated in both the depletion region and the n-side and p-side diffusion regions, whose equivalent width is of the order of the diffusion length in the two sides, $L_{np} = \sqrt{D_{np}\tau_n}$ for electrons in the p side (where D_{np} is the electron diffusivity, τ_n the electron lifetime), $L_{hn} = \sqrt{D_{hn}\tau_h}$ for holes in the n side (where D_{hn} is the hole diffusivity, τ_h the hole lifetime); see Section 1.8.2. Let us assume for simplicity that the optical generation rate is uniform and equal to G_o along the photodiode;[7] the total DC current (see (4.24), evaluated for $\omega = 0$ and with $G_o \approx G_{on} \approx G_{op}$) is

$$I_L = qA \int_W G_o \, dx + I_{Lp} + I_{Ln} \approx qAG_o(W + L_{np} + L_{hn}).$$

Apparently, carriers generated in the diffusion regions increase the device responsivity:

$$\Re = \frac{I_L}{P_{in}} = \frac{q}{hf}\alpha(W + L_{np} + L_{hn}), \tag{4.13}$$

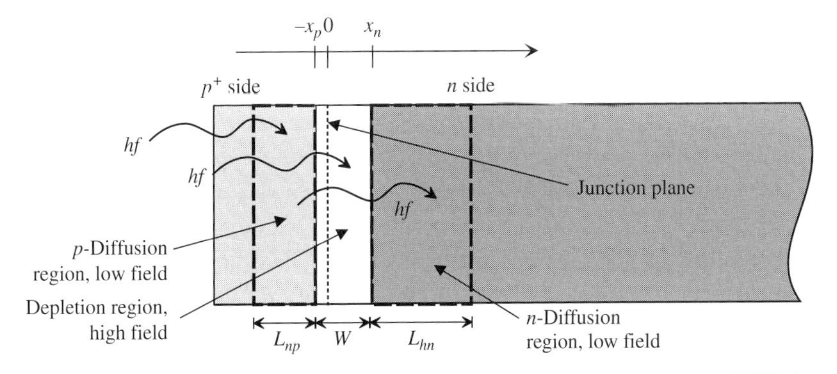

Figure 4.16 Generation of photocarriers in a pn photodiode: contribution of depletion and diffusion regions.

[7] This implies that the total thickness of the structure is smaller than the absorption length. This approximation, which will be removed in the treatment of pin diodes, helps to simplify the analysis of pn photodiodes while preserving the main points of the discussion.

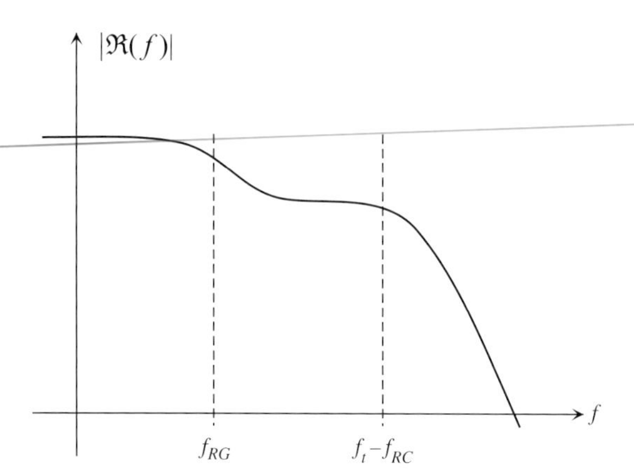

Figure 4.17 Frequency response of a *pn* photodiode: f_{RG} is the lifetime cutoff ($\approx$MHz), f_t and f_{RC} are the transit time and RC cutoff, both in the GHz range (typically, the RC cutoff dominates in *pn* photodiodes).

where W is the width of the depletion region.[8] However, assuming time-varying input optical power, a frequency-domain small-signal analysis shows (see (4.24), with $G_o \approx G_{on} \approx G_{op}$) that

$$I_L(\omega) = qAG_o(W + \widetilde{L}_{np} + \widetilde{L}_{hn}), \quad \widetilde{L}_{np} = \frac{L_{np}}{\sqrt{1 + \mathrm{j}\omega\tau_h}}, \quad \widetilde{L}_{hn} = \frac{L_{hn}}{\sqrt{1 + \mathrm{j}\omega\tau_n}}.$$

Thus, the additional response proportional to $\widetilde{L}_{np} + \widetilde{L}_{hn}$ exhibits a cutoff frequency of the order of the inverse of the lifetime, and the whole response has a double cutoff (low frequency, proportional to the inverse of the lifetime; high frequency, associated with transit time and capacitance effects), as shown in Fig. 4.17. From a system standpoint, this is quite an inconvenient response, and the device speed is practically limited by the device lifetime (with maximum cutoff frequencies of the order of 100–200 MHz)

4.7.1 Analysis of the *pn* photodiode response

To evaluate the *pn* photodiode total current, we assume that the diode has an abrupt doping profile, with junction plane at $x = 0$. The p-type side has uniform doping N_A and extends (ideally) from $-\infty$ to 0; the n-type side extends from 0 to $+\infty$. Across the junction plane, a depletion (space-charge) region hosts a dipole layer supporting, in equilibrium (i.e., for zero applied bias), the built-in voltage V_{bi}. The space-charge region extends from $-x_p$ (p side) to x_n (n side). By imposing equilibrium conditions and the Boltzmann statistics, the built-in voltage can be derived, as a function of the side dopings and of the intrinsic semiconductor concentration, as[9]

[8] Remember that (4.13) holds when the absorption length is larger than the total equivalent absorption width, i.e., when $\alpha(W + L_{np} + L_{hn}) < 1$. For arbitrary α, (4.13) would lead to an unphysical result, larger than the ideal limit in (4.9).

[9] The *pn* junction theory is covered in all textbooks on electron devices, see, e.g., [31].

$$V_{\text{bi}} = \frac{k_B T}{q} \log \left(\frac{N_D N_A}{n_i^2} \right). \tag{4.14}$$

Solution of the Poisson equation in the presence of an applied bias V_A ($V_A < 0$ in reverse bias) leads to the following expressions for the depletion region widths:

$$x_n = \sqrt{\frac{2\epsilon N_{\text{eq}} (V_{\text{bi}} - V_A)}{q N_D^2}} \tag{4.15}$$

$$x_p = x_n \frac{N_D}{N_A} = \sqrt{\frac{2\epsilon N_{\text{eq}} (V_{\text{bi}} - V_A)}{q N_A^2}}, \tag{4.16}$$

where $N_{\text{eq}}^{-1} = N_D^{-1} + N_A^{-1}$. The total width of the depletion region will be:

$$W = x_n + x_p = \sqrt{\frac{2\epsilon (V_{\text{bi}} - V_A)}{q N_{\text{eq}}}},$$

which increases with increasing reverse applied voltage.

To evaluate the photocurrent, assume for simplicity that the photogeneration term is constant in the two sides and in the depletion region and assumes, respectively, values G_{op} (p side), G_{on} (n side), and G_o (depletion region). This assumption will be conveniently removed in the analysis of the *pin* photodiode. Minority carriers in excess with respect to the thermodynamic equilibrium condition (corresponding to zero applied bias and no illumination), i.e., excess electrons in the p side, excess holes in the n side, follow the electron and hole continuity equations, respectively. In the two sides we assume quasi-neutrality (i.e., negligible electric field), and carrier recombination is modeled by the lifetime approximation. Assuming that the structure is one-dimensional along x, the time-domain continuity equations read

$$\frac{\partial n'}{\partial t} = -\frac{\partial}{\partial x}\left(-D_{np}\frac{\partial n'}{\partial x}\right) - \frac{n'}{\tau_n} + G_{op} \ (p \text{ side})$$

$$\frac{\partial p'}{\partial t} = \frac{\partial}{\partial x}\left(D_{hn}\frac{\partial p'}{\partial x}\right) - \frac{p'}{\tau_h} + G_{on} \ (n \text{ side}).$$

Transforming to the frequency domain, we have

$$\mathrm{j}\omega n' = D_{np}\frac{\mathrm{d}^2 n'}{\mathrm{d}x^2} - \frac{n'}{\tau_n} + G_{op} \ (p \text{ side})$$

$$\mathrm{j}\omega p' = D_{hn}\frac{\mathrm{d}^2 p'}{\mathrm{d}x^2} - \frac{p'}{\tau_h} + G_{on} \ (n \text{ side}),$$

Introducing the *complex diffusion lengths*

$$\frac{1}{\widetilde{L}_{np}} = \sqrt{\frac{1+\mathrm{j}\omega\tau_n}{\tau_n D_{np}}} = \frac{\sqrt{1+\mathrm{j}\omega\tau_n}}{L_{np}}$$

$$\frac{1}{\widetilde{L}_{hn}} = \sqrt{\frac{1+\mathrm{j}\omega\tau_h}{\tau_h D_{hn}}} = \frac{\sqrt{1+\mathrm{j}\omega\tau_h}}{L_{hn}},$$

where $L_{np} = \sqrt{D_{np}\tau_n}$, $L_{hn} = \sqrt{D_{hn}\tau_h}$ are the diffusion lengths for excess electrons in the p side and excess holes in the n side, we finally have

$$\frac{\mathrm{d}^2 n'}{\mathrm{d}x^2} = \frac{n'}{\widetilde{L}_{np}^2} - \frac{G_{op}}{D_{np}} \quad (p \text{ side}) \tag{4.17}$$

$$\frac{\mathrm{d}^2 p'}{\mathrm{d}x^2} = \frac{p'}{\widetilde{L}_{hn}^2} - \frac{G_{on}}{D_{hn}} \quad (n \text{ side}), \tag{4.18}$$

with boundary conditions:

$$n'(-x_p) = \frac{n_i^2}{N_A}\left[\exp\left(\frac{V_A}{V_T}\right) - 1\right], \quad n'(-\infty) = 0 \ \ (p \text{ side}) \tag{4.19}$$

$$p'(x_n) = \frac{n_i^2}{N_D}\left[\exp\left(\frac{V_A}{V_T}\right) - 1\right], \quad p'(+\infty) = 0 \ \ (n \text{ side}). \tag{4.20}$$

In each side, the first condition derives again from the Boltzmann statistics, while the second implies that very far away from the junction the excess carrier population should vanish.

Solving the linear set in (4.17) and (4.18) with boundary conditions (4.19) and (4.20) we obtain

$$n'(x) = \left[\frac{n_i^2}{N_A}\left(\mathrm{e}^{\frac{V_A}{V_T}} - 1\right) - \frac{\tau_n G_{op}}{1+\mathrm{j}\omega\tau_n}\right]\exp\left(\frac{x + x_p}{\widetilde{L}_{np}}\right) + \frac{\tau_n G_{op}}{1+\mathrm{j}\omega\tau_n} \quad (p \text{ side})$$

$$p'(x) = \left[\frac{n_i^2}{N_D}\left(\mathrm{e}^{\frac{V_A}{V_T}} - 1\right) - \frac{\tau_n G_{on}}{1+\mathrm{j}\omega\tau_h}\right]\exp\left(-\frac{x - x_n}{\widetilde{L}_{hn}}\right) + \frac{\tau_h G_{on}}{1+\mathrm{j}\omega\tau_h} \quad (n \text{ side}).$$

To evaluate the current, let us express the total current density $-J_{PD}$ (taken as positive outgoing from the p side) as the sum of drift $J_{n,\mathrm{dr}}$ and diffusion $J_{h,\mathrm{d}}$ currents of electrons and holes (taken as positive in the positive x direction):

$$-J_{PD} = J_{n,\mathrm{dr}} + J_{n,\mathrm{d}} + J_{h,\mathrm{dr}} + J_{h,\mathrm{d}}.$$

However, in the n side the drift current of the minority carriers (holes) is negligible, and so is the drift electron current density in the p side; we thus have

$$J_n \approx J_{n,\mathrm{d}} \quad \text{for } x \le -x_p \ (p \text{ side})$$

$$J_h \approx J_{h,\mathrm{d}} \quad \text{for } x \ge x_n \ (n \text{ side}).$$

In the depletion region, only the optical generation term is considered (thermal generation is neglected) and the continuity equations for the electron and hole current densities can be expressed as

$$\frac{\mathrm{d}J_n}{\mathrm{d}x} = -qG_o \tag{4.21}$$

$$\frac{\mathrm{d}J_h}{\mathrm{d}x} = -qG_o, \tag{4.22}$$

where G_o is the optical generation rate in the depletion region. Integrating (4.21) on the depletion region, we obtain

$$J_n(x_n) - J_n(-x_p) = q\int_{-x_p}^{x_n} G_o \, \mathrm{d}x \approx -qWG_o,$$

i.e.,

$$J_n(x_n) = J_n(-x_p) - qWG_o.$$

We can now conveniently express the total current (which is independent of x) in a section corresponding to the depletion region edges; e.g., in point x_n,

$$\begin{aligned} -J_{PD} &= J_n(x_n) + J_h(x_n) = J_n(-x_p) - qWG_o + J_h(x_n) \\ &= J_{n,\mathrm{d}}(-x_p) + J_{h,\mathrm{d}}(x_n) - qWG_o, \end{aligned}$$

since

$$J_n(-x_p) = J_{n,\mathrm{d}}(-x_p), \quad J_h(x_n) = J_{h,\mathrm{d}}(x_n).$$

We obtain the same result evaluating the current in $-x_p$. Expressing all current densities as a function of the excess charge gradients (see Section 1.8.1), we finally have

$$\begin{aligned} -J_{PD} &= J_{n,\mathrm{d}}(-x_p) + J_{h,\mathrm{d}}(x_n) + qWG_o \\ &= qD_{np} \left.\frac{\partial n'}{\partial x}\right|_{-x_p} - qD_{hn} \left.\frac{\partial p'}{\partial x}\right|_{x_n} - qWG_o = -J_d - J_L, \end{aligned}$$

where J_d is the dark current density and J_L is the photocurrent density. The corresponding currents are

$$I_d = AJ_d = qAn_i^2 \left(\frac{1}{\widetilde{L}_{np}} \frac{D_{np}}{N_A} + \frac{1}{\widetilde{L}_{hn}} \frac{D_{hn}}{N_D} \right) \left(\mathrm{e}^{\frac{V_A}{V_T}} - 1 \right) \tag{4.23}$$

$$I_L = AJ_L = qA \left(\widetilde{L}_{np} G_{op} + \widetilde{L}_{hn} G_{on} + WG_o \right), \tag{4.24}$$

where A is the detector area. The dark current follows the Shockley diode law and yields a positive contribution to the total detector current I_{PD} for large negative applied voltage V_A. From (4.24) the photocurrent I_L is found to include three contributions, the first two referring to the diffusion regions, the last (typically much smaller) to the depletion region. Although the diffusion contribution apparently enhances the photocurrent, in fact this contribution rapidly decreases with increasing speed of the input optical signal, with a characteristic time given by the lifetime (in a direct-gap semiconductor, of the order of 1 ns, much larger in indirect-gap semiconductors). As a result, the overall device response exhibits an early cutoff (due to lifetime) plus a very high-frequency transit-time or RC cutoff.

4.8 The *pin* photodiode

To improve the frequency response and the high-frequency efficiency, the depletion region width W should be made much larger than the width of the diffusion region. However, in the *pn* diode this would require impractically large reverse voltages. An obvious solution is to interpose a lightly doped or intrinsic region between the p and n layers that is completely depleted in reverse bias and whose electric field is almost constant (since the space charge is negligible). The resulting structure is the *pin* photodiode, see Fig. 4.18 (above); by proper design $W \gg L_{hn} + L_{np}$, making diffusion photocurrents negligible. Additionally, exploiting compound semiconductors and heterojunction devices, the slow part of the response can be suppressed altogether through NpP or NiP structures, where N or P as usual denote widegap semiconductors (e.g., AlGaAs/GaAs/AlGaAs or InP/InAlAs/InP); in this way, no absorption at all takes place in the external regions and the related current contribution vanishes.

Compound semiconductor *pin* structures (usually PiN) are today probably among the best component available for 10 Gbps and 40 Gbps systems, although for 10 Gbps application a competition exists with avalanche photodiodes. Cutoff frequencies in excess of 80 GHz have been demonstrated. Usually, a compromise must be reached between speed and responsivity (efficiency); the device speed is dominated by transit

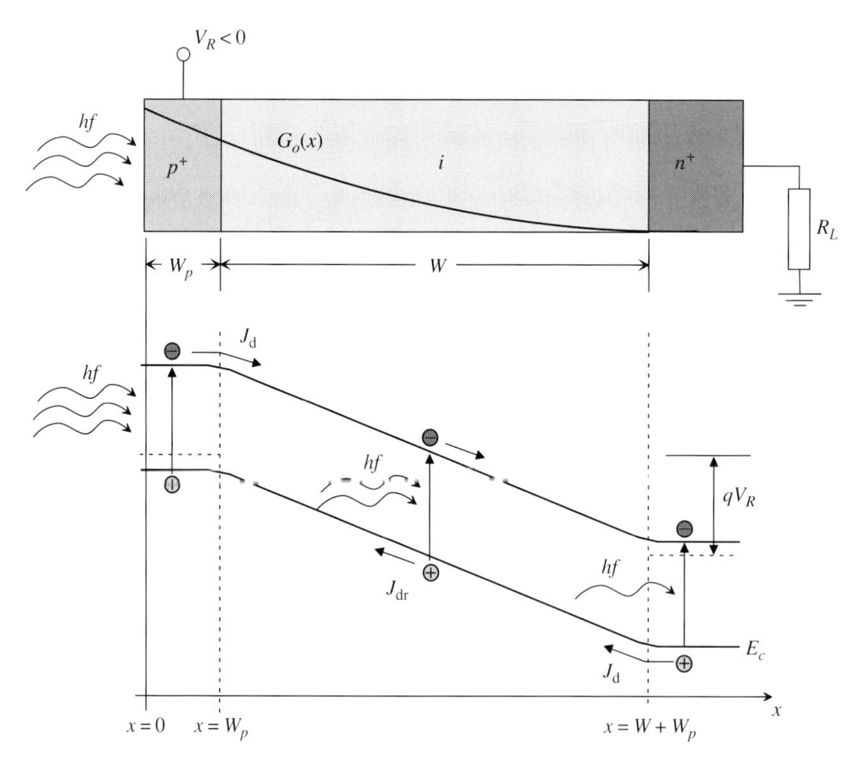

Figure 4.18 Schematic structure (above) and band diagram (below) of a *pin* photodiode in reverse bias. J_{dr} and J_d are drift and diffusion current densities, respectively.

time and the parasitic capacitance. Such limitations can be overcome in more advanced structures such as traveling-wave waveguide photodiodes; see Section 4.10.2.

4.8.1 The *pin* photocurrent, responsivity, and efficiency

The depletion region of a *pin* photodiode in reverse bias supports a constant electric field and linearly varying energy bands. In fact, from Poisson's equation we have

$$\frac{\mathrm{d}^2\phi}{\mathrm{d}x^2} = -\frac{q}{\epsilon}(N_D^+ - N_A^- + p - n) \approx 0$$

if we neglect the intrinsic region background doping and assume that the generated photocharge is negligible or quasi-neutral. This implies that ϕ is a linear function of position and, therefore, so is the conduction band edge $E_c = -q\phi + C$, where C is an arbitrary constant; see Fig. 4.18 (below). The main contribution to the photocurrent is given by the drift current associated with carriers generated inside the intrinsic depleted region; secondary contributions (which disappear in heterojunction devices) come from the diffusion regions. Due to the large width of the depletion region W, the optical generation rate will be nonuniform, according to the law

$$G_o(x) = \eta_Q \frac{P_{in}(1-R)}{Ahf} \alpha \mathrm{e}^{-\alpha x} = G_o(0)\mathrm{e}^{-\alpha x},$$

where η_Q is the intrinsic quantum efficiency, P_{in} the total incident power on the left (upper) photodiode facet, A is the detector area, hf is the photon energy, α is the material absorption, and R is the power reflectivity of the upper surface.

The analysis of the DC *pin* current I_{PD} (taken as positive outcoming from the p terminal; the sign is opposite to currents defined positive along the positive x axis) follows the same ideas as for the *pn* photodiode, i.e., the decomposition of the total current in electron (drift and diffusion) and hole (drift and diffusion) components. The device current is constant in x and can be expressed as

$$-I_{PD} = I_n(x) + I_h(x) = I_{n,\mathrm{dr}}(x) + I_{n,\mathrm{d}}(x) + I_{h,\mathrm{dr}}(x) + I_{h,\mathrm{d}}(x).$$

We consider a structure with a surface p^+ layer and an n^+ substrate. The depletion region begins approximately in $x = W_p$ and extends until $x = W + W_p$. In the p^+ region the electron (minority carrier) current is diffusion only, while in the n^+ region the hole current is diffusion only. Thus,

$$-I_{PD} = I_{n,\mathrm{d}}(W_p) + I_h(W_p) = I_n(W + W_p) + I_{h,\mathrm{d}}(W + W_p). \tag{4.25}$$

Multiplying the electron continuity equation (4.21) on the depletion region by the diode area A,

$$\frac{\mathrm{d}I_n}{\mathrm{d}x} = -qAG_o,$$

and integrating over the depletion region, we obtain

$$I_n(W + W_p) - I_n(W_p) \approx I_n(W + W_p) - I_{n,\mathrm{d}}(W_p) = I_i, \tag{4.26}$$

where

$$I_i = -qA \int_{W_p}^{W+W_p} G_o(x)\,\mathrm{d}x \tag{4.27}$$

is the photocurrent generated by the intrinsic absorption region. Substituting $I_n(W + W_p)$ from (4.26) into (4.25) we obtain

$$-I_{PD} = I_{n,\mathrm{d}}(W_p) + I_i + I_{h,\mathrm{d}}(W + W_p), \tag{4.28}$$

where the first contribution includes the diffusion current from the surface layer; the second contribution, the (photogenerated) drift current from the intrinsic layer; the third contribution, the diffusion current from the substrate. In the diffusion currents, we have both photocurrent and dark current components.

Carrying out the integration in (4.27) and expressing the optical generation rate G_o from (4.6), also accounting for surface reflection and the intrinsic quantum efficiency η_Q, we obtain

$$\begin{aligned} -I_i &= q\eta_Q \frac{P_{in}(1-R)}{hf} \int_{W_p}^{W_p+W} \alpha \mathrm{e}^{-\alpha x}\,\mathrm{d}x = q\eta_Q \frac{P_{in}(1-R)}{hf}\left[-\mathrm{e}^{-\alpha x}\right]_{W_p}^{W_p+W} \\ &= q\eta_Q \frac{P_{in}(1-R)}{hf} \mathrm{e}^{-\alpha W_p}\left(1-\mathrm{e}^{-\alpha W}\right). \end{aligned} \tag{4.29}$$

Concerning the diffusion contributions, we have to solve the same excess carrier equations as for the *pn* photodiode, but with a space-dependent generation term and slightly different boundary conditions.

In practical structures, the highly doped surface side p^+ is very thin, i.e., $W_p \ll L_\alpha$, $W_p \ll L_{np}$; this somewhat simplifies the solution of the diffusion equation:

$$\frac{\mathrm{d}^2 n'}{\mathrm{d}x^2} = \frac{n'}{L_{np}^2} - \frac{G_o(0)\mathrm{e}^{-\alpha x}}{D_{np}} \approx \frac{n'}{L_{np}^2} - \frac{G_o(0)}{D_{np}} \quad (p^+ \text{ side}), \tag{4.30}$$

with boundary conditions

$$n'(0) = 0, \quad n'(W_p) = n'_0. \tag{4.31}$$

The excess population n'_0 depends on the applied voltage in the same way as in the *pn* case (at least in reverse bias; in direct bias, the voltage drop over the intrinsic region should also be considered). The solution of (4.30) can be expressed as

$$n' = A\exp\left(\frac{x}{L_{np}}\right) + B\exp\left(-\frac{x}{L_{np}}\right) + \frac{L_{np}^2 G_o(0)}{D_{np}};$$

A and B can be obtained by imposing the boundary conditions (4.31); we obtain

$$n' = n'_0 \frac{\exp\left(-\dfrac{x}{L_{np}}\right) - \exp\left(\dfrac{x}{L_{np}}\right)}{\exp\left(-\dfrac{W_p}{L_{np}}\right) - \exp\left(\dfrac{W_p}{L_{np}}\right)}$$

$$+ \frac{L_{np}^2 G_o(0)}{D_{np}} \left\{ \frac{\left[\exp\left(\dfrac{x}{L_{np}}\right) + \exp\left(\dfrac{W_p - x}{L_{np}}\right)\right]\left[1 - \exp\left(-\dfrac{W_p}{L_{np}}\right)\right]}{\exp\left(-\dfrac{W_p}{L_{np}}\right) - \exp\left(\dfrac{W_p}{L_{np}}\right)} + 1 \right\}.$$

For $W_p \ll L_{np}$, the solution can be approximated as

$$n' \approx n'_0 \frac{x}{W_p},$$

i.e., the optical generation contribution is negligible.

We now consider the substrate side (assumed as n-type, see Fig. 4.18). The hole diffusion equation here can be written as

$$\frac{d^2 p'}{dx^2} = \frac{p'}{L_{hn}^2} - \frac{G_o(0) e^{-\alpha x}}{D_{hn}} \quad (n^+ \text{ side}),$$

with boundary conditions

$$p'(W_p + W) = p'_0, \quad p'(\infty) = 0, \tag{4.32}$$

where p'_0 is the excess concentration at the depletion region edge, which depends on the applied voltage and vanishes at zero applied bias. We now postulate the trial solution, vanishing for $x \to \infty$:

$$p'(x) = A \exp\left(-\frac{x - W_p - W}{L_{hn}}\right) + B \exp(-\alpha x).$$

By applying the boundary conditions (4.32) we obtain

$$p'(x) = \left[p'_0 - \frac{L_{hn}^2}{1 - \alpha^2 L_{hn}^2} \frac{G_o(0)}{D_{hn}} e^{-\alpha(W_p + W)}\right] \exp\left(-\frac{x - W_p - W}{L_{hn}}\right)$$

$$+ \frac{L_{hn}^2}{1 - \alpha^2 L_{hn}^2} \frac{G_o(0)}{D_{hn}} e^{-\alpha x}.$$

We can now evaluate the diffusion currents in W_p (electrons) and $W + W_p$ (holes) as

$$I_{n,d}(W_p) = qAD_{np} \left.\frac{\partial n'}{\partial x}\right|_{W_p} = \frac{qAD_{np}}{W_p} n'_0$$

$$I_{h,d}(W + W_p) = -qAD_{hn} \left.\frac{\partial p'}{\partial x}\right|_{W_p + W} = \frac{qAD_{hn}}{L_{hn}} p'_0 - \frac{qAG_o(0)L_{hn}}{1 + \alpha L_{hn}} e^{-\alpha(W_p + W)}.$$

The total photodiode current is therefore, from (4.28),

$$-I_{PD} = \underbrace{\frac{qAD_{np}}{W_p} n'_0 + \frac{qAD_{hn}}{L_{hn}} p'_0}_{-I_d} - \underbrace{\left[\frac{qAG_o(0)L_{hn}}{1 + \alpha L_{hn}} e^{-\alpha(W_p + W)} + I_i\right]}_{I_L},$$

where the first contribution is the dark current, the second the photocurrent. Expanding the generation rate G_o and using (4.29), the photocurrent can be expressed as[10]

$$I_L = \eta_Q \frac{q}{hf} P_{in} (1 - R) \mathrm{e}^{-\alpha W_p} \left(1 - \frac{\mathrm{e}^{-\alpha W}}{1 + \alpha L_{hn}}\right), \tag{4.33}$$

thus leading to the responsivity

$$\mathfrak{R} = \frac{I_L}{P_{in}} = \eta_Q \frac{q}{hf} (1 - R) \mathrm{e}^{-\alpha W_p} \left(1 - \frac{\mathrm{e}^{-\alpha W}}{1 + \alpha L_{hn}}\right),$$

and external quantum efficiency

$$\eta_x = \frac{I_L/q}{P_{in}/hf} = \eta_Q (1 - R) \mathrm{e}^{-\alpha W_p} \left(1 - \frac{\mathrm{e}^{-\alpha W}}{1 + \alpha L_{hn}}\right).$$

The diffusion contributions to currents are much slower than the drift contributions in dynamic operation, and should be reduced to optimize the high-speed response. This can be immediately achieved in heterojunction devices, where the substrate layer below the absorption region is widegap and therefore does not appreciably absorb light. To maximize η_x, we must also require that the thickness of the top layer be much smaller than α^{-1}, or that the top layer again be widegap, i.e., transparent to the incoming light. For high-speed, high-efficiency photodiodes $\alpha W_p \to 0$ and $\alpha L_{hn} \to 0$, so that the external device quantum efficiency and responsivity are

$$\eta_x \approx \eta_Q (1 - R) \left(1 - \mathrm{e}^{-\alpha W}\right), \tag{4.34}$$

$$\mathfrak{R} \approx \frac{q}{hf} \eta_Q (1 - R) \left(1 - \mathrm{e}^{-\alpha W}\right). \tag{4.35}$$

4.8.2 Conventional *pin* photodetector structures

Most high-performance *pin* detectors for long-wavelength communication systems exploit vertically illuminated structures with InGaAs absorption layers. An example of a diffused surface illuminated *pin* with planar layout and top ring electrodes is shown in Fig. 4.19 together with a mesa version where the lateral sides are etched and the free surface is properly passivated to avoid large dark currents. In the planar structure, the intrinsic or n^--type (also called ν-type) absorption layer is followed by an n-doped InP epitaxial buffer grown on the n-type InP substrate. In some cases, as in the structure shown in Fig. 4.20(b), the intrinsic absorption layer is followed by a nonabsorbing drift layer. The *pn* junction is obtained through diffusion, thus separating the junction from the hetero-interface; diffusion also allows the junction not to extend to the device surface. The ring electrode (p-contact) is clearly visible, while the n^+ layer is confined (rectangular footprint) to the device area; the external coplanar electrodes are deposited on a semi-insulating (S.I.) InP layer to decrease the parasitic capacitance and losses. The same device can also be realized with illumination from the bottom; the ring electrode

[10] Equation (4.33) holds when αW_p is small, since photocurrent generation in the top layer is explicitly neglected.

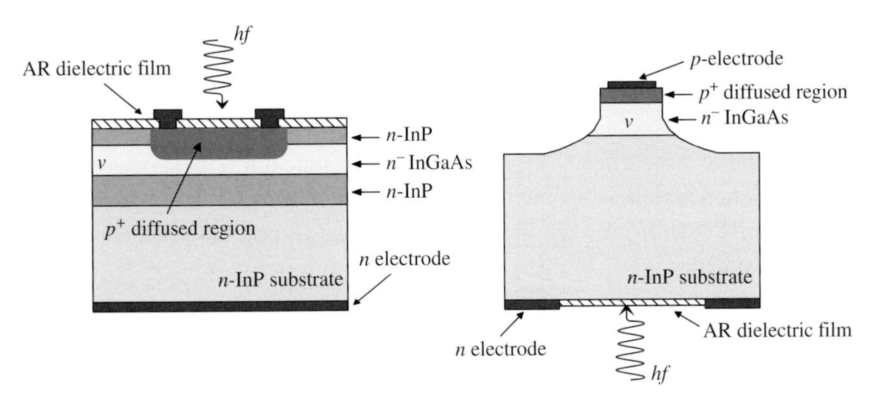

Figure 4.19 Planar double heterostructure *pin* (left); mesa single heterojunction *pin* (right).

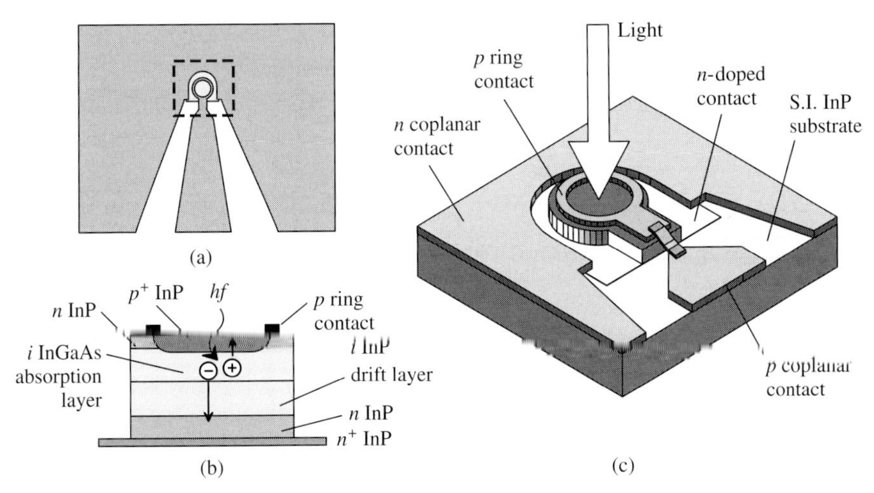

Figure 4.20 High-speed long-wavelength *pin* with coplanar waveguide output: (a) coplanar electrode layout; (b) section of the epitaxial structure; (c) detail of the active region (dashed box in (a)).

is then substituted by a circular electrode. In this case, for reasons related to the carrier speed and resulting transit time limited bandwidth, the preferred epitaxial structure would be with n^+ surface layer and p^+ bottom layer.

A guard ring, see Fig. 4.35, can be put around the *pn* junction to prevent breakdown due to the high field generated at the edge of the diffused region and reduce the device junction capacitance; however, this solution is more common in avalanche photodiodes due to the potentially high operating voltages and fields.

4.9 The *pin* frequency response

Four main mechanisms limit the speed of *pin* photodiodes under dynamic excitation:

1. the effect of the total diode capacitance, including the depleted region diode capacitance and any other external parasitic capacitance;
2. the transit time of the carriers drifting across the depletion layer of width W;

3. the diffusion time of carriers generated outside the undepleted regions (mainly in homojunction devices);
4. the charge trapping at heterojunctions (in heterojunction devices).

Transit time effects are negligible in *pn* junction photodiodes owing to the small depletion region width, but become a dominant mechanism in *pin* devices. Transit time and *RC* cutoff are thus the main limitations in practical, technology-optimized *pin* photodiodes.

4.9.1 Carrier diffusion and heterojunction charge trapping

Photogenerated carrier diffusion in the cladding layers introduces into the device response a slow component, already discussed in the treatment of the *pn* junction diode. In *pin* devices, such a component is potentially marginal due to the large value of W, but can be detected as a slow time constant in the time response; see Fig. 4.21. The diffusion current component can be minimized in homojunction devices by cladding the intrinsic layer between heavily doped layers; in this case, the carrier mobility μ decreases owing to strong impurity scattering and the diffusivity D decreases (remember $D/\mu = k_B T/q$). Addition of recombination centers leads at the same time to a decrease in the carrier lifetime τ, finally implying a decrease in the diffusion length $L = \sqrt{D\tau}$ (remember that the diffusion component depends on αL; see (4.33)). In compound semiconductor devices, making the top and/or bottom layers from a widegap material ultimately prevents absorption in the cladding layers.

Heterojunction *pin* detectors can be affected by charge trapping at heterojunctions, related to the valence and/or conduction band discontinuities introduced there. Consider, for instance, an InP/InGaAs/InP PiN photodiode, in which the active region is the ternary InGaAs layer. Charge trapping can be minimized by not letting the two junctions between doped and intrinsic materials (in particular the surface junction) coincide with the heterojunctions. In this way, for example, the Pi junction becomes a Ppi junction, where the homojunction and heterojunction are very close to the top window layer. From a technological standpoint, this can be obtained by diffusion of a p-type dopant into the intrinsic layer. Another possible technique, less common in practical

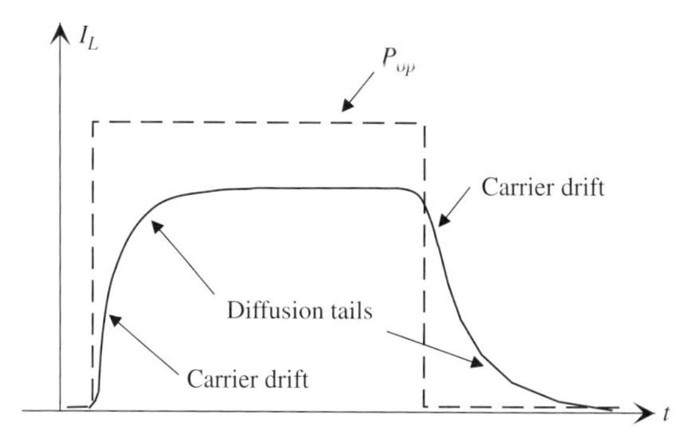

Figure 4.21 Effect of diffusion currents on the *pin* dynamic response.

devices, is inserting a thin quaternary layer or a superlattice graded bandgap layer at the hetero-interface. Another point to be considered is the need to remove the *pi* junction from the surface of the device, in order to avoid surface trapping effects; this can be obtained in mesa-type structures also by keeping the junction entirely inside the device volume.

4.9.2 Dynamic *pin* model and space-charge effects

In order to analyze the dynamics of the electrons and holes photogenerated in the intrinsic region, let us consider a time-domain model including the 1D continuity equations for electrons and holes:

$$\frac{\partial p}{\partial t} = -\frac{p - p_0}{\tau_b} + G_{op}(x, t) - \frac{1}{q}\frac{\partial J_h}{\partial x} \tag{4.36}$$

$$\frac{\partial n}{\partial t} = -\frac{n - n_0}{\tau_b} + G_{op}(x, t) + \frac{1}{q}\frac{\partial J_n}{\partial x}; \tag{4.37}$$

τ_b is the excess carrier lifetime, p_0 and n_0 are the equilibrium carrier densities, and the electron and hole current densities are

$$J_h = q v_h(\mathcal{E})\, p - q D_h \frac{\partial p}{\partial x}$$

$$J_n = q v_n(\mathcal{E})\, n + q D_n \frac{\partial n}{\partial x},$$

where v_n and v_h are the field dependent electron and hole drift velocities. The above equation must be then coupled to the Poisson equation:

$$\frac{\partial \mathcal{E}}{\partial x} = \frac{\rho}{\epsilon_s},$$

where ρ is the total charge density (evaluated neglecting the background doping in the intrinsic region, i.e., as $\rho = qp - qn$).

In the intrinsic region, photogenerated carriers, under the effect of the electric field induced by the applied bias, assume opposite drift velocities, so that electrons and holes separate. Since their drift velocities are in principle different in absolute value, some charge imbalance arises, which in turn perturbs (screens) the driving electric field.

We will assume that such space charge effects are negligible, i.e., that the driving electric field is the external field only, and that all diffusion currents are negligible. Among diffusion currents we also include those in the depletion region; in low-field conditions one has (e.g., for holes):

$$J_h = q v_h(\mathcal{E})\, p - q D_h \frac{\partial p}{\partial x} \approx q\mu_h \mathcal{E} p - q D_h \frac{\partial p}{\partial x}$$

$$\approx q\mu_h \frac{|V_A|}{W} p - q\frac{k_B T}{q}\mu_h \frac{\Delta p}{W} = \frac{q\mu_h}{W}\left(|V_A|\, p - \frac{k_B T}{q}\Delta p\right),$$

where we have assumed that the injected charge in the intrinsic region is approximately linear with slope $\Delta p / W$; considering that $\Delta p \approx p$ at worst, the diffusion current contribution is negligible when $|V_A| \gg k_B T/q = 26$ mV at ambient temperature, a condition that is again typically met in practical devices. Moreover, we will also assume that the transit time of drifting photogenerated carriers, τ_{dr} (defined as the average time photocarriers need to reach the external contact), is $\ll \tau_b$, so that carriers cannot recombine during their drift to the external circuit. Finally, the electric field is assumed high enough to saturate the carrier velocity (this approximately means $\mathcal{E} \geq 10$ kV/cm, i.e., $\mathcal{E} \geq 1$ V/μm for InGaAs, a condition that is easily obtained in practical devices where $W \approx$ μm).

Neglecting space charge effects is legitimate when the incident optical power is small; increasing P_{in} inevitably leads to an increase of the space charge ρ in the intrinsic region; the external electric field is screened by the space charge and becomes nonuniform, thus decreasing the photocarrier driving force; eventually, the photocurrent saturates. To approximately evaluate the impact of the space charge effect, assume that $v_n \approx v_h \approx v$ or, in the low-field approximation, $\mu_n \approx \mu_h \approx \mu$; moreover, $n \approx p$; it follows that

$$J_L = J_h + J_n \approx q\mu_h \mathcal{E}_0 p + q\mu_n \mathcal{E}_0 n \approx 2q\mu\mathcal{E}_0 n = 2qvn,$$

where $\mathcal{E}_0$ is the electric field induced by the applied bias. The charge density $|\rho| = qn = qp$ associated with the electrons or holes is

$$|\rho| \approx \frac{J_L}{2\mu\mathcal{E}_0} = \frac{J_L}{2v}.$$

If we assume, somewhat inconsistently, that, due to the displacement of photogenerated carriers, such a charge density also acts as an uncompensated density in the Poissons equation, we have

$$\frac{\partial \mathcal{E}}{\partial x} = \frac{\rho}{\epsilon_s} = \frac{J_L}{2\mu\epsilon_s\mathcal{E}_0} = \frac{J_L}{2\epsilon_s v},$$

where $\mathcal{E}$ is the extra field induced by uncompensated space charge. Integrating over the intrinsic layer (and assuming constant current density) we obtain

$$\mathcal{E} = \frac{J_L W}{2\mu\epsilon_s\mathcal{E}_0} = \frac{J_L W}{2\epsilon_s v}. \tag{4.38}$$

Therefore, the electric field induced by the space charge $\mathcal{E}$ is negligible with respect to the external field $\mathcal{E}_0$ if (*low-field case*):

$$\mathcal{E} = \frac{J_L W}{2\mu\epsilon_s\mathcal{E}_0} \ll \mathcal{E}_0 \rightarrow |\mathcal{E}_0| \gg \sqrt{\frac{J_L W}{2\mu\epsilon_s}},$$

or if (*velocity-saturation case*):

$$\mathcal{E} = \frac{J_L W}{2\epsilon_s v} \ll \mathcal{E}_0 \rightarrow |\mathcal{E}_0| \gg \frac{J_L W}{2\epsilon_s v}.$$

The above criteria permit estimation of the photodetector saturation power.

Example 4.2: Consider an AlGaAs/GaAs *pin* photodiode with area $A = 200\,\mu\text{m}^2$. The active region thickness is $W = 2\,\mu\text{m}$, and the device responsivity is $\mathfrak{R} = 0.5$. Assuming an applied reverse voltage of 4 V, a low-field mobility $\mu = 3000\,\text{cm}^2\text{V}^{-1}\text{s}^{-1}$ and a saturation velocity $v = 10^7$ cm/s, evaluate the optical saturation power as the input power for which the field induced in the intrinsic region by the external voltage is $\mathcal{E}_0 = 10\mathcal{E}$, where $\mathcal{E}$ is given by (4.38). Assume $\epsilon_s = 13\epsilon_0$.

An applied voltage of 4 V leads to an internal field on the intrinsic GaAs layer:

$$\mathcal{E}_0 = \frac{4}{2 \times 10^{-4}\,\text{cm}} = 2 \times 10^4\ \text{V/cm}$$

corresponding to a theoretical low-field velocity,

$$v = \mu\mathcal{E}_0 = 3000 \cdot 2 \times 10^4 = 6 \times 10^7\ \text{cm/s} \gg v.$$

Therefore we should use the expression for saturated velocity,

$$|\mathcal{E}_0| = 10 \cdot \frac{J_L W}{2\epsilon_s v} = 10 \cdot \frac{\mathfrak{R}\widetilde{P}_{in} W}{2\epsilon_s v} \rightarrow \widetilde{P}_{in} = \frac{2\epsilon_s v\,|\mathcal{E}_0|}{10\mathfrak{R}W}$$
$$= \frac{2 \cdot 13 \cdot 8.85 \times 10^{-12} \cdot 1 \times 10^7 \cdot 2 \times 10^6}{10 \cdot 0.5 \cdot 2 \times 10^{-6}} = 46.02 \times 10^7\ \text{W/m}^2,$$

leading to an input optical power

$$P_{in} = A\widetilde{P}_{in} = 46.02 \times 10^7 \cdot 200 \times 10^{-12} = 92\ \text{mW}.$$

4.9.3 Transit time analysis and transit time-limited bandwidth

Let us go back to the *pin* photocurrent analysis; neglecting the diffusion currents, as stated, and assuming constant electric field. System (4.36), (4.37) becomes

$$\frac{\partial p}{\partial t} = G_o(x,t) - \frac{1}{q}\frac{\partial J_h}{\partial x} \tag{4.39}$$

$$\frac{\partial n}{\partial t} = G_o(x,t) + \frac{1}{q}\frac{\partial J_n}{\partial x}, \tag{4.40}$$

where

$$J_h = q v_{h,\text{sat}} p, \quad J_n = q v_{n,\text{sat}} n. \tag{4.41}$$

For simplicity, the coordinate system is chosen so that $x = 0$ corresponds to the p^+i junction, $x = W$ to the in^+ junction. Since the system is linear, we can assume harmonic optical incident power at the modulation angular frequency ω,

$$p_{in}(t) = P_{in}(\omega)\text{e}^{\text{j}\omega t},$$

and work in the frequency domain. Substituting (4.41) in (4.39) and (4.40), we obtain

$$j\omega p(x) = G_o(x) - v_{h,\text{sat}}\frac{dp(x)}{dx} = G_o(0)e^{-\alpha x} - v_{h,\text{sat}}\frac{dp(x)}{dx} \tag{4.42}$$

$$j\omega n(x) = G_o(x) + v_{n,\text{sat}}\frac{dn(x)}{dx} = G_o(0)e^{-\alpha x} + v_{n,\text{sat}}\frac{dn(x)}{dx}, \tag{4.43}$$

with boundary conditions $p(W) = 0$, $n(0) = 0$. In fact, in the reverse-bias p^+i junction ($x = 0$), the total minority (electron) carrier density is zero, and so is the minority (hole) carrier density in the reverse bias in^+ junction in $x = W$. Since the system is also linear in x, we look for an exponential trial solution in the homogeneous associate (we start from the hole equation)

$$j\omega p'(x) = -v_{h,\text{sat}}\frac{dp'(x)}{dx} \tag{4.44}$$

under the form

$$p'(x) = Ae^{jkx}.$$

Substituting into (4.44), we obtain

$$j\omega Ae^{jkx} = -v_{h,\text{sat}}kAe^{jkx} \to k = -\frac{j\omega}{v_{h,\text{sat}}}.$$

We now express the solution of the complete equation (4.42) as $p_1 + p_2$, where p_2 is a particular solution of (4.42).

We assume as a trial solution $p_2 = B\exp(-\alpha x)$, so that

$$p(x) = p_1 + p_2 = Ae^{-\frac{j\omega x}{v_{h,\text{sat}}}} + Be^{-\alpha x},$$

where A and B are constants to be determined. Substituting and imposing the condition $p(W) = 0$ we obtain for the hole density

$$p(x) = \frac{G_o(0,\omega)}{j\omega - \alpha v_{h,\text{sat}}}e^{-\alpha W}\left[e^{-\alpha(x-W)} - e^{-\frac{j\omega(x-W)}{v_{h,\text{sat}}}}\right]. \tag{4.45}$$

For the electron density, we use for (4.43) the trial solution

$$n(x) = Ae^{\frac{j\omega x}{v_{n,\text{sat}}}} + Be^{-\alpha x}.$$

Substituting in the electron continuity equation and imposing condition $n(0) = 0$, we similarly have

$$n(x) = \frac{G_o(0,\omega)}{j\omega + \alpha v_{n,\text{sat}}}\left[e^{-\alpha x} - e^{\frac{j\omega x}{v_{n,\text{sat}}}}\right]. \tag{4.46}$$

From (4.45) and (4.46) the drift currents are

$$J_h(x) = qv_{h,\text{sat}}p(x) = \frac{qv_{h,\text{sat}}G_o(0,\omega)}{j\omega - \alpha v_{h,\text{sat}}}e^{-\alpha W}\left[e^{-\alpha(x-W)} - e^{-\frac{j\omega(x-W)}{v_{h,\text{sat}}}}\right]$$

$$J_n(x) = qv_{n,\text{sat}}n(x) = \frac{qv_{n,\text{sat}}G_o(0,\omega)}{j\omega + \alpha v_{n,\text{sat}}}\left[e^{-\alpha x} - e^{\frac{j\omega x}{v_{n,\text{sat}}}}\right].$$

The total current (constant in a 1D system) is the sum of $J_h(x)$, $J_n(x)$ and of the *displacement current*:

$$J_t(\omega) = J_h + J_n + \mathrm{j}\omega\epsilon_s E(x,\omega),$$

where $E(\omega)$ is the electric field harmonic component at ω. Integrating both sides from 0 to W we obtain

$$\begin{aligned}
\int_0^W J_t(\omega)\,\mathrm{d}x &= W J_t(\omega) = \int_0^W \left[J_h(x) + J_n(x) + \mathrm{j}\omega\epsilon_s E(x,\omega)\right]\mathrm{d}x \\
&= \frac{q v_{h,\mathrm{sat}} G_o(0,\omega)}{\mathrm{j}\omega - \alpha v_{h,\mathrm{sat}}} \mathrm{e}^{-\alpha W} \int_0^W \left(\mathrm{e}^{-\alpha(x-W)} - \mathrm{e}^{-\frac{\mathrm{j}\omega(x-W)}{v_{h,\mathrm{sat}}}}\right)\mathrm{d}x \\
&\quad + \frac{q v_{n,\mathrm{sat}} G_o(0,\omega)}{\mathrm{j}\omega + \alpha v_{n,\mathrm{sat}}} \int_0^W \left(\mathrm{e}^{-\alpha x} - \mathrm{e}^{\frac{\mathrm{j}\omega x}{v_{n,\mathrm{sat}}}}\right)\mathrm{d}x + \mathrm{j}\omega\epsilon_s \int_0^W E(x,\omega)\,\mathrm{d}x \\
&= \frac{q v_{h,\mathrm{sat}} G_o(0,\omega)}{\mathrm{j}\omega - \alpha v_{h,\mathrm{sat}}} \mathrm{e}^{-\alpha W} \left[\frac{\mathrm{e}^{-\alpha(x-W)}}{-\alpha} - \frac{\mathrm{e}^{-\frac{\mathrm{j}\omega(x-W)}{v_{h,\mathrm{sat}}}}}{-\mathrm{j}\omega/v_{h,\mathrm{sat}}}\right]_0^W \\
&\quad + \frac{q v_{n,\mathrm{sat}} G_o(0,\omega)}{\mathrm{j}\omega + \alpha v_{n,\mathrm{sat}}} \left[\frac{\mathrm{e}^{-\alpha x}}{-\alpha} - \frac{\mathrm{e}^{\frac{\mathrm{j}\omega x}{v_{n,\mathrm{sat}}}}}{\mathrm{j}\omega/v_{n,\mathrm{sat}}}\right]_0^W + \mathrm{j}\omega\epsilon_s \left[-V\right]_0^W .
\end{aligned}$$

Solving:

$$\begin{aligned}
J_t(\omega) = {} & \frac{1}{W} \frac{q v_{h,\mathrm{sat}} G_o}{\mathrm{j}\omega - \alpha v_{h,\mathrm{sat}}} \mathrm{e}^{-\alpha W} \left(\frac{\mathrm{e}^{\alpha W} - 1}{\alpha} + \frac{1 - \mathrm{e}^{\frac{\mathrm{j}\omega W}{v_{h,\mathrm{sat}}}}}{\mathrm{j}\omega/v_{h,\mathrm{sat}}}\right) \\
& + \frac{q v_{n,\mathrm{sat}} G_o}{\mathrm{j}\omega + \alpha v_{n,\mathrm{sat}}} \left(\frac{1 - \mathrm{e}^{-\alpha W}}{\alpha} + \frac{1 - \mathrm{e}^{\frac{\mathrm{j}\omega W}{v_{n,\mathrm{sat}}}}}{\mathrm{j}\omega/v_{h,\mathrm{sat}}}\right) + \mathrm{j}\omega \frac{\epsilon_s}{W} \left[V(0) - V(W)\right].
\end{aligned}$$

Introducing the electron and hole transit times,

$$\tau_{dr,n} = \frac{W}{v_{n,\mathrm{sat}}}, \quad \tau_{dr,h} = \frac{W}{v_{h,\mathrm{sat}}},$$

and the expression for the optical generation rate (accounting for surface reflection and the intrinsic quantum efficiency),

$$G_o(0,\omega) = \eta_Q \frac{(1-R)}{Ahf} \alpha P_{in}(\omega),$$

we can express the photodiode total current $I_t = A J_t$ as

$$
\begin{aligned}
I_t(\omega) &= \alpha W \frac{q}{hf} \eta_Q (1 - R) P_{in}(\omega) \\
&\times \left\{ \frac{e^{-\alpha W} - 1}{\alpha W \left(\alpha W - j\omega\tau_{dr,h}\right)} + e^{-\alpha W} \frac{e^{j\omega\tau_{dr,h}} - 1}{j\omega\tau_{dr,h} \left(\alpha W - j\omega\tau_{dr,h}\right)} \right. \\
&+ \left. \frac{1 - e^{-\alpha W}}{\alpha W \left(j\omega\tau_{dr,n} + \alpha W\right)} + \frac{1 - e^{j\omega\tau_{dr,n}}}{j\omega\tau_{dr,n} \left(j\omega\tau_{dr,n} + \alpha W\right)} \right\} + j\omega \frac{A\epsilon_s}{W} V_A(\omega) \\
&= -I_L(\omega) + j\omega C V_A(\omega).
\end{aligned}
$$

In the above equation, the term in braces is the small-signal short-circuit photocurrent $-I_L$, while the last term is the current absorbed by the intrinsic layer geometric capacitance, vanishing if a DC bias only is applied to the diode (since in this case the signal voltage $V_A(\omega) = 0\ \forall\omega$ and the small-signal load is a short circuit). Note that for $\omega \to 0$ the above equation reduces to

$$
I_t(0) = -I_L(0) = -\frac{q}{hf} \eta_Q (1 - R) P_{in}(0) \left[1 - \exp(-\alpha W)\right],
$$

which coincides with (4.33) on the assumption of negligible αW_p and αL_{hn}. In conclusion, the small-signal photocurrent component at the modulation frequency ω is

$$
\begin{aligned}
I_L(\omega) &= \alpha W \frac{q}{hf} \eta_Q (1 - R) P_{in}(\omega) \\
&\times \left\{ \frac{1}{\alpha W - j\omega\tau_{dr,h}} \left[\frac{1 - e^{-\alpha W}}{\alpha W} + e^{-\alpha W} \frac{1 - e^{j\omega\tau_{dr,h}}}{j\omega\tau_{dr,h}} \right] \right. \\
&- \left. \frac{1}{\alpha W + j\omega\tau_{dr,n}} \left[\frac{1 - e^{-\alpha W}}{\alpha W} + \frac{1 - e^{j\omega\tau_{dr,n}}}{j\omega\tau_{dr,n}} \right] \right\}.
\end{aligned}
\tag{4.47}
$$

The expression of the *pin* frequency response is rather complex, and the cutoff frequency can only be evaluated numerically. However, there are a number of particular cases in which the modulation bandwidth can be estimated analytically. Let us define the *pin* normalized responsivity:

$$
\begin{aligned}
\mathfrak{r}(\omega) = \frac{I_L(\omega)}{I_L(0)} &= \frac{1}{\alpha W - j\omega\tau_{dr,h}} \left[\frac{1}{\alpha W} + \frac{1 - e^{j\omega\tau_{dr,h}}}{j\omega\tau_{dr,h}} \frac{1}{e^{\alpha W} - 1} \right] \\
&- \frac{1}{\alpha W + j\omega\tau_{dr,n}} \left[\frac{1}{\alpha W} + \frac{1 - e^{j\omega\tau_{dr,n}}}{j\omega\tau_{dr,n}} \frac{e^{\alpha W}}{e^{\alpha W} - 1} \right].
\end{aligned}
\tag{4.48}
$$

The frequency response simplifies in the following limiting cases:

- The diode is *thick*, i.e., $\alpha W \gg 1$; in this case we have

$$
|\mathfrak{r}(\omega)| \approx \left| \frac{\sin\left(\dfrac{\omega\tau_{dr,n}}{2}\right)}{\dfrac{\omega\tau_{dr,n}}{2}} \right|.
$$

  The 3 dB bandwidth condition, with $\xi = j\omega\tau_{dr,n}/2$, is

$$
20 \log_{10} \left|\mathfrak{r}(\omega_{3dB,tr})\right| = 20 \log_{10} \left| \frac{\sin(\xi)}{\xi} \right| = -3,
$$

i.e.,

$$\frac{\omega_{3\text{dB},tr}\tau_{dr,n}}{2} \approx 1.391 \rightarrow f_{3\text{dB},tr} = \frac{2 \times 1.391}{2\pi}\frac{1}{\tau_{dr,n}} = 0.443\frac{v_{n,\text{sat}}}{W}.$$

Notice that $f_{3\text{dB},tr}$ is the so-called *transit time-limited cutoff frequency*, which depends, in this approximation, on the transit time of the minority carriers generated in the most illuminated part of the intrinsic region, i.e., close to the p^+ surface side; in the present case, such carriers are the electrons.

For a diode illuminated from the back (n^+ side), one would have instead

$$f_{3\text{dB},tr} = 0.443\frac{v_{h,\text{sat}}}{W}.$$

Since holes are slower than electrons, illumination should rather come from the p^+ side to maximize the device speed.

Assuming, on the other hand, that both carriers have the same transit time we obtain the approximate expression

$$f_{3\text{dB},tr} \approx \frac{1}{2.2\tau_t},$$

where τ_t is the electron or hole transit time.

- The diode is *thin* ($\alpha W \ll 1$); in this case, the generation of pairs along the i layer is almost uniform and the frequency response is limited by both carriers; an approximation of the cutoff frequency is given by [40]:

$$f_{3\text{dB},tr} = \frac{3.5\bar{v}}{2\pi W}, \quad \text{where} \quad \frac{1}{\bar{v}^4} = \frac{1}{2}\left(\frac{1}{v_{n,\text{sat}}^4} + \frac{1}{v_{h,\text{sat}}^4}\right).$$

In practice, the difference between the thin and thick diode approaches may be slight if the electron and hole saturation velocities are similar. Consider, for instance, the example in Fig. 4.22, where the active region width is 1.5 μm and the absorption varies from the thin to the thick diode condition; the transit-time cutoff frequency $f_{3\text{dB},tr}$, evaluated numerically from (4.48), undergoes a variation of 10% in the whole range, with a maximum around $\alpha W = 2$.

Independent of the kind of approximation made, the transit time-limited cutoff frequency increases with decreasing W. In the example shown in Fig. 4.23, where α is varied between 0.01 and 100 μm^{-1} and both the thin and the thick formulae are used, together with the general expression for the response, we see that the cutoff frequency is inversely proportional to W and almost independent of α. Values in excess of 100 GHz can be obtained for very thin diodes, but in that case the responsivity will be very low due to the low efficiency. Such a speed–efficiency trade-off will be discussed in Section 4.9.5.

4.9.4 Capacitance-limited bandwidth

The photocurrent (4.47) derived from the frequency-domain transit time analysis can be associated with the photocurrent generator in the detector equivalent circuit; see

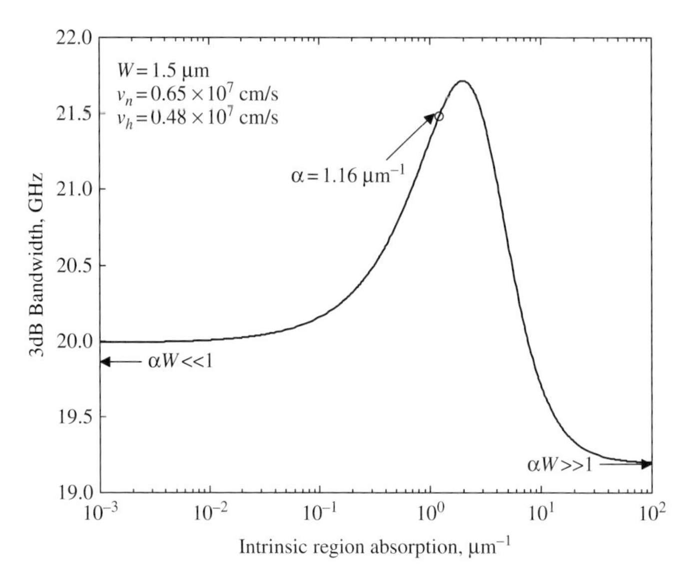

Figure 4.22 3 dB transit-time bandwidth $f_{3\mathrm{dB},tr}$ of pin photodiode versus the active region absorption.

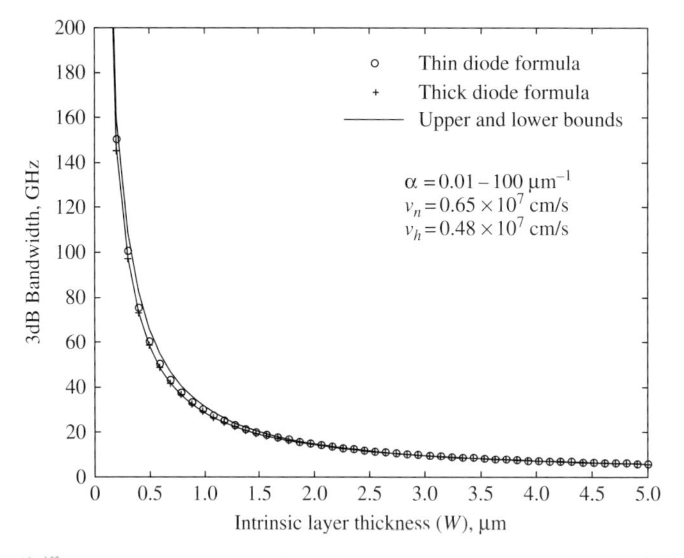

Figure 4.23 3 dB transit-time pin bandwidth $f_{3\mathrm{dB},tr}$ versus the active (intrinsic) layer thickness. The upper and lower bounds are evaluated from (4.48) on the whole absorption range.

Fig. 4.24. From the equivalent circuit we can also estimate the RC-limited cutoff frequency. Let C_p be the external diode parasitic capacitance and R_s the series parasitic diode resistance, R_D the parallel diode resistance, C_j the intrinsic capacitance (dominated by the intrinsic layer capacitance); we have, in the approximation

$$R_D \gg R_s, R_L,$$

that the 3 dB RC-limited photodiode bandwidth is given by

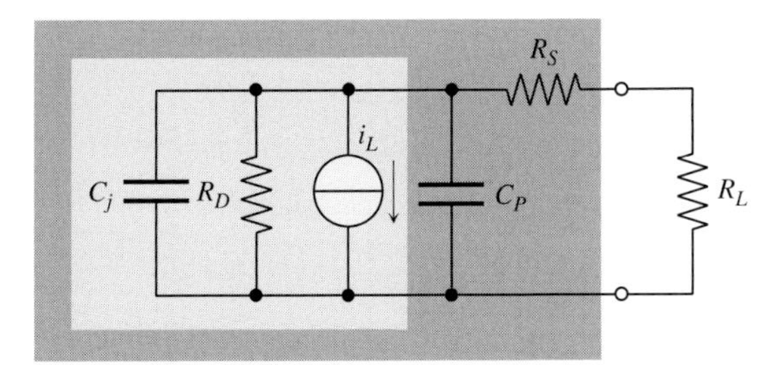

Figure 4.24 Simplified *pin* photodiode equivalent circuit; the light gray box is the intrinsic device, the dark gray box is the device with parasitics. The dark current is neglected.

$$f_{3\mathrm{dB},RC} \approx \frac{1}{2\pi RC},$$

where

$$R \approx R_S + R_L, \quad C \approx C_j + C_p, \quad C_j = \frac{\epsilon_s A}{W}.$$

The total cutoff frequency resulting from the transit time and RC effect can be evaluated at a circuit level; an approximate expression is

$$f_{3\mathrm{dB}} \approx \frac{1}{\sqrt{f_{3\mathrm{dB},RC}^{-2} + f_{3\mathrm{dB},tr}^{-2}}}. \tag{4.49}$$

4.9.5 Bandwidth–efficiency trade-off

In a vertically illuminated photodiode, optimization of the external quantum efficiency suggests $W \gg L_\alpha = 1/\alpha$; moreover, the detection area A should be large in order to improve coupling with the external source (e.g., an optical fiber). However, increasing W increases the RC-limited bandwidth (since it decreases the junction capacitance) but decreases the transit time-limited bandwidth. Increasing the device area has no influence on the transit time-limited bandwidth but makes the capacitance larger and therefore decreases the RC-limited bandwidth. Keeping the device area A constant, we therefore have $f_{3\mathrm{dB},RC} \propto W$ but $f_{3\mathrm{dB},tr} \propto 1/W$. Since $f_{3\mathrm{dB}} < \min(f_{3\mathrm{dB},RC}, f_{3\mathrm{dB},tr})$, the total bandwidth is dominated by $f_{3\mathrm{dB},RC} \propto W$ (low W) or $f_{3\mathrm{dB},tr} \propto 1/W$ (large W). The total bandwidth then first increases as a function of W, then decreases; $f_{3\mathrm{dB}}$ therefore has a maximum, which shifts toward smaller values of W and larger cutoff frequencies with decreasing A. At the same time, the efficiency always increases with W. As a consequence, high-frequency operation (high $f_{3\mathrm{dB}}$) requires small-area diodes, with small W and increasingly poor efficiency.

An example of the trade-off between speed and efficiency is presented in Fig. 4.25, showing the efficiency (we assume zero surface power reflectivity) and the bandwidth of an InGaAs *pin* photodiode with circular illuminated area. Only the intrinsic junction capacitance is considered; in a real high-speed device as much as 50% of the total capacitance may derive from parasitics. From Fig. 4.25, we see that 40 Gbps operation requires diodes with diameter as low as 20 μm; the resulting efficiency is lower than

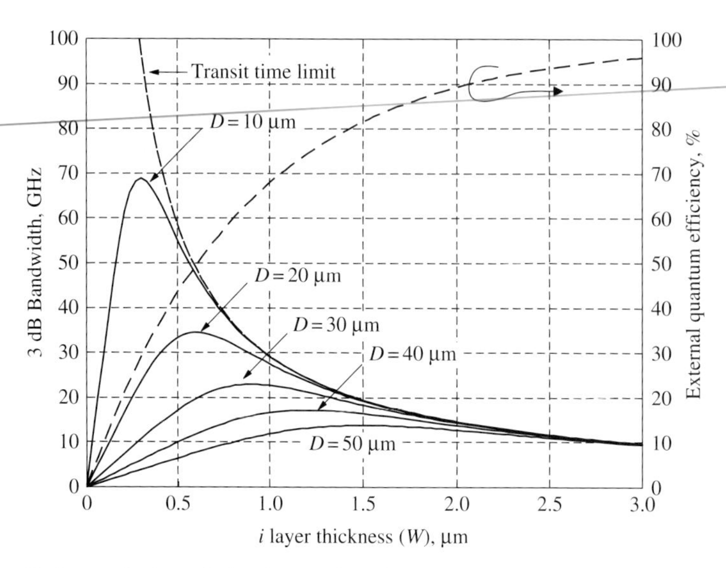

Figure 4.25 High-speed *pin* optimization: trade-off between speed and efficiency.

50%. Commercially available long-wavelength, high-speed diodes exhibit responsivities around 0.7–0.9 A/W and have active region thicknesses $W < 1\,\mu$m. Devices with top or bottom illumination are possible (bottom illumination allows increase in the active device area, since this is not partly covered by a ring contact).

4.10 Advanced *pin* photodiodes

Although conventional (vertical) *pin* photodiodes can today achieve bandwidths well in excess of 40 Gbps, such structures are close to the limit performances. Their main limitation, i.e., the efficiency–bandwidth trade-off, can be overcome by properly modifying the device design.

From a physical standpoint, the efficiency–bandwidth trade-off originates from the fact that photons and photogenerated carriers run parallel. This implies that increasing the thickness of the absorption layer (and, therefore, the efficiency) automatically leads to an increase of the transit time. However, if the angle between the power flux and the collected photocarrier path is made larger than zero, the absorption region can be made wider while preserving a low transit time. Alternatively, if photons are made to cross the active region several times, absorption can be high even with a low value of the photocarrier transit time.

The first idea is implemented in the refracting facet photodetector (RFPD, Fig. 4.26, inset) [41]. Using a refracting facet, the angle of the light crossing the active region, ϕ (Fig. 4.26, inset), is decreased from the value $\pi/2$ (typical of a conventional structure), and the efficiency can be optimized by increasing the length of the light path in the active region. Photocarriers to be collected travel vertically, following a path shorter than that of the photons. Moreover, photons are reflected by the top layer, thus doubling the

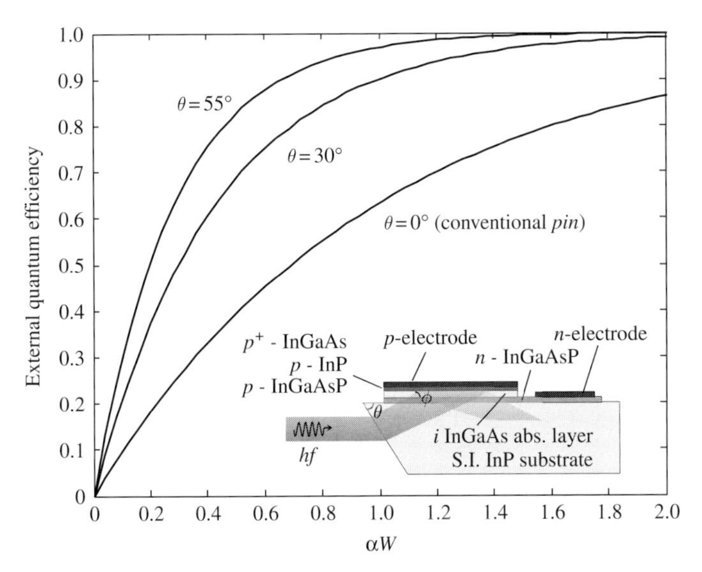

Figure 4.26 Efficiency of an RFPD compared to the conventional *pin* diode. The inset showing the RFPD structure is adapted from [41], Fig. 1.

absorption region's effective thickness. In the best case (100% metal reflectivity), given an active region thickness W, the effective absorption thickness W_{eff} will therefore be

$$W_{\text{eff}} = \frac{2W}{\cos\theta},$$

where $\theta = \pi/2 - \phi$, and the external efficiency will depend on W_{eff},

$$\eta_x \approx \eta_Q (1 - R) \left(1 - e^{-\alpha W_{\text{eff}}}\right), \tag{4.50}$$

while the carrier transit time will of course be related to W. The improvement in the efficiency vs. the angle θ following from (4.50) is shown in Fig. 4.26. In practice, the RFPD operation is made more critical by the coupling of the light source (e.g., a fiber) with a slanted surface; moreover, the efficiency is decreased by the finite reflectivity of the top metal layer. In the above remarks, we have neglected the effect of refraction at the heterojunctions.

In *resonant cavity detectors*, light is made to cross the absorption layer several times by inserting such a layer in a vertical optical resonator confined by front and back Bragg reflectors (a grating obtained by stacking several semiconductor layers of low and high refractive index, see Section 5.8.4). This solution increases the efficiency but, since the cavity is resonant, the optical bandwidth is narrow.

4.10.1 Waveguide photodiodes

In waveguide photodiodes, the photon flux and the photocarriers' motion are orthogonal. Light is guided by an optical waveguide made of an intrinsic narrow-gap semiconductor layer, sandwiched between two highly doped widegap layers

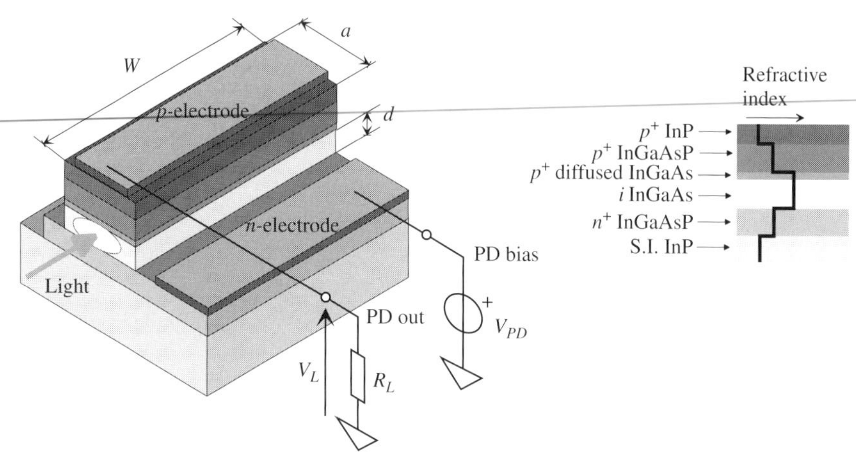

Figure 4.27 Schematic structure of an InGaAs waveguide photodiode (left) and details of the epitaxial structure (right) showing the guiding refractive index profile.

(see Fig. 4.27). Applying a reverse bias voltage, photocarriers are collected by the doped layers after a very short transit time (the waveguide height is typically $d < 1\,\mu\text{m}$); as the waveguide length can be designed so that $W \gg L_\alpha$, the majority of photons are absorbed, without affecting the transit time. Since the width a of the waveguide and of the top electrode is $a \approx 2-5\,\mu\text{m}$, the capacitance is low, so that high-frequency operation is not compromised by RC cutoff either. A disadvantage of the waveguide approach is the fact that the optical field is not completely confined by the narrowgap waveguide core, but partly extends into the widegap cladding, where absorption is negligible; the waveguide effective absorption is therefore a properly averaged value $\alpha_{\text{eff}} = \Gamma_{ov}\alpha$, where $\Gamma_{ov} < 1$ is the *overlap integral* and α is the core absorption. The overlap integral is a relative measure of the optical field energy within the absorbing part of the optical waveguide, vs. the total energy:

$$\Gamma_{ov} = \frac{\int_0^d |E_{op}|^2\,\mathrm{d}z}{\int_{-\infty}^{\infty} |E_{op}|^2\,\mathrm{d}z} < 1,$$

where the absorption layer extends from $z = 0$ to $z = d$ and E_{op} is the optical field. The same concept is exploited in laser modeling; see Section 5.5.1.

Denoting by d the active (absorption) region thickness, by a the junction width, and by W the optical waveguide length (see Fig. 4.27), the photodiode external device quantum efficiency and responsivity are:

$$\eta_x \approx \eta_Q (1 - R)\left(1 - \mathrm{e}^{-\Gamma_{ov}\alpha W}\right) \tag{4.51}$$

$$\mathfrak{R} = \frac{q}{hf}\eta_Q (1 - R)\left(1 - \mathrm{e}^{-\Gamma_{ov}\alpha W}\right), \tag{4.52}$$

while again $f_{3\text{dB},RC} = 1/2\pi RC$, where

$$R \approx R_S + R_L, \quad C \approx C_j + C_p, \quad C_j = \frac{\epsilon_s a W}{d}.$$

Finally, the transit time-limited cutoff frequency (assuming $\Gamma_{ov}\alpha d \ll 1$) will be

$$f_{3\mathrm{dB},tr} = \frac{3.5\bar{v}}{2\pi d}, \quad \text{where} \quad \frac{1}{\bar{v}^4} = \frac{1}{2}\left(\frac{1}{v_{n,\mathrm{sat}}^4} + \frac{1}{v_{h,\mathrm{sat}}^4}\right).$$

An example of waveguide photodiode structure is shown in Fig. 4.27. In practice, the intrinsic bandwidth of such structures can be designed well in excess of 40 GHz, but it is somewhat limited by parasitics; moreover, the total efficiency tends to decrease due to coupling losses between the external medium (e.g., the optical fiber) and the waveguide. Finally, such structures suffer from earlier power saturation with respect to conventional photodiodes, due to the very small cross section. Lateral waveguiding is usually obtained by including the active region in a ridge. A similar arrangement is exploited in electroabsorption modulators, which basically are waveguide photodetectors with variable absorption; see Section 6.8.

4.10.2 Traveling-wave photodetectors

The high-speed operation of waveguide photodiodes is mainly limited by RC cutoff; further bandwidth improvements can be obtained by turning the RF electrodes into a transmission line. This can be done by properly feeding the signal from one end of the structure (see Fig. 4.28) and by loading the other end with a matched load; the detector becomes a nonconventional *coplanar waveguide,* continuously loaded by the *pin* reverse-biased capacitance. The resulting *traveling-wave* or *distributed* photodetector (TWP) is able, at least in theory, to overcome the RC limitations of the simple lumped-parameter waveguide photodiode [42]. TWPs are also interesting as an example of devices wherein the interaction between the optical and RF signal is *distributed*; a similar principle will also be applied in distributed electrooptic and electroabsorption modulators.

To simplify the treatment, suppose that the transmission line is terminated, at both sides (generator and load), on the characteristic impedance. The distributed photodiode extends from $z = 0$ to $z = W$ (see Fig. 4.28, right); let us consider an infinitesimal slice of the photodiode between z and $z + \mathrm{d}z$. Such a slice excites an infinitesimal small-signal current at t, proportional to the optical power that crosses at t section z. However,

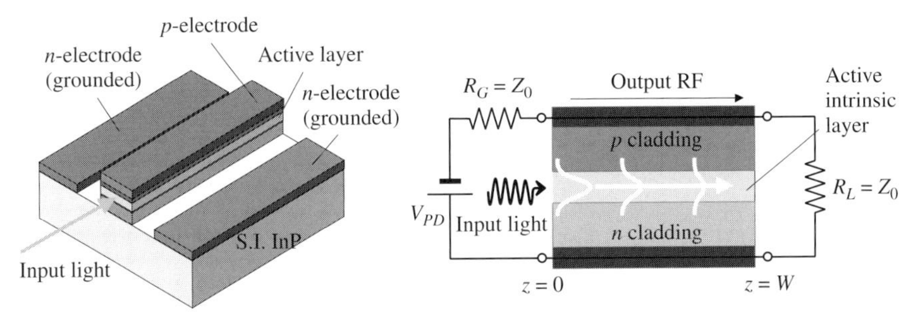

Figure 4.28 Distributed waveguide photodiode structure (left) and equivalent circuit in terms of optical and RF coupled waveguides (right) showing the generator and load conditions.

the optical signal entering the photodiode at $z = 0$ crosses section z after a delay $t_0 = z/v_o$, where v_o is the phase velocity of the optical signal. Such a delay translates, in the frequency domain, into a phase shift $\exp(-j\omega t_0) = \exp(-j\omega z/v_o) = \exp(-j\beta_o z)$, where $\beta_o = \omega/v_o$ (ω is the angular frequency of the optical signal modulation). Therefore, the parallel photocurrent excited by light at section z, taking into account power attenuation and phase shift, is

$$\begin{aligned} dI_L &= \frac{q\eta_Q}{hf}\left[P_{in}(0)e^{-j\beta_o z}e^{-\Gamma_{ov}\alpha z} - P_{in}(0)e^{-j\beta_o z}e^{-\Gamma_{ov}\alpha(z+dz)}\right] \\ &= \frac{q\eta_Q}{hf}\Gamma_{ov}\alpha P_{in}(0)e^{-\Gamma_{ov}\alpha z}e^{-j\beta_o z}\,dz, \end{aligned}$$

from which

$$\frac{dI_L}{dz} = \frac{q\eta_Q}{hf}\Gamma_{ov}\alpha P_{in}(0)e^{-\Gamma_{ov}\alpha z}e^{-j\beta_o z} = I_L'(0)e^{-(\Gamma_{ov}\alpha+j\beta_o)z}.$$

To further simplify the problem, let us suppose that the transmission line is lossless.

From the telegraphers' equations in the frequency domain, (3.10) and (3.11), we have, neglecting RF losses and introducing the source term related to the photocurrent, dI_L/dz:

$$\frac{dV}{dz} = -j\omega\mathcal{L}I \tag{4.53}$$

$$\frac{dI}{dz} = -j\omega\mathcal{C}V + \frac{dI_L}{dz}, \tag{4.54}$$

where $\mathcal{L}$ is the line inductance per unit length and $\mathcal{C}$ is the line capacitance per unit length. Differentiating (4.53) and substituting (4.54) into (4.53) we obtain

$$\begin{aligned} \frac{d^2V}{dz^2} &= -j\omega\mathcal{L}\frac{dI}{dz} = -j\omega\mathcal{L}\left(-j\omega\mathcal{C}V + \frac{dI_L}{dz}\right) \\ &= -\omega^2\mathcal{L}\mathcal{C}V - j\omega\mathcal{L}I_L'(0)e^{-(\Gamma_{ov}\alpha+j\beta_o)z}. \end{aligned}$$

We now use a trial solution of the form

$$V(z) = V_0^+e^{-j\beta_m z} + V_0^-e^{j\beta_m z} + V_Le^{-(\Gamma_{ov}\alpha+j\beta_o)z}.$$

Substituting, we obtain

$$V_L = -\frac{j\omega\mathcal{L}I_L'(0)}{(\Gamma_{ov}\alpha + j\beta_o)^2 + \beta_m^2},$$

where $\beta_m = \omega\sqrt{\mathcal{L}\mathcal{C}}$ is the transmission line propagation constant, i.e.,

$$V(z) = V_0^+e^{-j\beta_m z} + V_0^-e^{j\beta_m z} - \frac{j\omega\mathcal{L}I_L'(0)}{(\Gamma_{ov}\alpha + j\beta_o)^2 + \beta_m^2}e^{-(\Gamma_{ov}\alpha+j\beta_o)z}. \tag{4.55}$$

From the first telegraphers' equation (4.53) we then derive the current:

$$I = -\frac{1}{j\omega\mathcal{L}}\frac{dV}{dz} = \frac{V_0^+}{Z_0}e^{-j\beta_m z} - \frac{V_0^-}{Z_0}e^{j\beta_m z} - \frac{(\Gamma_{ov}\alpha + j\beta_o)\,I_L'(0)}{(\Gamma_{ov}\alpha + j\beta_o)^2 + \beta_m^2}e^{-(\Gamma_{ov}\alpha+j\beta_o)z}, \tag{4.56}$$

where $Z_0 = \sqrt{\mathcal{L}/\mathcal{C}}$ (see (3.5)), is the line characteristic impedance. The DC bias generator connected at the input behaves as a short for the sinusoidal photocurrent generated along the line. Taking into account the load (Z_0 at both sides), the boundary conditions (b.c.) in $x = 0$ and $x = W$ are therefore

$$V(0) = -Z_0 I(0), \quad V(W) = Z_0 I(W);$$

Substituting into the b.c. set the expressions for $V(z)$ (4.55) and $I(z)$ (4.56) we obtain the two constants V_0^+ and V_0^- as

$$V_0^+ = -\frac{1}{2} I_L'(0) \frac{\left[j\omega\mathcal{L} + Z_0 \left(\Gamma_{ov}\alpha + j\beta_o \right) \right]}{(\Gamma_{ov}\alpha + j\beta_o)^2 + \beta_m^2}$$

$$V_0^- = -\frac{1}{2} I_L'(0) \frac{\left[j\omega\mathcal{L} - Z_0 \left(\Gamma_{ov}\alpha + j\beta_o \right) \right]}{(\Gamma_{ov}\alpha + j\beta_o)^2 + \beta_m^2} e^{-(\Gamma_{ov}\alpha + j\beta_o + j\beta_m)W}.$$

The current on the load (i.e., the output photocurrent) is therefore, substituting $I_L'(0) = (q/hf)\eta_Q \Gamma_{ov}\alpha P_{in}(0)$,

$$I(W, \omega) = I_L(\omega) = \frac{1}{2} \frac{q}{hf} \eta_Q \Gamma_{ov}\alpha P_{in}(0) \frac{e^{-j\beta_m W} \left[1 - e^{-(\Gamma_{ov}\alpha + j\beta_o - j\beta_m)W} \right]}{\Gamma_{ov}\alpha + j\beta_o - j\beta_m}.$$

In the same way, we can evaluate the current at the device input, i.e., the current flowing into the matched load connected at $z = 0$:

$$I(0, \omega) = -\frac{1}{2} \frac{q}{hf} \eta_Q \Gamma_{ov}\alpha P_{in}(0) \frac{1 - e^{-(\Gamma_{ov}\alpha + j\beta_o + j\beta_m)W}}{\Gamma_{ov}\alpha + j\beta_o + j\beta_m}.$$

At zero frequency we have

$$I(W, 0) = \frac{1}{2} \frac{q}{hf} \eta_Q P_{in}(0) \left(1 - e^{-\Gamma_{ov}\alpha W} \right) = -I(0, 0),$$

which corresponds to *one half* of the conventional waveguide *pin* responsivity; see (4.52). Apart from the opposite current signs (deriving from the transmission line current sign convention) this simply means that in DC the distributed detector works like a lumped one, but with two loads (that are in parallel in DC), so that *half* of the photocurrent is absorbed by each load.

Assuming as the device photocurrent that absorbed by the output load in $z = W$, this can be cast into the equivalent form

$$|I_L(\omega)| = \frac{q}{hf} \eta_Q \frac{\Gamma_{ov}\alpha W P_{in}(0)}{2} \exp\left(-\frac{\Gamma_{ov}\alpha W}{2} \right) \left| \frac{\sinh \xi}{\xi} \right|,$$

where

$$\xi(\omega) = \frac{(\Gamma_{ov}\alpha + j\beta_o - j\beta_m)W}{2} = \frac{(\Gamma_{ov}\alpha + j\Delta\beta(\omega))W}{2},$$

$$\Delta\beta(\omega) = \beta_o - \beta_m = \omega \left(\frac{1}{v_o} - \frac{1}{v_m} \right).$$

The normalized frequency response is

$$
\begin{aligned}
|\mathfrak{r}(\omega)| &= \left|\frac{I_L(\omega)}{I_L(0)}\right| = \frac{\xi(0)}{\sinh \xi(0)}\left|\frac{\sinh \xi(\omega)}{\xi(\omega)}\right| \\
&= \left[1+\left(\frac{\Delta\beta}{\Gamma_{ov}\alpha}\right)^2\right]^{-1/2} \sqrt{\coth^2 \frac{\Gamma_{ov}\alpha W}{2} \sin^2 \frac{\Delta\beta W}{2} + \cos^2 \frac{\Delta\beta W}{2}}. \qquad (4.57)
\end{aligned}
$$

The maximum of $|\mathfrak{r}(\omega)|$ as a function of $\Delta\beta$ occurs when $\Delta\beta = 0$. This implies that if the optical mode and microwave velocities are the same (*synchronous condition*) the frequency response is flat and

$$
I_L(\omega) = \frac{q}{2hf}\eta_Q P_{in}(0)\left[1 - e^{-\Gamma_{ov}\alpha W}\right] \quad \forall\omega,
$$

coinciding of course with the DC photocurrent (one-half of the conventional *pin* photodiode output).

An example of the frequency response (4.57) as a function of the normalized frequency $\Delta\beta W$ is shown in Fig. 4.29; note that the sinc behavior appears only for very low values of $\Gamma_{ov}\alpha W$ (corresponding to low efficiency). Decreasing the velocity mismatch leads of course to lower values of $\Delta\beta W$ at a given frequency, i.e., to an increase of the bandwidth.

For exact *velocity matching*, the bandwidth is (theoretically) infinity, even though in practice RF losses limit the bandwidth in the synchronous condition as well. The 3 dB bandwidth can be expressed in a closed form in two cases:

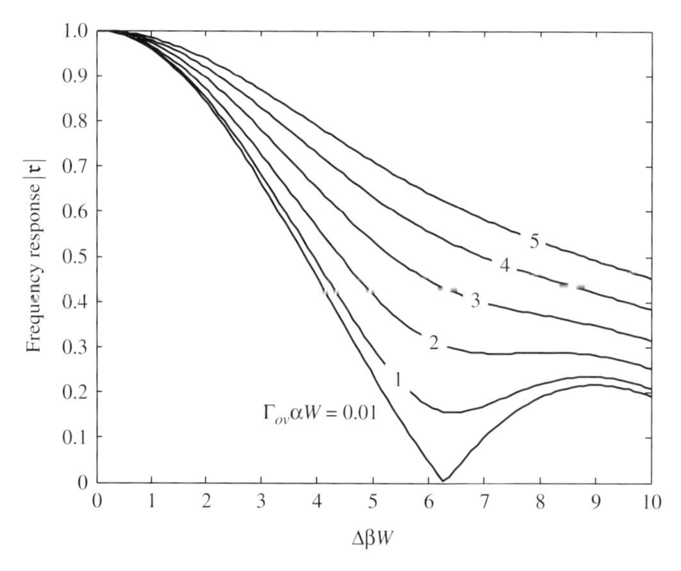

Figure 4.29 Frequency response of a distributed photodetector for different values of the absorption. RF losses have been neglected.

1. If the absorption region is short ($\Gamma_{ov}\alpha W \ll 1$) one has

$$|\mathfrak{r}(\omega)| \approx \left| \frac{\sin \dfrac{\Delta\beta W}{2}}{\dfrac{\Delta\beta W}{2}} \right|,$$

and the 3 dB bandwidth, which depends only on the velocity difference between the two modes, corresponds to the value $\Delta\beta W/2 \approx 1.391$, i.e.,

$$f_{3\text{dB}} = \frac{0.44}{W} \frac{v_o v_m}{|v_m - v_o|}.$$

In this case, the cutoff frequency is inversely proportional to the device length, and the resulting response is of $\sin x/x$ type.

2. For a (more realistic) case in which the absorption region is long ($\Gamma_{ov}\alpha W \gg 1$), it is convenient to write the response in a different form:

$$\begin{aligned} |I_L(\omega)| &= \frac{1}{2} \frac{q}{hf} \eta_Q \Gamma_{ov}\alpha P_{in}(0) \frac{\left|1 - \mathrm{e}^{-(\Gamma_{ov}\alpha + \mathrm{j}\beta_o - \mathrm{j}\beta_m)W}\right|}{\sqrt{\Gamma_{ov}^2\alpha^2 + (\beta_o - \beta_m)^2}} \\ &\approx \frac{1}{2} \frac{q}{hf} \eta_Q \Gamma_{ov}\alpha P_{in}(0) \frac{1}{\sqrt{\Gamma_{ov}^2\alpha^2 + (\beta_o - \beta_m)^2}}. \end{aligned}$$

The normalized frequency response is

$$|\mathfrak{r}(\omega)| = \left| \frac{I_L(\omega)}{I_L(0)} \right| = \frac{\Gamma_{ov}\alpha}{\sqrt{\Gamma_{ov}^2\alpha^2 + \Delta\beta^2}} = \frac{1}{\sqrt{1 + \dfrac{\omega^2}{\Gamma_{ov}^2\alpha^2}\left(\dfrac{1}{v_o} - \dfrac{1}{v_m}\right)^2}}.$$

In such a case, the photodetector 3 dB bandwidth is given by

$$f_{3\text{dB}} = \frac{\Gamma_{ov}\alpha}{2\pi} \frac{v_o v_m}{|v_m - v_o|} \approx \frac{\Gamma_{ov}\alpha}{2\pi} v \frac{v}{\Delta v}.$$

The approximation holds if $v_o \approx v_m \approx v$; $\Delta v = |v_m - v_o|$. Note that in this case ($\Gamma_{ov}\alpha W \gg 1$), $f_{3\text{dB}}$ is independent on W, because only the initial part of the device generates photocurrent, and the equivalent length whereon nonsynchronous coupling can play a role is of the order of only a few absorption lengths.

While the DC photocurrents on the generator and load impedances are equal, the frequency response on the generator impedance shows a faster decay. In fact, for a long photodiode, we have

$$|I(0)| = \frac{1}{2} \frac{q}{hf} \eta_Q \Gamma_{ov}\alpha P_{in}(0) \frac{1}{\sqrt{\Gamma_{ov}^2\alpha^2 + (\beta_o + \beta_m)^2}},$$

and the corresponding normalized frequency response is

$$\left|\mathfrak{r}'(\omega)\right| = \left| \frac{I(0,\omega)}{I(0,0)} \right| = \frac{1}{\sqrt{1 + \dfrac{\omega^2}{\Gamma_{ov}^2\alpha^2}\left(\dfrac{1}{v_o} + \dfrac{1}{v_m}\right)^2}},$$

with a 3 dB cutoff:

$$f'_{3\text{dB}} = \frac{\Gamma_{ov}\alpha}{2\pi}\frac{v_o v_m}{v_m + v_o} \ll f_{3\text{dB}}.$$

In fact, the interaction between the optical wave and the regressive RF wave entering the load in $x = 0$ is always *counterpropagating* (i.e., the two wave velocities are opposite in sign), and synchronous coupling is never achieved. In distributed photodiodes limited by velocity mismatch, the load in $z = W$ will draw the majority of the photocurrent; this conclusion is not necessarily true, however, if the speed is limited by transit time. In any case, matching the detector at both sides (generator and load) imposes a 50% penalty on the efficiency (although it increases the detector bandwidth).

Example 4.3: Consider a photodetector of length $W = 50\,\mu\text{m}$, active region thickness $d = 0.25\,\mu\text{m}$ (permittivity $\epsilon_r = 13$), width $a = 4\,\mu\text{m}$. The waveguide absorption length is $L_{\Gamma_{ov}\alpha} = 30\,\mu\text{m}$. Assume for the RF velocity $v_m = 1.2 \times 10^8$ m/s, for the optical velocity $v_o = 0.9 \times 10^8$ m/s; the photocarrier drift velocity is $\bar{v} = 1 \times 10^7$ cm/s. The characteristic impedance of the line is $Z_0 = 20\ \Omega$ and the photodetector is matched at both ends on Z_0. Evaluate the detector bandwidth in concentrated and distributed forms, taking also into account the effect of the carrier transit time. Evaluate the efficiency of the distributed and waveguide detectors and compare with a conventional, vertically illuminated detector.

Let us consider first the transit time cutoff frequency. Since light is illuminating the active region almost uniformly, the generation of e-h pairs along the i layer is approximately constant, and the frequency response is limited by both carriers; we can exploit the expression for the transit-time limited bandwidth:

$$f_{3\text{dB},tr} = \frac{3.5\bar{v}}{2\pi d} = \frac{3.5 \cdot 1 \times 10^5}{2\pi \cdot 0.25 \times 10^{-6}} = 223\,\text{GHz}.$$

Concerning the bandwidth of the distributed detector, we have (we assume from the data that the absorption region is long)

$$f_{3\text{dB},d} = \frac{\Gamma_{ov}\alpha}{2\pi}\frac{v_o v_m}{|v_m - v_o|} = \frac{1}{2\pi \cdot 30 \times 10^{-6}} \cdot \frac{0.9 \times 10^8 \cdot 1.2 \times 10^8}{1.2 \times 10^8 - 0.9 \times 10^8} = 1910\,\text{GHz}.$$

The cutoff frequency for the photocurrent on the generator internal impedance is instead

$$f'_{3\text{dB},d} = \frac{\Gamma_{ov}\alpha}{2\pi}\frac{v_o v_m}{|v_m + v_o|} = \frac{1}{2\pi \cdot 30 \times 10^{-6}} \cdot \frac{0.9 \times 10^8 \cdot 1.2 \times 10^8}{1.2 \times 10^8 + 0.9 \times 10^8} = 273\,\text{GHz}.$$

For the lumped detector, the capacitance is

$$C_j = \epsilon_0\epsilon_r\frac{aW}{d} = 8.86 \times 10^{-12} \cdot 13 \cdot \frac{4 \times 10^{-6} \cdot 50 \times 10^{-6}}{0.25 \times 10^{-6}} = 0.092\ \text{pF},$$

so that the RC cutoff ($R = Z_0/2 = 10\,\Omega$) is

$$f_{3\text{dB},RC} = \frac{1}{2\pi RC_j} = \frac{1}{2\pi \cdot 10 \cdot 0.092 \times 10^{-12}} = 173\,\text{GHz}.$$

The approximate total cutoff frequency for the distributed detector can be evaluated using the same approach as in (4.49):

$$f_{3\text{dB}}^{-2} \approx f_{3\text{dB},tr}^{-2} + f_{3\text{dB},d}^{-2}. \tag{4.58}$$

We obtain $f_{3\text{dB}} = 221\,\text{GHz}$ (distributed detector) and $f_{3\text{dB}} = 137\,\text{GHz}$ (waveguide detector). The total cutoff frequency for the photocurrent on the generator is instead $f_{3\text{dB}} = 172\,\text{GHz}$. Although $f'_{3\text{dB},d} \ll f_{3\text{dB},d}$, at high frequency also the current on the generator matched resistance is therefore far from being negligible. For the efficiency we have, neglecting reflections and coupling loss and considering only the photocurrent on the load:

$$\eta_x \approx \frac{1}{2}\left(1 - \text{e}^{-\Gamma_{ov}\alpha W}\right) = \frac{1}{2}\left(1 - \text{e}^{-50/30}\right) \approx 40.5\% \text{ distributed/waveguide}$$

$$\eta_x \approx \left(1 - \text{e}^{-\Gamma_{ov}\alpha d}\right) = \left(1 - \text{e}^{-0.25/30}\right) = 0.8\% \text{ conventional (vertical).}$$

In the vertical case we have considered the total photocurrent in both the generator and the load. In conclusion, the efficiency of the distributed and waveguide detectors is the same while the bandwidth of the distributed detector is almost twice that of the waveguide device. In both cases the efficiency is much larger than for the vertical case.

4.10.3 Velocity matched traveling-wave photodetectors

In practical implementations, the ideal version of the traveling-wave photodetector is affected by heavy RF losses; moreover, obtaining velocity matching is critical. The RF velocity depends in fact on the kind of quasi-TEM transmission line exploited in the distributed detector. While the optical refractive index is close to the active material index, the microwave effective index can be lower (due to quasi-TEM averaging between the air and the dielectric, see (4.59)) or, in some structures, much larger (and also strongly frequency-dependent) because of the slow-wave effect; see Section 6.10.

To decrease RF losses, probably the most important limitation to high-speed operation, periodically loaded distributed detectors have been proposed; see an example in Fig. 4.30, where the RF line is a quasi-TEM coplanar stripline (a TXL made of two parallel coplanar conductors, one acting as the ground plane) periodically loaded with *pin* photodiodes. Owing to the wider RF conductors vs. the design in Fig. 4.28 (where the signal conductor width is a few micrometers), the RF losses decrease dramatically and the effective length of the detector can be increased. However, over such a long structure, velocity matching becomes an issue. In fact, while the RF permittivity in a bulk semiconductor is larger than the optical permittivity, in the coplanar stripline exploited for the design of Fig. 4.30 the RF effective index is typically *lower* than the optical effective refractive index, due to the air–dielectric quasi-static averaging; see (4.59). Assume that the optical permittivity is ϵ_o, and that the RF permittivity of the semiconductor is $\epsilon_{RF} > \epsilon_o$. For a coplanar stripline, as for a coplanar waveguide, the effective line permittivity is, from (3.32)

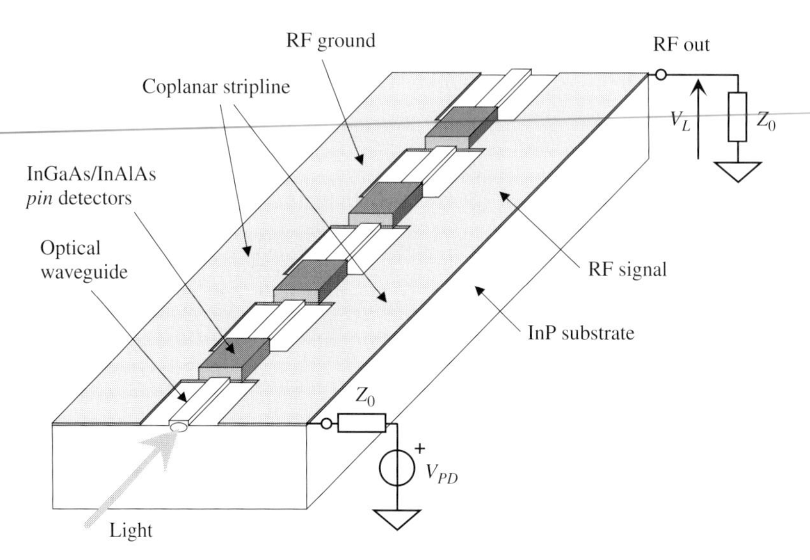

Figure 4.30 Distributed photodetector consisting of a RF waveguide periodically loaded with photodiodes.

$$\epsilon_{\text{eff}} \approx \frac{\epsilon_{RF} + 1}{2} < \epsilon_o. \tag{4.59}$$

However, the coplanar stripline in Fig. 4.30 is periodically loaded with the (shunt) lumped capacitances of the detectors. Such capacitances do not appreciably alter the low-frequency current density pattern in the RF electrodes and the associated magnetic field distribution, since they behave as open circuits. The p.u.l. line inductance $\mathcal{L}$ is therefore almost unchanged, and so is the *in vacuo* capacitance $\mathcal{C}_0 \propto \mathcal{L}^{-1}$ (3.27); at the same time, the average p.u.l. capacitance is increased due to the capacitive loading effect of the junctions. The effective permittivity of a uniform line of length W can be, in fact, expressed as the ratio between the total capacitance and the capacitance in air, the former including the shunt (junction) loading capacitance on the length W, C_j:

$$\epsilon_{\text{eff},l} = \frac{\mathcal{C}W + C_j}{\mathcal{C}_0 W} = \frac{\mathcal{C}}{\mathcal{C}_0} + \frac{C_j}{\mathcal{C}_0 W} = \epsilon_{\text{eff}} + \epsilon_l.$$

By properly selecting the amount of loading, the additional contribution ϵ_l can be chosen so that $\epsilon_{\text{eff}} + \epsilon_l = \epsilon_o$. In conclusion, capacitive loading decreases the RF velocity, so that velocity matching can be obtained, by careful design, over a wide frequency range. The resulting structure, called a velocity-matched traveling-wave detector (VMTWD) [43] as opposed to the simple traveling-wave detector (TWD), can achieve bandwidths well in excess of 100 GHz.

4.10.4 Uni-traveling carrier photodiodes

Uni-traveling carrier photodiodes (UTC-PDs) are a high-speed evolution of conventional, vertically illuminated *pin* photodiodes [44]. In the conventional *pin*, absorption takes place in the intrinsic layer, where electrons and holes are generated. The transit time is negatively affected by the lower velocity of holes vs. electrons; moreover, if the absorption layer thickness is reduced to improve the transit time, the efficiency

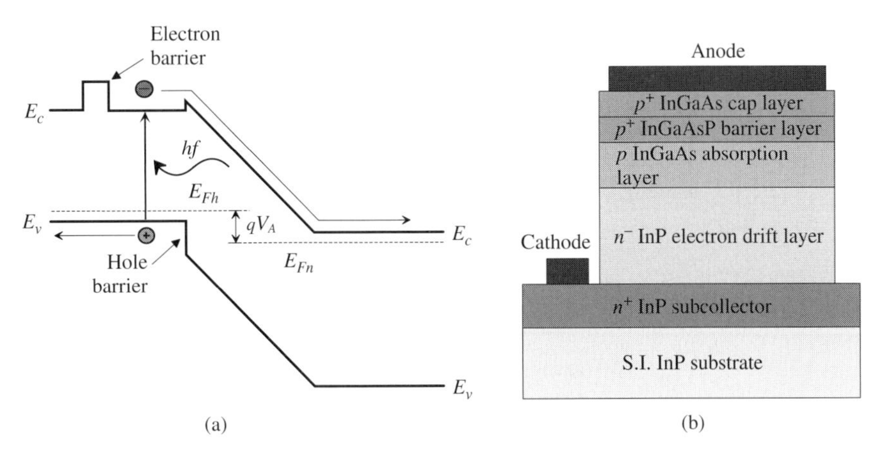

Figure 4.31 Simplified band diagram of uni-traveling carrier photodiode (UTC-PD) under reverse bias and backside illumination (a); layer structure of UTC-PD (b). Adapted from [44], Fig. 1.

and the RC-limited bandwidth decrease. UTC-PDs overcome such limitations in two ways. Firstly, photons are absorbed in a p layer where the field is low, but holes are collected by the p contact, while electrons quickly diffuse into an intrinsic or n^- high-field drift region, where they undergo quasi-ballistic (i.e., not affected by scattering events) motion to the collector. Due to the very thin absorption layer, the transit time depends only on the electron drift through the intrinsic layer, while holes do not play a role; however, due to quasi-ballistic motion, the electron transit time is low. Secondly, the electron drift layer is thick enough to limit the diode capacitance, thus increasing the RC-limited bandwidth.

Figure 4.31 shows the band diagram and layer structure of a long-wavelength UTC-PD grown on a InP substrate. Note the InGaAsP electron barrier layer and the hole barrier introduced by the InGaAs–InP valence band discontinuity. Typical absorption layers are thin in order not to decrease the response speed due to diffusion in the neutral absorption layer, with reported widths of the order of 100–200 nm, leading to rather poor responsivities (around 0.15 mA/mW); backside illumination is exploited to improve the photon collection and avoid photons being absorbed by the electron barrier layer. UTC-PDs developed by NTT Photonics Labs have demonstrated record performances for 1.55 μm detectors, with 3 dB bandwidth of 310 GHz [45]. The estimated average electron velocity is of the order of 3×10^7 cm/s, thus demonstrating velocity overshoot effects and quasi-ballistic transport in the drift region. With respect to distributed and waveguide photodetectors, UTC-PD have a much simpler, vertical structure and comparable speed; typical responsivities are lower, however.

4.11 Avalanche photodiodes

Avalanche photodiodes (APDs) exploit the avalanche multiplication of photogenerated carriers through *impact ionization* to amplify the detector current and improve the device sensitivity. A larger photocurrent (for the same illumination level) is obtained

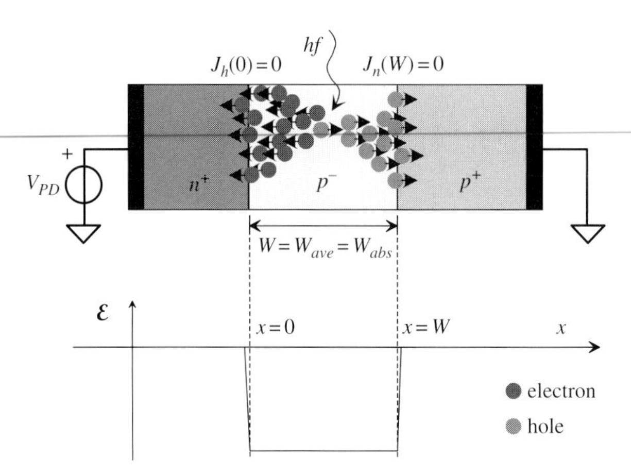

Figure 4.32 Conventional APD (above) and electric field profile (below).

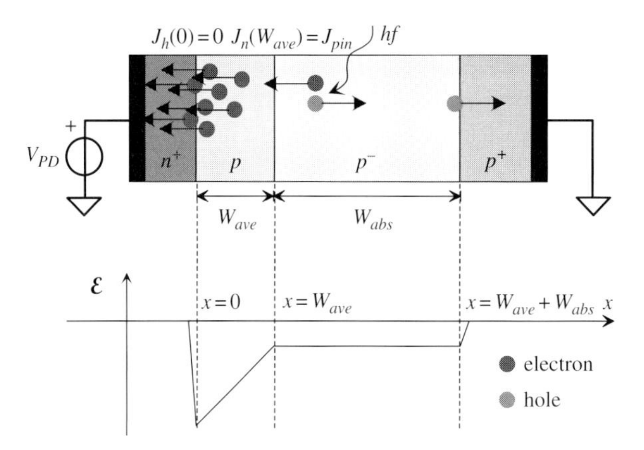

Figure 4.33 SAM-APD (electron-triggered avalanche) with electric field profile.

with respect to *pin* photodiodes, but also higher noise. In principle, the device includes two regions: the *generation region* (low to medium electric field) and the *multiplication region* (high field). The two regions can be physically the same, as in *conventional APDs* (see Fig. 4.32), or can be separated (SAM-APD: *separate absorption and multiplication APD*). In the APD shown in Fig. 4.33 the avalanche is triggered by electrons; the dual case (avalanche triggered by holes) is also possible; see Fig. 4.34.

The structure of the conventional APD is a *pin* diode in reverse bias, with negligible minority carrier diffusion currents at the edges of the depletion region, i.e., $J_h(0) = 0$ (n^+ side) and $J_n(W) = 0$ (p^+ side). The width of the avalanche and photogeneration region is $W \equiv W_{abs} \equiv W_{av}$; see Fig. 4.32. As already mentioned, photocarrier generation takes place in a high-field region, where, at the same time, avalanche multiplication by impact ionization occurs.

In the SAM-APD the absorption region is intrinsic, and multiplication takes place either in an n^+p junction at the device left-hand side, see Fig. 4.33, or in a np^+ junction at the device right-hand side, see Fig. 4.34. In the first case (Fig. 4.33), the hole current

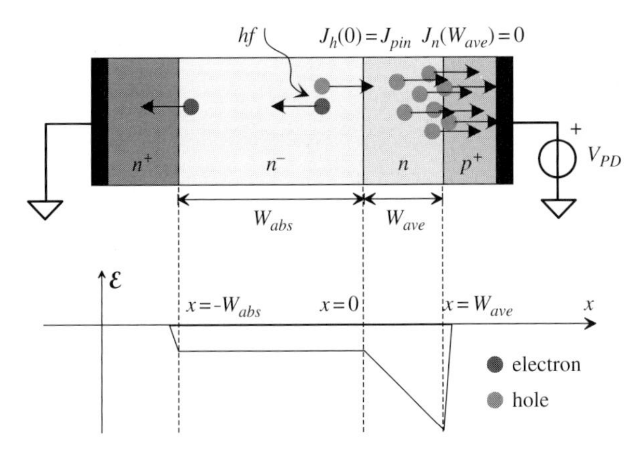

Figure 4.34 SAM-APD (hole-triggered avalanche) with electric field profile.

at the left-hand side depletion edge is negligible, $J_h(0) = 0$, while the electron current at the beginning of the avalanche region is $J_n(W_{av}) = J_{pin}$, where J_{pin} is the *primary current* photogenerated by the absorption region alone. In the second case (Fig. 4.34), the electron current at the right-hand side depletion edge is negligible, $J_n(W_{av}) = 0$, while the hole current at the beginning of the avalanche region is $J_n(0) = J_{pin}$, where J_{pin} is the primary current photogenerated by the absorption region alone.

Although the device analysis shows that the responsivity increases with increasing avalanche multiplication, which is, in turn, larger if the hole and electron ionization coefficients are similar (i.e., if both carriers contribute to the avalanche process), large values of the multiplication factor are inconvenient because they increase noise and decrease speed. Optimum noise and speed are achieved when the avalanche process is almost unipolar: this happens when the ionization coefficient of the *avalanche-triggering carrier* is much larger than the ionization coefficient of the other carrier. For some materials, this condition is met naturally; it can be artificially introduced through bandstructure engineering and the use of superlattices. From the technological standpoint, early APDs were Si- or Ge-based homojunction devices, with large operating voltages (e.g., in excess of 50 V). Today, InP-based long-wavelength APDs are available, and also the operating APD voltage has considerably decreased, thanks to device downsizing. Examples of APD structures are shown in Fig. 4.35; note the guard ring introduced into the junction. The structure is typically vertical (top or bottom illuminated), although waveguide APDs also have been proposed.

4.11.1 Analysis of APD responsivity

In order to evaluate the APD responsivity, we first consider the avalanche process in a region of width W, in which no photogeneration is present. Avalanche may originate from hole injection in $x = 0$ or electron injection in $x = W$; the result is exploited to analyze the two SAM-APD cases. The analysis of the conventional SAM (less important, since the structure does not allow the absorption and avalanche regions to be independently optimized and moreover, as discussed later, it exhibits larger noise) is

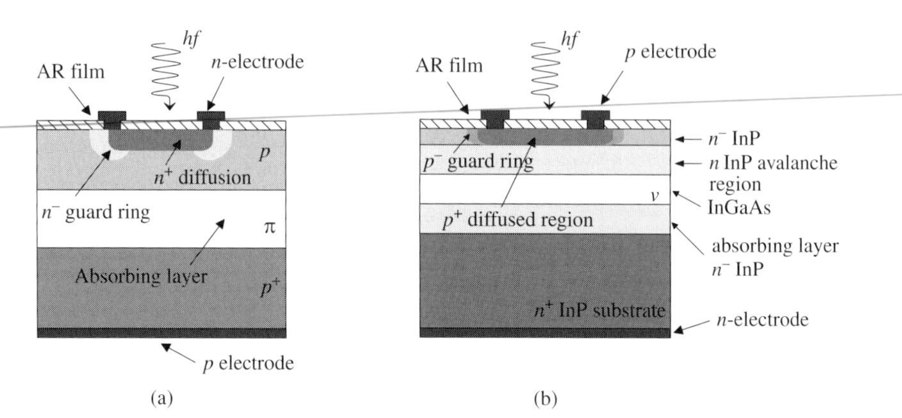

Figure 4.35 (a) Si-based APD; (b) InGaAs long-wavelength APD.

carried out assuming that photogeneration is present in the avalanche region, but no carrier injection in $x = W$ and $x = 0$.

Let us first consider a DC stationary condition, in which avalanche generation is triggered, in a region of length W and electric field $\mathcal{E}(x)$, by the injection of a hole current in $x = 0$ and an electron current in $x = W$ (i.e., the avalanche region extends from 0 to W).

From the electron and hole continuity equations, and neglecting all generation and recombination phenomena apart from avalanche multiplication, we have

$$\frac{\mathrm{d}J_n}{\mathrm{d}x} = -\alpha_h J_h - \alpha_n J_n \tag{4.60}$$

$$\frac{\mathrm{d}J_h}{\mathrm{d}x} = \alpha_h J_h + \alpha_n J_n. \tag{4.61}$$

For large electric fields, carrier currents are mainly drift currents, not diffusion currents. In this case, the electron and hole currents have the same sign, and (4.60) and (4.61) imply that J_h increases with increasing x (from left to right) while J_n increases with decreasing x (from right to left).

Taking into account that the current density $J = J_n(x) + J_h(x)$ is constant vs. x, we can derive a single equation, for example, in the hole current density. Eliminating the electron current density from (4.60), we obtain

$$\frac{\mathrm{d}J_h}{\mathrm{d}x} - (\alpha_h - \alpha_n)J_h = \alpha_n J. \tag{4.62}$$

As boundary conditions for system (4.60) and (4.61), we assign a hole current incident from 0^-, $J_h(0)$, and an electron current incident from W^+, $J_n(W)$. To solve (4.62), we actually only need the first boundary condition (as appropriate for a first-order equation).

Since α_n and α_h are generally functions of x in the presence of a nonuniform electric field, (4.62) is a first-order linear equation with variable coefficients of the kind

$$\frac{\mathrm{d}y}{\mathrm{d}x} + a(x)y = b(x), \tag{4.63}$$

with general solution

$$y(x) = \mathrm{e}^{-\int_0^x a(\xi)\,\mathrm{d}\xi}\left[y(0) + \int_0^x b(\xi)\mathrm{e}^{\int_0^\xi a(\eta)\,\mathrm{d}\eta}\,\mathrm{d}\xi\right].$$

Identifying $a(x) = -(\alpha_h - \alpha_n)$ and $b(x) = \alpha_n J$ (J constant), we obtain

$$J_h(x) = \mathrm{e}^{\int_0^x (\alpha_h-\alpha_n)\,\mathrm{d}\xi}\left[J_h(0) + J\int_0^x \alpha_n \mathrm{e}^{-\int_0^\xi (\alpha_h-\alpha_n)\,\mathrm{d}\eta}\,\mathrm{d}\xi\right]. \quad (4.64)$$

If $J_h(0)$ and $J_n(W)$ are assigned as boundary conditions, the total current density J can be expressed as

$$J = J_n(W) + J_h(W) = J_n(W) + \mathrm{e}^{\int_0^W (\alpha_h-\alpha_n)\,\mathrm{d}\xi} \times \left[J_h(0) + J\int_0^W \alpha_n \mathrm{e}^{-\int_0^\xi (\alpha_h-\alpha_n)\,\mathrm{d}\eta}\,\mathrm{d}\xi\right].$$

Solving with respect to J, we obtain

$$J = \frac{\mathrm{e}^{\int_0^W (\alpha_h-\alpha_n)\,\mathrm{d}\xi} J_h(0) + J_n(W)}{1 - \mathrm{e}^{\int_0^W (\alpha_h-\alpha_n)\,\mathrm{d}\xi}\int_0^W \alpha_n \mathrm{e}^{-\int_0^\xi (\alpha_h-\alpha_n)\,\mathrm{d}\eta}\,\mathrm{d}\xi} \equiv \frac{N}{D}$$
$$= M_h J_h(0) + M_n J_n(W), \quad (4.65)$$

where M_h and M_n are the *electron* and *hole multiplication factors*.

From (4.65) we see that, in the presence of a *finite* injected electron or hole current, the total current becomes larger (if M_h, $M_n > 1$), and can diverge due to the carrier avalanche generation. The condition $J \to \infty$ corresponds to $D \to 0$ in (4.65); however, since[11]

$$\int_0^W (\alpha_n - \alpha_h)\mathrm{e}^{-\int_0^\xi (\alpha_h-\alpha_n)\,\mathrm{d}\eta}\mathrm{d}\xi = \mathrm{e}^{-\int_0^W (\alpha_h-\alpha_n)\,\mathrm{d}\xi} - 1, \quad (4.66)$$

we immediately have

$$D = \mathrm{e}^{\int_0^W (\alpha_h-\alpha_n)\,\mathrm{d}\xi}\left[1 - \int_0^W \alpha_h \mathrm{e}^{-\int_0^\xi (\alpha_h-\alpha_n)\,\mathrm{d}\eta}\,\mathrm{d}\xi\right]. \quad (4.67)$$

Thus, the avalanche breakdown condition, corresponding to infinite current, is

$$\int_0^W \alpha_h \mathrm{e}^{-\int_0^x (\alpha_h-\alpha_n)\,\mathrm{d}\xi}\mathrm{d}x \equiv I_{h,\mathrm{ion}} = 1, \quad (4.68)$$

[11] In fact, defining

$$-\int_0^\xi (\alpha_h - \alpha_n)\,\mathrm{d}\eta = f(\xi),$$

Eq. (4.66) becomes

$$\int_0^W \mathrm{e}^{f(\xi)}\frac{\mathrm{d}f}{\mathrm{d}\xi}\,\mathrm{d}\xi = \int_{f(0)}^{f(W)} \mathrm{e}^f\,\mathrm{d}f = \mathrm{e}^{f(W)} - \mathrm{e}^{f(0)} = \mathrm{e}^{f(W)} - 1$$

since $f(0) = 0$ by definition.

where $I_{h,\mathrm{ion}}$ is the *hole ionization integral*, a function of the electron field profile and, therefore, of the applied voltage. Similarly, we can define the *electron ionization integral*:

$$I_{n,\mathrm{ion}} = \int_0^W \alpha_n \mathrm{e}^{\int_x^W (\alpha_h - \alpha_n)\,\mathrm{d}\xi}\,\mathrm{d}x. \tag{4.69}$$

From (4.65) and taking into account (4.67), (4.68) and (4.69), the multiplication factors can therefore be expressed as

$$M_h = \frac{1}{1 - \int_0^W \alpha_h \mathrm{e}^{-\int_0^\xi (\alpha_h - \alpha_n)\,\mathrm{d}\eta}\,\mathrm{d}\xi} = \frac{1}{1 - I_{h,\mathrm{ion}}}$$

$$M_n = \frac{1}{1 - \int_0^W \alpha_n \mathrm{e}^{\int_\xi^W (\alpha_h - \alpha_n)\,\mathrm{d}\eta}\,\mathrm{d}\xi} = \frac{1}{1 - I_{n,\mathrm{ion}}}.$$

The divergence condition can be equivalently formulated as $I_{n,\mathrm{ion}} = 1$ or $I_{h,\mathrm{ion}} = 1$; however, $I_{n,\mathrm{ion}}$ is more appropriate to describe electron injection from $x = W$. If one of the two injected currents vanishes, we say that the avalanche multiplication is *triggered by holes* ($J_h(0) \neq 0$, $J_n(W) = 0$) or *triggered by electrons* ($J_h(0) = 0$, $J_n(W) \neq 0$).

We consider now some simplifying cases. If the two avalanche ionization coefficients are equal ($\alpha_n = \alpha_h$), the ionization integrals become

$$I_{n,\mathrm{ion}} = \int_0^W \alpha_n\,\mathrm{d}x = I_{h,\mathrm{ion}} = \int_0^W \alpha_h\,\mathrm{d}x,$$

so that the avalanche condition is simply

$$\int_0^W \alpha_n\,\mathrm{d}x = \int_0^W \alpha_h\,\mathrm{d}x = 1. \tag{4.70}$$

Condition $\alpha_n \approx \alpha_h$ is approximately verified in some compound semiconductors, while in Si the ionization rates are significantly different. In many cases we can approximately assume that the avalanche ionization coefficients are different but *constant over the avalanche region*. In this case, the ionization integrals simplify as follows:

$$I_{n,\mathrm{ion}} = \frac{\alpha_n}{\alpha_h - \alpha_n}\left[\mathrm{e}^{(\alpha_h - \alpha_n)W} - 1\right]$$

$$I_{h,\mathrm{ion}} = \frac{\alpha_h}{\alpha_n - \alpha_h}\left[\mathrm{e}^{(\alpha_n - \alpha_h)W} - 1\right];$$

and the multiplication factors become

$$M_n = \frac{1}{1 - I_{n,\mathrm{ion}}} = \frac{\alpha_h - \alpha_n}{\alpha_h - \alpha_n \mathrm{e}^{(\alpha_h - \alpha_n)W}} = \frac{(1 - k_{hn})\,\mathrm{e}^{\alpha_n(1-k_{hn})W}}{1 - k_{hn}\mathrm{e}^{\alpha_n(1-k_{hn})W}} \tag{4.71}$$

$$M_h = \frac{1}{1 - I_{h,\mathrm{ion}}} = \frac{\alpha_n - \alpha_h}{\alpha_n - \alpha_h \mathrm{e}^{(\alpha_n - \alpha_h)W}} = \frac{k_{hn} - 1}{k_{hn}\mathrm{e}^{\alpha_n(1-k_{hn})W} - 1}, \tag{4.72}$$

where the ionization coefficient ratio

$$k_{hn} = \frac{\alpha_h}{\alpha_n}. \tag{4.73}$$

has been introduced.

We now apply the above approach to the analysis of the conventional APDs and SAM-APDs. For the electron-triggered SAM-APD shown in Fig. 4.33, we set $J_h(0) = 0$ in (4.65), and thus $J = M_n J_{pin}$, where $J_{pin} = J_n(W)$; thus, the equivalent *pin* current in multiplied by a factor $M_n > 1$,[12] so that the APD responsivity will be

$$\mathfrak{R}^n_{\text{SAM-APD}} = M_n \mathfrak{R}_{pin} = M_n \frac{q\eta_Q}{hf} (1 - R) \left(1 - e^{-\alpha W_{abs}}\right),$$

where $\mathfrak{R}_{pin}$ is the responsivity of the equivalent *pin*.

For the hole-triggered SAM-APD shown in Fig. 4.34, we set $J_n(W) = 0$ in (4.65), and thus $J = M_h J_{pin}$, where $J_{pin} = J_h(0)$; thus, the equivalent *pin* current is multiplied by a factor $M_h > 1$, so that the APD responsivity will be

$$\mathfrak{R}^h_{\text{SAM-APD}} = M_h \mathfrak{R}_{pin} = M_h \frac{q\eta_Q}{hf} (1 - R) \left(1 - e^{-\alpha W_{abs}}\right).$$

With respect to the *pin*, the avalanche photodiode is therefore characterized by a larger responsivity, possibly larger than the ideal, material-dependent, value defined in (4.9).

To model the *conventional APD*, see Fig. 4.32, we now finally consider the case where in the avalanche region photogeneration occurs, while there is no current injection in $x = 0$ and $x = W$. For simplicity, we assume that the optical generation rate is uniform; the resulting multiplication factor M_o will however be exploited (as an approximation), see (4.80), also in cases where the $W > L_\alpha$ and therefore the equivalent *pin* response is described by (4.35). Introducing the optical generation in (4.60) and (4.61),

$$\frac{dJ_n}{dx} = -\alpha_h J_h - \alpha_n J_n - qG_o \tag{4.74}$$

$$\frac{dJ_h}{dx} = \alpha_h J_h + \alpha_n J_n + qG_o, \tag{4.75}$$

we have, accounting for the constant total current,

$$\frac{dJ_h}{dx} - (\alpha_h - \alpha_n) J_h = \alpha_n J + qG_o. \tag{4.76}$$

As boundary conditions, we have $J_h(0) = 0$ and $J_n(W) = 0$. From the linear equation template (4.63), we identify $b(x) = \alpha_n J + qG_o$, thus obtaining the solution:

$$J_h(x) = e^{\int_0^x (\alpha_h - \alpha_n)\, d\xi} \times \left[J_h(0) + J \int_0^x \alpha_n e^{-\int_0^\xi (\alpha_h - \alpha_n)\, d\eta}\, d\xi + qG_o \int_0^x e^{-\int_0^\xi (\alpha_h - \alpha_n)\, d\eta}\, d\xi \right]. \tag{4.77}$$

Taking into account the boundary conditions, the total current density J can be expressed as

[12] For *equivalent pin* we mean a detector ideally obtained from the APD by suppressing the avalanche region and its effect.

$$J = J_h(W) = e^{\int_0^W (\alpha_h - \alpha_n)\,dx} \times \left[J \int_0^W \alpha_n e^{-\int_0^x (\alpha_h-\alpha_n)\,d\xi}\,dx + qG_o \int_0^W e^{-\int_0^x (\alpha_h-\alpha_n)\,d\xi}\,dx \right],$$

or, solving by J and taking (4.66) into account,

$$J = qG_o \frac{e^{\int_0^W (\alpha_h-\alpha_n)\,d\xi} \int_0^W e^{-\int_0^\xi (\alpha_h-\alpha_n)\,d\eta}\,d\xi}{1 - e^{\int_0^W (\alpha_h-\alpha_n)\,d\xi} \int_0^W \alpha_n e^{-\int_0^\xi (\alpha_h-\alpha_n)\,d\eta}\,d\xi} = qG_o W M_o = M_o J_{pin}, \quad (4.78)$$

where J_{pin} is the photocurrent density without avalanche generation, and M_o is the photocurrent multiplication factor:

$$M_o = \frac{1}{W} \frac{\int_0^W e^{-\int_0^\xi (\alpha_h-\alpha_n)\,d\eta}\,d\xi}{1 - \int_0^W \alpha_h e^{-\int_0^\xi (\alpha_h-\alpha_n)\,d\eta}\,d\xi}.$$

For $\alpha_h = \alpha_n$ the multiplication factor simplifies to

$$M_o = \frac{1}{1 - \int_0^W \alpha_h\,dx},$$

which again diverges if the avalanche breakdown condition (4.70) is met. For constant ionization coefficients we have instead:

$$M_o = \frac{1}{W} \frac{e^{(\alpha_n-\alpha_h)W} - 1}{\alpha_n - \alpha_h e^{(\alpha_n-\alpha_h)W}} = \frac{1}{W\alpha_n} \frac{e^{\alpha_n(1-k_{hn})W} - 1}{1 - k_{hn} e^{\alpha_n(1-k_{hn})W}}. \quad (4.79)$$

In the conventional APD, the equivalent *pin* responsivity is therefore multiplied by $M_o > 1$:

$$\mathfrak{R}_{\text{APD}} = M_o \mathfrak{R}_{pin}. \quad (4.80)$$

Thus, in both the conventional APD and the SAM-APD the responsivity is enhanced due to the effect of avalanche multiplication. From a technological standpoint, the SAM-APD allows for separate optimization of the avalanche region and of the absorption region, while in the conventional APD the high-field avalanche region width should be $W > L_\alpha$; as a result, the APD voltage may be very large, unless compound semiconductors are used. The separate optimization of the avalanche region thickness and of the absorption region length can be carried out in principle by exploiting waveguide photodetector structures; see Section 4.14.

In all cases considered, the APD multiplication factor (M_n, M_h or M_o) increases with increasing voltage; the resulting V–I characteristics are shown in Fig. 4.36, which can also be fitted by assuming for the multiplication factor (M_o, M_n or M_h) the empirical expression

$$M \approx \left[1 - \left(\frac{V_A - R_s I}{V_{br}} \right) \right]^{-n},$$

where R_s is the parasitic series resistance, V_{br} is the diode breakdown voltage, and V_A is the applied reverse voltage. Practical values for M typically are < 10 to decrease noise, lower the bias voltage, and improve the device reliability.

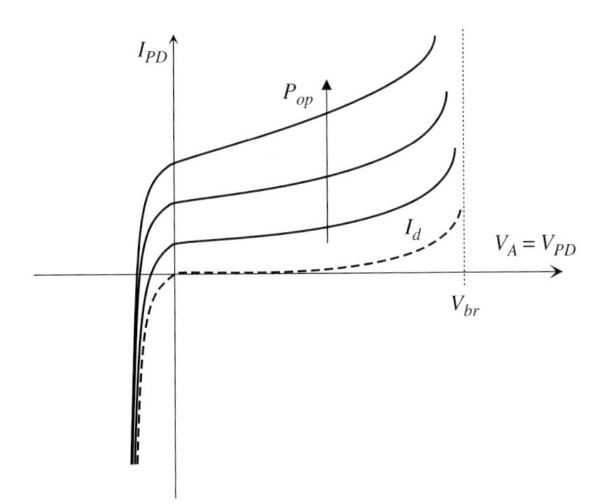

Figure 4.36 V–I characteristics of an APD.

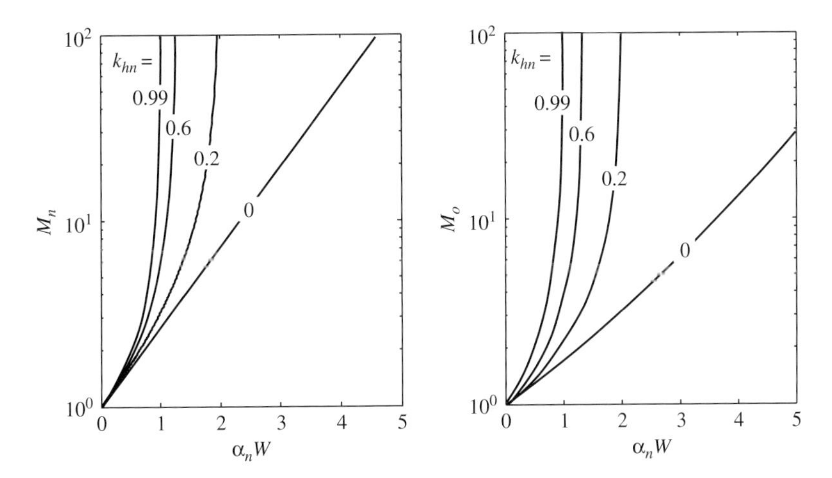

Figure 4.37 Ionization coefficients M_n and M_o as a function of $\alpha_n W$.

To compare, from the standpoint of the avalanche multiplication effectiveness, the conventional APD and the SAM-APD with electron injection, we plot the multiplication factors M_o and M_n as a function of the product $\alpha_n W$; see Fig. 4.37. The behavior of the multiplication factors is similar, although, for $k_{hn} = 0$, $M_n > M_h$. In fact, in this case one has

$$M_n = e^{\alpha_n W} > M_o = \frac{e^{\alpha_n W} - 1}{\alpha_n W},$$

while for $k_{hn} = 1$,

$$M_o = \frac{1}{1 - \alpha_n W} = M_n.$$

From Fig. 4.37 we see that for $k_{hn} \leq 1$ the electron-triggered SAM-APD multiplication factor is generally larger than for the conventional APD. From a physical standpoint this is because in the SAM-APD the whole primary photocurrent (from the most effective avalanching carrier, the electrons) is injected entirely at the edges of the avalanche region, while in the conventional APD the photocurrent injection takes place only gradually within the avalanche region (also acting as the absorption region). To optimize the multiplication factor in a SAM-APD the avalanche triggering (injected) carrier should be in principle the one with the larger ionization coefficient. If, on the other hand, $k_{hn} = 1$, we have $M_o = M_h = M_n$, i.e., all structures are equivalent concerning the multiplication factor.

4.12 Noise in APDs and *pins*

The short-circuit current fluctuation spectrum of a pin diode can, as already remarked, be modeled as shot noise according to (4.11). Neglecting the dark current, we have that the power spectrum of the noise current i_n (see Fig. 4.14) can be expressed as

$$S_{i_n,pin} = 2qI_L = 2q\mathfrak{R}_{pin}P_{in}.$$

Taking into account that in the APD the (primary) photocurrent generated by the equivalent pin is multiplied by $M = M_n$, M_h or M_o:

$$i_{L,\mathrm{APD}} = Mi_{L,pin},$$

we would expect fluctuations to be amplified according to the same law. Since the power spectrum of the process $y = Ax$ is $S_y = |A|^2 S_x$, this would imply the power spectrum of the APD current fluctuations to be $S_{i_n,\mathrm{APD}} = M^2 S_{i_n,pin}$. Unfortunately, the avalanche multiplication of carriers generates *additional noise* with respect to the simple multiplication of the shot noise associated with the primary photocurrent. We can therefore write

$$S_{i_n,\mathrm{APD}} = M^2 F S_{i_n,pin} = 2qM^2F\mathfrak{R}_{pin}P_{in} = 2qMF\mathfrak{R}_{\mathrm{APD}}P_{in},$$

where $F = F(M) > 1$ is the *excess noise factor*. The evaluation of the excess noise factors in the conventional and SAM-APD cases is carried out in detail below in Section 4.12.1;[13] we here summarize the main results and discuss the noise behavior vs the material parameters, in particular the ratio k_{hn} (4.73) and $k_{nh} = 1/k_{hn}$.

The short-circuit noise currents for the three cases (conventional APD, SAM-APD electron or hole triggered), neglecting the dark current, can be expressed as follows. For the **APD** we have

$$S^o_{\delta i} = 2qI_L M_o F_o(M_o) \tag{4.81}$$

[13] We follow the classical approach of McIntyre [46]. The resulting analysis is quasi-static, i.e., it assumes that carriers injected in a high-field region immediately ionize according to the local field; this approximation is inaccurate when considering extremely thin avalanche regions, (see, e.g., [47]), where the presence of a *dead space* becomes significant.

$$M_o = \frac{1}{W\alpha_n}\frac{e^{\alpha_n(1-k_{hn})W} - 1}{1 - k_{hn}e^{\alpha_n(1-k_{hn})W}} \quad (4.82)$$

$$F_o = \frac{(1+\alpha_h W M_o)(1+\alpha_n W M_o)}{M_o}. \quad (4.83)$$

For the **SAM-APD**, *electron triggered*:

$$S_{\delta i}^{\mathrm{SAM}_n} = 2qI_L M_n F_n(M_n) \quad (4.84)$$

$$M_n = \frac{(1-k_{hn})\,e^{\alpha_n(1-k_{hn})W_{av}}}{1 - k_{hn}e^{\alpha_n(1-k_{hn})W_{av}}} \quad (4.85)$$

$$F_n = k_{hn}M_n + (1-k_{hn})\left(2 - M_n^{-1}\right) = M_n\left[1 - (1-k_{hn})\left(1 - M_n^{-1}\right)^2\right]. \quad (4.86)$$

And for the **SAM-APD**, *hole triggered*:

$$S_{\delta i}^{\mathrm{SAM}_h} = 2qI_L M_h F_h(M_h) \quad (4.87)$$

$$M_h = \frac{k_{hn} - 1}{k_{hn}e^{\alpha_n(1-k_{hn})W_{av}} - 1} \quad (4.88)$$

$$F_h = k_{nh}M_h + (1-k_{nh})\left(2 - M_h^{-1}\right) = M_h\left[1 - (1-k_{nh})\left(1 - M_h^{-1}\right)^2\right] \quad (4.89)$$

A comparison of the different APD solutions from the standpoint of noise shows that the excess noise factor increases if the electrons and holes have similar ionization coefficients. Let us concentrate on the electron-triggered avalanche case; for $k_{hn} = 0$, the excess noise factor is minimum, while it increases for increasing k_{hn} and for increasing M_n; see Fig. 4.38. Concerning the conventional APD case, the behavior is similar, though the excess noise factor F_o tends to be larger, with the same multiplication factor, than the excess noise factor F_n. In general, for low values of k_{hn}, the SAM-APD is able to reach low values of the excess noise factor independent of gain, while this is not the case for conventional APDs; see Fig. 4.39, where the F_o and F_n are plotted together as a function of $\alpha_n W$. The larger excess noise factor F_o for low k_{hn} is evident.

SAM-APDs therefore, beside allowing to optimize the width of the absorption region so that $W_{abs} > L_\alpha$ without requiring unrealistically large driving voltages, are also potentially less noisy than conventional APDs; moreover, in SAM-APDs the avalanche region can be implemented with a larger gap material.

Concerning materials for APDs, the suitability of compound semiconductors for which $k_{hn} \approx 1$ should be discussed. Asymptotic high-field values for GaAs are $\alpha_n = 1.9 \times 10^5\ \mathrm{cm}^{-1}$, $\alpha_h = 2.22 \times 10^5\ \mathrm{cm}^{-1}$, making the ratio close to one; GaAs and also AlGaAs therefore are not particularly well suited to APD development. $\mathrm{In_{0.53}Ga_{0.47}As}$ has the asymptotic values of $\alpha_n = 2.27 \times 10^6\ \mathrm{cm}^{-1}$, $\alpha_h = 3.95 \times 10^6\ \mathrm{cm}^{-1}$; this would suggest that for very high field, hole injection devices are a more convenient choice.

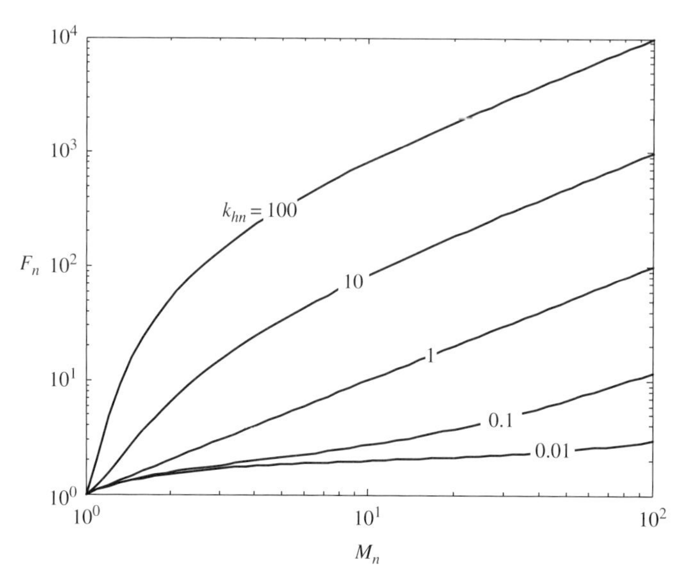

Figure 4.38 Excess noise factor F_n as a function of the electron multiplication coefficient M_n; the parameter is k_{hn}.

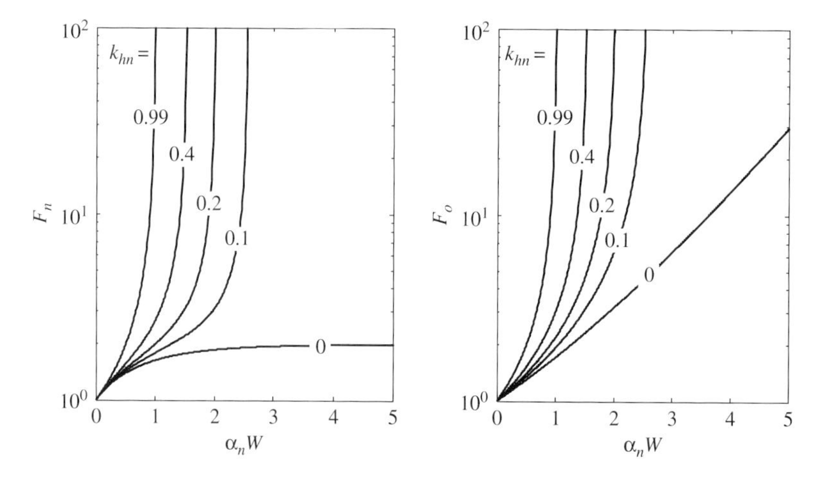

Figure 4.39 Excess noise factor for a SAM-APD (electron triggered) and conventional APD as a function of $\alpha_n W$

Independent of the material, however, low-voltage operation leads to low multiplication factor, but also to low excess noise.

4.12.1 Analysis of APD noise

Estimating the excess noise factor requires a microscopic-level analysis. We start by remarking that both optical and avalanche generation lead to fluctuations in the carrier population, and therefore in the current density. Fluctuations originating in section x of the avalanche photodetector are propagated to the external circuit through a multiplication process, leading to their amplification. We first have to characterize the local fluctuations from a statistical standpoint and, then to evaluate their propagation to the overall device current fluctuations.

Consider, first, the statistical characterization of fluctuations in section x. From the hole continuity equation we have, introducing fluctuations as random source terms according to the Langevin approach (see Section 5.13)

$$\frac{dJ_h}{dx} = \alpha_h J_h + \alpha_n J_n + qG_o + \gamma_h + \gamma_n + \gamma_o,$$

where γ_h, γ_n, γ_o are *zero average random processes* that express the fluctuations in the avalanche generation of holes and electrons, and in the optical generation, respectively. Since the total current is constant, the electron current continuity equation does not add independent information, also considering that the noise source terms correspond to direct (interband) process, and that, as a consequence, population fluctuations in the conduction and valence bands are fully correlated (i.e., the same fluctuation source appears in both continuity equations). Note that we have here made an approximation in neglecting the time dynamics of the avalanche process; since the microscopic noise sources are modeled by a white process, the resulting current fluctuations will also be white.

To model the second-order statistical properties of γ_h, γ_n, γ_o, we assume that (particle) generation and recombination can be described by a *Poissonian process* $x(t) = X + \delta x(t)$ in the particle number, where X is the average value and $\delta x(t)$ the zero-average fluctuation; the related (charge) GR process will be $qx(t)$. The power spectrum of (particle) number fluctuations $\delta x(t)$ will follow the Campbell theorem, i.e., $S_{\delta x} = 2X$. More specifically, consider now that fluctuations γ_h, γ_n, γ_o in the generation rate refer to generated *charge* per unit time and volume; the corresponding (average) *particle* generation rates are $(\alpha_h J_h)/q$, $(\alpha_n J_n)/q$, $(qG_o)/q$, respectively. Since the power spectrum of $q\delta x(t)$ is $q^2 S_{\delta x}$, we have[14]

$$\begin{aligned}\overline{\gamma_h(x)\gamma_h^*(x')} &= S_{\gamma_h}(\omega, x, x') = q^2 \cdot 2\left(\frac{\alpha_h J_h}{q}\right)\delta(x - x') \\ &= 2q\alpha_h J_h(x)\delta(x - x') \equiv K_{\gamma_h}(\omega, x)\delta(x - x')\end{aligned} \tag{4.90}$$

$$\begin{aligned}\overline{\gamma_n(x)\gamma_n^*(x')} &= S_{\gamma_n}(\omega, x, x') = q^2 \cdot 2\left(\frac{\alpha_n J_n}{q}\right)\delta(x - x') \\ &= 2q\alpha_n\left[J - J_h(x)\right]\delta(x - x') \equiv K_{\gamma_n}(\omega, x)\delta(x - x')\end{aligned} \tag{4.91}$$

$$\begin{aligned}\overline{\gamma_o(x)\gamma_o^*(x')} &= S_{\gamma_o}(\omega, x, x') = q^2 \cdot 2\left(\frac{qG_o}{q}\right)\delta(x - x') \\ &= 2q^2 G_o\delta(x - x') \equiv K_{\gamma_o}(\omega, x)\delta(x - x'),\end{aligned} \tag{4.92}$$

[14] Given a process $x(t)$ with Fourier transform $X(\omega)$, we formally denote the power spectrum of x, S_x, as a spectral ensemble average $\overline{XX^*}$.

where the Dirac delta $\delta(x - x')$ accounts for spatial uncorrelation of the noise sources in points x and x', and K_{γ_i} denotes the *local noise source.*

Since the continuity equations are linear (also with respect to forcing terms and boundary conditions), we can exploit superposition and analyze a template continuity equation pair of the kind

$$\frac{\mathrm{d}J_h}{\mathrm{d}x} = \alpha_h J_h + \alpha_n J_n + \gamma \tag{4.93}$$

$$\frac{\mathrm{d}J_n}{\mathrm{d}x} = -\alpha_h J_h - \alpha_n J_n - \gamma, \tag{4.94}$$

where γ is a *unit amplitude impulsive source* located in x', $\gamma(x) = \delta(x - x')$, and there is no injection at the region ends ($J_h(0) = J_n(W) = 0$). The two equations are not independent, since the total current is constant; however, introducing the electron continuity equation makes it easier to impose the boundary conditions. Thus, imposing constant total current J and with boundary conditions $J_h(0) = J_n(W) = 0$ and assuming $\gamma(x) = \delta(x - x')$, we have the set

$$\frac{\mathrm{d}J_h(x, x')}{\mathrm{d}x} = (\alpha_h - \alpha_n)\, J_h(x, x') + \alpha_n J + \delta(x - x') \tag{4.95}$$

$$\frac{\mathrm{d}J_n(x, x')}{\mathrm{d}x} = (\alpha_h - \alpha_n)\, J_n(x, x') - \alpha_h J - \delta(x - x'). \tag{4.96}$$

Integrating (4.95) from x'^- to x'^+ we obtain a *jump condition* for the hole current density that will be exploited further on:

$$\int_{x'^-}^{x'^+} \frac{\mathrm{d}J_h(x, x')}{\mathrm{d}x} = J_h(x'^+, x') - J_h(x'^-, x') = 1. \tag{4.97}$$

We can now solve (4.95) and (4.96) for $x \neq x'$ (i.e., an equation without the impulsive forcing term) according to the approach already developed in Section 4.11.1.

Let us split the domain into two parts, $x < x'$ and $x > x'$; for $x < x'$ solution of (4.95) with boundary conditions $J_h(0) = 0$ yields

$$J_h(x) = J \mathrm{e}^{\int_0^x (\alpha_h - \alpha_n)\,\mathrm{d}\xi} \int_0^x \alpha_n \mathrm{e}^{-\int_0^\xi (\alpha_h - \alpha_n)\,\mathrm{d}\eta}\, \mathrm{d}\xi, \quad x < x'. \tag{4.98}$$

On the other hand, for $x > x'$ we have, solving (4.96)[15]

$$J_n(x) = \mathrm{e}^{-\int_x^W (\alpha_h - \alpha_n)\,\mathrm{d}\xi} J_n(W) + J \mathrm{e}^{\int_0^x (\alpha_h - \alpha_n)\,\mathrm{d}\xi} \int_x^W \alpha_n \mathrm{e}^{-\int_0^\xi (\alpha_h - \alpha_n)\mathrm{d}\eta}\, \mathrm{d}\xi,$$

that is, with, $J_n(W) = 0$ and $J_h = J - J_n$,

$$J_h(x) = J\left[1 - \mathrm{e}^{\int_0^x (\alpha_h - \alpha_n)\,\mathrm{d}\xi} \int_x^W \alpha_h \mathrm{e}^{-\int_0^\xi (\alpha_h - \alpha_n)\,\mathrm{d}\eta}\, \mathrm{d}\xi\right], \quad x > x'. \tag{4.99}$$

[15] We can in fact express the solution of the template equation as:

$$y(x) = \mathrm{e}^{\int_x^W a(\xi)\,\mathrm{d}\xi} y(W) - \mathrm{e}^{-\int_0^x a(\xi)\,\mathrm{d}\xi} \int_x^W b(\xi) \mathrm{e}^{\int_0^\xi a(\eta)\,\mathrm{d}\eta}\, \mathrm{d}\xi.$$

The impulsive term in (4.93) and (4.94) causes a variation of the total current density J; since J is constant, such a variation will be position-independent. Let us denote the total current density induced by a unit source in $x = x'$ as $J \equiv M(x')$; this, in fact, coincides with the multiplication factor already discussed in the APD responsivity analysis.[16]

To simplify the treatment, from now on we will assume that ionization coefficients are *constant*. In this case, the expressions of the hole current density in (4.98) and (4.99) reduce to

$$J_h(x, x') = J\frac{\alpha_n}{\alpha_n - \alpha_h}\left[1 - \mathrm{e}^{-(\alpha_n - \alpha_h)x}\right], \quad x < x',$$
$$J_h(x, x') = J\frac{\alpha_n}{\alpha_n - \alpha_h}\left[1 - k_{hn}\mathrm{e}^{(\alpha_n - \alpha_h)(W - x)}\right], \quad x > x'.$$

From the jump condition (4.97) we can now derive the value of the total current density fluctuation $J \equiv M(x')$ induced by an unit impulsive current density source in $x = x'$:

$$M(x') = \frac{(\alpha_n - \alpha_h)\,\mathrm{e}^{(\alpha_n - \alpha_h)x'}}{\alpha_n - \alpha_h \mathrm{e}^{(\alpha_n - \alpha_h)W}} = \frac{(1 - k_{hn})\,\mathrm{e}^{\alpha_n(1 - k_{hn})x'}}{1 - k_{hn}\mathrm{e}^{\alpha_n(1 - k_{hn})W}}. \tag{4.100}$$

For electron injection in W or hole injection in 0, $M(x')$ reduces to the already introduced electron and hole multiplication factors defined in (4.71) and (4.72), i.e., $M(W) = M_n$, $M(0) = M_h$.

We now evaluate the output current fluctuations as a function of the local noise source $K_\gamma(\omega, x')$ associated with γ. The total fluctuation induced by a generic source $\gamma(x)$ will be

$$\delta J = \int_0^W M(x')\gamma(x')\,\mathrm{d}x',$$

where spatial superposition (allowed by spatial uncorrelation) was used. From the frequency-domain average we obtain for the noise current power spectrum $S_{\delta J}$:

$$\begin{aligned} S_{\delta J} = \overline{\delta J \delta J^*} &= \int_0^W \int_0^W M(x')M(x'')\overline{\gamma(x')\gamma^*(x'')}\,\mathrm{d}x'\,\mathrm{d}x'' \\ &= \int_0^W \int_0^W M(x')M(x'')K_\gamma(\omega, x')\delta(x' - x'')\,\mathrm{d}x'\,\mathrm{d}x'' \\ &= \int_0^W M^2(x')K_\gamma(\omega, x')\,\mathrm{d}x'. \end{aligned} \tag{4.101}$$

We now apply (4.101) separately to discuss three cases: the *conventional APD*, the SAM-APD with *electron-triggered avalanche,* and the SAM-APD with *hole-triggered avalanche.*

[16] From the dimensional standpoint, a δ source in the continuity equation has the value of 1 A m^2 m^{-1}; since the resulting induced current variation will be $M \cdot 1$ A/m^2, M is actually adimensional.

In the *conventional APD*, both avalanche and optical generation take place in the absorption region; in this case one has, exploiting (4.101) and superposition:

$$\begin{aligned} S_{\delta J} &= \int_0^W M^2(x)\left[K_{\gamma_h}(\omega,x)+K_{\gamma_n}(\omega,x)+K_{\gamma_o}(\omega,x)\right]\mathrm{d}x \\ &= 2q\int_0^W M^2(x)\left\{\alpha_h J_h(x)+\alpha_n\left[J-J_h(x)\right]+qG_o\right\}\mathrm{d}x \\ &= 2q\left(\alpha_h-\alpha_n\right)\int_0^W M^2(x)J_h(x)\mathrm{d}x+2q\left(\alpha_n J+qG_o\right)\int_0^W M^2(x)\,\mathrm{d}x. \end{aligned}$$

The total current in the conventional APD is, from (4.78),

$$J = qWG_oM_o,$$

while the hole current density is derived from (4.77) and (4.78), assuming position-independent ionization coefficients, as

$$J_h(x) = \frac{qG_o}{\alpha_n-\alpha_h}\left(1+W\alpha_n M_o\right)\left[1-\mathrm{e}^{-\alpha_n(1-k_{hn})x}\right].$$

With some manipulations we have

$$\begin{aligned} S_{\delta J} &= 2q\left(\alpha_h-\alpha_n\right)\int_0^W M^2(x)J_h(x)\,\mathrm{d}x+2q\left(\alpha_n J+qG_o\right)\int_0^W M^2(x)\,\mathrm{d}x \\ &= 2q^2G_o\left(1+W\alpha_n M_o\right)\int_0^W M^2(x)\mathrm{e}^{-\alpha_n(1-k_{hn})x}\,\mathrm{d}x \\ &= 2qJ\left[\left(1+W\alpha_n M_o\right)\left(1+W\alpha_h M_o\right)\right], \end{aligned} \qquad (4.102)$$

since, from (4.100),

$$\begin{aligned} \int_0^W M^2(x)\mathrm{e}^{-\alpha_n(1-k_{hn})x}\,\mathrm{d}x &= \frac{1}{\alpha_n}\frac{\left(1-k_{hn}\right)\left(\mathrm{e}^{\alpha_n(1-k_{hn})W}-1\right)}{\left(1-k_{hn}\mathrm{e}^{\alpha_n(1-k_{hn})W}\right)^2} \\ &= \frac{1}{\alpha_n}W\alpha_n M_o\left(1+W\alpha_h M_o\right), \end{aligned} \qquad (4.103)$$

where we have taken into account (see (4.79)) the transformation

$$W\alpha_n M_o = \frac{\mathrm{e}^{\alpha_n(1-k_{hn})W}-1}{1-k_{hn}\mathrm{e}^{\alpha_n(1-k_{hn})W}} \rightarrow \mathrm{e}^{\alpha_n(1-k_{hn})W} = \frac{1+W\alpha_n M_o}{1+W\alpha_n M_o k_{hn}}.$$

From (4.102), the short-circuit current fluctuation spectrum of the conventional APD, $S^o_{\delta i}$, can be finally expressed as

$$S^o_{\delta i} = AS_{\delta J} = 2qIM_oF_o(M_o),$$

where A is the detector area,[17] with excess noise factor given by

$$F_o(M_o) = \frac{(1 + \alpha_h W M_o)(1 + \alpha_n W M_o)}{M_o}.$$

In the *electron-triggered SAM-APD* there is no optical generation in the multiplication region, only avalanche generation; however, there is injection of a noisy primary photocurrent from the absorption region. This corresponds to an equivalent generation noise source in $x = W$, such as

$$qG_o(x') = J_n(W)\delta(x' - W) = J_{pin}\delta(x' - W) = \frac{J}{M_n}\delta(x' - W).$$

Introducing all relevant noise sources into (4.101) we have

$$\begin{aligned} S_{\delta J} &= \int_0^W M^2(x)\left[K_{\gamma_h}(\omega, x) + K_{\gamma_n}(\omega, x) + \frac{J}{M_n}\delta(x' - W)\right] \mathrm{d}x \\ &= 2q \int_0^W M^2(x)\left\{\alpha_h J_h(x) + \alpha_n \left[J - J_h(x)\right]\right\} \mathrm{d}x + 2q\frac{J}{M_n}M^2(W) \\ &= 2q\left(\alpha_h - \alpha_n\right) \int_0^W M^2(x) J_h(x)\,\mathrm{d}x + 2q\alpha_n J \int_0^W M^2(x)\,\mathrm{d}x + 2qJM_n. \end{aligned}$$

The hole current density for the electron injection case can be evaluated, e.g., using (4.64) with $J_h(x) = 0$ and constant ionization coefficients, as

$$J_h(x) = J\frac{1 - \mathrm{e}^{-\alpha_n(1-k_{hn})x}}{1 - k_{hn}}.$$

Thus,

$$\begin{aligned} S_{\delta J} &= -2q\alpha_n(1 - k_{hn}) \int_0^W M^2(x) J_h(x)\,\mathrm{d}x + 2q\alpha_n J \int_0^W M^2(x)\,\mathrm{d}x + 2qJM_n \\ &= 2q\alpha_n J \int_0^W M^2(x)\left[\mathrm{e}^{-\alpha_n(1-k_{hn})x} - 1\right] \mathrm{d}x + 2q\alpha_n J \int_0^W M^2(x)\,\mathrm{d}x + 2qJM_n \\ &= 2q\alpha_n J \int_0^W M^2(x)\mathrm{e}^{-\alpha_n(1-k_{hn})x}\,\mathrm{d}x + 2qJM_n. \end{aligned}$$

Using (4.71),

$$M_n = \frac{(1 - k_{hn})\,\mathrm{e}^{\alpha_n(1-k_{hn})W}}{1 - k_{hn}\mathrm{e}^{\alpha_n(1-k_{hn})W}} \rightarrow \mathrm{e}^{\alpha_n(1-k_{hn})W} = \frac{M_n}{1 + k_{hn}\left(M_n - 1\right)}$$

and (4.103) we have

$$\int_0^W M^2(x)\mathrm{e}^{-\alpha_n(1-k_{hn})x}\,\mathrm{d}x = \frac{1}{\alpha_n}\left[k_{hn}M_n^2 + (1 - 2k_{hn})\,M_n - (1 - k_{hn})\right].$$

[17] Multiplication by the area A results from the fact that all current density fluctuations in the device cross section are *spatially uncorrelated*.

We thus obtain

$$\begin{aligned} S_{\delta J} &= 2q\alpha_n J \int_0^W M^2(x) \mathrm{e}^{-\alpha_n(1-k_{hn})x}\, \mathrm{d}x + 2qJM_n \\ &= 2qJ\left[k_{hn}M_n^2 + (1-2k_{hn})\,M_n - (1-k_{hn}) + M_n\right] \\ &= 2qJM_n\left[k_{hn}M_n + (1-k_{hn})\left(2-\frac{1}{M_n}\right)\right]. \end{aligned}$$

The spectrum of the short-circuit current fluctuations δi for electron-triggered avalanche, $S_{\delta i}^{\mathrm{SAM}_n}$, can be finally recovered by multiplying $S_{\delta J}$ by the diode area A as

$$S_{\delta i}^{\mathrm{SAM}_n} = AS_{\delta J} = 2qIM_nF_n(M_n),$$

where the excess noise factor $F_n(M_n)$ is given by

$$F_n(M_n) = k_{hn}M_n + (1-k_{hn})\left(2-\frac{1}{M_n}\right) = M_n\left[1-(1-k_{hn})\left(\frac{M_n-1}{M_n}\right)^2\right].$$

It is useful to analyze the asymptotic behavior of the excess noise factor. For small and large M_n one has the following limits:

$$F_n(M_n) \xrightarrow[M_n\to\infty]{} k_{hn}M_n, \quad F_n(M_n) \xrightarrow[M_n\to 1]{} 1, \quad F_n(M_n) \xrightarrow[k_{hn}\to 1]{} M_n.$$

In the *SAM-APD with hole-triggered avalanche*, as in the dual electron-triggered case, there is no optical generation contributing to noise, only avalanche, and injection of a noisy hole current in $x = 0$; this corresponds to an equivalent generation noise source in $x = 0$ such as

$$qG_o(x') = J_h(0)\delta(x') = J_{pin}\delta(x') = \frac{J}{M_h}\delta(x').$$

Instead of explicitly solving this case, we can derive the expression by duality from the electron injection case, using $M_n \to M_h$, $k_{hn} = k_{nh} = 1/k_{hn}$; we obtain that the short-circuit current fluctuation spectrum $S_{\delta i}^{\mathrm{SAM}_h}$ for hole-triggered avalanche can be expressed as

$$S_{\delta i}^{\mathrm{SAM}_h} = 2qIM_hF_h(M_h),$$

where the excess noise factor $F_h(M_h)$ is given by

$$F_h(M_h) = \frac{M_h}{k_{hn}} + \left(1-\frac{1}{k_{hn}}\right)\left(2-\frac{1}{M_h}\right) = M_h\left[1-\left(1-\frac{1}{k_{hn}}\right)\left(\frac{M_h-1}{M_h}\right)^2\right].$$

4.13 The APD frequency response

Let us consider for simplicity a SAM-APD with electron-triggered avalanche, absorption region width W_{abs}, avalanche region width W_{av}. We assume with this structure $k_{hn} \le 1$. The frequency response can be approximately expressed in terms of the delay that carriers undergo before being collected. Such delays are related to several factors (note that we mention only intrinsic effects, leaving extrinsic RC-cutoff to a circuit-level analysis):[18]

- The electron transit time in the absorption region:

$$\tau_{tr} = W_{abs}/v_{n,\text{sat}}.$$

- The avalanche buildup time (or avalanche delay), which can be expressed as

$$\tau_A \approx \frac{r(k_{hn}) M_n W_{av} k_{hn}}{v_e}, \tag{4.104}$$

  where $r(k_{hn})$ is a slowly varying function of k_{hn} such as $r(1) = 1/6$, $r(0) = 1$, and v_e is an effective velocity defined as

$$v_e = \frac{v_{n,\text{sat}} v_{h,\text{sat}}}{v_{n,\text{sat}} + v_{h,\text{sat}}}.$$

  Expression (4.104) is actually valid if $M_n k_{hn} > 1$, see [48], [49], [50].
- The transit time of avalanche-generated holes:

$$\tau_{tr,Ah} = \frac{W_{abs} + W_{av}}{v_{h,\text{sat}}}.$$

From the above partial delays, we can approximately derive the total delay of a SAM-APD as

$$\tau = \tau_{tr} + \tau_A + \tau_{tr,Ah} = \frac{W_{abs}}{v_{n,\text{sat}}} + \frac{M_n W_{av} k_{hn}}{v_{n,\text{sat}}} + \frac{W_{abs} + W_{av}}{v_{h,\text{sat}}}$$

implying that the total delay is minimized if $k_{hn} \ll 1$, i.e., if the ionization coefficient of the avalanche triggering carrier is much larger than the ionization coefficient of the other carrier. On this basis, the intrinsic cutoff frequency can be approximately expressed as

$$f_T \approx \frac{1}{2.2\tau}.$$

Due to the additional delay mechanisms, the APD is typically slower than the *pin*; moreover, the device response is faster when only the avalanche triggering carriers (e.g., electrons) ionize. In fact, if both electrons and holes have the same ionization probability, an electron pulse injected into the high-field region generates holes which, instead of simply being collected in the p-side, in turn generate secondary electrons, whose delay before collection is of course larger than the delay of the electrons directly generated by the initial pulse. Secondary electrons generate new holes, and so on; if the multiplication factor is finite, the process finally dies out with a slow tail. As a result, a larger number of carriers is collected with respect to the $k_{hn} = 0$ case (and, indeed the multiplication factor is larger), but such carriers will be collected over a longer time interval,

[18] The analysis of the APD avalanche region frequency response is straightforward but lengthy, and the formulae presented are just an approximation of a transit time-like cutoff. For a complete discussion see [48], [49], [50].

leading to an output current pulse with a large spread in time. This finally amounts to a slower response.

Example 4.4: Consider a hole-triggered InGaAs SAM-APD with $W_{abs} = 1\,\mu\text{m}$, $W_{av} = 0.03\,\mu\text{m}$, $\alpha_n = 2.27 \times 10^5\,\text{cm}^{-1}$, $\alpha_h = 3.95 \times 10^5\,\text{cm}^{-1}$. Assuming $v_{n,\text{sat}} = v_{h,\text{sat}} = 10^7$ cm/s, evaluate the multiplication factor, excess noise factor and detector intrinsic bandwidth in a hole-triggered APD.

We have

$$k_{hn} = \frac{\alpha_h}{\alpha_n} = \frac{3.95}{2.27} = 1.74$$

and therefore

$$\begin{aligned} M_h &= \frac{k_{hn} - 1}{k_{hn} \exp\left[\alpha_n (1 - k_{hn}) W_{av}\right] - 1} \\ &= \frac{1.74 - 1}{1.74 \cdot \exp\left[2.27 \times 10^5 \cdot (1 - 1.74) \cdot 0.03 \times 10^{-4}\right] - 1} = 14.44 \end{aligned}$$

and

$$\begin{aligned} F_h(M_h) &= \frac{M_h}{k_{hn}} + \left(1 - \frac{1}{k_{hn}}\right)\left(2 - \frac{1}{M_h}\right) \\ &= \frac{14.44}{1.74} + \left(1 - \frac{1}{1.74}\right)\left(2 - \frac{1}{14.44}\right) = 9.11. \end{aligned}$$

Delay times can be derived exploiting dual relations with respect to those for the electron-triggered SAM-APD. We obtain (with $v_e = v_{h,\text{sat}}/2 = 0.5 \cdot 10^7$ cm/s):

$$\begin{aligned} \tau_{tr} &= \frac{W_{abs}}{v_{h,\text{sat}}} = \frac{10^{-4}}{10^7} = 10^{-11}\ \text{s} \\ \tau_A &= \frac{r\,(k_{nh})\, M_h W_{av} k_{nh}}{v_e} \approx \frac{M_h W_{av}}{v_e k_{hn}} = \frac{14.44 \cdot 0.03 \times 10^{-4}}{0.5 \cdot 10^7 \cdot 1.74} = 0.498 \times 10^{-11}\ \text{s} \\ \tau_{tr,An} &= \frac{W_{abs} + W_{av}}{v_{n,\text{sat}}} = \frac{10^{-4} + 0.03 \times 10^{-4}}{10^7} = 1.03 \times 10^{-11}\ \text{s} \\ \tau &= (1 + 0.498 + 1.03) \cdot 10^{-11}\ \text{s} = 2.528 \times 10^{-11}\ \text{s}, \end{aligned}$$

so that the cutoff frequency will be

$$f_T \approx \frac{1}{2.2\tau} = \frac{1}{2.2 \cdot 2.528 \times 10^{-11}} = 17.98\,\text{GHz}$$

while the corresponding *pin* transit time-limited cutoff frequency (assuming, somewhat arbitrarily, the long diode expression) is

$$f_{T,pin} \approx \frac{1}{2.2\tau} = \frac{1}{2.2 \cdot 10^{-11}} = 45.45\,\text{GHz}.$$

However, the APD responsivity is 14.44 times larger than the *pin* responsivity.

4.14 Advanced APD structures

Due to internal gain, APDs are able to reach spectacular efficiencies; however, the additional delay introduced by the avalanche process typically leads, at least in conventional (vertically illuminated) APDs, to lower cutoff frequency vs. *pin* photodiodes. The APD speed can be improved by applying some of the solutions already investigated in the development of high-speed *pin*s, namely, the use of resonant structures [51], [52] and of waveguide photodiodes [53], [54]. In both cases, the absorption layer can be made thinner with respect to vertically illuminated structures, thus reducing the transit time and globally improving the device speed (which, however, remains inferior to that of the equivalent *pin*).

In [51] a resonant-cavity AlGaAs–GaAs–InGaAs SAM-APD (RCAPD) is presented with a peak $\eta_x = 80\%$, achieved with a 35 nm thick $In_{0.1}Ga_{0.9}As$ absorption region; the low-gain bandwidth was 20 GHz; operation is in the first window. An optimized RCAPD structure developed by the same research group is presented in [52]. The structure has a number of additional layers besides the thin (35 nm) $In_{0.1}Ga_{0.9}As$ absorption layer and the 80 nm $Al_{0.2}Ga_{0.8}As$ undoped avalanche layer, to improve the field profile (see Fig. 4.40(a)); bottom reflection is provided by a $\lambda/4$ AlGaAs/GaAs 20-layer stack, while the top reflector is a p^+ $Al_{0.2}Ga_{0.8}As$ layer grown to a suitable thickness to make the cavity resonant at the desired wavelength (not explicitly specified, but around 0.8 μm). The measured 3 dB bandwidth in the low-gain regime (multiplication factor down to 2, achieved with a reverse bias around 10 V) is 33 GHz, while

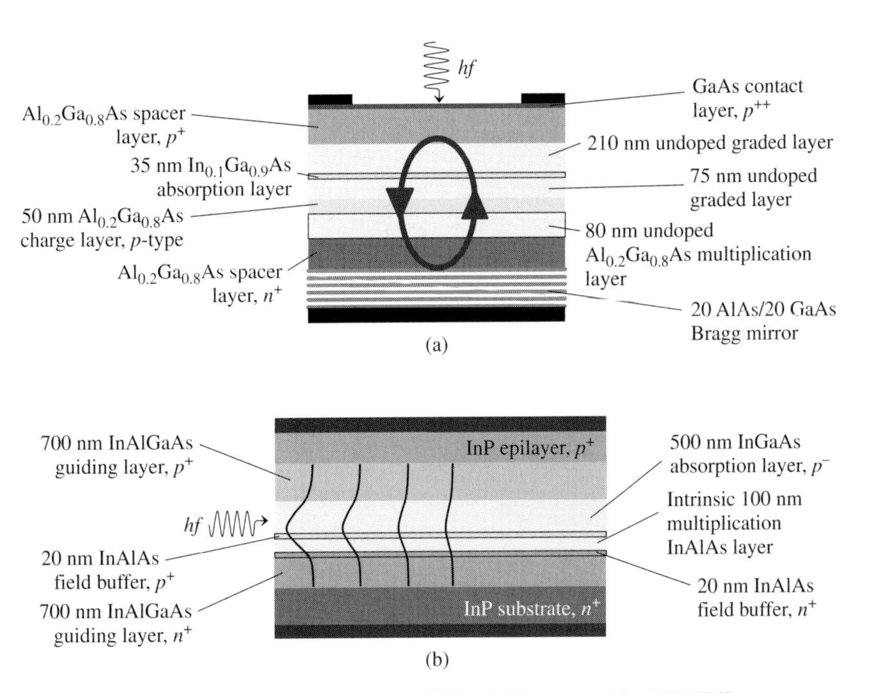

Figure 4.40 Layer structure of (a) reflecting-facet APD [52] and (b) waveguide APD [54].

the gain–bandwidth product in the high-gain regime is 290 GHz. The resulting external quantum efficiency in the low-gain, high-speed regime is not specified.

Waveguide avalanche photodiodes (WAPD) can be obtained both in conventional and in the SAM version. In conventional WAPDs photons are injected in the absorption region, which also operates as the avalanche multiplication region, parallel to the junction plane. High efficiency can therefore be obtained, as in waveguide *pins*, without increasing the photogenerated and multiplied carriers' transit time. Better results can be obtained in SAM-WAPDs, where the waveguide absorption layer is followed by an avalanche layer. A long-wavelength (1.55 μm) SAM-WAPD was proposed in [53], exploiting a 200 nm InGaAs absorption layer sandwiched between two 100 nm InAlGaAs guiding layers (the waveguide) and an external 150 nm InGaAs avalanche layer lattice-matched to the InP substrate. The detector is built on a mesa structure with length 10–100 μm. The low-gain bandwidth was reported as 28 GHz, with a maximum external quantum efficiency of 16% and a maximum gain–bandwidth product of 320 GHz. However, the maximum η_x was not in fact achieved with the high-bandwidth device, for which coupling losses reduced the efficiency down to 5%.

More recently, an InAlAs/InGaAs WAPD was presented in [54], see Fig. 4.40(b); the waveguide structure consists of an InAlAs/InGaAs core region sandwiched between upper p^+ and a lower n^+ InAlGaAs guiding layers. The InGaAs absorption layer is 500 nm thick and is separated from the 100 nm thick InAlAs multiplication layer by a 20 nm InAlAs field buffer layer (intended to improve the RF field distribution). The overall thickness of the optical guiding layers is 700 nm and was designed to improve the multimode coupling efficiency. The waveguide mesa was 6 μm wide and 20 μm long. The measured low-gain detector bandwidth was 35 GHz with responsivities in the range 0.73–0.88 A/W at 1.3 μm and 1.55 μm, respectively, corresponding to external efficiencies around 70% in both cases; the overall receiver sensitivity (the receiver also included a transimpedance front-end amplifier) was −19 dBm at 40 Gbps.

In conclusion, high-speed APDs still exhibit frequency responses that are not comparable with fast *pins*; however, the gain–bandwidth products obtained are promising and further optimization could make APDs the devices of choice also above 10 Gbps.

4.15 Concluding remarks on high-speed PDs

Standard high-speed photodetectors are vertically illuminated structures whose performances are limited by the bandwidth–efficiency product: high speed is obtained at the expense of reduced responsivity. In waveguide photodetectors the speed and efficiency issues are decoupled, but the speed limitation in terms of RC and transit time cutoff remains; efficiency can be high, though in practice coupling losses are critical due to the small waveguide cross section, but cutoff frequencies of the order of 100 GHz can be obtained with proper design. To completely overcome the RC-cutoff limitation, traveling-wave detectors have been proposed, in which the main limitation is velocity mismatch; however, (vertical) transit time may also again become important at extremely high speed. Due to the high losses and poor velocity match of simple

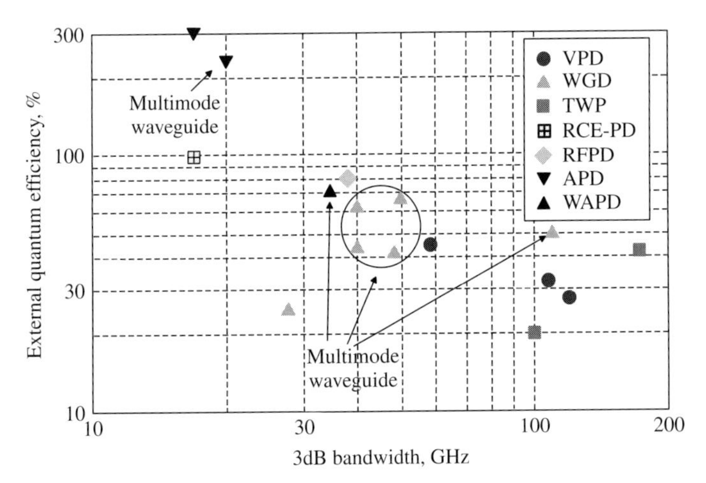

Figure 4.41 Efficiency–bandwidth trade-off in some high-speed detectors. Partly adapted from [40], Fig. 1 (©1999 IEEE); the WAPD point data are from [54].

distributed detectors, periodically loaded distributed detectors have been developed to further increase the speed. Such detectors also have the advantage of a larger absorption volume and therefore of lower saturation power.

High-speed APDs are based on vertical geometries similar to those exploited in *pins*. Despite being slower than *pins* of similar structure, during the last few years compound semiconductor APDs have gained popularity for applications in the 10 Gbps range, due to the better sensitivity when included in a receiver dominated by the front-end amplifier white noise (usually referred to as "front-end thermal noise"). Advanced waveguide APDs have also been developed, with performances in the 40 Gbps range.

In Fig. 4.41 we show a few examples of the performances of high-speed detectors [40]. VPD is the conventional vertically illuminated *pin*; WGD is the waveguide photodetector; TWP the traveling-wave photodetector; RCE-PD is the resonant cavity enhanced photodetector; RFPD the reflecting-facet photodetector; WAPD the waveguide APD. Two vertical APD examples are also shown for comparison. For some structures, a multimode waveguide was exploited as a guiding device. Top results in terms of speed are achieved by TWPs, with efficiencies of the order of 40%; however, both waveguide detectors and VPDs (with a somewhat decreased efficiency) have comparable performances. WAPDs show promising performances in terms of efficiency and speed. It may be concluded that the real need for traveling-wave detectors is still to come in the application field, while waveguide detectors may be an asset for photonic integrated circuits bringing together passive components based on waveguiding structures and detectors.

4.16 The photodiode front end

The (weak) photodetector output signal current can be converted into a voltage through a load resistor and then amplified by a *front-end preamplifier stage*; in high-speed receivers, conversion and amplification are typically integrated in a *transimpedance amplifier* (TIA) providing both power gain and current–voltage conversion. The usual photodiode bias point is at negative voltage and negative current, to optimize linearity, dynamic range and dark current. Detectors can be capacitively coupled to the load or TIA front-end to suppress the dark current and other DC current components. Before discussing some possible front-end choices, we consider the equivalent signal and noise circuit exploited in the detector model.

4.16.1 Photodetector and front-end signal and noise model

A photodiode equivalent circuit has already been introduced, see Fig. 4.14. The internal admittance is mainly reactive, due to the photodiode capacitance C_D. The deterministic current generator includes the photocurrent i_L (in small-signal phasor form I_L) and the dark current I_d (often neglected), while the random generator i_n models the current fluctuations. We will here confine the treatment to the *pin* and APD cases.

As already discussed, the noise in *pin* detectors is intrinsic shot noise, plus some thermal noise arising from parasitics that we will neglect for simplicity. One has, therefore,

$$S_{i_n,pin} = 2q\left(I_L + I_d\right).$$

In APD photodiodes, we have multiplied and excess shot noise; for the SAM-APD with electron-triggered avalanche the noise current power spectrum is, from (4.84),

$$S_{i_n,\mathrm{APD}} = 2qI_{\mathrm{APD}}M_nF_n$$

where the multiplication factor M_n and the excess noise factor F_n are reported in (4.85) and (4.86). Similar expressions hold for the other cases.

The front-end amplifier input can be modeled in turn as an impedance Z_i; the front-end noise can be approximately represented through a parallel current noise generator according to the scheme in Fig. 4.14 where $R_L \to Z_i$.

4.16.2 High- and low-impedance front ends

Let us initially consider a front end made of a load resistor with resistance R_i, connected to a high-input-impedance voltage amplifier. In such a case, we can approximate $Z_i = R_i$. Suppose, for simplicity, that the amplifier noise is negligible and that only the load resistor (thermal) noise is significant. It can easily be shown that the choice of the load resistance directly impacts on the receiver performance in terms of *bandwidth*, *noise*, and *input linear dynamic range*.

In fact, high load resistance implies low noise but low bandwidth B; low input resistance implies high bandwidth but also high noise. Concerning the input dynamic range in which the device operates linearly, this is wide for low input resistance and narrow for large input resistance.

To simplify the discussion, we will exploit a simple bias scheme in which the DC bias resistance is the same as the signal resistance. The two resistances can actually be decoupled by use of proper bias Ts (see Section 3.4.1).

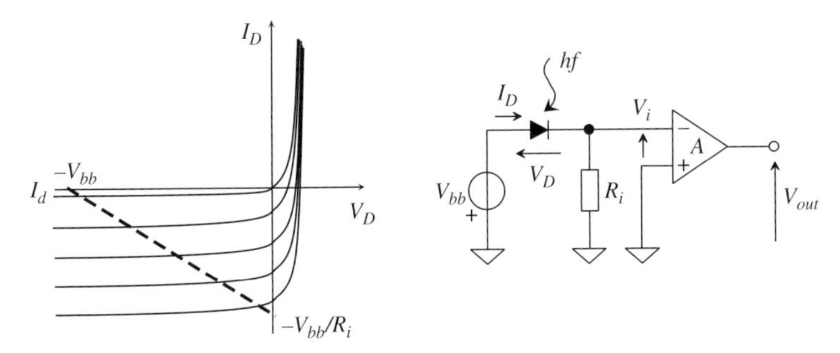

Figure 4.42 Low-impedance front-end amplifier: load line (left) and stage simplified structure (right).

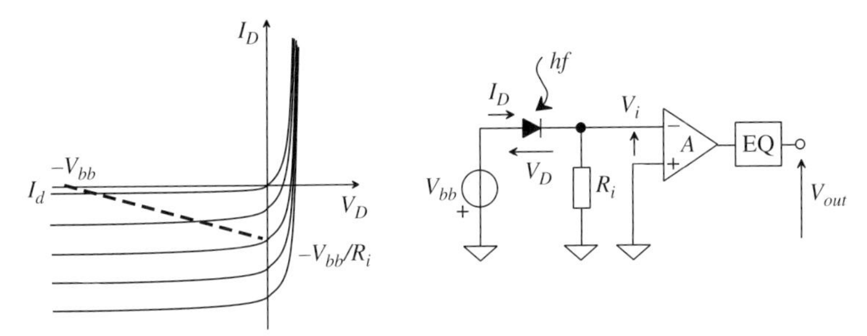

Figure 4.43 High-impedance front-end amplifier: load line (left) and stage simplified structure with equalizer (right).

In *low-impedance front ends* (see Fig. 4.42) the load resistance R_i, and therefore the time constant $R_i C_D$, are small, leading to a wide RC-limited bandwidth (remember that other, *intrinsic*, limitations to bandwidth exist, such as the transit time). However, the Nyquist noise generated by the load is large, since it is proportional to $G_i = 1/R_i$. As a result, this configuration is rather noisy. However, the large slope of the load line of the photodiode allows for a wide linear dynamic range of input optical power (intrinsic saturation effects may also limit the receiver dynamics).

In *high-impedance front ends* (see Fig. 4.43) the load resistance R_i is high, and therefore the time constant $R_i C_D$ is large, leading to narrow RC-limited bandwidth. The front-end stage may behave like an integrator, thus requiring an equalization stage to be inserted after the amplifier to reshape the waveform. The Nyquist noise generated by the load is small, however, since it is proportional to $G_i = 1/R_i$. As a result, this configuration is slow, but has low noise. A further disadvantage is related to the small slope of the load line, which leads to a limited linear dynamic range in the input optical power.

The frequency response of the low- and high-input impedance stages can be analyzed from the equivalent circuit of Fig. 4.44, in which the high-impedance amplifier is modeled as an input capacitance and an output voltage-controlled voltage generator. Define as $\mathfrak{R}(\omega)$ the small-signal responsivity resulting from intrinsic mechanisms (such

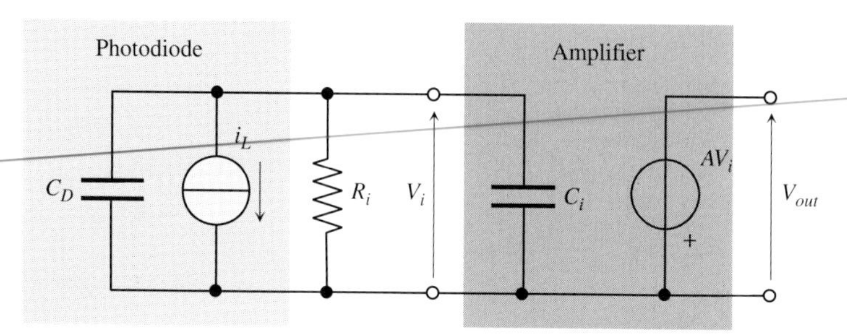

Figure 4.44 Equivalent circuit for the analysis of the low- and high-impedance front end.

as transit time and avalanche buildup), and approximate such a frequency behavior by a single pole response, as

$$\Re(\omega) = \frac{\Re(0)}{1 + j\omega/\omega_{Ti}},$$

where $\omega_{Ti} = 2\pi f_{Ti}$ is the intrinsic PD cutoff frequency. The total PD cutoff (including RC-cutoff) also results from extrinsic elements, and is circuit dependent. From the amplifier small-signal circuit one has

$$V_i = -\Re(\omega) P_{in} \frac{R_i}{1 + j\omega(C_D + C_i)R_i}$$

$$V_{out} = -AV_i = \frac{A\Re(\omega) P_{in} R_i}{1 + j\omega(C_D + C_i)R_i};$$

solving, the overall response (output voltage vs. input optical power) is

$$\frac{V_{out}}{P_{in}} = AR_i\Re(0)\frac{1}{1 + j\omega/\omega_{Ti}}\frac{1}{1 + j\omega/\omega_{Tx}},$$

with *extrinsic* cutoff frequency

$$\omega_{Tx} = 2\pi f_{Tx}, \quad f_{Tx} = \frac{1}{2\pi(C_D + C_i)R_i}.$$

The extrinsic cutoff frequency usually dominates, leading to the behavior shown in Fig. 4.45 for low- and high-input-impedance front ends, respectively. In the high-impedance case, the amplifier behaves like an integrator, and a derivative equalizing block is needed in cascade to the amplifier to restore the signal.

4.16.3 Transimpedance amplifier front ends

A far better compromise between bandwidth, dynamics, and noise requirements is achieved through the *transimpedance amplifier (TIA) front end*. The purpose of the TIA is to turn an input current into an output voltage; due to the input current drive, the input impedance is typically low and the gain (the transimpedance Z_m) has the dimension of an impedance. TIAs can be implemented through voltage amplifier stages with parallel resistive feedback; see Fig. 4.46 (right). Let A be the open-loop voltage amplification; from the analysis, the TIA input impedance is

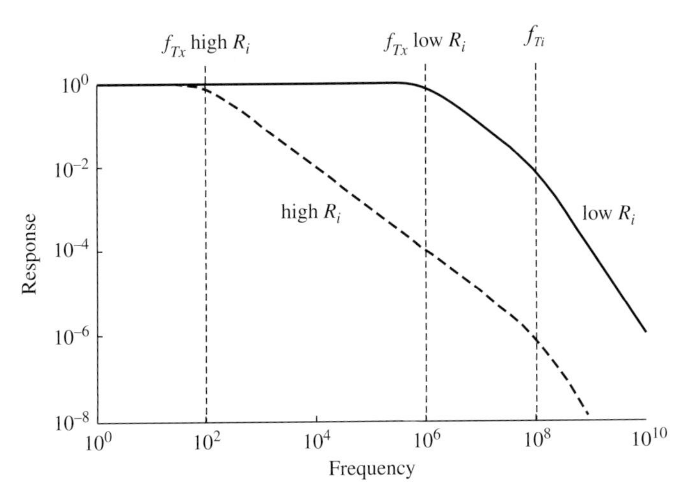

Figure 4.45 Frequency response of the low- and high-impedance front-end stage. f_{Tx} is the extrinsic cutoff frequency, f_{Ti} the intrinsic cutoff frequency.

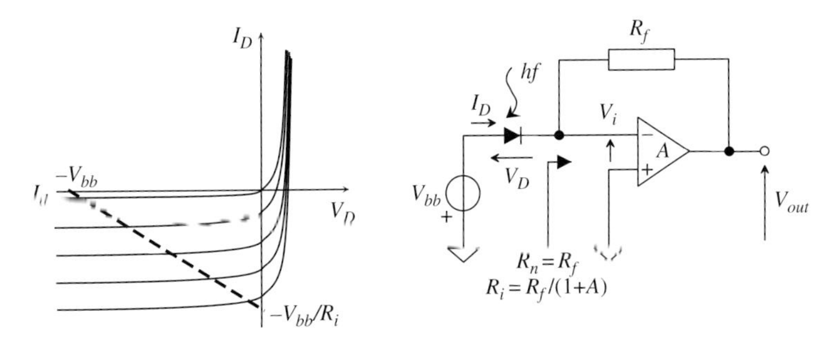

Figure 4.46 Transimpedance front end: load line (left) and stage simplified structure (right).

$$R_i \approx R_f/A,$$

where R_f is the feedback resistance. For large gain, the input resistance is low, yielding high linear dynamic range and high bandwidth. However, the input noise resistance can be shown to be equal to R_f, thus yielding low parallel thermal noise ("cold resistance stage").[19] Let us analyze first the transimpedance cutoff frequency and response. From Fig. 4.47, one has the nodal (input and output) equations:

$$\Re(\omega)P_{in} + \left[j\omega\left(C_i + C_D\right)\right]V_i + \frac{1}{R_f}\left(V_i - V_{out}\right) = 0$$

$$V_{out} = -AV_i.$$

[19] The name "cold resistance" comes from the fact that the r.m.s. noise current can be expressed as $\sqrt{4k_BTG_n}$ where $G_n = 1/R_f$; alternatively we can associate the noise current to the input conductance $G_{in} \approx A/R_f$ but with a noise temperature $T_n = T/A$ as $\sqrt{4k_BT_nG_{in}}$. In this second case we can imagine that the input conductance has low noise because its noise temperature is low (from which comes the idea of "cold resistance").

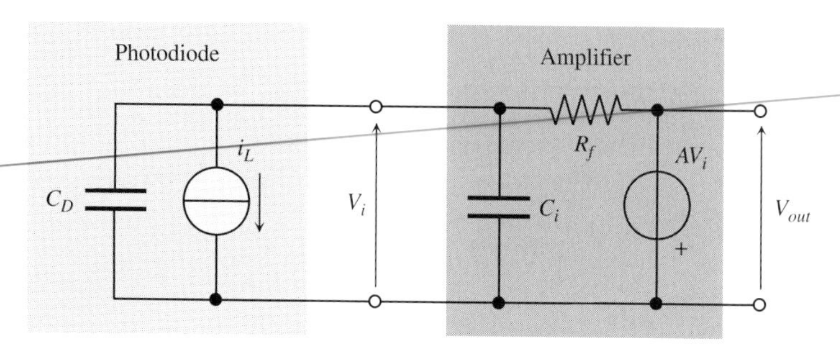

Figure 4.47 Transimpedance stage analysis: equivalent circuit.

Substituting V_{out} into the first equation and solving by V_i, we obtain

$$V_i = \frac{-\Re(\omega) P_{in}}{\dfrac{1+A}{R_f}} \frac{1}{1 + j\omega \dfrac{C_i + C_D}{\dfrac{1+A}{R_f}}} \underset{A \gg 1}{\approx} \underbrace{-\Re(\omega) P_{in}}_{I_{in}} \cdot \frac{R_f}{A} \frac{1}{1 + j\omega \dfrac{(C_i + C_D) R_f}{A}}$$

and

$$V_{out} = \frac{\Re(\omega) P_{in} A}{\dfrac{1+A}{R_f}} \frac{1}{1 + j\omega \dfrac{C_i + C_D}{\dfrac{1+A}{R_f}}} \underset{A \gg 1}{\approx} \underbrace{-\Re(\omega) P_{in}}_{I_{in}} \cdot R_f \frac{1}{1 + j\omega \dfrac{(C_i + C_D) R_f}{A}}.$$

Thus, the input impedance is

$$Z_{in} = \frac{V_i}{I_{in}} = \frac{R_f}{A} \frac{1}{1 + j\omega \dfrac{(C_i + C_D) R_f}{A}} \underset{\omega \to 0}{\approx} \frac{R_f}{A} = R_{in}.$$

The *transimpedance* Z_m of the stage (i.e., the ratio between the output voltage and the input current) can therefore be evaluated as

$$Z_m = \frac{V_{out}}{I_{in}} = -\frac{A}{\dfrac{1+A}{R_f}} \frac{1}{1 + j\omega \dfrac{C_i + C_D}{\dfrac{1+A}{R_f}}} \underset{A \gg 1}{\approx} -\frac{R_f}{1 + j\omega \dfrac{(C_i + C_D) R_f}{A}}, \quad (4.105)$$

That is, at low frequency,

$$Z_m = -R_f.$$

The stage frequency response is therefore

$$\frac{V_{out}}{P_{in}} = \frac{\Re(\omega) R_f}{1 + j\omega \dfrac{(C_i + C_D) R_f}{A}},$$

with extrinsic cutoff frequency

$$f_{Tx} = \frac{A}{2\pi (C_D + C_i) R_f},$$

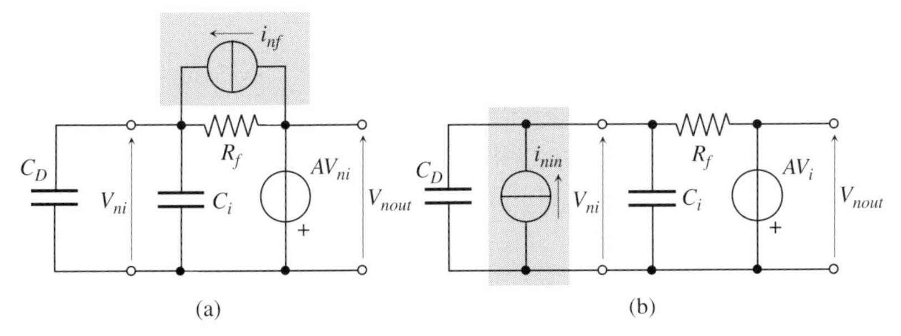

Figure 4.48 Noise analysis of transimpedance stage: (a) feedback with noisy resistor; (b) the noise generator is brought to the input.

which is typically large owing to the amplification factor. In conclusion, the transimpedance stage exhibits low input impedance and high extrinsic bandwidth, without the trade-off required by high- or low-impedance stages.

To justify the noise properties of transimpedance stages, we evaluate the effect of the noisy feedback resistor by computing the total noise voltage at the output and interpreting it as the result of an input noise generator i_{nin} (the "input referred current noise generator") to be compared to the input noise generators in the high- and low-impedance cases. For simplicity, we assume that the amplifier introduces no additional noise due to active elements, though in fact this may be significant, as discussed later, (see Section 4.16.4).

In order to carry out the computation, we consider the circuit as excited by the noise generator of R_f, I_{nf} (power spectrum $4k_BT/R_f$), the corresponding nodal equations of the circuit in Fig. 4.48 (all other deterministic or noise current generators have been disconnected, i.e., replaced by an open circuit) are

$$\left[j\omega\left(C_i + C_D\right)\right] V_{ni} + \frac{1}{R_f}\left(V_{ni} - V_{nout}\right) = I_{nf}$$

$$V_{nout} = -AV_{ni}.$$

Substituting V_{nout} into the first equation and solving by V_{ni} and V_{nout}, we obtain

$$V_{ni} = \frac{I_{nf}}{\dfrac{1+A}{R_f}} \frac{1}{1 + j\omega\dfrac{C_i + C_D}{\dfrac{1+A}{R_f}}} \underset{A\gg 1}{\approx} \frac{I_{nf}\dfrac{R_f}{A}}{1 + j\omega\dfrac{(C_i + C_D)\,R_f}{A}}$$

$$V_{nout} = \frac{-AI_{nf}}{\dfrac{1+A}{R_f}} \frac{1}{1 + j\omega\dfrac{C_i + C_D}{\dfrac{1+A}{R_f}}} \underset{A\gg 1}{\approx} -\frac{I_{nf}R_f}{1 + j\omega\dfrac{(C_i + C_D)\,R_f}{A}}.$$

We have evaluated the input and output noise voltages induced by the feedback noise generator. Now we want to compute the input noise generator I_{nin} needed to produce the same noise output voltage V_{nout}.

The generator I_{nin} should satisfy the equation

$$V_{nout} = -\frac{R_f}{1 + j\omega \frac{(C_i + C_D) R_f}{A}} I_{nf} = Z_m I_{nin},$$

but, taking into account (4.105), we obtain that the input generator coincides with the feedback generator:

$$I_{nin} = I_{nf},$$

i.e., the input noise generator has r.m.s. short-circuit current equal to

$$\overline{i_{nin}^2} = S_{i_{nin}} \times B = 4k_B T \frac{B}{R_f} = 4k_B T \frac{B}{A R_{in}}.$$

Thus, the noise input resistance is equal to the low-frequency input impedance R_{in}, multiplied by the amplification A; in other words, the low input impedance is *not* obtained at the price of a larger noise conductance.

4.16.4 High-speed transimpedance stages

In high-speed (high-frequency) transimpedance amplifier (TIA) stages, noise from the amplifier active elements cannot be neglected with respect to feedback resistor noise. We assume that the TIA noise can be referred to the input as a noise current generator in parallel to the photodetector. This is just an approximation, since a complete input-referred model can be shown to require two correlated series voltage and a parallel current input generators, (see e.g. [36]); however, the series voltage generator is less important due to the low input impedance of the stage, and will be neglected.

The power spectrum of the input-referred current noise generator $i_{n,TIA}$ will be shown to include an f^2 component, see (4.106), besides a $1/f$ (flicker) component which will be neglected in the analysis. The f^2 component can be attributed to the drain or collector noise of the front-end amplifier referred to the input.

To evaluate the power spectrum of $i_{n,TIA}$ we shall separately take into account the effect of the amplifier and of the feedback resistor. We will consider a simple one stage FET amplifier whose noise behavior is dominated by the drain (output) current generator.

The FET equivalent circuit considered is shown in Fig. 4.49, where C_i is the gate–source capacitance, and $g_m V_{in}$ is the short-circuit drain current. Note that this form of the TIA actually requires a finite load to operate as a transimpedance amplifier, since the output circuit is a current generator. The drain noise current generator i_{nD} has power spectrum $S_{i_{Dn} i_{Dn}} \approx 4k_B T g_m P$; see (3.56). We want now to identify the input generator $i_{n,TIA}$ leading to the same short-circuit noise current as the output generator i_{nD}. From Fig. 4.49(above), we have $i_{nout} = i_{nD}$ while from Fig. 4.49(below), we obtain in the frequency domain, considering only the input generator $i_{n,TIA}$ and suppressing the output generator i_{nD},

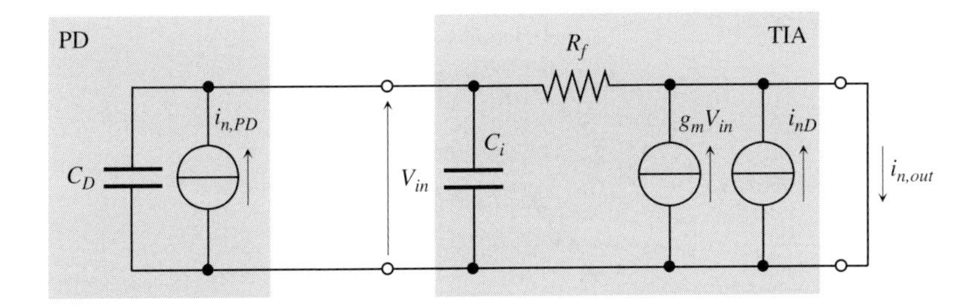

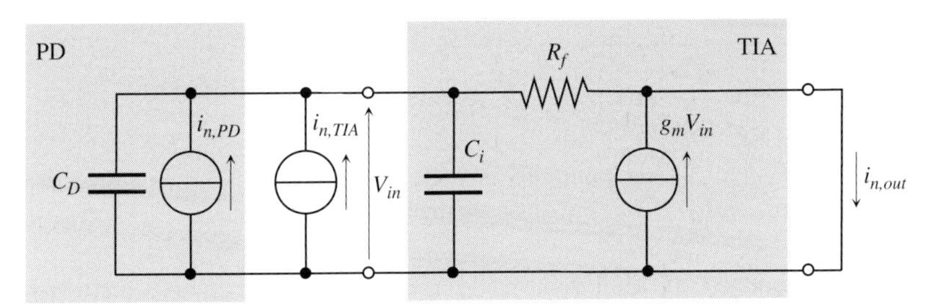

Figure 4.49 Equivalent circuit of the FET TIA (above); equivalent circuit for the evaluation of the input-referred current generator with a FET TIA (below).

$$I_{nout} = g_m V_{in} = g_m I_{n,TIA} \frac{R_f}{1 + \mathrm{j}\omega C_T R_f},$$

where $C_T = C_i + C_D$. Since we must impose that $i_{nout} = i_{nD}$ and find $i_{n,TIA}$ as a consequence, we have

$$I_{n,TIA} = \frac{1 + \mathrm{j}\omega C_T R_f}{g_m R_f} I_{nD}.$$

The power spectrum of the input-referred TIA noise generator will therefore be

$$S_{i_{n,TIA}} = \frac{1 + (\omega C_T R_f)^2}{(g_m R_f)^2} S_{i_{nD}} = \frac{1 + (\omega C_T R_f)^2}{(g_m R_f)^2} 4k_B T g_m P. \tag{4.106}$$

Note the high-pass behavior with cutin frequency:

$$f_c = \frac{1}{2\pi C_T R_f}.$$

In conclusion, the total, input-referred TIA noise power spectrum, including the feedback resistor contribution, will be

$$S_{n,in} = 4k_B T \frac{1}{R_f} + 4k_B T \frac{P}{g_m R_f^2} + 4k_B T \frac{P (2\pi C_T)^2}{g_m} f^2,$$

with a highpass behavior whose characteristic frequency is typically *lower* than the TIA cutoff. In fact we have, for the TIA extrinsic cutoff frequency:

$$f_{Tx,TIA} = \frac{A}{2\pi (C_D + C_i) R_f} = \frac{A}{2\pi C_T R_f} = A f_c, \quad A \gg 1.$$

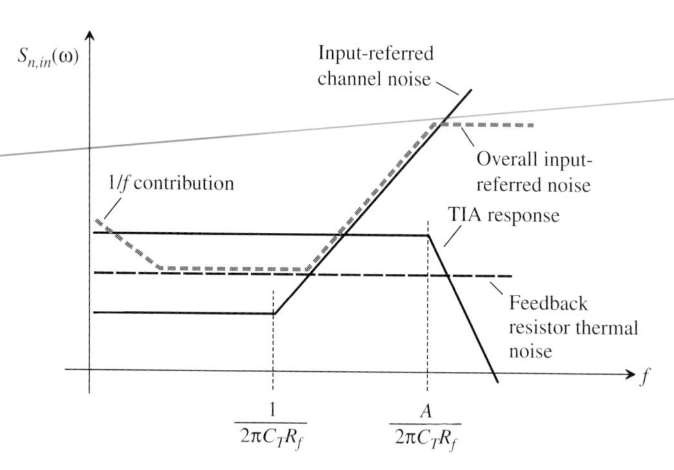

Figure 4.50 Noise overshoot in the TIA input-referred noise generator; the dark gray dashed line is the total noise spectral density. Notice the $1/f$ contribution at low frequency.

As a consequence, the total input-referred noise generator exhibits an overshoot at a frequency below the amplifier cutoff frequency, as shown in Fig. 4.50 (the $1/f$ component, not reported in the formulae, has been added to the plot).

Finally, the total input noise current including the detector contribution will be

$$S_{n,in}^{tot} = S_{n,in} + 2qI_L = 2qI_L + 4k_B T \frac{1}{R_f} + 4k_B T \frac{P}{g_m R_f^2} + 4k_B T \frac{P\,(2\pi C_T)^2}{g_m} f^2.$$

The input-referred noise current spectrum overshoot can be minimized by reducing the total capacitance C_T at high frequency through narrowband compensation by a series inductor. Having somewhat reduced C_T, some further optimization can be carried out by taking into account that the FET cutoff frequency is $f_T = g_m/2\pi C_i$; we have

$$S_{n,in}^{tot} = 2qI_L + 4k_B T \frac{1}{R_f} + 4k_B T P \frac{1}{g_m R_f^2} + 4k_B T P g_m \underbrace{\frac{(C_D + C_i)^2}{C_i^2}}_{f(C_i)} \left(\frac{f}{f_T}\right)^2,$$

but $f(C_i)$ can be minimized by making $C_i = C_D$ through a proper choice of the FET device periphery. Therefore, careful noise matching of the detector and input stage capacitance can lead to a minimization of the f^2 component of the spectrum.[20]

4.17 Front-end SNR analysis and *pin*–APD comparison

In the following discussion, we will model the input-referred front-end noise current with a noise conductance g_n. Moreover, we will approximately consider all spectra to be white, and make reference to the quadratic values of the photodiode and front-end (FE)

[20] The flicker noise component initially considered is in fact neglected in the final result.

stage noise and signal currents. Assuming a receiver with a photodiode of responsivity $\mathfrak{R}$ and input optical power P_{in}, the FE output signal to noise ratio (SNR) is given by

$$\mathrm{SNR} = \frac{\overline{i_s^2}}{\overline{i_{PD}^2} + \overline{i_{FE}^2}} = \frac{\mathfrak{R}^2 P_{in}^2}{2q\mathfrak{R}P_{in}B + 4k_B T g_n B},$$

where $\overline{i_s^2}$ is the quadratic mean of the signal current, $\overline{i_{PD}^2}$ is the quadratic mean of the photodiode (shot) noise current, and $\overline{i_{FE}^2} = 4k_B T g_n B$ is the quadratic mean of the FE input referred noise current. For PD noise, reference is made for the moment to a conventional *pin*. For low input optical power, the photodiode noise is negligible and the SNR is in the *thermal noise limit*, where it increases as the square of the optical power:

$$\mathrm{SNR} \approx \frac{\mathfrak{R}^2 P_{in}^2}{4k_B T g_n B}.$$

On the other hand, at high input optical power, the FE noise becomes negligible and the SNR is in the *shot noise limit*, where it depends only linearly on the optical power:

$$\mathrm{SNR} \approx \frac{\mathfrak{R}^2 P_{in}^2}{2q\mathfrak{R}P_{in}B} = \frac{\mathfrak{R}P_{in}}{2qB} = \frac{I_L}{2qB}.$$

For extremely high P_{in}, the laser relative intensity noise (RIN) dominates. However, most links work (in practice) in the thermal noise limit or, at best, near the onset of the shot noise-limited regime.[21] An example of the overall behavior of the link SNR is shown in Fig. 4.51. For simplicity the laser RIN has been kept constant and independent of the input power.

The *receiver sensitivity* S is the minimum input power (averaged between 0s and 1s for a digital link) required to yield, with a given detection scheme (e.g., intensity modulation, direct detection: IM-DD), an SNR or bit error rate (BER) complying with the system specifications (e.g., $\mathrm{BER} \le 1 \times 10^{-9}$ or $\mathrm{BER} \le 1 \times 10^{-12}$). Given

$$\mathrm{BER} = \frac{1}{2}\mathrm{erfc}\left(\frac{Q}{\sqrt{2}}\right),$$

where

$$Q = \frac{\sqrt{\mathrm{SNR}}}{2} = \frac{1}{2}\sqrt{\frac{(\mathfrak{R}P_{in1})^2}{\overline{i_n^2}}} = \frac{\mathfrak{R}S}{\sqrt{\overline{i_n^2}}},$$

P_{in1} being the high level optical power in a digital scheme, one has,

$$S = \frac{Q\sqrt{\overline{i_n^2}}}{\mathfrak{R}},$$

[21] In a fiber link, an upper limitation to the spurious-free dynamic range also comes from the effect of the fiber nonlinearity, which we completely neglect here.

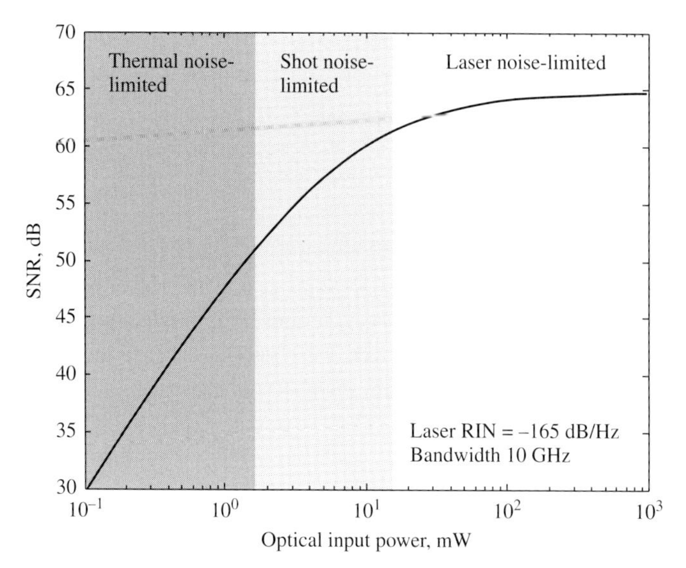

Figure 4.51 SNR of a link as a function of the received power: thermal noise-limited, shot noise-limited, and RIN-limited regions.

where i_n is the total input noise current (including PD and FE noise) (see, e.g., [55]). For a shot noise-limited *pin* receiver and a given Q, i.e., a given SNR or BER:

$$S_{pin} = Q\frac{\sqrt{2q\left(\Re_{pin}S_{pin}\right)B}}{\Re_{pin}} \longrightarrow S_{pin} = \frac{2qQ^2B}{\Re_{pin}} \approx \frac{2qQ^2R_b}{\Re_{pin}}.$$

That is, for a given BER, *S increases linearly with the bit rate* R_b.

Let us consider now the sensitivity in a FE with an APD photodetector. With reference (as usual) to the SAM-APD case with electron-triggered avalanche we obtain

$$S_{\text{APD}} = Q\frac{\sqrt{2q\left(\Re_{\text{APD}}S_{\text{APD}}M_nF_n\right)B}}{\Re_{\text{APD}}} \longrightarrow S_{\text{APD}} = \frac{2qQ^2BM_nF_n}{\Re_{\text{APD}}}.$$

Comparing with the *pin* case, we have $\Re_{\text{APD}} = \Re_{pin}M_n$, i.e.,

$$S_{\text{APD}} = \frac{2qQ^2BM_nF_n}{\Re_{\text{APD}}} = \frac{2qQ^2BM_nF_n}{M_n\Re_{pin}} = F_nS_{pin}.$$

Apparently, the APD sensitivity is always worse than the *pin* sensitivity since the excess noise factor F_n (see Fig. 4.52) is always $F_n > 1$ (the lowest values are obtained for $k_{hn} \to 0$). However, this result holds only if both detectors work in the shot noise regime.

A more complete analysis shows that the APD may be superior if the detector is working in a *thermal noise-limited* link.

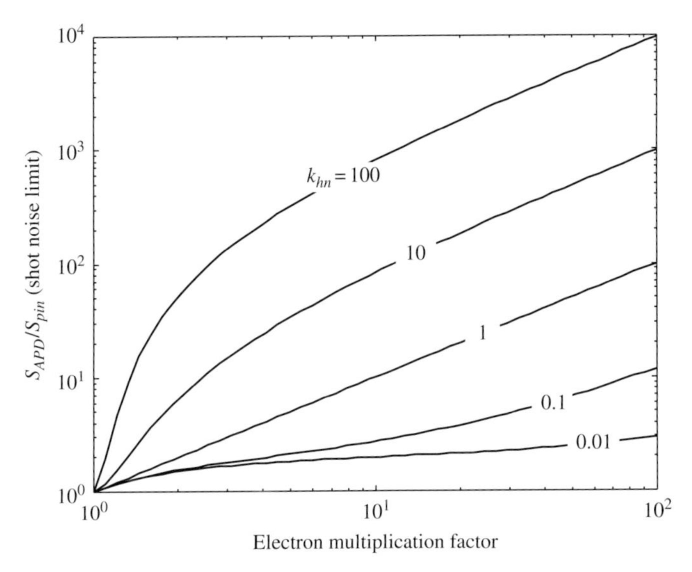

Figure 4.52 Ratio between the APD and the *pin* sensitivities in the shot noise limit; k_{hn} is the α_h/α_n ratio.

Going back to the APD SNR in the general case, one has

$$\begin{aligned}\text{SNR} &= \frac{\overline{i_s^2}}{\overline{i_{\text{APD}}^2} + \overline{i_{FE}^2}} = \frac{\Re_{\text{APD}}^2 P_{in}^2}{2qM_n F_n \Re_{\text{APD}} P_{in} B + 4k_B T g_n B} \\ &= \frac{M_n^2 \Re_{pin}^2 P_{in}^2}{2qM_n^2 F_n \Re_{pin} P_{in} B + 4k_B T g_n B} \\ &= \frac{\Re_{pin}^2 P_{in}^2}{2q\Re_{pin} F_n P_{in} B + 4k_B T \dfrac{g_n B}{M_n^2}}.\end{aligned}$$

Thus, an APD is equivalent to a *pin* receiver where *the weight of receiver noise is decreased by* M_n^2 while *the pin noise is increased by* F_n.

In particular, if the receiver operates in the thermal noise limit, one has

$$\text{SNR} \approx \frac{\Re_{\text{APD}}^2 S_{\text{APD}}^2}{4k_B T g_n B} = \frac{\Re_{pin}^2 M_n^2 S_{\text{APD}}^2}{4k_B T g_n B} = \frac{\Re_{pin}^2 S_{pin}^2}{4k_B T g_n B} \rightarrow S_{\text{APD}} = \frac{S_{pin}}{M_n}.$$

In the thermal noise limit the APD is therefore *always superior*. However, with increasing M_n one ultimately has $F_n \approx k_{hn} M_n$, i.e., for the APD,

$$\begin{aligned}\text{SNR}|_{\text{APD}} &\underset{M_n \to \infty}{\approx} \frac{\Re_{\text{APD}}^2 S_{\text{APD}}^2}{2qk_{hn} M_n^2 \Re_{\text{APD}} S_{\text{APD}} B + 4k_B T g_n B} \\ &\approx \frac{\Re_{\text{APD}} S_{\text{APD}}}{2qk_{hn} M_n^2 B} = \frac{\Re_{pin} S_{\text{APD}}}{2qk_{hn} M_n B},\end{aligned}$$

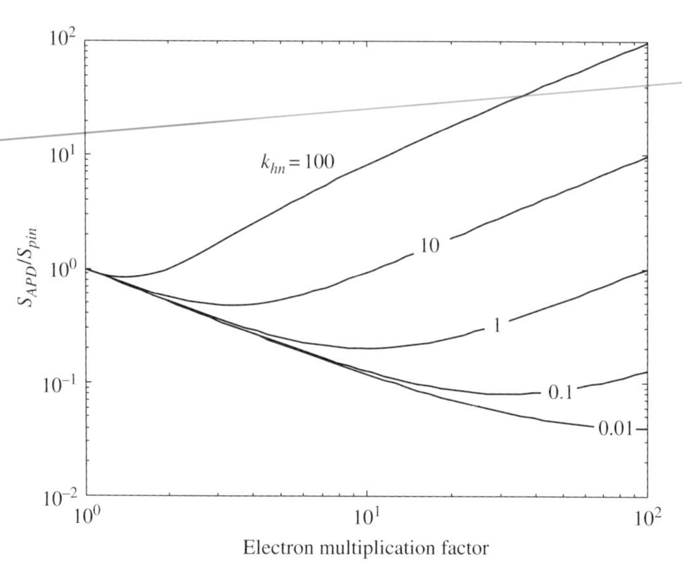

Figure 4.53 APD versus *pin* sensitivity as a function of the multiplication factor for several values of k_{hn}. We assume that the front-end thermal noise is 100 times the *pin* noise.

while, for the *pin*,

$$\text{SNR}|_{pin} = \frac{\mathfrak{R}^2_{pin} S^2_{pin}}{2q\mathfrak{R}_{pin} S_{pin} B + 4k_B T g_n B} \underset{M_n \to \infty}{\approx} \frac{\mathfrak{R}_{pin} S_{pin}}{2qB}.$$

That is, with the same SNR,

$$\frac{S_{\text{APD}}}{S_{pin}} \underset{M_n \to \infty}{\approx} k_{hn} M_n.$$

Therefore, in the limit of a large multiplication factor the APD sensitivity finally becomes worse than the *pin* sensitivity; again, the avalanche coefficient ratio should be kept as small as possible. Since, for intermediate values of the multiplication factor, thermal noise prevails and the APD is increasingly better for increasing M_n, while for extremely large values we are again in the shot noise limit, an optimum value of the multiplication factor exists that yields the minimum value of the S_{APD}/S_{pin} ratio. As shown in Fig. 4.53, which was derived assuming a specific value for thermal noise (FE noise equal to 100 times the *pin* noise for input power equal to the *pin* sensitivity), the sensitivity ratio becomes < 1 (i.e., favorable to the APD) and exhibits a well-defined minimum that is deeper for low value of k_{hn}. Of course, the physical realization of extremely low k_{hn} values may be difficult.

In conclusion, the *pin* has better sensitivity than the APD in the shot noise limit; in the thermal noise limit or in intermediate cases, APDs have better sensitivity, despite the higher noise, due to the larger responsivity. A theoretical optimum value for the multiplication rate exists, and more favorable conditions are reached anyway when decreasing k_{hn}. Large multiplication factors should, however, be avoided because they may lead to a reduction of the bandwidth.

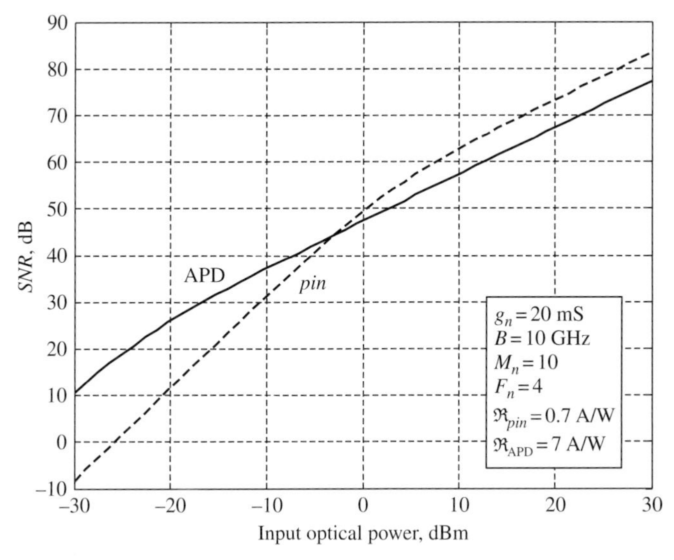

Figure 4.54 SNR for an APD and a *pin* as a function of the input power. The receiver parameters are shown in the inset.

An example is shown in Fig. 4.54, where the SNR of a link with APD and *pin* receiver is plotted as a function of the input power. For low power (thermal noise limit) the APD is superior, while for high power (shot noise limit) the *pin* finally prevails (i.e., has a larger SNR).

APD sensitivities are typically better than *pin* sensitivities, but the device is slower. The choice may also be dictated by the application: the SNR required by digital vs. analog systems (e.g., CATV, microwave photonics) systems is much lower (e.g., ≈ 17 dB vs. ≈ 40 dB), so that digital systems may work in the thermal noise limits with lower SNR, while analog systems, which require higher SNR, may work at or near the shot noise limit and therefore can better exploit *pin* devices.

4.18 Front-end examples

Several hybrid and OEIC (optoelectronic integrated circuit) TIA implementations have been presented so far with different technologies (conventional bipolar transistors, HBTs, FETs, HEMTs, PHEMTs, more recently MOS). Alternative solutions to the front-end stage are the hybrid solution (photodiode externally connected to an integrated TIA) and the OEIC solution where both the photodiode and the TIA are integrated on the same chip. Moreover, the detector gain can be introduced, not only through APDs, but also by amplifying the input optical signal through a SOA. Figure 4.55 reports the performances of a number of detector-TIA receivers with different technologies appearing mainly before 2000 [56]; see also [57], Table 1, 2 and 3, [58] and [59]. The figure of merit F_{OEIC} is a gain–bandwidth product defined as

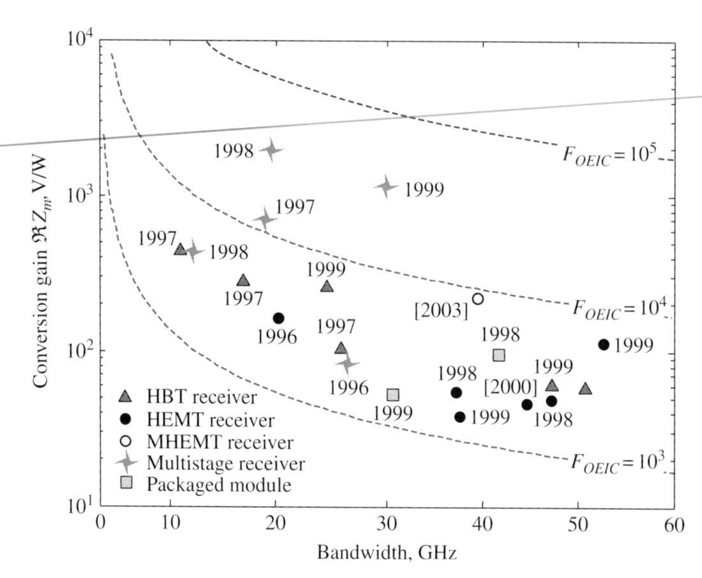

Figure 4.55 Some published receivers and their conversion gain, bandwidth, and F_{OEIC}. Adapted from [56], Fig. 1 (©2000 IEEE). For further details on the receiver data, see [56] and references therein. The points labeled [2000] and [2003] are from [58] and [59], respectively. MHEMT stands for metamorphic HEMT.

$$F_{OEIC} = \Re \left| Z_m(0) \right| B,$$

where $\Re$ is the detector responsivity.

A few commercial TIA products are reported in [60], Table 5.2, with a single-ended or *differential* topology. The products (which appeared before 2005) are listed according to the speed (2.5, 10, and 40 Gbps). Technologies include for the 2.5 Gbps products Si BJTs, SiGe HBTs, and GaAs MESFETs and HEMTs; at higher speed HBTs (SiGe and GaAs) dominate, with some examples of GaAs HEMT. The r.m.s. noise current $\sqrt{i^2_{n,TIA}}$ obviously increases with the receiver bandwidth and the values reported are 200–500 $n\text{A}/\sqrt{\text{GHz}}$ electrical bandwidth. More recently, both 10 Gbps and 40 Gbps transceivers in silicon-on-insulator (SOI) CMOS 0.13 μm technology were presented [61], showing that nanometer-scale NMOS devices can also play a role in high-speed optical receivers.

Although the monolithic integration of a detector with the transimpedance amplifier appears to be a promising process, in practice this approach is fraught with a number of technological problems – mainly compatibility between the electron and optoelectronic device process flow. An example of monolithic 40 Gbps InP-based *pin*-HBT integration is shown in Fig. 4.56, with the layer breakdown shown in Fig. 4.57 [56]. Note that the transistor collector layer is also exploited for the *pin* active region. The HBT preamplifier includes a low-input-impedance common-base stage, followed by a TIA closed on a common emitter buffer; two designs were implemented, with bandwidth of 30 GHz or 50 GHz, respectively, with a gain difference of about 7 dB.

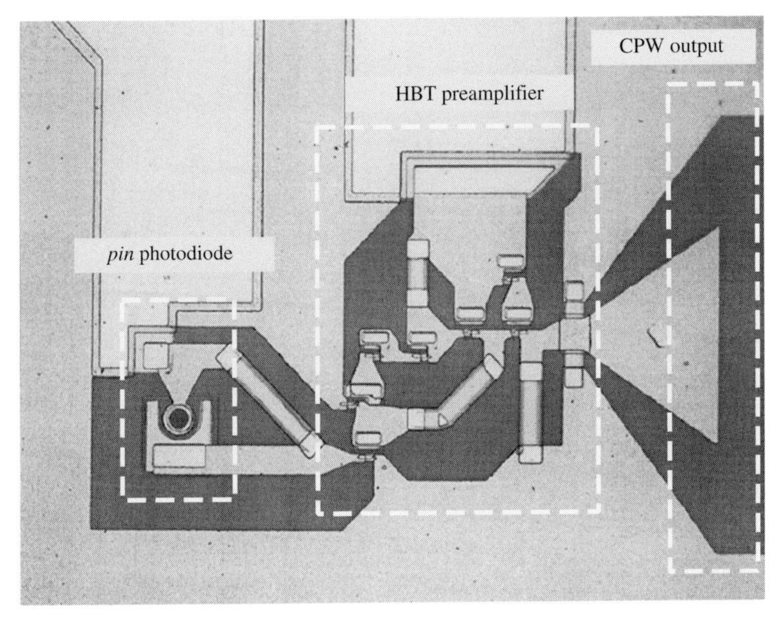

Figure 4.56 50 GHz integrated *pin*-HBT receiver. Adapted from [56], Fig. 12 (©2000 IEEE).

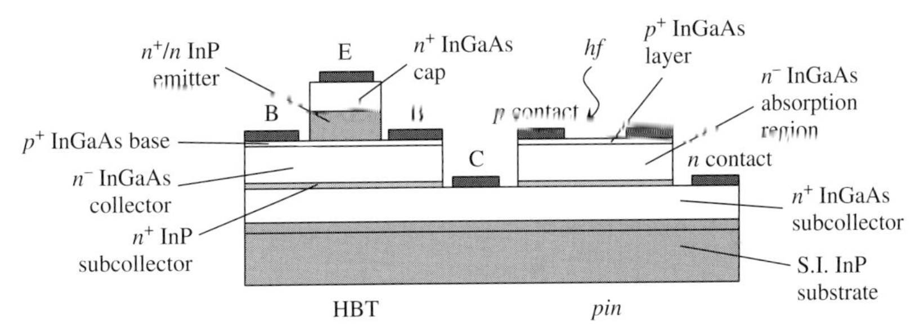

Figure 4.57 Integrated *pin*-HBT receiver structure: the HBT collector acts as the *pin* absorption layer. Adapted from [56], Fig. 2 (©2000 IEEE).

An example of the InP-based layer structure of the 40 Gbps receiver stage, integrating a coplanar MIM (metal–insulator–metal) detector with an InAlAs–InGaAs PHEMT front-end amplifier, is shown in Fig. 4.58 [62]. Epitaxial regrowth was exploited in this case to realize the MSM photodiode. The circuit layout is shown in Fig. 4.59.

More recently, monolithic integration of long-wavelength receivers explored all-Si solutions, based on combining nanometer CMOS processes with Ge-based photodiodes monolithically grown on a Si substrate. Germanium-on-Si photodetectors have been demonstrated in the past, see [63], [38], and integration of a Ge detector with the already mentioned 130 nm CMOS process was achieved [64] for a 10 Gbps transceiver, with a sensitivity around −15 dBm. The all-Si transceiver implementation also requires that a Si-based modulator be available.

In general, monolithic integration should introduce benefits for high-speed receivers, where removing inductive and capacitive parasitics associated with interconnections

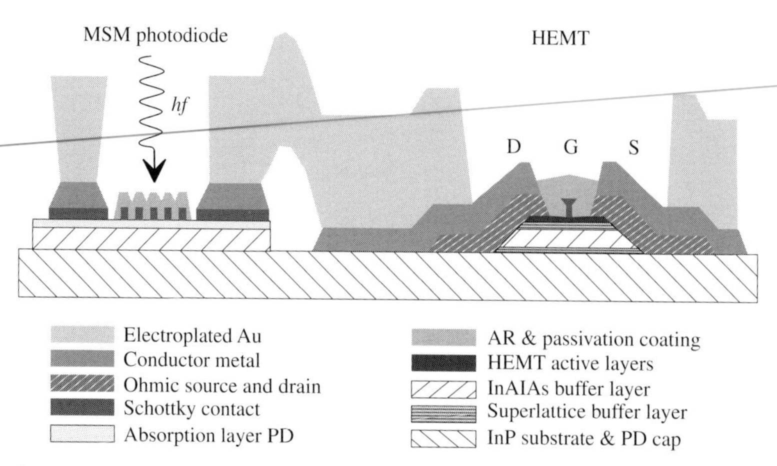

Figure 4.58 Layer structure of MSM-PHEMT integrated receiver. From [62], Fig. 4 (©1999 IEEE).

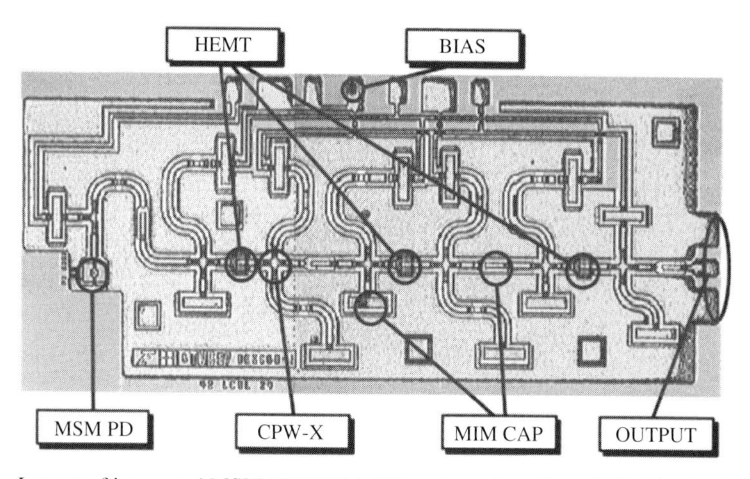

Figure 4.59 Layout of integrated MSM-PHEMT InP-based receiver. From [62], Fig. 8 (©1999 IEEE).

allows for bandwidth improvements. For applications below 10 Gbps, performance advantages are probably low, but cost benefits associated with large volume production could be an asset of monolithic integration.

4.18.1 Hybrid and monolithic front-end solutions

Figure 4.60 shows the sensitivity vs. bit rate of a number of hybrid and integrated receiver solutions. The hybrid, OEIC and APD best fits are from [57], Fig. 3, and refer to pre-1998 results; they suggest that hybrid receivers are superior to integrated receivers for speeds lower than 10 Gbps, inferior for high speed; APD-based hybrid receivers nevertheless, show sensitivities 10 dB better (typically) than *pin* based receivers. More recent results reported in Fig. 4.60 point out that high-speed OEIC receivers may outperform the 1998 best fit, but also that impressive results can be obtained through photo-HBT based OEICs and APD hybrids. In [59] a 40 Gbps *pin*-MHEMT receiver

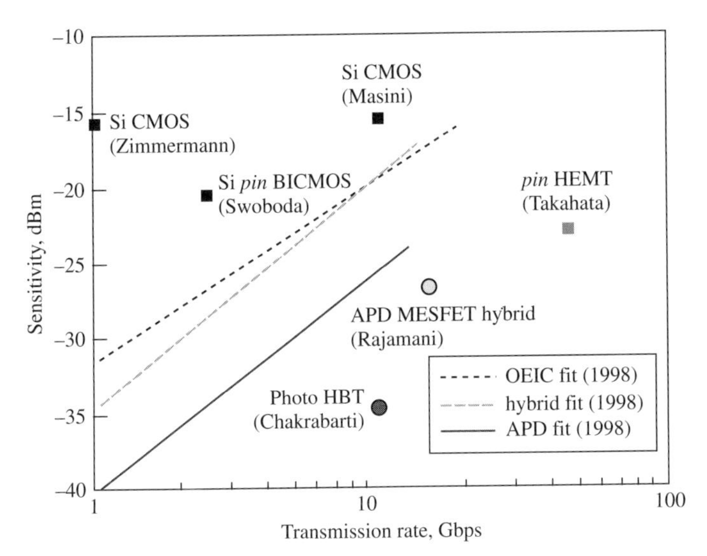

Figure 4.60 Integrated, hybrid, APD receiver sensitivity vs. bit rate. Best fits are from [57], while the other post-2000 data are from Zimmermann [65], Swoboda [66], Masini [67], Rajamani [68], Chakrabarti [69], and Takahata [58].

is implemented in both hybrid and monolithic form, and the OEIC implementation is found to have a larger gain (7 dB) and a wider bandwidth (6 GHz), confirming the conclusions in [57] at least for speeds well in excess of 10 Gbps.

In concluding the comparison between the *pin* and the APD photodiode, we should mention that other possible competitors to these solutions exist. Amplification of the primary photocurrent can take place by an avalanche mechanism, but another possibility is amplification within a phototransistor. HBT-based phototransistors have shown promising performances in the laboratory in terms of speed, but of course such devices require a compromise between transistor and photodetector active region design. A completely different approach is to move amplification from the electrical to the optical domain, inserting an optical amplifier (e.g., a SOA, semiconductor optical amplifier) between the fiber output and the *pin* input. The resulting solution can be competitive with respect to the APD in terms of sensitivity.

4.19 Questions and problems

4.19.1 Questions

1. What is the photodetector *responsivity*? Explain the behavior of responsivity as a function of the photon energy.
2. Define the internal (η_Q) and external (η_x) quantum efficiency for a photo detector.
3. Define the dark current of a detector.

4. Choose a material system for (a) long-wavelength detection in a high-speed, long-haul fiber link; (b) detection in the first window in a LAN; (c) far infrared detection for temperature mapping.
5. Discuss the role of Si and Ge as possible detector materials.
6. Explain the operation of a photoresistor and of a photodiode (*pn*).
7. What is a phototransistor? Explain why this device has internal gain.
8. Explain why the performances of a *pin* photodiode are superior to those of a *pn* photodiode.
9. Explain the operation of the avalanche photodiode (APD). What are the advantages of the SAM-APD with respect to the conventional APD?
10. Explain why the order of magnitude of the responsivity of an ideal semiconductor photodiode at 1 μm wavelength is close to 1.
11. Sketch the equivalent circuit of a photodiode.
12. What is the photovoltaic operation region of a photodiode? Is this bias condition convenient for a digital receiver?
13. Discuss the frequency response of a *pn* photodiode.
14. What is the difference between the *intrinsic* and the *extrinsic* cutoff frequency of a photodiode? Explain the role of the device capacitance in the extrinsic cutoff frequency.
15. Discuss the frequency response and modulation bandwidth of a *pin* photodiode.
16. Discuss the modulation bandwidth of an avalanche photodiode. Why is the SAM-APD faster than the simple APD? Why is the APD slower than the *pin*?
17. Select a suitable device for a high-speed communication system among *pin* diodes, photoresistors, APDs, phototransistors.
18. Suggest a reason why heterostructure InGaAs *pin* detectors typically exploit a *diffused pi* junction.
19. Explain why a distributed PD can overcome the *RC* limitation on bandwidth.
20. Explain the physical origin of photodetector gain in an APD.
21. Describe the disadvantages of a homojunction photodiode, and how they can be overcome by using a heterojunction photodiode.
22. Sketch a heterojunction photodiode suitable for 1.55 μm wavelength detection. What is the typical material system for this application?
23. Discuss the bandwidth–efficiency trade-off in vertical *pin* photodiodes.
24. Explain why waveguide detectors offer improved bandwidth–efficiency product with respect to vertically illuminated PDs.
25. Describe the bandwidth limitation factors in traveling-wave photodetectors.
26. Describe the structure of a velocity-matched traveling-wave detector.
27. Describe some possible solutions for high-speed APDs.
28. Define the *sensitivity* of a photodetector.
29. Discuss the *thermal limit* and the *shot noise limit* to the photodetector sensitivity.
30. Define the *quantum limit* to the photodetector sensitivity.
31. Explain why an avalanche photodiode can be superior to a *pin* diode despite having a higher noise than the latter.
32. Explain the pros and cons of the low-impedance, high-impedance, and trans-impedance front-end stages.

33. Define the noise equivalent power (NEP) in a photodiode.
34. What is the optimum condition in an APD for the ratio of the avalanche ionization coefficients of electrons and holes (supposing electrons are the avalanche-triggering carriers)?
35. Discuss noise optimization in a high-speed transimpedance stage.

4.19.2 Problems

1. A photodiode has an intrinsic cutoff frequency of 10 MHz and an extrinsic *RC* cutoff of 100 kHz. What is the cutoff frequency?
2. A photodiode has a capacitance of 50 fF and is closed on a 50 Ω resistance. Evaluate the extrinsic cutoff frequency.
3. A *pin* photodiode has an intrinsic layer of thickness $d = 1.5\,\mu m$. The absorption coefficient for the active region is 10^3 cm^{-1}. (1) Calculate the external quantum efficiency if only light absorbed in the undoped region contributes to the photocurrent. Assume no reflection losses and unit internal quantum efficiency. (2) Calculate the intrinsic layer thickness d needed to ensure an external efficiency of 0.9.
4. A GaAs *pn* photodiode has the following parameters at 300 K: side dopings $N_D = 10^{17}\,\text{cm}^{-3}$, $N_A = 10^{17}\,\text{cm}^{-3}$, $D_n = 20\,\text{cm}^2/\text{s}$, $D_h = D_n/2$; $\tau_n = \tau_h = 100\,\text{ns}$; $\alpha = 300\,\text{cm}^{-1}$. Taking into account the GaAs intrinsic concentration (assume $n_i = 2.1 \times 10^6\,\text{cm}^{-3}$), evaluate the depletion region width at a reverse voltage of 10 V and the diffusion lengths (assume $\epsilon_r = 13$). Evaluate the maximum theoretical device responsivity in DC (for $\hbar\omega \approx E_g$) and sketch the modulation response of the diode. Suppose that the diode sides are long (with respect to the diffusion length). Is the assumption of constant optical power vs. x realistic with the data provided?
5. A *pin* photodiode has an area of $0.05\,\text{mm}^2$ and the absorption region is $4\,\mu m$ thick. Supposing that the carriers travel at saturation velocity, evaluate the overall frequency response from the capacitance and transit time contributions and the photodiode external efficiency. Assume $\epsilon_r = 12$, $R_L = R_G = 50\ \Omega$, $\alpha = 10^4\,\text{cm}^{-1}$, 100% internal quantum efficiency and 30% surface power reflectivity.
6. We want to design a Si *pin* photodiode with a (minimum) 300 MHz bandwidth at 20 V reverse bias. The device area is $0.2\,\text{mm}^2$; the (thin) surface side is p^+ while the substrate is n^+. The efficiency (considering only the intrinsic layer) should be at least 50% (with unit internal quantum efficiency and zero surface reflectivity). Repeat the computation by considering also diffusion currents in the substrate. What is the (approximate) frequency response in this case? Assume $\alpha = 200\,\text{cm}^{-1}$, $\epsilon_r = 12$, $\mu_n = 1000\,\text{cm}^2\,\text{V}^{-1}\text{s}^{-1}$, $\mu_h = 500\,\text{cm}^2\,\text{V}^{-1}\text{s}^{-1}$, $R_L = 50\,\Omega$, $\tau_h = 100\,\text{ns}$.
7. The quantum efficiency of an InGaAsP/InP avalanche photodiode is 80% when detecting a radiation with $\lambda = 1.3\,\mu m$. With an incident optical power of $1\,\mu W$ the output current is $20\,\mu A$. Calculate the avalanche gain or current multiplication factor of the device.
8. An AlGaAs detector is designed to have a cutoff wavelength of $0.68\,\mu m$. Estimate the Al content of the detector. (Hint: for the AlGaAs energy gap

in the direct-bandgap region assume $E_g = 1.42 + 1.25x$ where x is the Al fraction.)

9. A Ge-based *pin* photodiode has the following parameters: $W = 50\,\mu\text{m}$, $\alpha = 500\,\text{cm}^{-1}$, $A = 1\ \text{mm}^2$, $\epsilon_r = 11$. Evaluate the external quantum efficiency (assume unit internal efficiency and zero power surface reflection coefficient) and the total bandwidth as a function of the applied voltage assuming $\mu_n \approx \mu_h \approx 1000\,\text{cm}^2\ \text{V}^{-1}\text{s}^{-1}$ and a carrier saturation velocity of $v_{\text{sat}} = 10^7$ cm/s. Assume a piecewise field–velocity curve for electrons and holes. The load resistance is $R_L = 100\,\Omega$.
10. A waveguide *pin* photodetector has an InGaAs active region thickness of $d = 0.5\,\mu\text{m}$. The device width is $W = 2\,\mu\text{m}$. Evaluate the device length L so as to achieve $\eta_x = 0.9$ (neglect coupling losses) and estimate the resulting bandwidth. Assume an absorption length of $5\,\mu\text{m}$, $\epsilon_r = 12$, and a carrier saturation velocity $v_{\text{sat}} = 10^7$ cm/s. Suppose a load resistance $R_L = 20\ \Omega$.
11. The avalanche region of a SAM avalanche photodiode is $0.8\,\mu\text{m}$ thick. Assuming $\alpha_n = \alpha_h = 10^4\,\text{cm}^{-1}$, discuss the multiplication factor of the device for electron triggered and hole triggered avalanche. Repeat for $\alpha_n = 10^4\,\text{cm}^{-1}, \alpha_h = 10^2\,\text{cm}^{-1}$.
12. The avalanche region of an avalanche photodiode is $2.5\,\mu\text{m}$ thick. Suppose $\alpha_n = 10^4\,\text{cm}^{-1}$, $\alpha_h = 10^2\ \text{cm}^{-1}$. Evaluate the excess noise factors of the electron-triggered SAM-APD and of the conventional APD.
13. A shot noise-limited detector has 10 GHz bandwidth. Evaluate the photocurrent needed to ensure a SNR of 60 dB.
14. A photodiode is connected to a front-end amplifier with a r.m.s. noise current of 14 pA/$\sqrt{\text{Hz}}$. Supposing that the PD responsivity is 0.8 mA/mW, evaluate the optical power for which the diode noise equals the front-end noise.
15. What is the input resistance of a transimpedance amplifier with a feedback resistance of $500\ \Omega$ and an open-loop amplification of 1000? What is the input noise resistance?
16. A photodiode operates at 10 Gbps and has a responsivity of 0.4 mA/mW. Evaluate the NEP neglecting the dark current.
17. Consider a *pin* diode with photocurrent $I_{ph} = 1$ mA and a SAM-APD with photocurrent $I_{ph}M_n$ (where $M_n = 10$ is the avalanche multiplication factor, the electrons being the avalanche triggering carriers). (a) Compare the shot noise r.m.s. currents for the two cases on a 10 GHz bandwidth assuming $k_{hn} = 0.2$. (b) Assuming that the thermal noise current from the front end is equal to ten times the *pin* shot noise current, evaluate the SNR in the two cases.
18. A *pin* photodiode has an internal capacitance $C_j = 40$ fF and is connected to a transimpedance amplifier with an input resistance $R_{in} = 50\ \Omega$ and an input capacitance $C_{in} = 10$ fF. The equivalent photodiode transit time is $\tau_t = 10$ ps. Evaluate the maximum bit rate for the photodiode.
19. A silicon *pin* diode is illuminated by an optical power $P_{in} = 100$ nW at $\lambda = 1\,\mu\text{m}$. The external quantum efficiency of the device is 55% and the dark current at the operating bias is negligible. Calculate the photocurrent and the r.m.s. shot noise current if $B = 500$ MHz and evaluate the output SNR.

5 Sources

5.1 Optical source choices

Two solid-state optical sources are currently available, the light-emitting diode (LED) and the laser diode (LASER stands for light amplification by the stimulated emission of radiation). LEDs have poor spectral purity and low speed under direct modulation, and are therefore unsuited to long-haul or high-speed communication systems.

Lasers, on the other hand, have at least one order of magnitude narrower linewidth than LEDs and, under direct modulation, can reach bit rates in excess of 10 Gbps (albeit with some deterioration of the spectral purity due to the spurious frequency modulation or chirp associated with the amplitude modulation). Lasers are therefore well suited for long-haul and high-speed applications, although indirect modulation is required to achieve bit rates in excess of 10 Gbps over long distances (e.g., $L > 10$ km).

5.2 Light-emitting diodes

Light-emitting diodes (LEDs) are based on electron–hole (e-h) pair recombination in a forward-biased *pn* junction or heterojunction, leading to spontaneous photon emission. Due to the emission mechanism, the LED output spectrum is comparatively broad (with a total width at half power of the order of $1.8k_BT \approx 47$ meV at ambient temperature), or, in terms of wavelength, about 50 nm (or 500 Å) with respect to a central wavelength around 1 μm. The spectral purity of LEDs is therefore low with respect to lasers, whose linewidth is at least one order of magnitude narrower. At the same time, the LED maximum modulation speed is of the order of 100 Mbps, about two orders of magnitude slower than the laser direct modulation response. From the standpoint of optical communication systems (operating in the near infrared), the LED is therefore confined to short-distance, low-speed links. However, a number of important applications exist for visible and also UV light LEDs, such as displays and lighting (automotive and domestic). Developments in the latter area have been fostered by GaN-based blue-light LEDs, which have in turn allowed for the development of white-light LED-based sources.

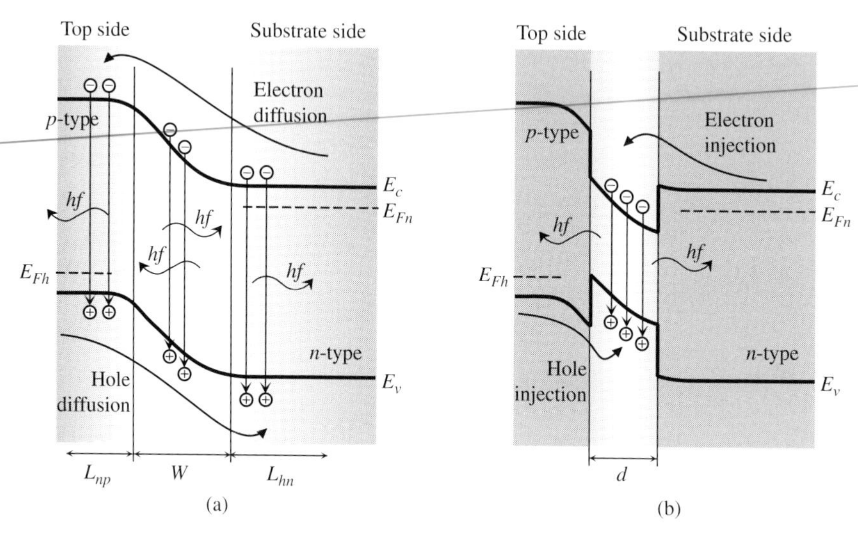

Figure 5.1 Principle of homojunction (a) and heterojunction (b) LEDs.

5.2.1 LED structures

Homojunction LEDs exploit a directly biased *pn* junction. In forward bias, excess carrier populations are injected into the two junction sides; carriers recombine (radiatively or nonradiatively) both in the two diffusion, low-field, regions and in the space-charge region; see Fig. 5.1(a). In the homojunction LED the emitted photons can, however, be absorbed back by the material, since the photon energy is larger than the material gap. As a consequence, photons emitted toward the substrate are mostly absorbed, while those emitted toward the device upper surface can escape and be coupled to an external medium (if they are not absorbed first or reflected by the upper device surface). To increase the device efficiency, therefore, photons must be mostly emitted near the surface; this can be obtained by means of an asymmetric junction where the (thin) surface layer is less doped than the substrate; in this way, excess carrier injection takes place mainly from the substrate into the surface layer. In the structure shown in Fig. 5.1(a), pn^+ doping allows surface generation to be maximized. Homojunction LEDs can be diffused or epitaxial.

In heterojunction LEDs carriers are injected, in direct bias, into a narrowgap material sandwiched between two widegap layers, e.g., an AlGaAs/GaAs/AlGaAs heterojunction or InAlAs/InGaAs/InP heterojunction. Injected excess carriers radiatively recombine, emitting photons; see Fig. 5.1(b). In heterojunction LEDs the photons emitted by the narrowgap layer cannot be absorbed by the widegap cladding; photons emitted toward the substrate can be recovered, e.g., through mirrors, thus improving the device efficiency.

LEDs can also be classified according to the direction along which the output beam is emitted with respect to the junction plane. *Vertical emission* LEDs, also called Burrus LEDs [70] are well suited to coupling with wide-core multimode optical fibers (with core widths of the order of 50–100 μm) due to the large emission area (see Fig. 5.2(a)).

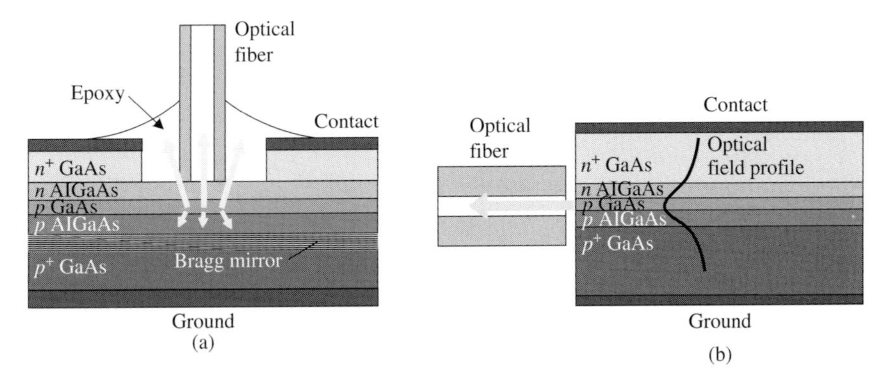

Figure 5.2 Heterojunction LEDs: (a) vertical AlGaAs/GaAs/AlGaAs LED with a substrate Bragg mirror; (b) edge-emission LED.

Many high-performance LEDs are, on the other hand, *edge-emitting* structures (see Fig. 5.2(b)) needing lateral coupling and alignment with the optical fiber. Edge-emitting LEDs closely resemble Fabry–Perot lasers in which the cavity quality factor is low due to the lack of mirrors. In some heterojunction edge-emitting LEDs, high injection conditions enable the onset of gain in the narrowgap material, thus allowing emitted photons to undergo multiplication through stimulated emission. The resulting device, called a *superradiant LED* (or superluminescent LED [71]), shows an increase in the optical power and also spectral narrowing (due to the fact that the gain spectrum is narrower than the spontaneous emission spectrum).

5.2.2 Homojunction LED power–current characteristics

The homojunction LED is a pn junction where radiative recombination dominates over nonradiative effects. Following the Shockley model, the total junction current can be written as the sum of three contributions, arising from recombination in the two injection sides (to define the structure, we suppose that the p side is on the surface, while the substrate is n^+) and in the space charge region. Let N_A be the doping of the (surface) p side, N_D the doping of the (substrate) n side, and L_{np} and L_{hn} the diffusion lengths of electrons in the p side and holes in the n side, respectively:

$$L_{np} = \sqrt{D_n \tau_n}, \quad L_{hn} = \sqrt{D_h \tau_h},$$

where D_n and D_h are the electron and hole diffusivities, τ_n and τ_h are the carrier total lifetimes. The current contribution from electron injection from the substrate into the surface layer is denoted as I_{n0}, while the current contribution from hole injection from the top layer into the substrate is denoted as I_{h0}. Finally, the current arising from generation and recombination in the space-charge region is I_{GR}. The total diode current is

$$I_D = I_{n0} + I_{h0} + I_{GR},$$

where

$$I_{n0} = \frac{qAD_n n_i^2}{L_{np} N_A} \left[\exp\left(\frac{V_A}{V_T}\right) - 1 \right] \tag{5.1}$$

$$I_{h0} = \frac{qAD_h n_i^2}{L_{hn} N_D} \left[\exp\left(\frac{V_A}{V_T}\right) - 1 \right] \tag{5.2}$$

$$I_{GR} = \frac{q n_i W}{2\tau} \left[\exp\left(\frac{V_A}{2V_T}\right) - 1 \right]. \tag{5.3}$$

In the above expressions, A is the device area, W is the space charge region width, τ is an equivalent lifetime for carriers in the space charge region, n_i is the semiconductor intrinsic density, and V_A is the applied voltage. The GR component is small, due to the reduced width of the space charge region in direct bias, and will be neglected in the rest of the discussion, i.e.,

$$I_D \approx I_{n0} + I_{h0}.$$

Since photons emitted in the substrate side are typically lost by absorption, only I_{n0} is a useful LED current component. We define the LED *injection efficiency* η_i as the ratio of the *useful* current (injected into the p side) to the total current:

$$\eta_i \approx \frac{I_{n0}}{I_{n0} + I_{h0}} \approx \frac{I_{n0}}{I_D} = \frac{1}{1 + \sqrt{\frac{D_h}{D_n}} \sqrt{\frac{\tau_n}{\tau_h}} \frac{N_A}{N_D}} \leq 1 \rightarrow I_{n0} = \eta_i I_D. \tag{5.4}$$

The injection efficiency η_i can be maximized making the junction asymmetric, with $N_D \gg N_A$; of course the p and n sides can be interchanged (i.e., the surface side can be low-doping n-type, the substrate high-doping p-type).

Let us now evaluate the optical generation associated with the current component I_{n0}. Since the photon generation rate (per unit time and volume) coincides with the (excess) minority carrier radiative recombination rate (per unit time and volume), we have, in the lifetime approximation and considering injected electrons in the surface p layer:

$$\left. \frac{\mathrm{d}}{\mathrm{d}t} \frac{n_{ph}}{V} \right|_{\text{sp. em.}} = \frac{n'(x)}{\tau_{n,r}} = \frac{n'(-x_p)}{\tau_{n,r}} \exp\left(\frac{x - x_p}{L_{np}}\right),$$

where $\tau_{n,r}$ is the radiative electron lifetime, $-x_p$ is the p-side boundary of the depletion layer, the surface p layer extends from $x = -W_p$ to $x = 0$ (corresponding to the junction plane) with (for simplicity) $W_p \gg L_{np}$, and the excess carrier concentration in the depletion layer boundary is given by the junction law as

$$n'(-x_p) = \frac{n_i^2}{N_A} \left[\exp\left(\frac{V_A}{V_T}\right) - 1 \right],$$

where V_A is the applied voltage.

The total generated optical power can be recovered by integrating the photon generation rate per unit time and volume from $-W_p \approx -\infty$ to $-x_p$ and over the junction area A, and multiplying the result by the photon energy $\hbar\omega \approx E_g$; we obtain

$$P_{out} = A\hbar\omega \int_{-\infty}^{-x_p} \frac{n'(-x_p)}{\tau_{n,r}} \exp\left(\frac{x - x_p}{L_{np}}\right) dx = AL_{np}\hbar\omega \frac{n'(-x_p)}{\tau_{n,r}}$$
$$= \hbar\omega \frac{AL_{np}}{\tau_{n,r}} \frac{n_i^2}{N_A} \left[\exp\left(\frac{V_A}{V_T}\right) - 1\right].$$

Taking into account the expression of I_{n0} from (5.1) and (5.4) we can also write

$$P_{out} = \frac{\hbar\omega}{q} \frac{\tau_n}{\tau_{n,r}} \eta_i I_D.$$

The total lifetime τ_n can be derived (by applying a Matthiessen rule) from the lifetimes associated with radiative recombination ($\tau_{n,r}$) and with nonradiative mechanisms, such as thermal and Auger recombination ($\tau_{n,nri}$), as

$$\frac{1}{\tau_n} = \frac{1}{\tau_{n,r}} + \sum_i \frac{1}{\tau_{n,nri}} = \frac{1}{\tau_{n,r}} + \frac{1}{\tau_{n,nr}}.$$

The ratio

$$\eta_r = \frac{\tau_n}{\tau_{n,r}} = \frac{\tau_{n,nr}}{\tau_{n,r} + \tau_{n,nr}} = \frac{\tau_{n,r}^{-1}}{\tau_{n,r}^{-1} + \tau_{n,nr}^{-1}} \tag{5.5}$$

is the *radiative efficiency*, i.e., the ratio between the optical power generated by the LED and the optical power the device would generate (with the same bias current I_D) if recombination were entirely radiative. Finally, photon losses also arise because photons are reflected by the semiconductor–air interface with power reflection coefficient R. The *transmission efficiency* is the ratio between the transmitted and the generated optical power and therefore coincides with the transmission coefficient $T = 1 - R$:

$$\eta_t = 1 - R.$$

For *normal incidence* on a semiconductor–dielectric interface, the power reflection coefficient is

$$R = \left(\frac{n_{r1} - n_{r2}}{n_{r1} + n_{r2}}\right)^2,$$

where n_{r1} is the semiconductor refractive index, n_{r2} is the external dielectric refractive index. We can thus express the power–current relation of the LED as

$$P_{out} = \frac{\hbar\omega}{q} \eta_t \eta_r \eta_i I_D \approx E_g\big|_{\text{eV}}\, \eta_x I_D, \tag{5.6}$$

where $\eta_x = \eta_t \eta_r \eta_i$ is the total LED *external efficiency*. The parameter

$$\frac{dP_{out}}{dI_D} = \eta_x\, E_g\big|_{\text{eV}} \tag{5.7}$$

is also called the LED *slope efficiency*.[1] For high power level, the linear relation no longer holds, mainly due to device heating (which in turn decreases the thermal lifetime) and, in high injection, to Auger recombination. The power–current LED characteristics therefore saturates at high current and power.

5.2.3 Charge control model and modulation bandwidth

The asymmetric homojunction LED can be conveniently modeled through a more intuitive charge-control approach. Introducing the excess electron charge $-Q_n$ stored in the p side of the diode, we have[2]

$$Q_n = qA\int_{-\infty}^{-x_p} n'(-x_p)\exp\left(\frac{x-x_p}{L_{np}}\right)\mathrm{d}x = qA\sqrt{D_n\tau_n}\frac{n_i^2}{N_A}\left[\exp\left(\frac{V_A}{V_T}\right)-1\right],$$

and, from (5.1)

$$I_{n0} = \eta_i I_D = qA\sqrt{\frac{D_n}{\tau_n}}\frac{n_i^2}{N_A}\left[\exp\left(\frac{V_A}{V_T}\right)-1\right] = \frac{Q_n}{\tau_n}, \tag{5.8}$$

i.e., the current I_{n0} supplies the charge flow needed to replace the total charge recombined per unit time in the diffusion region. However, the total charge is also related to the radiated power (we neglect reflections) as

$$P_{out} = \frac{\hbar\omega}{q}\frac{Q_n}{\tau_{n,r}}.$$

Thus, including reflection losses and taking into account (5.8), (5.4) and (5.5), we obtain again the power–current relation in (5.6).

The charge control approach can be exploited to evaluate the LED modulation response. For simplicity we assume $I_D \approx I_{n0}$; we can extend the static charge control equation to time-varying conditions as

$$\frac{\mathrm{d}q_n(t)}{\mathrm{d}t} = i_D(t) - \frac{q_n(t)}{\tau_n}.$$

Introducing for all quantities a DC and small-signal component,

$$i(t) = I_D + \widehat{I}_D\exp(\mathrm{j}\omega_m t), \quad q_n(t) = Q_n + \widehat{Q}_n\exp(\mathrm{j}\omega_m t)$$
$$p_{out}(t) = P_{out} + \widehat{P}_{out}\exp(\mathrm{j}\omega_m t),$$

we have, for the signal components and transforming to the frequency domain,

$$\mathrm{j}\omega_m\widehat{Q}_n = \widehat{I}_D - \frac{\widehat{Q}_n}{\tau_n} \rightarrow \widehat{Q}_n = \frac{\tau_n}{1+\mathrm{j}\omega_m\tau_n}\widehat{I}_D,$$

[1] Note the analogy between (5.7) with $\eta_x = 1$ and the expression for the (ideal) photodiode responsivity $\mathfrak{R} = dI/dP_{in} = 1/\left.E_g\right|_{\mathrm{eV}}$.

[2] For simplicity we define such charge as $-Q_n$, $Q_n > 0$.

that is,

$$\widehat{P}_{out} = \eta_r \frac{\hbar\omega}{q} \frac{\widehat{Q}_n}{\tau_{n,r}} = \frac{\hbar\omega}{q} \frac{\tau_n}{\tau_{n,r}} \frac{1}{1 + \mathrm{j}\omega_m \tau_n} \widehat{I}_D.$$

Therefore, the modulation response $m(\omega_m)$ will be a low-pass transfer function[3]

$$m(\omega_m) = \left| \frac{\widehat{P}_{out}(\omega_m)}{\widehat{P}_{out}(0)} \right| = \frac{1}{\sqrt{1 + \omega_m^2 \tau_n^2}}, \tag{5.9}$$

with cutoff frequency associated with the carrier total lifetime:

$$f_{3\mathrm{dB}} = \frac{1}{2\pi \tau_n}.$$

Since the radiative lifetime decreases with increasing charge injection to reach an asymptotic value τ_o (the spontaneous radiative lifetime), the modulation bandwidth depends on the LED bias. Assuming the dependence

$$\tau_n \approx \tau_{n,r} = \frac{K}{Q_n},$$

where K is a proper constant, we have

$$I_D = \frac{Q_n}{\tau_n} = \frac{Q_n^2}{K} \rightarrow Q_n = \sqrt{KI}, \quad \tau_n = \sqrt{\frac{K}{I_D}}, \quad f_{3\mathrm{dB}} \propto \sqrt{I_D}.$$

The LED bandwidth therefore increases as the square root of current, to reach the limiting value $f_{3\mathrm{dB},M} = 1/(2\pi\tau_o)$. For GaAs, the maximum modulation bandwidth is of the order of 300 MHz. Although the limiting value is material dependent, LEDs are anyway confined to applications below 1 Gbps.

5.2.4 Heterojunction LED analysis

The charge control model can be applied to the heterojunction LED as follows. Assuming an injected electron and hole density $n \approx p$, we have for the current and the generated optical power P_{op}:

$$I_L \approx qA \int_0^d \frac{n(x)}{\tau_n} \mathrm{d}x = \frac{Q_n}{\tau_n}$$

$$P_{op} = \hbar\omega A \int_0^d \frac{n(x)}{\tau_{n,r}} \mathrm{d}x = \frac{\hbar\omega}{q} \frac{Q_n}{\tau_{n,r}} = \frac{\hbar\omega}{q} \frac{\tau_n}{\tau_{n,r}} I_L,$$

[3] As discussed in detail in Section 6.2.3, the modulation response can make reference to the signal optical output power (optical definition) or to the electrical power originated by detecting the signal output power through a photodetector, proportional to the square of the optical power (electrical definition). The "optical" response is therefore $|m(\omega)|$ and in dB (called dBo), $10 \log_{10}(|m(\omega)|)$; the "electrical" response is $|m(\omega)|^2$ and in dB (called dBe), $20 \log_{10}(|m(\omega)|)$. The 3 dB bandwidth corresponds to $|m(\omega)| = 1/2$ for the optical definition and to $|m(\omega)| = 1/\sqrt{2}$ for the electrical definition.

where d is the active layer thickness and $-Q_n$ is the total electron charge stored in the narrowgap active layer, equal to the total hole charge ($Q_n = Q_h$) for quasi-neutrality. The total LED current is approximated with the GR contribution in the narrowgap layer, while diffusion currents in the cladding layers have been neglected.

The output optical power will be, taking into account (5.5) and (5.8):

$$P_{out} = \eta_t \frac{\hbar\omega}{q} \frac{Q_n}{\tau_{n,r}} = \eta_t \frac{\hbar\omega}{q} \frac{\tau_n}{\tau_{n,r}} I_D = \frac{\hbar\omega}{q} \eta_t \eta_r I_D \approx \eta_x \left. E_g \right|_{\text{eV}} I_D,$$

where $\eta_x = \eta_t \eta_r$ is the device external efficiency. In heterojunction LEDs the injection efficiency (as defined in homojunction devices) is ideally 1, but in this case also photons emitted toward the substrate can be lost, thus decreasing the efficiency. Backside mirrors (cleaved or based on Bragg reflectors, see Section 5.8.4) can be exploited to recover the photon flux directed toward the substrate.

As in homojunction LEDs, the charge control analysis can be extended to the dynamic regime, obtaining the same results for the modulation response (5.9) and bandwidth.

Quantum well (QW) LEDs are heterojunction LEDs with $d \ll 1\,\mu\text{m}$; the charge control analysis still holds, with an injected electron charge:

$$Q_n = qAn_s,$$

where n_s is the sheet carrier density in the QW. The charge-control model enables us to directly relate the output power to the bias current. The voltage–current characteristic in heterojunction LEDs still qualitatively follows the Shockley law; however, the diode current in direct bias is mainly recombination current in the narrowgap material rather than a diffusion current as in homojunction LEDs, with implications for the diode threshold and ideality factor.

5.2.5 LED emission spectrum

The spontaneous emission spectrum $r_o^{sp}(\hbar\omega)$ of a direct-bandgap semiconductor is reported in (2.34). For low injection conditions (and nondegenerate material), (2.34) simplifies into (2.44); introducing the normalized variable $x = (\hbar\omega - E_g)/k_B T$ and taking into account (2.47), $r_{o,\text{ND}}^{sp}$ can be approximated as a function of x as

$$\begin{aligned} r_{o,\text{ND}}^{sp}(\hbar\omega) &= \frac{1}{\tau_0} \frac{np}{n_i^2} \frac{(2m_r^* k_B T)^{3/2}}{2\pi^2\hbar^3} \exp\left(-\frac{E_g}{k_B T}\right) \left(\frac{k_B T x + E_g}{E_g}\right) \sqrt{x} \exp(-x) \\ &\approx \frac{1}{\tau_0} \frac{np}{n_i^2} \frac{(2m_r^*)^{3/2} (k_B T)^{5/2}}{2\pi^2\hbar^3} \exp\left(-\frac{E_g}{k_B T}\right) \sqrt{x} \exp(-x) \\ &= K\sqrt{x} \exp(-x), \quad x > 0. \end{aligned}$$

for $\hbar\omega - E_g \ll E_g$. The behavior is shown in Fig. 5.3. To derive the full spectral width at half maximum (FWHM), we notice that the function $y = \sqrt{x}\exp(-x)$ has a maximum at $x = 1/2$ with value $y_M = \text{e}^{-1/2}/\sqrt{2}$. Solving the equation $y = \sqrt{x}\exp(-x) = \text{e}^{-1/2}/\left(2\sqrt{2}\right)$, we find $x_1 = 5.091 \times 10^{-2}$, $x_2 = 1.847$, $\Delta x = 1.795$. The low-injection LED FHWM is therefore $\Delta E \approx 1.8 k_B T$.

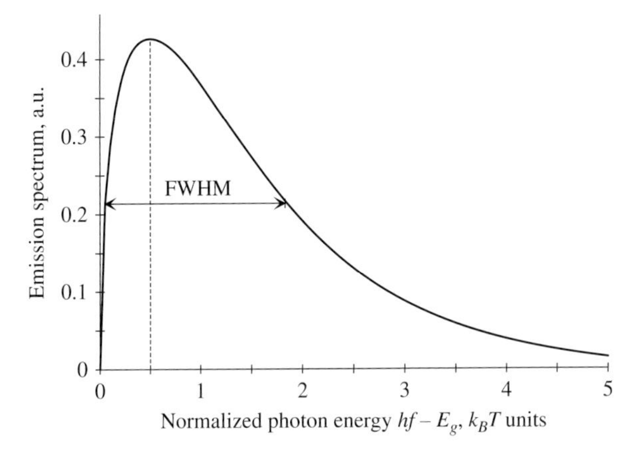

Figure 5.3 Normalized emission spectrum (arbitrary units, a.u.) as a function of the normalized photon energy x.

Example 5.1: Evaluate the LED FWHM for an emission wavelength of $\lambda = 1.3\,\mu\text{m}$ in terms of wavelength and frequency.

We have, with wavelengths in μm and energies in eV:

$$\frac{\Delta E}{E} = \frac{1.8 k_B T}{1.24/\lambda} = \frac{\Delta\lambda}{\lambda} \rightarrow \Delta\lambda = 63.8\,\text{nm}.$$

The spectral width in Hz is given by $\Delta E = \Delta hf = 1.8 k_B T$, i.e., at $T = 300\,\text{K}$:

$$\Delta f = \frac{1.8 k_B T}{h} = \frac{1.8 \cdot 1.381 \times 10^{-23} \cdot 300}{6.626 \times 10^{-34}} = 11.255\ \text{THz}.$$

In high injection, the linewidth increases due to the onset of degeneracy. In general the emission spectrum must be evaluated numerically; a high-injection trend can, however, be recovered from the limit at 0 K. The high-injection (degenerate) emission spectrum at 0 K, assuming quasi-neutrality, can be obtained as the function under the integral in (2.55); for $\hbar\omega - E_g \ll E_g$ one has

$$r^{sp}_{o,D} \approx \frac{2\sqrt{2} q^2 n_r m_r^{*3/2} E_g \mathfrak{p}_{cv}^2}{3\pi^3 \hbar^3 m_0^2 c_0^3 \epsilon_0} \sqrt{\hbar\omega - E_g}, \qquad E_g < \hbar\omega < E_{\max}$$

where (see (2.54) and (2.56) with $E_{\max} - E_g = E_{Fn} - E_c$):

$$E_{\max} = E_g + \frac{\hbar^2 \left(3\pi^2\right)^{2/3}}{2m_r^*} n^{2/3}.$$

In high-injection conditions, we approximate $\Delta E = E_{\max} - E_g$, i.e.,

$$\Delta E \approx \frac{\hbar^2 \left(3\pi^2\right)^{2/3}}{2m_r^*} n^{2/3}.$$

Using (for GaAs) $m_r^* = 0.057m_0$ we obtain, e.g., for $n = 10^{19}\,\mathrm{cm}^{-3}$:

$$\Delta E|_{\mathrm{eV}} \approx \frac{\hbar^2 \left(3\pi^2\right)^{2/3} n^{2/3}}{2qm_r^*} = \frac{\left(1.05 \times 10^{-34}\right)^2 \cdot \left(3\pi^2\right)^{2/3} \cdot \left(10^{25}\right)^{2/3}}{1.6 \times 10^{-19} \cdot 2 \cdot 0.057 \cdot 9.11 \times 10^{-31}} = 294\,\mathrm{meV},$$

much larger than the low-injection value. At high injection, however, the LED may become superradiant or superluminescent, thus implying that part of the spontaneous emission spectrum is affected by gain, and that the resulting total spectrum will follow the narrower gain spectrum. Other effects, such as nonparabolic bands, can play a role in narrowing the LED emission. As an example, in the superradiant LED analyzed in [71] the linewidth reduces from 20 nm to 9 nm when increasing the LED current from 750 mA to 1.5 A (emission is around 0.87 μm); see [71], Fig. 12.

In QW heterojunction LEDs the emission spectrum follows a staircase density of states, where exciton peaks are also present; in fact, in low-injection conditions the Roosbroeck–Shockley relation (2.41) suggests that the emission spectrum is proportional to the absorption spectrum, where resonance peaks appear:

$$r_0^{sp} \approx \frac{2n_r^2}{\pi h c_0^2}(\omega)^2 \alpha(\hbar\omega) \exp\left(-\frac{h\omega}{k_B T}\right), \quad h\omega \gg k_B T.$$

A similar behavior, with marked emission peaks, is observed in quantum wire or quantum dot (QD) LEDs. An example of a QD LED emission spectrum, with broadened peaks related to the δ-like density of states, is shown in Fig. 5.4 [72].

5.2.6 LED materials

Since the LED emission spectrum is concentrated around the semiconductor energy gap, the material of choice depends on the application. Visible LEDs (with emission between 400 and 700 nm) exploit materials with a comparatively large gap. Examples of visible LED materials are the $\mathrm{GaAs}_{1-x}\mathrm{P}_x$ alloy, covering red, orange,

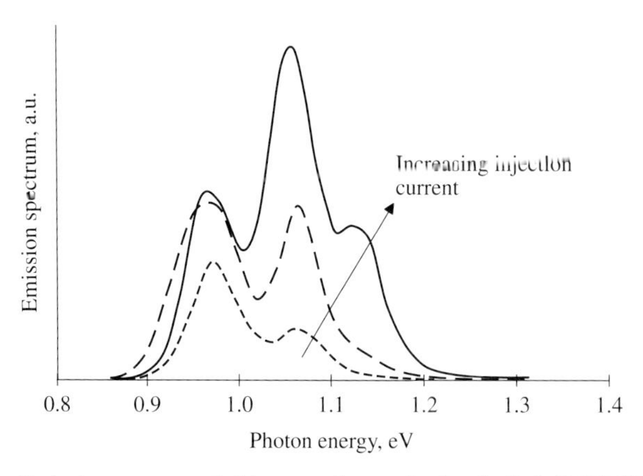

Figure 5.4 Emission spectrum (arbitrary units, a.u.) of an InGaAsP QD LED with increasing injection current. Adapted from [72], Fig. 1 (inset).

and yellow colours, and GaN based alloys, covering green, blue, and violet. Since $GaAs_{1-x}P_x$ is indirect bandgap for $x > 0.5$, the problem arises of exploiting indirect-bandgap materials for spontaneous emission. Due to the high radiative lifetime and low radiative efficiency, indirect-bandgap materials are characterized by very low external efficiencies, unless the material is doped with an impurity (e.g., nitrogen) called an *isoelectronic impurity,* whose effect is to improve radiative processes without leading to acceptor or donor behavior (isoelectronic means that the number of external electrons of the impurity is the same as for the hosting material). Isoelectronic doping improves the radiation efficiency of indirect-bandgap compound semiconductors by about two orders of magnitude, although this is always much lower than in direct-bandgap alloys. While the InGaN alloy covers the visible range from green to blue, UV LEDs require AlGaN, with a wider bandgap. Applications of UV LEDs include monitoring of contaminants and ambient-temperature portable water purification systems.

LEDs for optical communications operating in the first window exploit AlGaAs/GaAs/AlGaAs heterostructures. Long-wavelength operation is possible with InP-based alloys, such as InGaAsP; the resulting LEDs, often edge-emitting and superradiant, offer the best performance in this device class with 1.3 μm and 1.5 μm emission.

5.3 From LED to laser

Both the LED and the laser diodes exploit radiative recombination of electron–hole pairs in a forward-biased junction to emit light. However, LEDs are based on *spontaneous emission*, implying broad linewidth (of the order of $2k_BT$) and narrow modulation bandwidth, well below 1 GHz. The dominant emission mechanism in lasers is *stimulated emission*: photons of a specific energy and wavenumber stimulate the emission of *coherent* photons (with the same energy and wavenumber), thus leading to EM wave amplification; all e-h pairs recombine to generate coherent photons, and narrow linewidth results. At the same time, the lifetime associated with stimulated emission can be shorter, for high photon density, than the spontaneous radiative lifetime τ_0; thus, the laser can achieve modulation bandwidths as wide as 20–30 GHz.

In summary, *stimulated emission* is the key to improving the LED spectral purity and modulation bandwidth or speed. To turn a LED into a laser, however, we need a mechanism able to foster stimulated rather than spontaneous recombination at a certain photon energy, i.e., a frequency-selective structure such as an optical resonator or cavity. The optical cavity is compatible with a discrete set of photon states, whose density is large within the cavity, and which operate as positive feedback with respect to stimulated emission.[4]

[4] The coupling of a *gain block* (the junction where stimulated emission occurs) and a *frequency-selective feedback block* makes the laser somewhat similar to an electronic oscillator, whose basic components are an amplifier and a frequency-selective feedback loop.

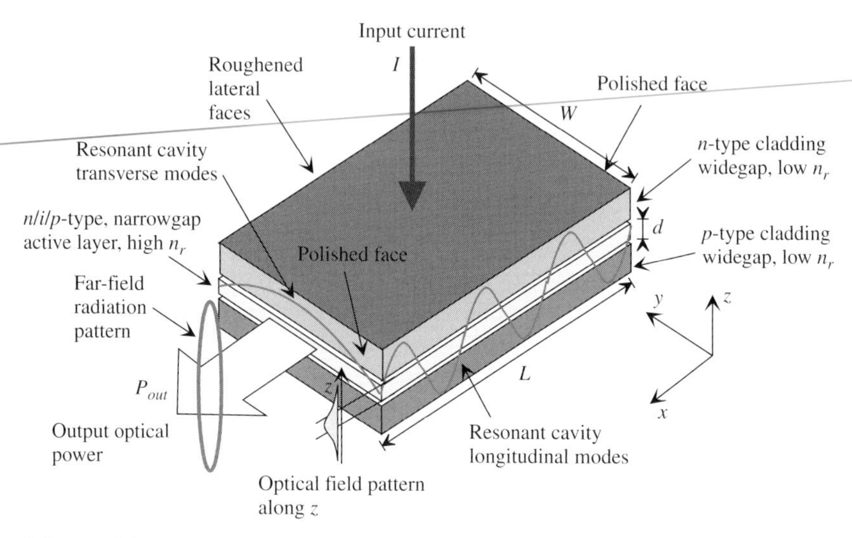

Figure 5.5 Scheme of the Fabry–Perot cavity laser.

In the simplest cavity lasers, the Fabry–Perot (FP) lasers, the cavity and the junction are completely integrated, as shown in Fig. 5.5.[5] The *pn* junction is typically implemented through a *PiN* heterostructure, in which the narrowgap central portion acts as the active layer (where stimulated emission occurs through e-h pair recombination), while the two external widegap (cladding) layers, having slightly lower refractive index than the active region, contribute to the vertical confinement of the optical field. The cavity optical field can be described as a superposition of waves guided (in the direction orthogonal to the junction plane) by the narrowgap layer, and reflected back and forth from the two polished faces (lateral faces are typically roughened to suppress the so-called transversal resonances). A forward-bias current is fed into the junction and injected carriers recombine, sustaining the photon density inside the cavity and compensating for photon losses due to absorption in the cladding and, above all, to permeable (i.e., non-perfectly-reflecting) mirrors. The mirror power reflectivity (r is the field reflectivity), defined at normal incidence by the refractive index step with respect to air:

$$R = |r|^2 = \left(\frac{n_r - 1}{n_r + 1}\right)^2$$

is typically $\ll 1$ (for example, $R = (\sqrt{13} - 1)^2/(\sqrt{13} + 1)^2 = 0.32$ for a GaAs–air interface); therefore, many photons escape the cavity, allowing the laser field to be coupled with an external propagation medium (e.g., an optical fiber). Since the field distribution on the mirror faces is narrow across and broad parallel to the heterojunction, the far-field laser radiation pattern (proportional to the spatial Fourier transform of the aperture field distribution) will be narrow in the horizontal plane and broad in

[5] The Fabry–Perot cavity is originally made of two permeable metal mirrors, while the semiconductor implementation mirrors are obtained by cleaving the facets of a semiconductor chip.

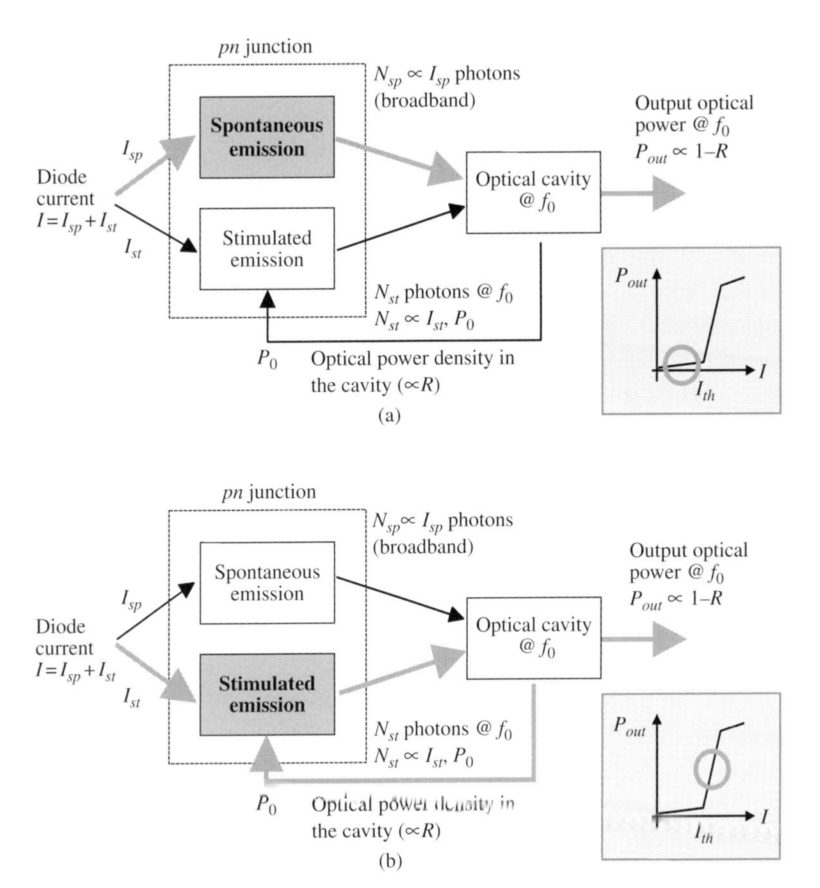

Figure 5.6 Laser functional scheme below threshold (a) and above threshold (b). R is the mirror power reflectivity; N_{sp} and N_{st} are the photon densities emitted via spontaneous and stimulated emission.

the vertical plane. In summary, the double heterojunction enables, in the vertical direction, carrier confinement (due to the heterojunction potential wells in the conduction and valence bands) and photon confinement (due to the higher refractive index of the active layer vs. the cladding).

The cavity feedback makes stimulated emission dominant only when the photon density in the cavity is large enough. Consider the laser block scheme in Fig. 5.6. The junction hosts two radiative recombination mechanisms, spontaneous and stimulated; we neglect for the moment other competing nonradiative processes, such as thermal and Auger recombination. The diode current injects carriers recombining in either radiative mechanism according to the relative lifetimes. For low current, the carrier density in the junction is small, and so is the phonon density in the cavity. In this region, spontaneous emission dominates over stimulated emission and the laser works as a LED, according to the flow scheme shown by the thick arrows in Fig. 5.6(a). The output power near the laser emission wavelength is small, because most photons are emitted on a broad spectrum.

Increasing the current density, the photon density in the cavity allowed modes increases. Correspondingly, the stimulated lifetime (inversely proportional to the cavity mode photon density) decreases, becoming smaller than the spontaneous lifetime. The feedback loop identified by the thick arrow in Fig. 5.6(b) now becomes dominant and typical laser operation starts. Above the laser threshold – corresponding to self-sustaining oscillations where the photons lost in one cycle are replaced by the photons generated by stimulated emission – the laser spectrum narrows. A small fraction of carriers still recombine spontaneously, emitting incoherent photons and thus contributing to the laser phase noise and finite linewidth.

Above threshold, the number of carriers injected per unit time (i.e., the junction current) is approximately equal (with unit quantum efficiency, i.e., neglecting nonradiative recombination processes) to the number of photons generated per unit time within the cavity; in steady-state conditions the photon density N remains constant because generated photons are lost at a rate N/τ_{ph} where τ_{ph} (the photon lifetime) is the average survival time of a photon before it leaves the cavity through the mirrors or is absorbed; we assume here that the first mechanism, called mirror or end loss, is dominant. Mirror loss is the source of the output optical power, which is therefore proportional to the input electric current. At very high input current densities, however, the radiative efficiency of the laser decreases, due to nonradiative (e.g., Auger) recombination processes competing with the stimulated recombination, self-heating, and gain compression, so that the laser output power finally saturates.

Taking into account that the total number of photons generated is anyway proportional to the input current both in the LED-like and in the laser regime (below and above threshold), it is clear that the different slope of the laser power-current characteristics (see the insets in Fig. 5.6), also called slope efficiency, depends on the fact that the laser concentrates, around a narrow linewidth all the photons generated, which are emitted on a broader bandwidth in the LED. FP lasers have linewidths of the order of 2–3 nm (20–30 Å), at least one order of magnitude narrower than typical LED linewidths.

5.4 The Fabry–Perot cavity resonant modes

Let us consider in more detail the operation of the FP cavity laser. The cavity length is L and the cavity width W (along x and y axes, respectively, see Fig. 5.5). The active region thickness is d; the front and back cavity edges are cleaved so as to act as a pair of flat parallel mirrors, while the lateral faces (across y) are treated to minimize lateral reflections. The optical field can be TE or TM (electric field parallel or perpendicular to the stratification; see Fig. 5.7). Due to the refractive index step between the active layer and the surrounding widegap cladding layers, the optical field is confined in the active region, and extends into the cladding with exponentially decreasing tails. Resonant modes are characterized by a longitudinal (x) field pattern, similar to that shown in Fig. 5.5, i.e., the cavity length L is an integer multiple of a half-wavelength (we assume that the cavity facets are ideal mirrors). In the y direction the optical field may exhibit

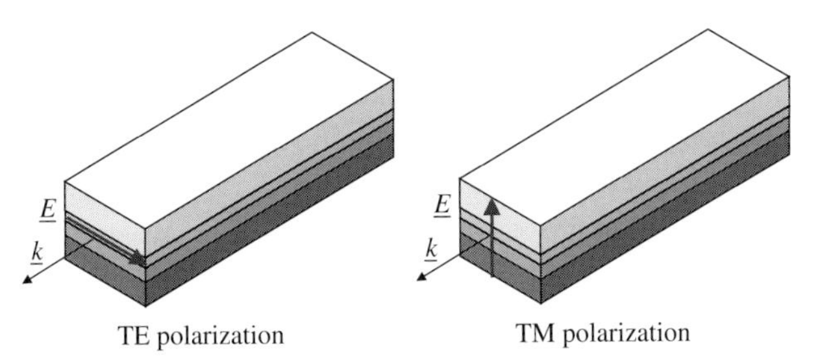

Figure 5.7 TE (left) and TM (right) polarization in a Fabry–Perot cavity.

transversal resonances, depending on the cavity width W and on the treatment of the lateral facets.

Concerning the cavity mode polarization, TE modes typically have better overlap integral Γ_{ov} (see Section 5.5.1) and better vertical (z) confinement of the optical field; the mirror reflectivity is also slightly better for TE polarization.

The FP cavity is 3D; however, in the vertical (z) direction, only the fundamental mode (with one maximum) is excited. Since the field distribution can be factorized as $X(x)Y(y)Z(z)$, we refer to *transversal* and *longitudinal* modes with reference to functions X and Y, respectively; in particular, modes with many oscillations in the transverse direction (x) are *higher-order transversal modes*.

5.4.1 Analysis of the TE slab waveguide fundamental mode

The laser-active region, sandwiched between two cladding layers, is an example of three-layer slab waveguide. In most advanced laser structures, however, the confinement of the optical field occurs not only in the vertical (z) direction but also in the lateral (y) direction, due to either a refractive index variation (index-guided lasers) or to a variation of gain (gain-guided lasers). The closed-form analysis of the propagation in the slab waveguide is, however, useful to introduce a number of important parameters and to provide explicit analytical approximations for them. Propagation in more complex structures where the refractive index varies in the cavity cross section as $n(y, z)$ often has to be analyzed by numerical techniques; however, the propagation parameters can always be expressed through the help of a modal effective refractive index, n_{eff}, already introduced in the analysis of quasi-TEM waveguides; see Section 3.2.

Let us consider now the analysis of the slab waveguide optical field in terms of TE and TM modes. For the more important TE case, the electric field can be expressed as

$$E_y(x, z) = E(z)\exp(-\mathrm{j}\beta x),$$

i.e., the field is polarized in the stratification plane. The corresponding wave equation yields

$$\frac{\mathrm{d}^2 E(z)}{\mathrm{d}z^2} - \left[\beta^2 - n^2(z)k_0^2\right] E(z) = 0, \tag{5.10}$$

where k_0 is the free-space propagation constant and n is the refractive index. Suppose now, for simplicity, that the slab waveguide is symmetrical, with active region refractive index n_1 and cladding index $n_2 < n_1$; the active region extends between $z = -d/2$ and $z = d/2$ and the cladding is unbounded. The fundamental mode electric field is even in z; since the wave equation is linear, we look for exponential or sinusoidal solutions:

$$\begin{aligned} E(z) &= A\exp(-\alpha_2 z), & z &> d/2 \\ E(z) &= B\cos(\alpha_1 z), & -d/2 &< z < d/2 \\ E(z) &= A\exp(\alpha_2 z), & z &< -d/2. \end{aligned}$$

Substituting into (5.10) and simplifying, we obtain

$$\alpha_2 = \sqrt{\beta^2 - n_2^2 k_0^2}, \quad \alpha_1 = \sqrt{n_1^2 k_0^2 - \beta^2}. \tag{5.11}$$

Imposing now that the TE field be continuous, with continuous first derivative, in $z = d/2$ (continuity conditions in $z = -d/2$ follow from symmetry) and eliminating the arbitrary constants A and B we obtain, using (5.11),

$$\sqrt{n_1^2 k_0^2 - \beta^2}\tan\left(\frac{\sqrt{n_1^2 k_0^2 - \beta^2}\,d}{2}\right) = \sqrt{\beta^2 - n_2^2 k_0^2}. \tag{5.12}$$

The dispersion relation (5.12) can be expressed in a more suitable way by introducing the normalized parameters:

$$u = \frac{\sqrt{n_1^2 k_0^2 - \beta^2}\,d}{2} = \frac{k_0 d}{2}\sqrt{n_1^2 - n_{\text{eff}}^2}$$

$$w = \frac{\sqrt{\beta^2 - n_2^2 k_0^2}\,d}{2} = \frac{k_0 d}{2}\sqrt{n_{\text{eff}}^2 - n_2^2},$$

where the propagation constant is $\beta = k_0 n_{\text{eff}}$ (n_{eff} mode effective refractive index, $\epsilon_{\text{eff}} = n_{\text{eff}}^2$ mode effective permittivity). The normalized frequency v and propagation constant b ($0 \le b \le 1$) are defined as

$$v = \sqrt{u^2 + w^2} = \frac{k_0 d}{2}\sqrt{n_1^2 - n_2^2} = \sqrt{2}\pi\frac{d}{\lambda_0}\sqrt{\overline{n}\Delta n} \tag{5.13}$$

$$b = \frac{n_{\text{eff}}^2 - n_2^2}{n_1^2 - n_2^2}, \tag{5.14}$$

where $\overline{n} = (n_1 + n_2)/2$, $\Delta n = n_1 - n_2$. From the above results we obtain

$$w^2 = \left(\frac{k_0 d}{2}\right)^2\left(n_{\text{eff}}^2 - n_2^2\right) = b\left(\frac{k_0 d}{2}\right)^2(n_1^2 - n_2^2) = bv^2$$

$$u^2 = \left(\frac{k_0 d}{2}\right)^2\left(n_1^2 - n_{\text{eff}}^2\right) = \left(\frac{k_0 d}{2}\right)^2\left[\left(n_1^2 - n_2^2\right) - \left(n_{\text{eff}}^2 - n_2^2\right)\right] = (1-b)v^2.$$

Therefore, the dispersion relation (5.12) for the fundamental mode becomes

$$\tan\left(\sqrt{1-b}v\right) = \sqrt{\frac{b}{1-b}} \to v = \frac{1}{\sqrt{1-b}}\arctan\sqrt{\frac{b}{1-b}}. \tag{5.15}$$

In the dispersion relation the parameter v is defined from the structure data (dimension and refractive index profile) and the operating frequency (or wavelength); b is then derived from the solution of (5.15). From b, the effective refractive index is

$$n_{\mathrm{eff}} = \sqrt{bn_1^2 + (1-b)n_2^2}.$$

An approximate explicit solution for b can be expressed as

$$b \approx 1 - \frac{1}{2v^2}\log(1+2v^2). \tag{5.16}$$

Example 5.2: Assuming $n_1 = 3.6$, $n_2 = 3.2$, $d = 0.5\,\mu\mathrm{m}$, $\lambda = 1.55\,\mu\mathrm{m}$, evaluate the effective permittivity according to the exact and approximate approaches.

The normalized frequency is

$$v = \frac{k_0 d}{2}\sqrt{n_1^2 - n_2^2} = \frac{\pi d}{\lambda}\sqrt{n_1^2 - n_2^2} = \frac{\pi \cdot 0.5}{1.55}\sqrt{3.6^2 - 3.2^2} = 1.671.$$

Solving the dispersion relation (5.15) we find $b = 0.703$, i.e.,

$$n_{\mathrm{eff}} = \sqrt{bn_1^2 + (1-b)n_2^2} = \sqrt{0.703 \cdot 3.6^2 + (1-0.703)\cdot 3.2^2} = 3.486.$$

The approximate explicit solution is

$$b \approx 1 - \frac{1}{2 \cdot 1.671^2}\log(1 + 2\cdot 1.671^2) = 0.662,$$

leading to the approximate effective index

$$n_{\mathrm{eff}} \approx \sqrt{0.662 \cdot 3.6^2 + (1-0.662)\cdot 3.2^2} = 3.470.$$

As an example of the diffraction index step Δn that may arise in the active region of InP based lasers, Fig. 5.8 shows the InP and InGaAsP refractive index as a function of wavelength for different quaternary alloys supporting emission at 1.1, 1.33, and 1.55 μm. The refractive index of $\mathrm{In}_{1-x}\mathrm{Ga}_x\mathrm{As}_y\mathrm{P}_{1-y}$ lattice-matched to ($x = 0.46y$) can be approximated, as a function of the emission energy $E_p \approx E_G$ and of the photon energy E_{ph}, as [73]

$$n_2 = 1 + \sum_{k=1}^{2} A_k\left[1 - \left(\frac{E_{ph}}{E_p + E_k}\right)^2\right]^{-1},$$

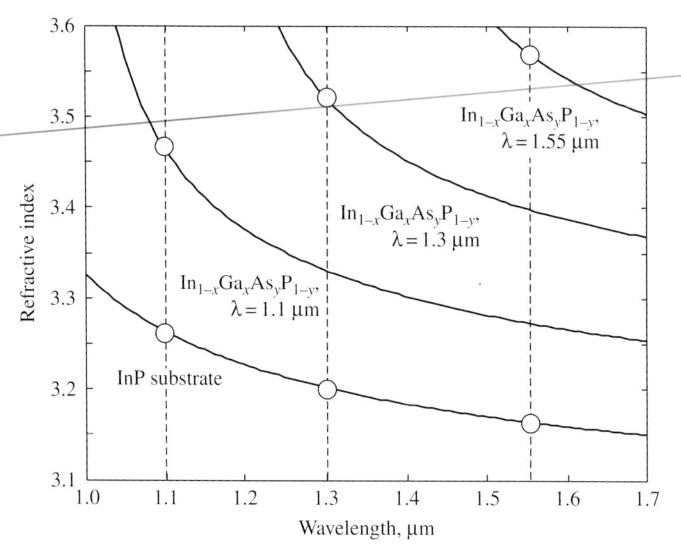

Figure 5.8 Refractive index of InP and of several quaternary alloys suited for emission at $\lambda = 1.1, 1.3, 1.55$ μm as a function of wavelength.

where (energies in eV):

$$A_1 = 13.3510 - 5.4555E_p + 1.2332E_p^2$$
$$A_2 = 0.7140 - 0.3606E_p$$
$$E_1 = 2.5048 \text{ eV}$$
$$E_2 = 0.1638 \text{ eV}.$$

5.4.2 Longitudinal and transversal cavity resonances

Let us denote the cavity side lengths as $L = L_x$ and $W = L_y$; assume an oscillating (TE or TM) field pattern:

$$E(x, y, z) = E_0(z) \sin(k_x x) \sin(k_y y).$$

The modal distribution $E_0(z)$ was already analyzed in detail for the TE polarization. Assuming for simplicity $\Gamma = -1$ for the mirror reflectivity, the total electric field on the mirrors is zero and the resonance condition follows as

$$k_x L_x = n\pi, \quad k_y L_y = m\pi,$$

where m and n are integers. For modes with effective refractive index n_{eff}, the following dispersion relation arises (the field is assumed as a plane wave in the (x, y) plane):

$$k_x^2 + k_y^2 = \left(\frac{n\pi}{L_x}\right)^2 + \left(\frac{m\pi}{L_y}\right)^2 = k_0^2 = \left(\frac{2\pi n_{\text{eff}}}{\lambda_{0nm}}\right)^2.$$

The resonant wavelengths λ_{0nm} become

$$\frac{n_{\text{eff}}}{\lambda_{0nm}} = \sqrt{\left(\frac{n}{2L_x}\right)^2 + \left(\frac{m}{2L_y}\right)^2}, \tag{5.17}$$

where typically $n \gg m$ since many oscillations arise in the longitudinal direction but only a few (or, hopefully, just one) in the transversal direction. Indeed, let us initially assume that no oscillations exist in the transversal direction; the electric field is constant along y, a condition compatible with roughened lateral surfaces imposing homogeneous Neumann boundary conditions on the field. Condition (5.17) then reduces to

$$\frac{n_{\text{eff}}}{\lambda_{0n}} = \frac{n}{2L_x} \rightarrow n\frac{\lambda_{0n}}{2n_r} = L_x \rightarrow \lambda_{0n} = \frac{2n_{\text{eff}}L_x}{n}.$$

Therefore, having assigned an "excited interval" around λ_0, where the lasing mode will fall, the mode index will be

$$n \approx \frac{2n_{\text{eff}}L_x}{\lambda_0}.$$

For a cavity with typical length of 100 μm, effective refractive index $n_{\text{eff}} \approx 3$, emission wavelength in vacuo $\lambda_0 \approx 1$ μm, this amounts to $n \approx 600$. A high-order longitudinal mode is therefore excited in common (edge-emitting) lasers, while for vertical emission lasers (VCSELs) the mode order can be much lower due to the reduced cavity length. Correspondingly, the resonant frequencies are

$$f_n = n\frac{c_0}{2n_{\text{eff}}L_x}.$$

It is important to evaluate the spacing between two successive resonant wavelengths. We have

$$\Delta\lambda_{0n} = \lambda_{0n} - \lambda_{0(n+1)} = \left(\frac{1}{n} - \frac{1}{n+1}\right)2n_{\text{eff}}L_x = \frac{2n_{\text{eff}}L_x}{n\,(n+1)} \approx \frac{\lambda_{0n}}{n},$$

and therefore

$$\Delta\lambda_{0n} \approx \frac{2n_{\text{eff}}L_x}{n^2}.$$

The corresponding, uniform, frequency spacing will be

$$\Delta f_n = f_{n+1} - f_n = [(n+1) - n]\frac{c_0}{2n_{\text{eff}}L_x} = \frac{c_0}{2n_{\text{eff}}L_x}.$$

Since $\lambda_{0n} f_n = c_0$, by differentiation we obtain

$$\frac{\Delta f_n}{f_n} \approx \frac{\Delta\lambda_{0n}}{\lambda_{0n}}.$$

Let us introduce again transversal modes, but assuming $n \gg m$; in this case, the resonant condition of the transversal modes can be evaluated perturbatively from the corresponding longitudinal mode as

$$\frac{n_{\text{eff}}}{\lambda_{0nm}} = \sqrt{\left(\frac{n}{2L_x}\right)^2 + \left(\frac{m}{2L_y}\right)^2} \approx \left(\frac{n}{2L_x}\right)\left[1 + \frac{1}{2}\left(\frac{m}{n}\right)^2\left(\frac{L_x}{L_y}\right)^2\right],$$

leading to

$$\lambda_{0nm} \approx \lambda_{0n}\left[1 + \frac{1}{2}\left(\frac{m}{n}\right)^2\left(\frac{L_x}{L_y}\right)^2\right]^{-1} \approx \lambda_{0n} - \lambda_{0n}\frac{1}{2}\left(\frac{m}{n}\right)^2\left(\frac{L_x}{L_y}\right)^2,$$

that is,

$$\frac{\lambda_{0nm} - \lambda_{0n}}{\lambda_{0n}} \approx -\frac{1}{2}\left(\frac{m}{n}\right)^2\left(\frac{L_x}{L_y}\right)^2.$$

Example 5.3: Suppose $L = 200\,\mu\text{m}$, $W = 50\,\mu\text{m}$, $n_{\text{eff}} = 3.3$, excitation wavelength around $0.8\,\mu\text{m}$; evaluate the spacing between longitudinal and transversal modes.

The modal order is

$$n \approx \frac{2n_{\text{eff}}L_x}{\lambda_0} = \frac{2 \times 3.3 \times 200}{0.8} \approx 1650 \rightarrow \lambda_{0n} = 800\,\text{nm},$$

leading exactly to the prescribed excitation wavelength. The longitudinal mode line spacing is

$$\Delta\lambda_{0n} \approx \frac{\lambda_{0n}}{n} = \frac{800}{1650} = 0.48\,\text{nm},$$

which corresponds to a frequency spacing

$$\Delta f_n = \frac{c_0}{2n_{\text{eff}}L_x} = \frac{3 \cdot 10^8}{2 \times 3.3 \times 200 \cdot 10^{-6}} = 227\,\text{GHz}.$$

The longitudinal mode spacing $\Delta\lambda_{0n}$ is typically smaller than the FP laser linewidth; thus, more than one longitudinal mode is usually excited. Consider now some transversal modes around the longitudinal mode selected; we have, in general,

$$\lambda_{0,1650,m} = \frac{n_{\text{eff}}}{\sqrt{\left(\frac{n}{2L_x}\right)^2 + \left(\frac{m}{2L_y}\right)^2}} = \frac{3.3}{\sqrt{\left(\frac{1650}{2 \times 200}\right)^2 + \left(\frac{m}{2 \times 50}\right)^2}},$$

which corresponds, e.g., to

$$\lambda_{0,1650,0} = 800\,\text{nm}, \quad \lambda_{0,1650,1} = 799.998\,\text{nm}, \quad \lambda_{0,1650,2} = 799.991\,\text{nm}.$$

Transversal modes are therefore close to the longitudinal (fundamental) one, and are usually all excited, contributing to undesirable features in the laser emission and coupling to external media. Higher-order transversal modes are inconvenient from the standpoint of the laser linewidth, but also of the far-field radiation pattern. Odd-order

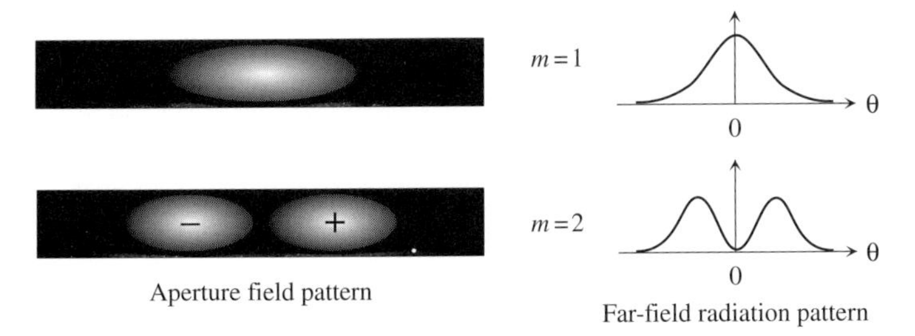

Figure 5.9 Aperture and radiation patterns of fundamental and higher-order transversal Fabry–Perot cavity modes. $\theta = 0$ corresponds to the paraxial direction.

transversal modes have a paraxial zero in the far-field radiation pattern, and therefore are poorly coupled with an optical fiber. Higher-order even modes tend to be less directional and therefore their coupling to an optical fiber deteriorates; see Fig. 5.9. transversal mode suppression can be achieved by making the cavity width W narrow, ideally $\approx \lambda_0/n_{\text{eff}}$.

5.5 Material and cavity gain

Stimulated emission, and therefore cavity gain, is the key to laser operation. However, the optical field extends also to the cladding, i.e., outside the active region of thickness d where gain is present; see Fig. 5.10. To account for this, an effective gain, called the *cavity gain* g_c, can be introduced. The cavity gain is related to the material gain in the active region by the relation

$$g_c = \Gamma_{ov} g,$$

where the parameter $\Gamma_{ov} < 1$ is the overlap integral (see Section 5.5.1):

$$\Gamma_{ov} = \frac{\int_0^d |E_{op}|^2 \, dz}{\int_{-\infty}^{\infty} |E_{op}|^2 \, dz} < 1. \tag{5.18}$$

E_{op} is the optical electric field.

5.5.1 Analysis of the overlap integral

Assuming a symmetrical slab waveguide with fundamental TE mode, the overlap integral can be evaluated from the field distribution. Exploiting the field symmetry, we have

$$\Gamma_{ov} = \frac{\int_0^{d/2} |E_{op}|^2 \, dz}{\int_0^{d/2} |E_{op}|^2 \, dz + \int_{d/2}^{\infty} |E_{op}|^2 \, dz} = \left(1 + \frac{\int_{d/2}^{\infty} |E_{op}|^2 \, dz}{\int_0^{d/2} |E_{op}|^2 \, dz}\right)^{-1}, \tag{5.19}$$

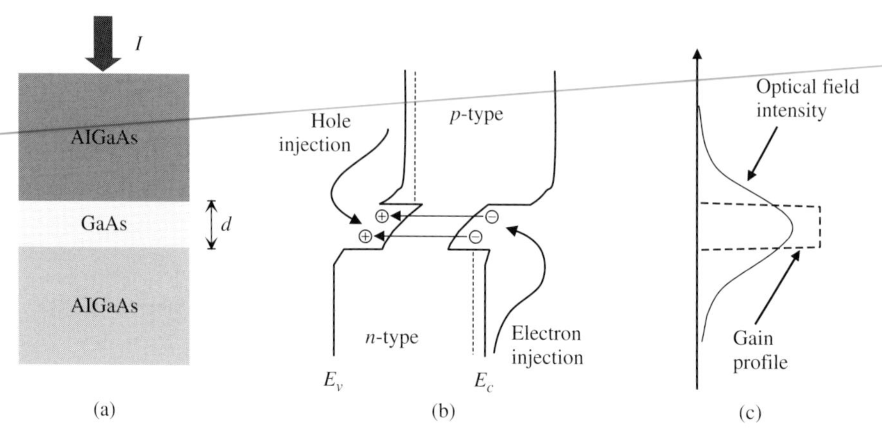

Figure 5.10 (a) AlGaAs/GaAs/AlGaAs double heterojunction laser structure; (b) band diagram under direct bias, and (c) gain and optical field intensity.

but

$$\begin{aligned}\frac{\int_{d/2}^{\infty}\left|E_{op}\right|^2 \mathrm{d}z}{\int_0^{d/2}\left|E_{op}\right|^2 \mathrm{d}z} &= \frac{A^2}{B^2}\frac{\int_{d/2}^{\infty}\exp(-2\alpha_2 z)\,\mathrm{d}z}{\int_0^{d/2}\cos^2(\alpha_1 z)\,\mathrm{d}z} = \frac{\alpha_1}{\alpha_2}\frac{1+\cos d\alpha_1}{d\alpha_1+\sin d\alpha_1}\\
&= \sqrt{\frac{1-b}{b}}\frac{1+\cos\left(2\sqrt{1-b}v\right)}{2\sqrt{1-b}v+\sin\left(2\sqrt{1-b}v\right)}\\
&= \frac{1}{\sqrt{b}}\frac{\sqrt{1-b}}{v\sqrt{1-b}\left[\tan^2\left(v\sqrt{1-b}\right)+1\right]+\tan\left(v\sqrt{1-b}\right)}\\
&= \frac{1}{\sqrt{b}}\frac{(1-b)}{v+\sqrt{b}},\end{aligned} \tag{5.20}$$

where we have exploited the relations

$$\frac{\alpha_1}{\alpha_2} = \frac{\sqrt{n_1^2k_0^2-\beta^2}}{\sqrt{\beta^2-n_2^2k_0^2}} = \frac{u}{w} = \sqrt{\frac{1-b}{b}}$$

$$d\alpha_1 = d\sqrt{n_1^2k_0^2-\beta^2} = 2u = 2\sqrt{1-b}v,$$

and the dispersion relation (5.15). We finally obtain from (5.19) and (5.20):

$$\Gamma_{ov} = \frac{\sqrt{b}\left(v+\sqrt{b}\right)}{\sqrt{b}v+1}.$$

Using the approximate expression (5.16) for b, an explicit expression can be derived as a function of the normalized frequency v; the resulting behavior is shown in Fig. 5.11.

For large values of d (and therefore of v) the overlap integral saturates to 1, while for small d (and therefore small v), we have from (5.16) the asymptotic behavior

$$b \underset{v\to 0}{\approx} v^2,$$

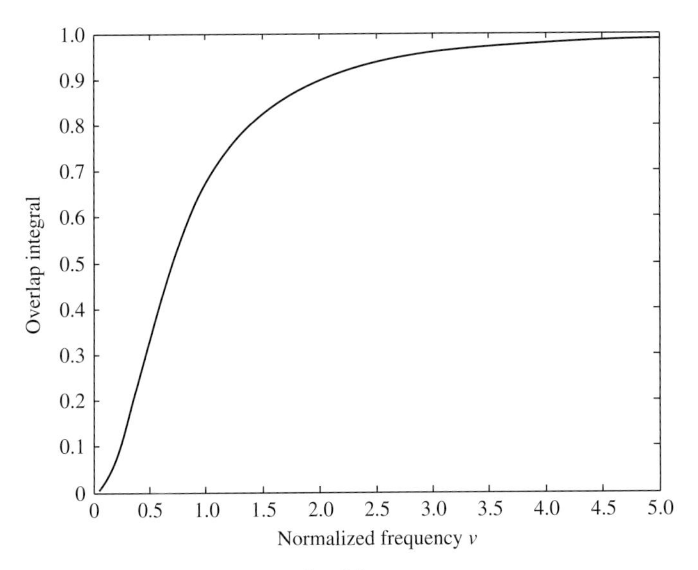

Figure 5.11 Overlap integral versus normalized frequency.

leading to

$$\Gamma_{ov} \underset{v \to 0}{\approx} \frac{2v^2}{v^2+1} \approx 2v^2 = 4\pi^2 \frac{d^2}{\lambda_0^2} \overline{n} \Delta n,$$

where $\overline{n}$ is the average cladding and active region index, and Δn is the index difference between the active region and the cladding.

If n_1 undergoes a variation Δn_1, the variation of the effective propagation index can be expressed through Γ_{ov} in a somewhat intuitive manner. In fact, from the dispersion relation (5.15), rewritten here for convenience,

$$f(b, v) = \tan\left(\sqrt{1-b}v\right) - \sqrt{\frac{b}{1-b}} = 0,$$

we can express the derivative

$$\frac{\mathrm{d}b}{\mathrm{d}v} = -\frac{\partial f/dv}{\partial f/db} = 2\frac{\sqrt{b}(1-b)}{\sqrt{b}v+1},$$

and rewrite Γ_{ov} in terms of it:

$$\Gamma_{ov} = \frac{\sqrt{b}\left(v+\sqrt{b}\right)}{\sqrt{b}v+1} = b + \frac{v}{2} \cdot 2\frac{\sqrt{b}(1-b)}{\sqrt{b}v+1} = b + \frac{v}{2}\frac{\mathrm{d}b}{\mathrm{d}v}. \tag{5.21}$$

Expressing v and b from (5.13) and (5.14) in terms of permittivities (rather than of refractive indices) we obtain

$$v^2 = \left(\frac{k_0 d}{2}\right)^2 (\epsilon_1 - \epsilon_2), \quad b = \frac{\epsilon_{\mathrm{eff}} - \epsilon_2}{\epsilon_1 - \epsilon_2}.$$

It follows that

$$v^2 b = \left(\frac{k_0 d}{2}\right)^2 (\epsilon_1 - \epsilon_2) \cdot \frac{\epsilon_{\text{eff}} - \epsilon_2}{\epsilon_1 - \epsilon_2} = \left(\frac{k_0 d}{2}\right)^2 (\epsilon_{\text{eff}} - \epsilon_2).$$

But from (5.21), taking into account that ϵ_1 is varying, inducing a change in ϵ_{eff}:

$$\Gamma_{ov} = b + \frac{v}{2}\frac{\mathrm{d}b}{\mathrm{d}v} = b + v^2 \frac{\mathrm{d}b}{\mathrm{d}v^2} = \frac{\mathrm{d}\left(v^2 b\right)}{\mathrm{d}v^2} = \frac{\mathrm{d}\left(\epsilon_{\text{eff}} - \epsilon_2\right)}{\mathrm{d}\left(\epsilon_1 - \epsilon_2\right)} = \frac{\mathrm{d}\epsilon_{\text{eff}}}{\mathrm{d}\epsilon_1} \approx \frac{\Delta n_{\text{eff}}}{\Delta n_1}, \quad (5.22)$$

if the index variation between the active region and the cladding is small. Thus, the variation of the effective index is simply the variation of the active region refractive index, weighted by the overlap integral:

$$\Delta n_{\text{eff}} \approx \Gamma_{ov} \Delta n_1.$$

If material gain is small, this can be interpreted as a (small) perturbation of the imaginary part of the active region complex refractive index $n_r = n_{r1} - \mathrm{j} n_{r2}$ from 0 to $n_{r2} \equiv \Delta n_1$:

$$\frac{g}{2} = \frac{2\pi}{\lambda_0} n_{r2} \equiv \frac{2\pi}{\lambda_0} \Delta n_1;$$

the related perturbation of the modal gain (i.e., the cavity gain g_c) will be proportional to the variation Δn_{eff} in the imaginary part of the (complex) effective index, as

$$\frac{g_c}{2} = \frac{2\pi}{\lambda_0} \Delta n_{\text{eff}} = \frac{2\pi}{\lambda_0} \Gamma_{ov} \Delta n_1 = \Gamma_{ov} \frac{g}{2}.$$

Thus,

$$g_c = \Gamma_{ov} g. \quad (5.23)$$

5.6 The FP laser from below to above threshold

Let us now summarize a few points from the previous discussion. In the FP cavity, a total current I is injected in forward bias, corresponding to a current density $J = I/A$, A being the junction area. Holes and electrons build up in the active layer with densities $n \approx p$ (because of quasi-neutrality), and *positive net gain* results over a certain energy interval. At the same time, photon losses occur due to absorption in the cladding (denoted as α_{loss}), and to the effect of mirrors (referred to as *mirror loss* or *end loss*). Mirror loss, however concentrated in space, can in turn be expressed in terms of an equivalent absorption α_m, as discussed later. Within the cavity, a suitable modal spectrum exists, with longitudinal modes and (if any) clusters of closely spaced transversal modes around them.

Suppose that the current density is much lower than the threshold density required for lasing, $J \ll J_{th}$; also the electron (hole) concentration in the active region will be lower than the threshold value, $n \ll n_{th}$, and the gain will be lower than the total loss $\alpha_{loss} + \alpha_m$. In such conditions, the laser operates like a LED, with a broad emission spectrum dominated by spontaneous emission; see Fig. 5.12(a).

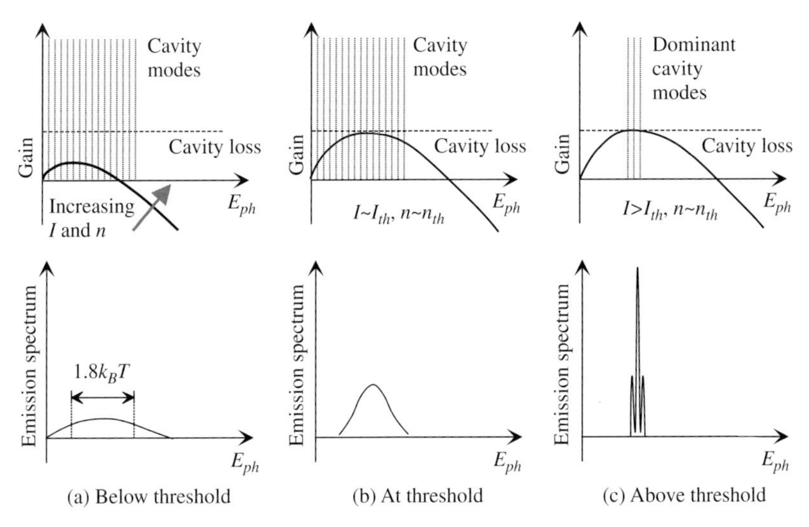

Figure 5.12 The laser gain profile and photon emission from below to above threshold.

Increasing the current density will cause the electron density (and, as a consequence, the material and cavity gain) to increase. When the gain almost compensates for the total losses, the photons belonging to the cavity modes close to the gain maximum begin to experience positive feedback from stimulated emission. The corresponding lifetime decreases, so that more and more e-h pairs recombine by stimulated emission. At threshold, losses and gain are equal, and the output spectrum narrows; see Fig. 5.12(b).

Above threshold, increasing current injection does not cause an appreciable increase of the electron density above n_{th}; this happens because the corresponding increase of the optical power density in the cavity reduces the carriers' stimulated lifetime. A certain set of modes (among which one can be dominant: in FP lasers, however, many longitudinal modes are typically excited) stimulates the e-h pair recombination, thus subtracting photons to the other modes and to spontaneous recombination, which remains as an incoherent noise background. In this way, only a few FP cavity lines are above threshold (with different amplitudes) while the others are quenched and below threshold; see Fig. 5.12(c).

5.6.1 The threshold condition

The FP laser threshold condition can be derived by considering that, at threshold, laser oscillations are self-sustaining. This means that two conditions are simultaneously met:

- The optical wave recovers the initial phase after a round-trip in the cavity, corresponding to cavity resonance.
- The optical wave recovers the initial amplitude after a round-trip in the cavity; this means that the cavity net gain is zero, i.e., the cavity gain equals the cavity losses (mirror or end loss, loss external to the active region or at least not accounted for in the active region material net gain).

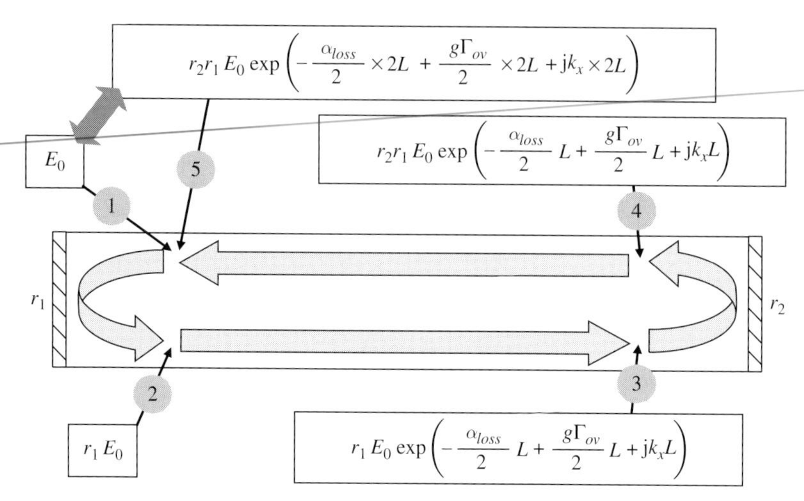

Figure 5.13 Round-trip phase and amplitude evaluation of FP cavity threshold condition.

We can follow the electric field through a cavity round-trip, as shown in Fig. 5.13. We first consider a cavity with unequal field mirror reflectivities r_1 and r_2.[6] The round-trip begins at point 1, immediately before the mirror. Following reflection from the mirror, the field amplitude is multiplied by the mirror reflectivity r_1 (point 2). The field then travels through the cavity, experiencing a total field attenuation and gain $\frac{1}{2}(\alpha_{loss} - g)L$ (remember that the power absorption and gain must be divided by 2 to obtain the corresponding field parameters). At the same time, a phase delay builds up, proportional to $k_x L$. After reflection from the second mirror (point 3 to 4 in Fig. 5.13), the round-trip is closed at point 5 with a further amplitude and phase variation. We now require that the amplitude and phase should be the same at points 1 (beginning of round-trip) and 5 (end of round-trip).

For simplicity, we will assume now that the mirror reflectivities are equal ($r_1 = r_2 = r$) and that $r^2 = |r|^2 = R$ (real, power reflectivity; we assume that the phase of r is π – a different phase would lead to a small variation of the resonance condition).[7] The round-trip consistency condition yields

$$E_0 = R E_0 \exp\left(-\alpha_{loss} L + g\Gamma_{ov} L + j2k_x L\right), \tag{5.24}$$

but the phase term vanishes, since

$$k_x L = \frac{2\pi n_r}{\lambda_0} \times n \frac{\lambda_0}{n_r} = n \times 2\pi.$$

Therefore, from (5.24) we obtain

$$1 = \exp\left(-\alpha_{loss} L + g\Gamma_{ov} L + \log R\right) \rightarrow -\alpha_{loss} L + g\Gamma_{ov} L + \log R = 0;$$

[6] This situation is common in practice, since one of the facets is treated so as to emit a low signal, exploited for monitoring purposes.

[7] The case of unequal complex reflectivities is trivial, since the phases are included in the resonance conditions, while the equivalent power reflectivity is $R_{eq} = |r_1 r_2|$.

i.e., introducing the threshold cavity gain $\Gamma_{ov} g_{th}$,

$$g_{th}(\hbar\omega)\Gamma_{ov} = \alpha_{loss} + \frac{1}{L}\log\frac{1}{R} = \alpha_{loss} + \alpha_m = \alpha_t, \tag{5.25}$$

where α_t is the total (cavity) loss, α_m is the mirror (or end) equivalent loss:

$$\alpha_m = \frac{1}{L}\log\frac{1}{R}, \tag{5.26}$$

and α_{loss}, as already stated, is the absorption in the cladding. The mirror loss or end loss acts like a *concentrated loss* equivalent to a *distributed absorption* such as $-\alpha_m L = \log R$. Note also that the *transparency condition* $g(\hbar\omega)\Gamma_{ov} = 0$ (often close to the threshold in the presence of very low mirror loss, as in VCSELs).

5.6.2 The emission spectrum

As already mentioned, below threshold the laser emission is similar to the LED broadband spectrum. Increasing the device current initially brings the device into a superradiant LED regime, with some spectral narrowing due to the onset of stimulated emission. Above threshold, the spectrum is narrowed but, still, a few FP cavity lines are excited. Some data relative to a GaAs/AlGaAs based FP laser are shown in Fig. 5.14 [74].

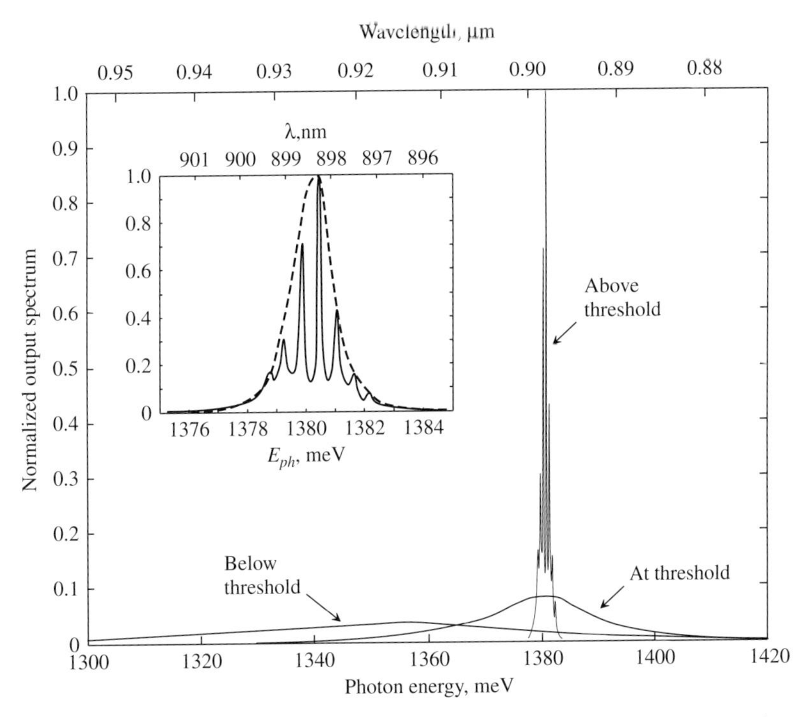

Figure 5.14 Evolution of the FP laser linewidth from below to above threshold. Inset: expansion of the behavior above threshold near the maximum. Adapted from [74], Fig. 4.14.

Below threshold a LED-like behavior dominates, with a FWHM linewidth of the order of 50 meV or 60 nm (around 1.3 μm). Near threshold, we have superradiant LED behavior, with a shift of the maximum around the gain maximum and linewidth around 20 nm (200 Å). Above threshold, we find typical laser behavior, with envelope linewidth (see inset in Fig. 5.14) around 1 nm or 10 Å; note, however, that the *single-line linewidth* is of the order of 1 Å. We shall return later to the cause of the finite linewidth of the laser, even in the case where multiple longitudinal resonances are suppressed.

5.6.3 The electron density and optical power

Below threshold, increasing the current increases the junction electron density n (and therefore the gain); recombination is, however, dominated by the spontaneous lifetime, which decreases with increasing injected charge to finally reach a limiting value τ_0. Above threshold, stimulated recombination prevails for some cavity modes; this leads to smaller stimulated lifetimes vs. spontaneous recombination, which is therefore reduced to a background noise. Always above threshold, the current increases but the junction injected charge density remains approximately constant to the threshold value ($n \approx n_{th}$); however, the increase of the optical power leads to a decrease of the stimulated lifetime; this allows the optical output power and the current density to increase, even with a junction charge density approximately constant to the threshold value. Finally, the slope efficiency $\mathrm{d}P_{out}/\mathrm{d}I$ increases after threshold because the emission spectrum becomes narrowband. Such behavior is summarized in Fig. 5.15.

In practice, the behavior of the laser spectrum above threshold is considerably more involved. Laser heating leads to a variation of the gain spectrum (e.g., due to a variation of the energy gap vs. temperature) and also to a variation of the cavity size and refractive index (with a shift of the cavity resonant modes). As a consequence, the emission spectrum shifts and mode hopping can occur (i.e., the dominant mode can change with increasing optical power).

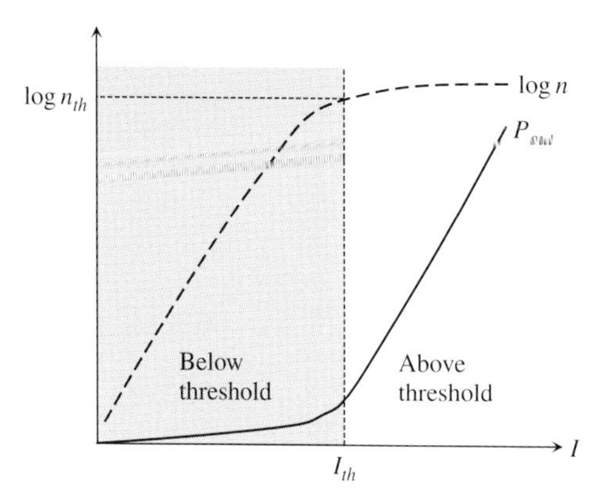

Figure 5.15 Power–current (continuous line) and electron density–current (dashed line) characteristics for a laser.

5.6.4 The power–current characteristics

To approximate the power–current characteristics above threshold, we express the laser current as a *recombination current* in the active region. Neglecting nonradiative effects we obtain from a charge-control model,

$$I = \frac{\mathrm{d}Q_n}{\mathrm{d}t} = q \int_{\text{active}} U_r \, \mathrm{d}V = qA \left(\frac{1}{\tau_n^{sp}} + \frac{1}{\tau_n^{st}} \right) nd, \tag{5.27}$$

where $-Q_n$ is the active-region electron charge, U_r is the radiative recombination rate, n is the electron density in the active region of thickness d, and A is the junction area. The spontaneous and stimulated lifetimes are denoted by τ_n^{sp} and τ_n^{st}, respectively. At threshold (5.27) becomes

$$I_{th} = qA \frac{n_{th} d}{\tau_{n,th}} = qA \left(\frac{1}{\tau_{n,th}^{sp}} + \frac{1}{\tau_{n,th}^{st}} \right) n_{th} d. \tag{5.28}$$

The total lifetime at threshold $\tau_{n,th}$ can be approximately assumed of the order of $4\tau_0$ (see [4], Section 4.7 (iv)). Above threshold, stimulated recombination becomes dominant; since, from (2.46) and (2.57) one has (both in low and high injection) that τ_n^{st} is inversely proportional to the optical power, the corresponding carrier lifetime can be expressed as

$$\tau_n^{st} \approx \tau_{n,th}^{st} \frac{P_{out,th}}{P_{out}}. \tag{5.29}$$

On the other hand, above threshold $n \approx n_{th}$ and $\tau_{n,th}^{sp} \approx \tau_n^{sp}$; we therefore have, from (5.28) and (5.29),

$$\begin{aligned} I &= qA \left(\frac{1}{\tau_{n,th}^{sp}} + \frac{1}{\tau_n^{st}} \right) n_{th} d = qA \left(\frac{1}{\tau_{n,th}^{sp}} + \frac{1}{\tau_{n,th}^{st}} + \frac{P_{out} - P_{out,th}}{\tau_{n,th}^{st} P_{out,th}} \right) n_{th} d \\ &= I_{th} + qA \frac{n_{th} d}{\tau_{n,th}^{st}} \frac{P_{out} - P_{out,th}}{P_{out,th}}, \end{aligned}$$

and, solving for the optical power,

$$P_{out} - P_{out,th} = \underbrace{\frac{P_{out,th}}{A} \frac{\tau_{n,th}^{st}}{q n_{th} d}}_{\text{slope efficiency}} (I - I_{th}). \tag{5.30}$$

5.6.5 The photon lifetimes

A useful parameter in laser modeling is the photon lifetime τ_{ph}, that is, the average lifetime of a photon between its emission and loss (due to external loss or to mirror loss). We defined the equivalent total loss of the cavity in (5.25) and (5.26) as

$$\alpha_t = \alpha_{loss} + \frac{1}{L} \log R^{-1} = \alpha_{loss} + \alpha_m.$$

However, the 1D continuity equation for the photon density N yields, in DC steady-state conditions,

$$\left.\frac{\mathrm{d}N}{\mathrm{d}t}\right|_{\mathrm{tot}} = \frac{\mathrm{d}}{\mathrm{d}x}\left[\frac{c_0}{n_r}N(x)\right] = \frac{c_0}{n_r}\frac{\mathrm{d}\left[N(0)\exp(-\alpha_t x)\right]}{\mathrm{d}x} = -\alpha_t\frac{Nc_0}{n_r} = -\frac{N}{\tau_{ph}},$$

where we have introduced the *photon total lifetime*:

$$\frac{1}{\tau_{ph}} = \frac{c_0}{n_r}\alpha_t = \frac{c_0}{n_r}\left(\alpha_{loss} + \frac{1}{L}\log R^{-1}\right) = \frac{1}{\tau_{loss}} + \frac{1}{\tau_m}. \tag{5.31}$$

The total lifetime can readily be split into the *photon (external) loss lifetime* and *mirror loss lifetime* (or end loss lifetime):

$$\frac{1}{\tau_{loss}} = \frac{c_0}{n_r}\alpha_{loss} \tag{5.32}$$

$$\frac{1}{\tau_m} = \frac{c_0}{n_r}\frac{1}{L}\log R^{-1}. \tag{5.33}$$

Introducing the photon cavity volume V such that the photon number is $n_{ph} = VN$, we can express the output power in terms of the photon number and of the mirror loss as

$$P_{out} = \hbar\omega\frac{n_{ph}}{\tau_m}. \tag{5.34}$$

Combining (5.30) and (5.34) we also obtain that the photon number increases with the driving current with the same law as the output power:

$$n_{ph} = n_{ph,th} + \frac{\tau_m}{\hbar\omega}\frac{P_{out,th}}{A}\frac{\tau^{st}_{n,th}}{qn_{th}d}(I - I_{th}) \approx \frac{\tau_m}{\hbar\omega}\frac{P_{out,th}}{A}\frac{\tau^{st}_{n,th}}{qn_{th}d}(I - I_{th}).$$

Note that, in this approximation, the output power vanishes at threshold (i.e., the spontaneous generation of photons is neglected).

5.6.6 Power–current characteristics from photon lifetimes

An alternative expression for the power–current characteristics of the laser above threshold (5.30) can be derived in terms of the photon lifetimes. Let us consider as the output power the power associated with the cavity modes above threshold; such power will be negligible at threshold and below. Assuming that all carriers injected into the junction recombine, emitting coherent photons, and that such photons are either absorbed in the cladding or emitted through the mirrors, we have the balance equation for the excess current vs. the threshold current:

$$\frac{I - I_{th}}{q} \approx \frac{n_{ph}}{\tau_{ph}}. \tag{5.35}$$

However, the output power corresponds to photons disappearing through mirrors with a lifetime τ_m:

$$P_{out} \approx \hbar\omega\frac{n_{ph}}{\tau_m}.$$

Eliminating the photon number n_{ph} we obtain

$$P_{out} \approx \frac{\hbar\omega}{q}\frac{\tau_{ph}}{\tau_m}(I - I_{th}) = \frac{\hbar\omega}{q}\frac{\frac{1}{L}\log R^{-1}}{\alpha_{loss} + \frac{1}{L}\log R^{-1}}(I - I_{th}). \quad (5.36)$$

Equations (5.36) and (5.30) are equivalent, see Example 5.7.

A best-case estimate of the laser slope efficiency $\mathrm{d}P_{out}/\mathrm{d}I$ can be derived by considering an ideal situation, in which each e-h pair recombining in the active region above threshold leads to the stimulated emission of a photon, which is then radiated out of the cavity. Assuming that such e-h pairs are related to the extra current vs. the threshold, $I - I_{th}$, one has

$$E_{ph} \cdot \frac{I - I_{th}}{q} = \text{photon energy} \times \text{photons generated per unit time} = P_{out},$$

$$\text{or, approximating } \hbar\omega \approx E_g, \quad P_{out} \approx \frac{E_g}{q}(I - I_{th}), \quad (5.37)$$

which yields an ideal slope efficiency, independent of the specific laser structure and only dependent on material properties. For low external loss, $\tau_{ph} \approx \tau_m$ and (5.36) coincides (assuming $\hbar\omega \approx E_g$) with (5.37). Notice that, in all the expressions considered, we have of course neglected power saturation, thus yielding a constant slope efficiency.

5.7 The laser evolution: tailoring the active region

The evolution of the laser active region is briefly summarized in Fig. 5.16. Early semiconductor lasers exploited homojunctions,[8] providing little confinement to carriers (whose concentration decreases away from the junction because of diffusion). The increase in the carrier concentration around the junction, due to carrier injection in direct bias, leads to a small variation of the refractive index (weakly proportional to the free carrier concentration). Low carrier and photon confinement required extremely high threshold currents and led to very short laser average time to failure, even at low operating temperature.

A fundamental step toward the development of practical semiconductor lasers was achieved with the introduction of heterojunctions (initially simple, and then double), providing strong confinement of injected carriers together with photon confinement through the refractive index increase in the low-gap (active) region. To decrease the threshold current I_{th} with the same amount of threshold charge density, the thickness d of the active region was progressively decreased. Below a certain thickness (a few hundred nanometers), however, two effects arise. First, the low-gap active region becomes a

[8] The first demonstration of the junction laser was in 1962.

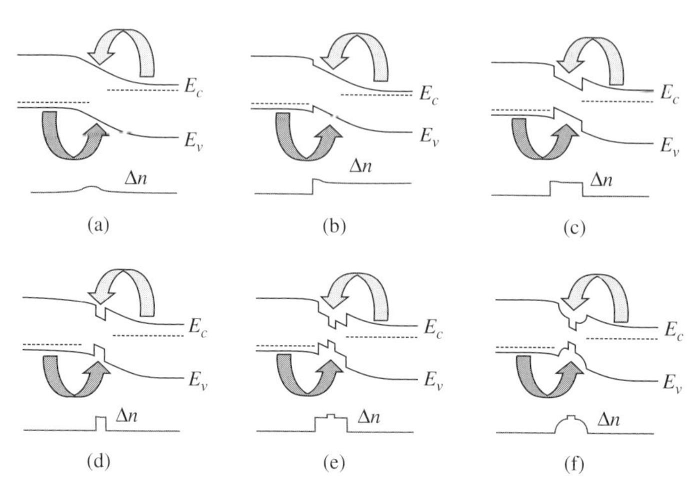

Figure 5.16 Evolution of the active-region design from homojunction (a) to single (b) and double (c) heterojunction, quantum well (d), quantum well with separate carrier and photon confinement (e), graded index separate confinement heterostructure (GRINSCH) lasers (f).

quantum well (QW), with a change (usually an improvement) in the bandstructure and optical properties. On the other hand, the overlap integral becomes increasingly small, due to the poor photon confinement provided by the increasingly thin QW. To improve photon confinement, the QW was embedded in another double heterojunction, realized with a material of intermediate gap between the active layer and the cladding (e.g., the active layer is GaAs, the optical confinement layer is $Al_{0.1}Ga_{0.9}As$, and the cladding is $Al_{0.3}Ga_{0.7}As$, all lattice matched); this provided *separate confinement* to carriers and the optical field. As a further step, nonuniform refractive index profiles were exploited in the so-called GRINSCH (*gr*aded *in*dex *s*eparate *c*onfinement *h*eterostructure) lasers, so as to optimize the overlap integral. Furthermore, multiple QW (MQW) lasers were introduced, typically providing separate confinement to electrons and photons, to increase the output laser power while preserving the same advantages as provided by simple QW structures.

In summary, the evolution of the laser active region was in the direction of improving the vertical confinement of electrons, holes, and photons, at the same time exploiting the potential advantages or features offered by QWs in terms of optical properties. Before discussing this point, however, we have to mention that lateral confinement should also be improved with respect to the conventional FP solution. The two main approaches (gain-guided and index-guided lasers) will be discussed in Section 5.8.2 and Section 5.8.3.

5.7.1 Quantum-well lasers

Introducing a QW leads to a change in the material gain properties, which become polarization dependent. Moreover, in the QW the degeneracy between the two topmost hole bands (HH and LH) is removed. Typically, the HH band is topmost in unstrained wells; tensile strain can reestablish degeneracy and even bring the LH band on top: this

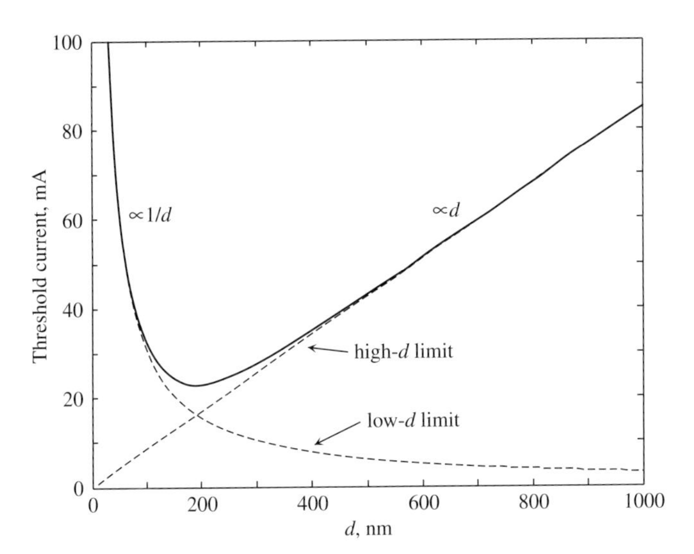

Figure 5.17 Behavior of threshold current vs. active region thickness (Example 5.4).

would decrease the hole effective mass and density of states, thus leading to an improvement in gain, and therefore in the threshold current. However, compressive strain is also beneficial since it reduces the valence band discontinuity and the hole escape lifetime, thus improving the injection efficiency. Another advantage of the QW approach is a reduction of Auger rates [4]. Finally, QW lasers are the ultimate step in decreasing the active-region laser thickness to reduce the threshold current density.

Decreasing the active region thickness in a conventional or QW laser initially leads to a decrease of the threshold current. Assume, as an approximation, that the material gain depends linearly on the carrier density; the threshold gain can be approximated as $g_{th} = K\Gamma_{ov} n_{th}$, where K is a suitable constant. Using the classical expression for the threshold current we obtain

$$I_{th} \approx qA\frac{n_{th}d}{4\tau_0} = qA\frac{g_{th}}{4\tau_0 K}\frac{d}{\Gamma_{ov}},$$

meaning that, for large d, the threshold current decreases with decreasing d, while for small d we ultimately have

$$\Gamma_{ov} \propto d^2 \rightarrow I_{th} \propto qA\frac{g_{th}}{4\tau_0 K}\frac{d}{d^2} \propto \frac{1}{d},$$

i.e., the threshold current increases for decreasing d, as shown in Fig. 5.17.

Example 5.4: A double heterojunction laser with $\lambda_0 = 1.3\,\mu\text{m}$ has an InP cladding with $n_1 = 3.2$ and an InGaAsP intrinsic layer, thickness d, with refractive index $n_2 \approx 3.55$. Assuming a threshold gain $g_{th} = 4000\,\text{cm}^{-1}$ and a threshold electron density $n_{th} = 1 \times 10^{18}\,\text{cm}^{-3}$, junction area $A = 300 \times 5\,\mu\text{m}^2$, estimate the threshold current as a function of d. Assume $\tau_0 = 0.7$ ns.

For large values of d, $\Gamma_{ov} \approx 1$ and we can express the threshold current as

$$I_{th} = qA\frac{n_{th}d}{4\tau_0} = 1.6 \times 10^{-19} \cdot 300 \cdot 5 \times 10^{-12} \cdot \frac{1 \times 10^{24} \cdot d|_{\text{nm}} \times 10^{-9}}{4 \cdot 0.7 \times 10^{-9}}$$
$$= 0.085 \cdot d|_{\text{nm}} \text{ mA}.$$

For small d, the overlap integral can be expressed as

$$\Gamma_{ov} \approx 4\pi^2\frac{d^2}{\lambda_0^2}\overline{n}\Delta n = 4\pi^2\frac{(d|_{\text{nm}})^2 \times 10^{-18}}{(1.3 \times 10^{-6})^2}\frac{3.2 + 3.55}{2} \cdot 0.35$$
$$= 2.76 \times 10^{-5} \cdot (d|_{\text{nm}})^2 .$$

However, $g_{th} \approx n_{th}K\Gamma_{ov} \approx n_{th}K$ for large d; thus $g_{th}/K \approx n_{th}$ and, for small d:

$$I_{th} = qA\frac{g_{th}}{4\tau_0 K}\frac{d}{\Gamma_{ov}} = qA\frac{n_{th}}{4\tau_0}\frac{d}{\Gamma_{ov}}$$
$$= \frac{1.6 \times 10^{-19} \cdot 300 \cdot 5 \times 10^{-12} \cdot 1 \times 10^{24} \times 10^{-9}}{4 \cdot 0.7 \times 10^{-9} \cdot 2.76 \times 10^{-5} \cdot d|_{\text{nm}}} = \frac{3105}{d|_{\text{nm}}} \text{ mA}.$$

Blending the two asymptotic behaviors we obtain the result shown in Fig. 5.17. The minimum threshold current is around 25 mA corresponding to a current density of 1.6 kA/cm^2 for an optimum thickness around 200 nm.

In the above analysis, we have assumed that the active layer sheet density can be given a quasi-3D expression as nd, where n is the active layer 3D concentration. In fact, the electron density of a QW exhibits quantum features, such as a density profile (orthogonal to the stratification) proportional to the squared magnitude of the subband envelope wavefunction. The electron sheet density relative to each subband can be denoted by n_{2D}, and, assuming for simplicity that only the first subband is involved in the laser action, the threshold laser current can be now more conveniently expressed as

$$I_{th} = qA\frac{n_{th,2D}}{\tau_{n,th}} = qA\left(\frac{1}{\tau_{n,th}^{sp}} + \frac{1}{\tau_{n,th}^{st}}\right)n_{th,2D} \approx qA\frac{n_{th,2D}}{4\tau_0}.$$

The already discussed behavior is, however, found when decreasing the QW thickness. In typical QW lasers, the emitted power density is increased through the use of MQW structures (but this, also leads to a higher current density).

The gain spectra of QWs exhibit slightly different features with respect to the bulk, due to the different joint density of states (JDOS), which is a step-like function, each step corresponding to an allowed transition between subbands. The energy behavior of gain in bulk was shown in Fig. 2.16 (for GaAs); the gain envelope globally follows the JDOS profile. In a QW the gain profile vs. energy, shown in Fig. 2.17, clearly exhibits features inherited from the QW JDOS; notice that increasing the electron density leads to a deeper modification of the gain profile, which exhibits peaks. Quantum wire and

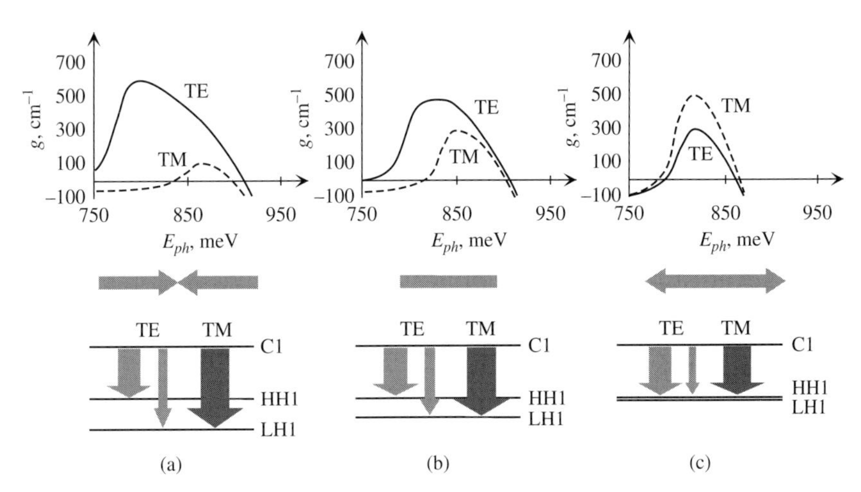

Figure 5.18 Gain profile in strained and unstrained QWs. The structure is an InGaAs/InGaAsP QW. Compressive strain (a), unstrained (b), tensile strain (c). The gain profiles are from [11], Fig. 10.31 (©1995, John Wiley & Sons); this material is reproduced with permission of John Wiley & Sons, Inc.

quantum dot gain spectra are even more deeply modified due to the strongly resonant JDOS behavior found in such structures.

A further degree of freedom in gain profile engineering finally originates from the use of strained or pseudomorphic QWs, which may allow for polarization-independent emission. In fact, strained QWs can lead to an alignment or disalignment of the valence subband level, as shown in Fig. 5.18. The interaction strength of the light holes (LH) and heavy holes (HH) with the electron subband is different, however, and turns out to be polarization dependent (from the computation of the corresponding matrix elements). The relative interaction strengths are (see, e.g., [11]):

	HH→C	LH→C
TE	3/4	1/4
TM	0	1

Such behavior is clearly visible in Fig. 5.18, which refers to an InGaAs/InGaAsP QW with sheet carrier density $n_s = 3 \times 10^{12}\,\mathrm{cm}^{-2}$ [11]. Three cases are considered: (a) QW with compressive strain, Ga mole fraction $x = 0.41$, well width $L = 4.5\,\mathrm{nm}$; (b) lattice-matched QW, Ga mole fraction $x = 0.47$, well width $L = 6\,\mathrm{nm}$; (c) QW with tensile strain, Ga mole fraction $x = 0.53$, well width $L = 1.5\,\mathrm{nm}$. In all cases the structure was designed so as to obtain the lowest band-edge transition wavelength close to $1.55\,\mu\mathrm{m}$ (energy ≈ 0.8 eV). The gains were normalized to a total length of 20 nm including the QW width L and the InGaAsP barrier widths [11]. In cases (a) and (b) the C1-HH1 transition is dominant, and the TE polarization gain prevails. In case (c) tensile strain equalizes the HH1 and LH1 levels, and the TM polarization gain is slightly prevalent (note that, with the same interaction strength, light holes are characterized by a lower density of states effective mass and therefore more easily reach the population inversion condition). Equalization of the gain spectra in a strained QW structure also corresponds (in the absence of current injection) to the equalization of absorption profiles, which can be exploited, as already remarked, in the realization of polarization-independent electroabsorption modulators or waveguide photodiodes.

5.7.2 Laser material systems

The material choice for the laser active region is related to the emission wavelength and therefore to the specific application. Material systems for the laser are somewhat similar to those exploited in the LED. The early development of lasers in the 1970s was mainly driven by the AlGaAs/GaAs material system, with emission wavelengths around 0.8 μm. Long-haul fiber communications stimulated the development in the 1980s of long-wavelength lasers, emitting at 1.3 and 1.55 μm in the regions of minimum fiber dispersion and attenuation, respectively. Such lasers were based on InP technology and, in particular, on the InGaAsP material system. In the same years, the introduction of CD players brought GaAs-based lasers into the consumer market, with a drive toward reducing the wavelength into the visible region (e.g., red) in DVD players to increase storage density. In the 1990s, InP-based lasers gradually became commercially available, while new material systems were investigated to achieve emission in the blue (0.4 μm wavelength). This was finally achieved by exploiting the widegap InGaN/GaN material system, opening the path to new applications, such as DVD systems with dramatically increased capacity (25 Gb per single-layer disc). On the other hand, extralong-wavelength emission was sought by exploiting, rather than band-to-band transitions, intersubband effects in multiquantum well (the so-called quantum cascade lasers, also suited for terahertz emission). As a last point, we may mention the quest for the Si laser – announced in recent years, but in the form, for the moment, of an optically pumped device based on the Raman effect [75]. For datacom and telecom applications, however, AlGaAs/GaAs and InGaAsP/InP today remain the main players.

5.8 The laser evolution: improving the spectral purity and stability

Early semiconductor lasers exploited Fabry–Perot cleaved cavities with rather large lateral dimension W. Several longitudinal and transversal modes were excited, making the linewidth one order of magnitude better than in LEDs, but still poor for long-haul applications.

Partial suppression of the higher-order transversal modes was first achieved by injecting current only in a stripe, thus causing a reduction of the lateral cavity size due to the so called *gain guiding* effect (only part of the laser offered gain and was therefore able to support the lasing action). Gain guiding was then replaced by more efficient *index-guided* structures, with just one transversal mode.

On the other hand, single longitudinal mode lasers required either the coupling of two Fabry–Perot cavities (as in cleaved coupled cavity or C^3 lasers),[9] or making use of a different mirror approach, based on distributed reflection from a diffraction grating, rather than on concentrated reflection by a cleaved mirror. Distributed reflectors have nonflat reflection coefficients vs. wavelength, and therefore more effectively suppress spurious longitudinal modes.

Gratings can be either integrated into the cavity as in DFB (distributed-feedback) lasers, or can be placed outside of the cavity, as in DBR (distributed Bragg reflector) lasers. External distributed mirrors involve additional complexities in the technology, but their reflecting properties are immune from the change in the laser bias current. DBR lasers therefore have better line stability and, if the external mirrors are suitably polarized with currents, the emitted wavelength can be changed, leading to *tunable lasers*.

Transition from the FP to the DFB and DBR lasers also leads to an improvement of the temperature stability of the emitted wavelength, but, if stringent requirements are imposed on this parameter, the laser has to be thermally stabilized through a Peltier cell, leading to expensive packaging and high-cost devices, such as those exploited in long-haul WDM (wavelength division multiplexing) systems. In recent years, however, market demand has been pushing more on short-length links supported through low-cost, integrated modules not needing temperature stabilization; this has in turn fostered the development of uncooled lasers exploiting material systems with reduced temperature sensitivity, like AlGaInAs alloys [76].

5.8.1 Conventional Fabry–Perot lasers

Conventional Fabry–Perot lasers have a simple manufacturing process; a *pn* heterojunction is grown epitaxially, and then cut so as to create bars approximately as long as the final cavity. End-facet cleaving along proper crystal planes allows the creation of the end mirrors, while the lateral sides are roughened so as to obtain a non-mirroring surface. However, due to the impossibility of laterally cleaving the cavity down to a few micrometers, the FP cavity is wide enough to excite many transversal modes (besides many longitudinal modes); see Fig. 5.19(a). Transversal mode excitation leads to poor coupling with the fiber, while the excitation of many longitudinal modes causes poor spectral purity (typically of the order of 1–3 nm) and current- and temperature-dependent spectral emission.

5.8.2 Gain-guided FP lasers

A first step toward improving the FP cavity laser resulted from confining the current injection to a narrow stripe. In its simplest form, the resulting *stripe laser* has a

[9] In C^3 lasers two unequal-length cavities act according to a vernier principle to allow oscillation only for a few isolated modes; see e.g., [4].

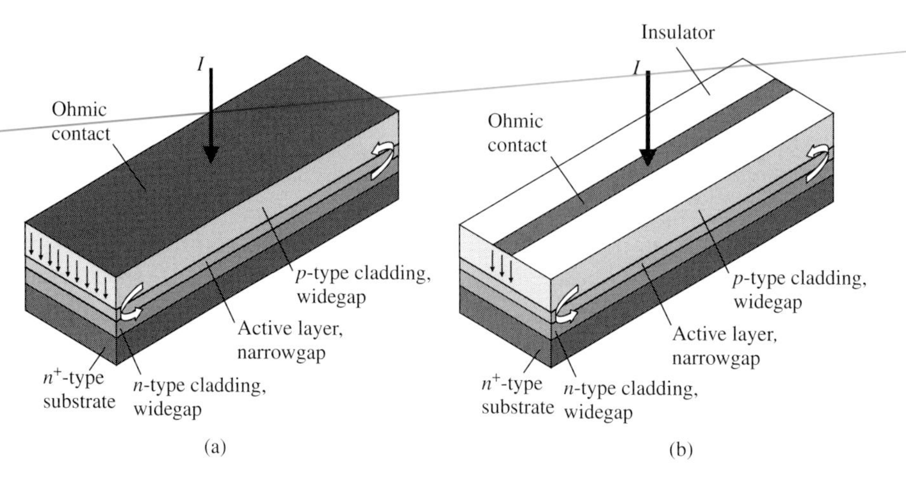

Figure 5.19 Fabry–Perot cavity laser (a) and gain-guided stripe laser (b).

passivated upper surface, while current flow is limited to the unpassivated metal stripe; see Fig. 5.19(b). Owing to the current injection, the region lying below the ohmic contact has higher carrier concentration than the rest of the device and exhibits gain. Since the current density also spreads laterally, the boundary between the gain region and the region without gain is not sharp. Moreover, the gain is not laterally uniform due to the inhomogeneous current injection. Lasing takes place only in the central region with gain (gain guiding) but, due to the nonuniform gain and refractive index, the optical field wavefronts bend so as to give rise to multiple longitudinal modes (while higher-order transversal modes are effectively suppressed). No strong lateral confinement is provided by this structure to the optical field or to carriers. Due to the multiple longitudinal modes excited, gain-guided lasers often exhibit kinks in the power–current characteristic, caused by the mode transition (Fig. 5.20); this also leads to a change in the emission wavelength.

5.8.3 Index-guided FP lasers

Further improvements to the laser structure were introduced by lateral confinement of photons through a refractive index step (index-guided laser) and/or of the carriers through an insulating interface and/or a potential well associated with a heterojunction. A first example of index-guided structure is the *ridge laser* (Fig. 5.21), in which the current density is injected through a ridge structure that increases the effective refractive index of the slab dielectric waveguide section lying below it, so as to confine the field in the region below the ridge. Although the gain profile is laterally not uniform, the consequence is less dramatic due to the improved lateral optical field confinement.

Better results can be obtained by creating a nonuniform lateral structure, either by passivation, or by epitaxial regrowth of a widegap material around the central active region. In both cases the structure is referred to as the *buried laser* (or *buried*

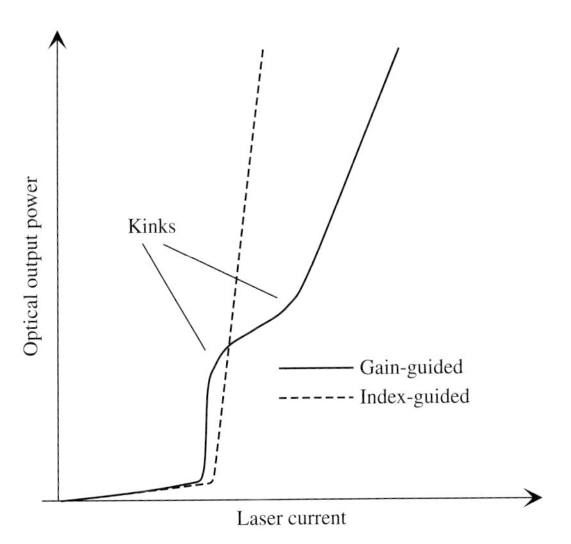

Figure 5.20 Power–current characteristics of gain-guided lasers (with kink effect) and of index-guided lasers.

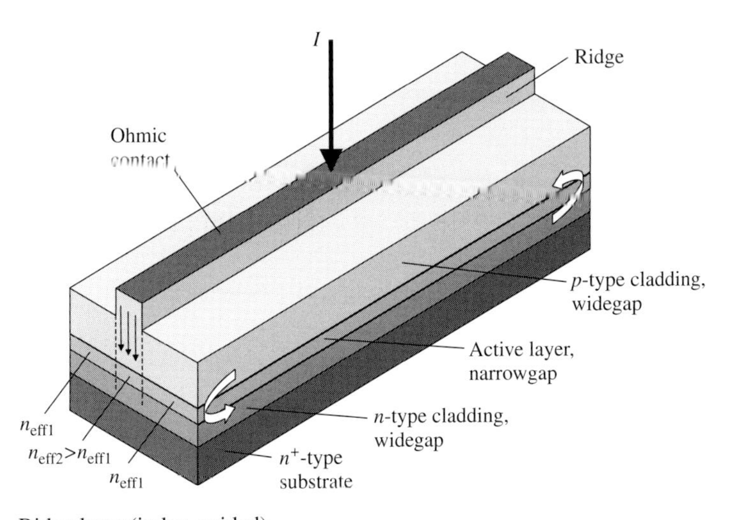

Figure 5.21 Ridge laser (index-guided).

heterostructure, BH, laser); see Fig. 5.22(a) for a passivated example and Fig. 5.22(b) for an epitaxially regrown device.

In index-guided structures, the active region is limited by barriers that laterally confine both carriers and photons. In such structures, the carrier concentration and, to some extent, also the photon density, can be considered as uniform in the lateral direction. The access region is often wedge-shaped in order to decrease the parasitic series resistance of the diode. Although the laser has more than one longitudinal mode excited, mode hopping as a function of the excitation current and of the temperature is avoided (Fig. 5.20). Despite the absence of mode hopping, the output wavelength changes with increasing current, for example, due to the change in refractive index associated with

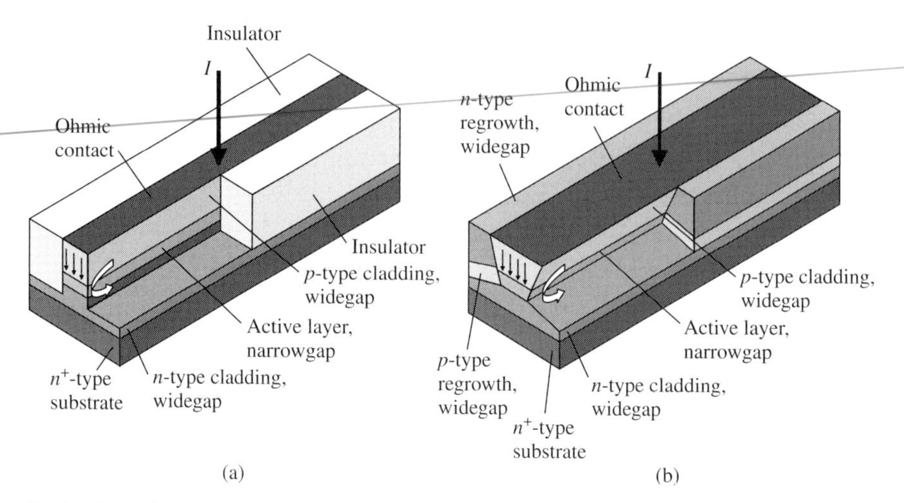

Figure 5.22 Buried laser, index guided, with oxide isolation (a) and epitaxially regrown buried laser (b).

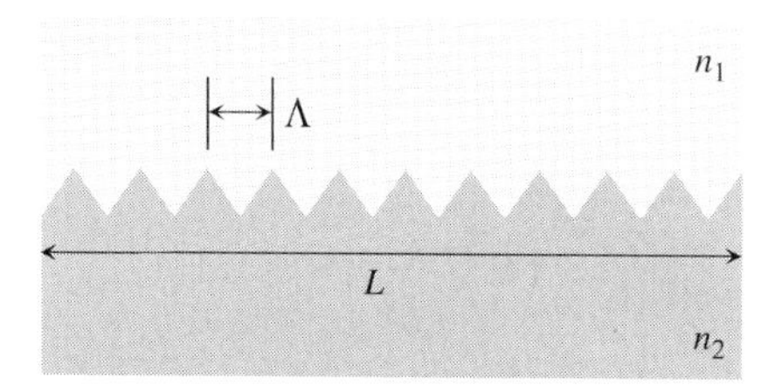

Figure 5.23 Diffraction grating and grating parameters.

the increasing temperature, but the current and temperature sensitivity of the emitted spectrum is smaller than for gain-guided lasers.

5.8.4 Distributed-feedback (DFB and DBR) lasers

Single (longitudinal) mode lasers are achieved with the help of mirror (diffraction) gratings (see Fig. 5.23), which substitute cleaved mirrors, or provide additional feedback. In DFB or *distributed-feedback* lasers (see Fig. 5.24(a)), the grating is integrated with the gain region, while gratings are external in DBR or *distributed Bragg reflector* lasers (see Fig. 5.24(b)).

The distributed mirror concept was introduced in the early 1970s [77]. While a cleaved mirror acting on the basis of the refractive index step has a reflection coefficient that is (almost) independent of the operating wavelength, the grating reflection coefficient can be strongly frequency dependent. If the distributed reflector is properly designed, side modes see a larger mirror or end loss with respect to the central mode, and therefore are below threshold.

A diffraction grating (see Fig. 5.23) can be created within or outside the active region of the laser by properly etching a periodic corrugation (with a submicrometer step related to the reflection peak) through a photoresist mask exposed by electron-beam

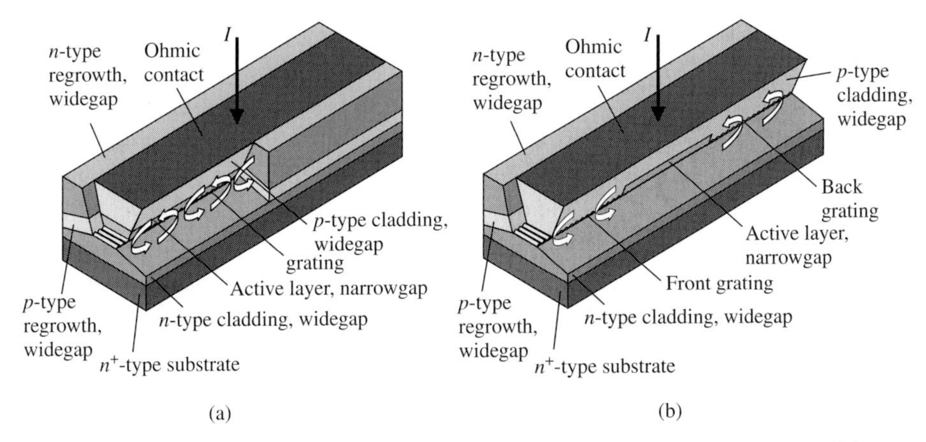

Figure 5.24 Distributed feedback (DFB) laser (a), distributed Bragg reflector (DBR) laser (b). Part of the lateral and front structure has been removed to expose the grating.

lithography (EBL) or other deep-UV processes. Interference-based techniques can be used to avoid the use of deep submicrometer lithography. Once the surface of the widegap material has been suitably etched, the epitaxial growth continues with a narrowgap material, thus directly creating a grating adjacent to the laser active region, which acts as a periodic perturbation of the refractive index within the active region.

To introduce the grating operation, suppose that a slab waveguide mode interacts with a grating of periodicity Λ (Fig. 5.23); each step of the grating causes a small amount of reflection. Reflections from two following grating periods have a phase difference, $\Delta\phi$; this leads to constructive interference from all reflected contributions (and therefore large overall reflection) only if

$$\Delta\phi = k\Delta l = 2k\Lambda = 4\pi\frac{\Lambda}{\lambda} = 2n\pi \rightarrow \lambda = 2\frac{\Lambda}{n}.$$

The above condition, with $n = 1$, is referred to as the *Bragg condition*, and dictates that the grating period must be one-half of the laser (guided) wavelength; $\lambda_B = 2\Lambda$ is also called the Bragg wavelength of the grating.[10] Higher-order reflections with $n > 1$ are sometimes used to relax lithography requirements, which can become very severe, e.g., for blue lasers, where the wavelength is around 400 nm. Two successive Bragg conditions (e.g., with $n = 1$ and $n = 2$) correspond to a large wavelength interval. Compare the Fabry–Perot resonance condition $n\lambda/2 = L$ with the Bragg condition $n\lambda/2 = \Lambda$; in both cases we obtain that

$$\frac{|\Delta\lambda|}{\lambda} \approx \frac{1}{n},$$

[10] Another interpretation of the grating operation sees the grating as a scatterer that imparts to the photons a momentum k_g, converting the forward wave into a reflected wave:

$$-k = k - k_g \rightarrow k_g = \frac{2\pi}{\Lambda} = 2k = \frac{4\pi}{\lambda} \rightarrow \lambda \equiv \lambda_B = 2\Lambda.$$

Of course, this again corresponds to the definition of the Bragg wavelength.

but since $L \gg \Lambda$, the corresponding value of n is about 1000 for the FP cavity, while it is only 1 for the grating. The spacing between two consecutive Bragg conditions is therefore several orders of magnitude larger than the spacing between two FP cavity modes. This, however, does not necessarily imply that grating-based lasers are safely single-mode, since the detailed analysis of a grating reveals that, around each of its Bragg resonances, the mirror reflectivity may have a comparatively wide bandwidth.

For the sake of simplicity, consider first the grating as a *distributed mirror* external to the gain region (the corresponding structure is the DBR laser). The analysis shows that, around a Bragg condition (e.g., that corresponding to $n = 1$), the grating reflection coefficient can be expressed as [37]

$$r = \frac{-\mathrm{j}\kappa L \sinh(\gamma L)}{\gamma L \cosh(\gamma L) + \mathrm{j}\Delta\beta L \sinh(\gamma L)}, \tag{5.38}$$

where κ is the coupling coefficient between the forward and backward waves, defined by the equations (E is the electric field phasor)

$$\left.\frac{\mathrm{d}E^+(x)}{\mathrm{d}x}\right|_{\text{grating}} = -\mathrm{j}\kappa E^-(x), \qquad \left.\frac{\mathrm{d}E^-(x)}{\mathrm{d}x}\right|_{\text{grating}} = \mathrm{j}\kappa E^+(x)$$

and

$$\gamma^2 = |\kappa|^2 - (\Delta\beta)^2 .$$

The *detuning* $\Delta\beta$ is the difference in propagation constant with respect to the Bragg condition; $\Delta\beta$ is related to the wavelength variation $\Delta\lambda$ as

$$\Delta\beta = \frac{2\pi}{\lambda_B} - \frac{2\pi}{\lambda_B + \Delta\lambda}. \tag{5.39}$$

Figure 5.25 shows a plot of the grating reflectivity as a function of detuning for several values of the normalized coupling κL. It can be seen that the region of high reflectivity

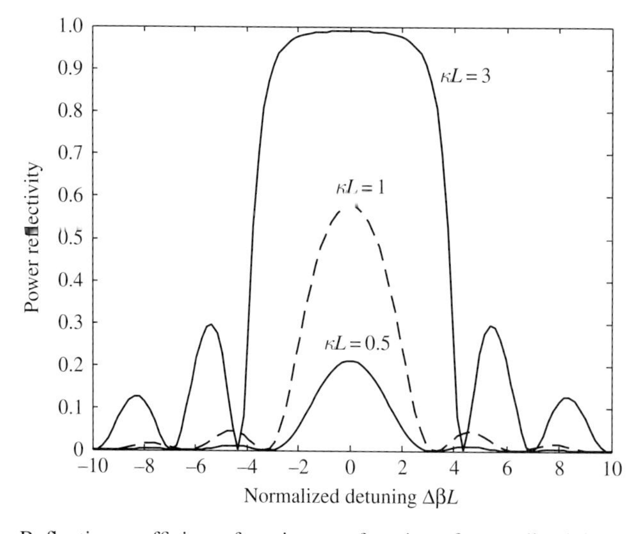

Figure 5.25 Reflection coefficient of grating as a function of normalized detuning.

around the zero detuning condition has the approximate width $|\Delta\beta L| \approx 2\kappa L$; for $|\Delta\beta L| > 2\kappa L$ the reflectivity drops and the mirror loss increases.

If the grating is present together with a cleaved cavity (as in DBR lasers, see Section 5.8.5), proper design of the grating is able to suppress lasing of all FP cavity modes but one, thus making laser operation single-mode. In fact, for a FP cavity mode we have

$$2kL = 2n\pi.$$

Therefore, if k_n corresponds to the central FP mode, the nearby FP side modes have normalized detuning $\pm\Delta\beta L = k_{n\pm1}L - k_n L = \pm\pi$. To decrease the grating reflectivity of the side modes (thus increasing their end loss), we therefore need to impose $2\kappa L_g < \pi$, where L_g is the grating length.

In DFB lasers, gain and distributed reflection occur in the same region, and the analysis of the threshold condition is more involved. In fact, the threshold condition now corresponds to the solution of the following transcendental equation [37]:

$$\gamma L \cosh(\gamma L) + \mathrm{j}\left(\Delta\beta L + \mathrm{j}\frac{g_{th}L}{2}\right)\sinh(\gamma L) = 0,$$

where g_{th} is the (cavity) threshold power net gain and

$$\gamma^2 = |\kappa|^2 - \left(\Delta\beta + \mathrm{j}\frac{g_{th}}{2}\right)^2.$$

Note that the threshold net gain corresponds to the sum of the external and mirror loss (or *end loss*, as it is more appropriately called in this context), approximately equivalent to the end loss. As usual, $\Delta\beta$ is the detuning with respect to the Bragg condition; see (5.39). Although one might have expected that a solution exists for zero detuning (i.e., for the Bragg wavelength), this is not the case for the simple uniform grating introduced here. In fact, with zero detuning we should have

$$\tanh\left(\sqrt{|\kappa|^2 + \left(\frac{g_{th}}{2}\right)^2}L\right) = \sqrt{|\kappa|^2 + \left(\frac{g_{th}}{2}\right)^2}L \Bigg/ \frac{g_{th}L}{2} > 1,$$

but the hyperbolic tangent of a real argument is always < 1. Therefore, no solution exists for zero detuning; however, having established, for example, the coupling κ, we can numerically analyze the threshold condition so as to obtain the detuning–gain pairs that satisfy it. The evolution of the solutions in the plane $(g_0 L, \Delta\beta L)$ for different values of the normalized coupling κL is shown in Fig. 5.26; the (cavity) field gain is $g_0 = g_{th}/2$ and the normalized coupling corresponding to dots is $\kappa L = 0.5, 1, 2, 5$. Two symmetric solutions exist for positive and negative detuning, having the same possibility for lasing.

Note that this result has been obtained by assuming that the FP mirror reflection is entirely suppressed (i.e., that the cavity facets are transparent, only distributed reflection takes place). From Fig. 5.26, it is clear that for high gain, low coupling the separation between the two solutions is the same as for Fabry–Perot side modes; besides, the

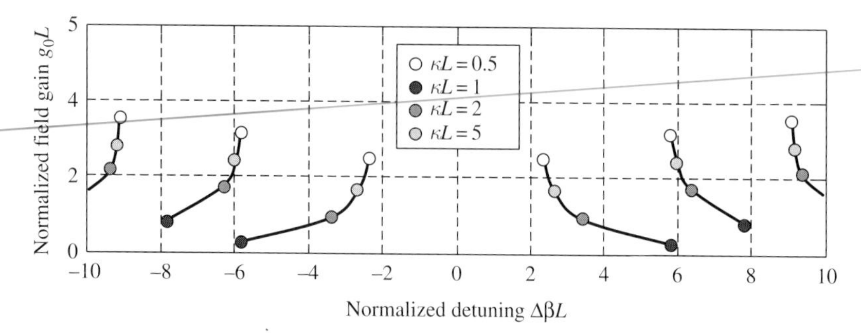

Figure 5.26 Normalized threshold field gain $g_0 L$ in a DBF laser as a function of the normalized detuning $\Delta\beta L$ for different values of the normalized coupling κL.

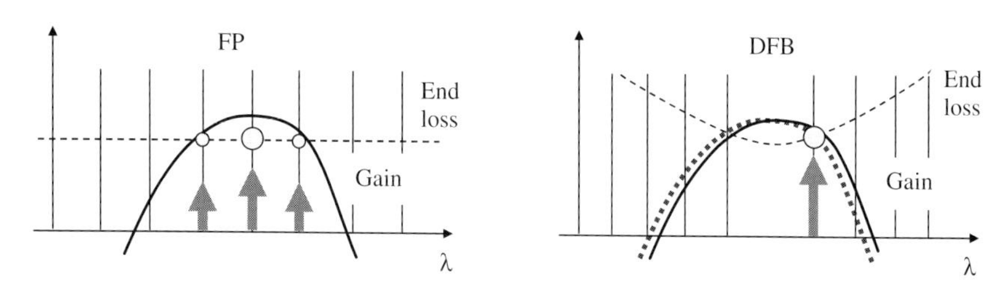

Figure 5.27 FP laser spectrum, left (constant end loss) and DFB laser spectrum, right (side modes have higher end loss). In the DFB case the dashed line refers to a strictly symmetrical structure in which two modes are above threshold, the continuous line to a structure in which asymmetries are introduced so as to bring only one mode above threshold.

additional side solutions, with larger detuning, have a small gain penalty.[11] In such conditions, the DFB laser is not markedly superior to the FP, even if the introduced gain penalty somewhat suppresses higher-order longitudinal modes. However, increasing the coupling, the two main lines increase their distance and the gain penalty (related to the larger end loss) of higher-order longitudinal modes becomes large enough to effectively suppress secondary-mode lasing.

In conclusion, while in FP lasers the mirror loss is the same for all longitudinal modes (Fig. 5.27), in the DFB laser side modes have a larger mirror loss and therefore cannot lase. The two symmetrical lasing modes are, however, practically inconvenient; one of them can be effectively suppressed either by introducing dissymmetries into the gain profile or in the reflecting FP-like facets, or by introducing an appropriate phase shift in the center of the grating ($\lambda/4$ shift DFB lasers). Quarter-wavelength shifting also leads to a dominant lasing mode with zero detuning.

[11] It can be shown that, for low coupling κL, the solution is asymptotically

$$\Delta\beta L = \left(n - \frac{1}{2}\right)\pi + \tan^{-1}\left(\frac{\Delta\beta L}{g_0 L}\right).$$

Neglecting the last term (solutions with high gain $g_0 L$ or low detuning $\Delta\beta L$), the mode spacing is the same as for the FP laser (this corresponds to the vertical asymptotes in Fig. 5.26). For high coupling ($\kappa L \gg 1$), on the other hand, $\Delta\beta L \approx \pi/\sqrt{g_0 L}$, corresponding to the external curves in Fig. 5.26.

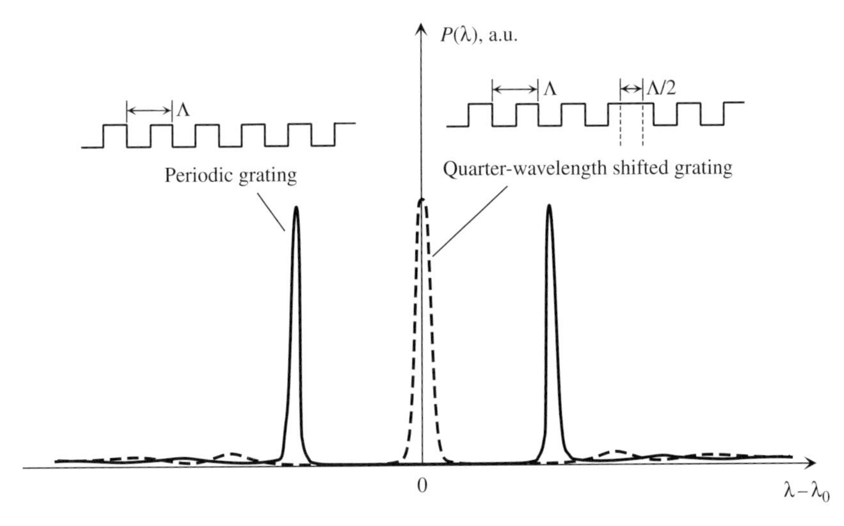

Figure 5.28 Output spectrum of a DFB laser with conventional grating and quarter-wavelength shifted grating.

A schematic example of a DFB laser is shown in Fig. 5.24 (a); the template is the buried heterostructure device already considered for the Fabry–Perot cavity laser; besides the cavity, distributed feedback is achieved through a grating in the active region interface. The qualitative behavior of the emission spectrum of a $\Lambda/2 = \lambda_g/4$ shifted DFB laser is compared in Fig. 5.28 (dashed line) to the spectrum of a conventional, ideal DFB laser (continuous line). With the quarter-wavelength grating, we obtain a dominating central mode with zero detuning and good suppression of additional transversal (side) modes.

5.8.5 DBR and tunable DBR lasers

In DBR (distributed Bragg reflector) lasers, the distributed mirrors are external to the active region. Figure 5.24(b) shows the principle of the implementation. In a DBR laser (or, at least, in its simplest form) gratings operate as mirrors, with no gain. The reflection coefficient (5.38) of the grating is frequency-dependent, and a proper choice of the grating length and coupling coefficient allows for selecting a single longitudinal mode. In certain cases, gratings can be totally external and extremely long (e.g., in an optical fiber), thus leading to very small linewidths. With respect to the DFB implementation, DBR lasers are more critical to manufacture and less compact, but external gratings are more stable than internal ones (not being influenced by bias or modulation current), thus leading to improved line stability. Finally, the grating reflection coefficient can be modulated by an external current, thus making the laser tunable. Tunable lasers typically have a more complex structure, including tunable mirror sections and phase sections; one of the laser facets can be a conventional cleaved mirror; see Fig. 5.29. Tunable lasers have several applications, not only telecom, but also in measurement systems; their tuning range varies according to the

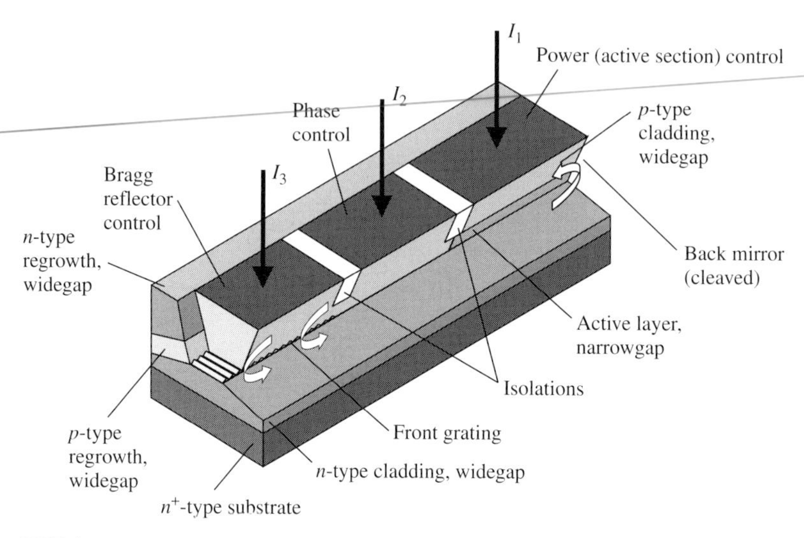

Figure 5.29 DBR laser structure with gain, phase, and Bragg mirror sections.

implementation but (with the combination of several sources in parallel) can be as wide as 40 nm.

5.8.6 Vertical cavity lasers

The laser structures discussed so far are *edge emitters* – light is emitted in the junction plane. Due to the aperture field pattern, the radiation diagram is not circular; this makes coupling with a fiber more critical. Vertical emission lasers, similar to vertical emission (or Burrus) LEDs, have better fiber coupling owing to the circular emitting area, which translates into a circular far-field beam. However, vertical emission requires mirrors parallel to the junction plane, which can be manufactured, according to the DBR approach, as diffraction gratings made by stacks of quarter-wavelength layers. The resulting cavity and mirror properties, however, are more critical than in conventional edge-emitting lasers. In fact, vertical cavity surface-emitting lasers (VCSELs) have a short cavity that requires *large gain* and *very low mirror loss* (e.g., 99.9%) to achieve lasing. Highly reflecting mirrors require, in turn, thick semiconductor stacks, with large series parasitic resistance. Because of the above issues, the development of VCSELs has been fraught with a number of technological and design problems. A comparative simplified picture of the typical layout of an edge-emitting Fabry–Perot laser and of an array of VCSELs is shown in Fig. 5.30; note that the small size of a VCSEL makes it well suited to manufacturing arrays of emitters.

In the VCSEL, the divergence of the laser beam is inversely proportional to the beam size at the source, i.e., the smaller the source, the larger the divergence. The typical length of the cavity is 1–3 λ; due to this, the Fabry–Perot cavity mode spacing is very large, and the cavity reflectivity is entirely dominated by DBR mirrors, which are

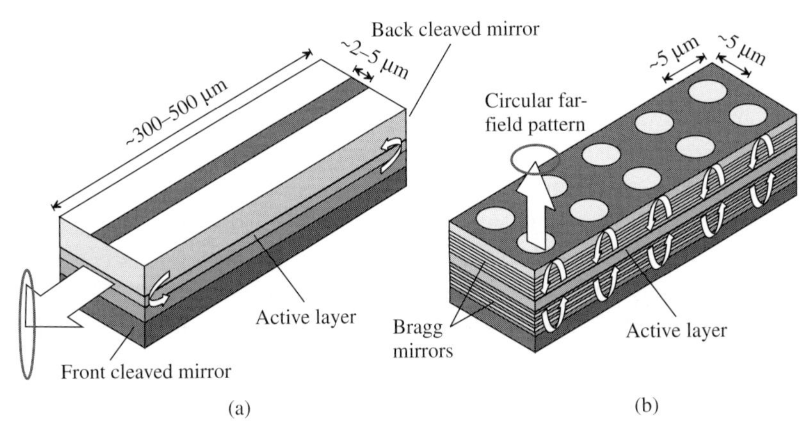

Figure 5.30 Qualitative layout of a Fabry–Perot cavity edge laser (a) and of a VCSEL array (b); note the different far-field radiation patterns. The two drawings are not to scale.

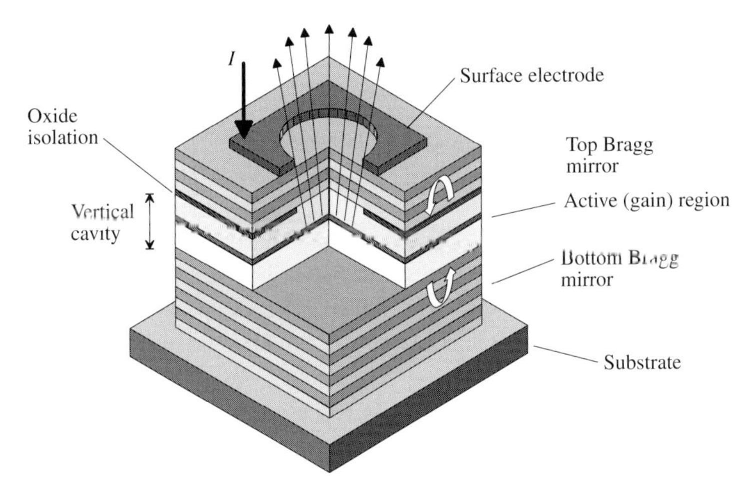

Figure 5.31 VCSEL structure with semiconductor distributed Bragg reflectors.

formed by depositing alternating layers of semiconductor or dielectric materials with a difference in refractive index (e.g., a heterostructure MQW). VCSELs are currently manufactured in several wavelengths (1300, 1550, and 850 nm; mature VCSELs are, however, in the short wavelength range) and exhibit some appealing features, such as an easier integrated technology with respect to edge emitters, easier testability, and therefore lower cost.

An example of VCSEL mirror and cavity structure is reported in Fig. 5.31; both dielectric and semiconductor stacks can be used as external Bragg reflectors. Note that, unlike in edge-emitting DBR lasers, the mirror characteristics potentially change as a function of the emitted power, since, as in DFB lasers, the bias current is, in fact, injected through the reflector stacks.

5.8.7 Quantum dot lasers

The laser structures discussed so far exploit bulk semiconductors or quantum wells. A number of advantages are expected from the use of zero-dimensionality structures, i.e., quantum dots (QDs). Since the density of states of a QD is a set of delta functions of the carrier energy, see Section 1.32, we also expect the absorption and gain profiles to be pulse-like or, allowing for some linewidth broadening mechanism, a set of broad pulses centered around each of the QD energies. When compared with conventional or QW structures, QDs have many favorable characteristics, such as higher differential gain (i.e., variation of gain with respect to the carrier density), and thus a larger modulation bandwidth, lower chirp, better spectral purity, low threshold current, and low sensitivity to the operating temperature. In order to be effective, the size of QDs should be as small as 10–100 nm. Nanometer-scale dots exhibit an atom-like behavior that makes them quite similar to the molecules exploited in gas and solid-state conventional lasers.

In practice, however, the introduction of QDs into a manufacturable and effective structure is fraught with many difficulties, see [78]. QDs can be implemented through a potential well confined in all directions; nevertheless, confinement should not prevent current injection into the QD, but should be strong enough (with potential barriers of the order of 100 meV or more) not to allow carriers to escape the dots by thermalizing into the bulk. Finally, since a single QD cannot provide significant output power, QDs have to be suitably manufactured into QD arrays. QD arrays are a kind of 2D or 3D superlattice (according to whether QDs are distributed in a plane or in a volume); however, the QD array should be extremely uniform in terms of QD size and spacing, since in nonuniform or random QD arrays, with a broad distribution of QD sizes, the density of states is smeared and the advantages of the QD optical properties are lost. QD arrays made through etching, as sets of circular pillars exhibit good uniformity but are affected by heavy surface effects.

The present technology of QD lasers is based on the self-assembling of 2D arrays of dots, made of a narrowgap material grown on a widegap substrate (e.g., InAs on GaAs). InAs nanodots are grown by molecular beam epitaxy on a GaAs substrate; in a first stage of the epitaxial deposition, InAs (lattice mismatched vs. GaAs) aggregates in a set of pyramidal structures whose height and width are in the nanometer range (e.g., 10–20 nm base, 2–3 nm height, with a spacing around 100 nm between structures). InAs growth is stopped at this early nucleation stage and an epitaxial growth of GaAs is performed, which is able to planarize the structure leaving, at the end, a set of almost uniformly sized square pyramidal dots of narrowgap InAs embedded into a thin layer of GaAs. Several techniques have been developed to improve the uniformity of QD spacing and also to allow for the growth of regular QD stacks making a 3D QD array; see Fig. 5.32.

One of the major interests in QD lasers is in the area of low-cost directly modulated lasers for 10 Gbps (and maybe above) applications not needing temperature stabilization and therefore expensive packaging. Due to the reduced chirp, such devices could be competitive with respect to lasers integrated with electrooptic or electroabsorption modulators. An example of recent QD laser development is a temperature-insensitive

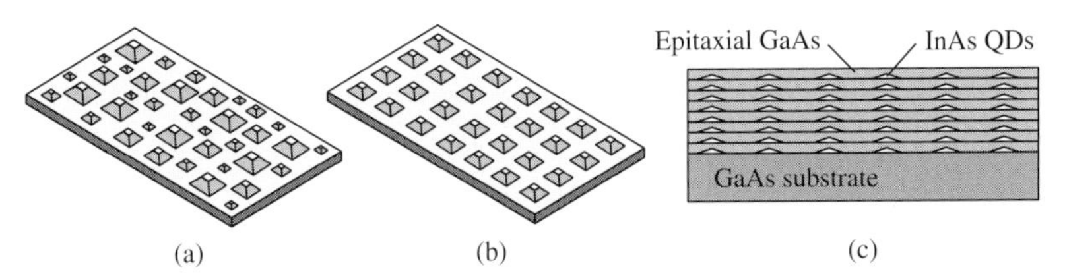

Figure 5.32 InAs quantum dots grown on a GaAs substrates by MBE: (a) disordered arrangement of QDs with irregular size; (b) regular arrangement of QDs with regular size; (c) 3D array of self-arranged QDs.

(20°C to 70°C) 1.3 μm QD laser with active region made of a multilayer InAs on GaAs QD array, including δ-doped p-type GaAs layers, with satisfactory performances in the 10 Gbps range [79].

5.9 The laser temperature behavior

The laser emission is temperature dependent: gain spectra generally depend on the active material temperature: increasing the ambient temperature, but also the dissipated power in the junction, leads to a shift in the gain spectrum. As a consequence, cavity modes experience a shift, due to the temperature dependence of the effective refractive index and, in a Fabry–Perot cavity, to the change in size of the cavity. In Fabry–Perot cavities, this leads both to temperature sensitivity of the emission spectrum and to the so-called *mode hopping* (i.e., the dominant mode changes with changing temperature). This increases the temperature sensitivity of the emitted wavelength up to values of the order of 4 Å/K. Mode hopping, on the other hand, is avoided in single-longitudinal-mode lasers such as DFB lasers, leading to a reduction of the temperature sensitivity to values of the order of 1 Å/K (see e.g., [4], Section 10.7).

Another important issue is the variation of the threshold current with temperature. Since the temperature increase favors nonoptical recombination processes, gain decreases, leading to larger threshold currents and to a corresponding shift of the power–current relation. The temperature dependence of the threshold current follows an Arrhenius-like law with characteristic temperature T_0:

$$I_{th}(T) \approx I_{th}(T_r) \exp\left(\frac{T - T_r}{T_0}\right);$$

T_r is the reference temperature. Particular material systems (including quantum dot-based-structures) can be implemented to decrease the temperature sensitivity.

Example 5.5: The threshold current of a laser is measured at different temperatures:

$$I_{th}(T_1) = 61\,\text{mA},\ T_1 = 20\,°\text{C} \qquad I_{th}(T_2) = 67\,\text{mA},\ T_2 = 35\,°\text{C}$$
$$I_{th}(T_3) = 75\,\text{mA},\ T_3 = 50\,°\text{C} \qquad I_{th}(T_4) = 85\,\text{mA},\ T_4 = 65\,°\text{C}.$$

Estimate the characteristic temperature T_0.

From the Arrhenius plot expression we derive, taking the logarithm of both sides,

$$T_0 \approx \frac{T - T_r}{\log\left(\dfrac{I_{th}(T)}{I_{th}(T_r)}\right)},$$

from which we obtain an approximate least-squares fit of $T_0 \approx 146\,\mathrm{K}$.

5.10 Laser linewidth

The laser linewidth can be decreased by suppression of extra transversal and longitudinal modes, but, ultimately, single-mode lasers also have a finite linewidth. Single-mode emission is, in fact, affected by phase fluctuations (*phase noise*), which broaden the laser output spectrum.

Fluctuations in the laser intensity and phase are caused by fluctuations in the cavity photon number and carrier active region population. Phase fluctuations seem to be particularly related to the spontaneous emission of noncoherent photons; in fact, while stimulated emission generates photons in phase with incident photons, spontaneous emission within the laser bandwidth creates photons with random phase. However, population fluctuations leading to fluctuations in the number of *coherent* photons, besides causing intensity fluctuations, also play a role in originating phase fluctuations, since they cause fluctuations in the cavity gain and therefore, via the Kramers–Kronig relations, in the cavity effective refractive index n_{eff}. This causes in turn the cavity resonant frequency (and therefore the emission frequency) to fluctuate, thus broadening the output spectrum. Such an *indirect* broadening mechanism is proportional to a factor α_H defining the relative variation of n_r versus the gain; α_H is called the *linewidth enhancement factor* or the *Henry alpha factor*, and reads

$$\alpha_H = \frac{\mathrm{d}n_r}{\mathrm{d}n_i} \approx \frac{\Delta n_r}{\Delta n_i}, \tag{5.40}$$

where n_r and n_i are the real and imaginary parts of the refractive index, respectively (n_i can be associated with gain or with losses). The same factor also occurs in the analysis of the frequency chirp of lasers under amplitude modulation and in the chirp modeling of electrooptic and electroabsorption modulators, from which comes the alternative name of *Henry chirp factor*. The relevant analysis is presented in Section 5.10.1; a more complete treatment (based on the Langevin approach), discussed in Section 5.13.5, leads to the same results.

The laser linewidth resulting from the two mechanisms outlined above (emission of incoherent photons and linewidth enhancement caused by population fluctuations), also called *intrinsic* laser linewidth, can be shown to be (see (5.74) in Section 5.10.1)

$$\Delta f \approx \frac{1}{2\pi\tau_{ph}} \frac{hf}{2\tau_{ph} P_{out}} (1 + \alpha_H^2),$$

where P_{out} is the total output laser power.

Apparently, a further, extrinsic limitation to the laser linewidth is imposed by the resonator (i.e., the laser cavity). The resonator-limited linewidth, also called *cavity linewidth*, is, however, typically larger than the intrinsic linewidth, due to the low quality factor Q of semiconductor laser cavities. In fact, the cavity linewidth can be derived from the definition of the cavity quality factor $Q = \Delta f / f$, where Δf is the cavity FWHM bandwidth. One has immediately

$$\Delta\omega = \frac{\omega}{Q}. \tag{5.41}$$

However, the cavity quality factor is defined as

$$Q = \omega \times \frac{\text{Energy stored}}{\text{Energy lost per unit time}} = \omega \times \hbar\omega n_{ph} \times \left(\frac{\hbar\omega n_{ph}}{\tau_{ph}}\right)^{-1} = \omega\tau_{ph}, \tag{5.42}$$

where n_{ph} is the total photon number in the cavity and τ_{ph} is the photon lifetime, i.e., the average time between the photon emission and its absorption or escape from the cavity through the mirrors; see (5.33) and Section 5.6.5. From (5.41) and (5.42) we obtain

$$\Delta\omega = \frac{\omega}{Q} = \frac{1}{\tau_{ph}},$$

and, defining such a bandwidth as the *cavity linewidth* Δf_c:

$$\Delta f_c = \frac{1}{2\pi\tau_{ph}}.$$

The cavity linewidth can be extremely narrow if the cavity length is very large or the mirrors are of high reflectivity. However, in semiconductor lasers the quality factor of the cavity is low enough (low reflectivity mirrors and short cavity) to make the laser linewidth dependent on or limited by intrinsic mechanisms; see Example 5.6.[12]

Laser linewidth requirements vary significantly according to the application. In the datacom and telecom field, fiber dispersion limits the length–bandwidth product of the link; moreover, the signal bandwidth is usually dominated by the source linewidth, unless very pure laser sources are used. According to the application, the required laser linewidth may vary. For single-wavelength local area network (LAN) links, the 20–30 Å (2–3 nm) linewidth typical of FP lasers is adequate and sometimes

[12] An intuitive explanation of the linewidth narrowing with respect to the cavity linewidth can be derived by taking into account that in the presence of gain we have the approximate dynamic equation for the photon density:

$$\frac{dN}{dt} = \frac{c_0 g_c}{n_r} N - \frac{N}{\tau_{ph}} \equiv \frac{N}{\tau_{eq}},$$

where the equivalent lifetime is much larger than τ_{ph}. Thus, the effective cavity Q is actually *increased* by the gain mechanism.

the 300–500 Å (or more) linewidth typical of LEDs can suffice. Long-haul single-frequency links (with spans up to 100 km) cannot be implemented through FP lasers but require 1–2 Å (0.1–0.2 nm) linewidths that can be achieved by DFB lasers. Finally, wavelength division multiplexing (WDM) long-haul applications exploiting several carriers impose further requirements on the line stability, thus needing DFB or DBR lasers with temperature control.

Example 5.6: Consider a cavity with $L = 300\,\mu\text{m}$, effective refractive index $n_r = 3.3$, mirror power reflectivity of 30%, power 1 mW, $\alpha_H = 5$. Evaluate the cavity and intrinsic laser linewidth. The photon energy is $hf = 1$ eV.

Neglecting additional losses, we have

$$\tau_{ph} \approx \frac{n_r L}{c_0 |\log R^{-1}|} = \frac{3.3 \cdot 300 \times 10^{-6}}{3 \times 10^8 \cdot \log 0.3^{-1}} = 2.74 \times 10^{-12}\ \text{s},$$

$$\Delta f_c = \frac{1}{2\pi\tau_{ph}} = \frac{1}{2\pi \cdot 2.74 \times 10^{-12}} = 58\ \text{GHz}.$$

On the other hand, the laser linewidth is

$$\Delta f = \frac{1}{2\pi\tau_{ph}} \frac{hf}{2\tau_{ph} P_{out}} (1 + \alpha_H^2)$$

$$= \frac{1}{2\pi \cdot 2.74 \times 10^{-12}} \frac{1 \cdot 1.69 \times 10^{-19}}{2 \cdot 2.74 \times 10^{-12} \cdot 10^{-3}} (1 + 5^2) = 47\ \text{MHz}.$$

The cavity linewidth is, therefore, much larger than the actual laser linewidth.

5.10.1 Linewidth broadening analysis

We present here a classical treatment of the laser phase noise, leading to expression (5.74) for the laser linewidth.[13] The analysis is based on a number of steps, which we will briefly summarize here. We start by postulating that the optical field intensity I and phase ϕ are affected by fluctuations, δI and $\delta\phi$, respectively. Phase fluctuations originate both *directly* from spontaneous emission events, and *indirectly* from intensity fluctuations.[14] Intensity fluctuations are, in fact, related to *population fluctuations* in the cavity, which in turn cause a fluctuation in the gain and, because of the Kramers–Kronig relations, in the cavity effective index and field phase. Population fluctuations are not introduced directly, but, rather, a dynamic model for the fluctuating field is developed in order to relate $\delta\phi$ to δI through the α_H parameter. Then, the phase and amplitude

[13] The treatment closely follows the Henry papers [80], [81]. An alternative discussion of laser noise according to the Langevin approach is presented in Section 5.13.5.

[14] In the Henry theory of linewidth broadening, intensity fluctuations caused by stimulated emission are not mentioned explicitly; the final result obtained is, however, consistent with the Langevin model in which fluctuations are caused by both spontaneous and stimulated emission. It should be taken into account that the spontaneous and stimulated emission rates are anyway proportional through the photon number.

fluctuations derived by adding to the field a single photon with random phase are evaluated, and it is shown that the average rate of this photon emission should coincide with the spontaneous emission rate. Finally, the total $\delta\phi$ deriving from direct and indirect phase fluctuations is obtained and its second-order statistical properties are evaluated, assuming that $\delta\phi$ has Gaussian statistics. This allows the autocorrelation and power spectrum of the fluctuating field to be estimated in the form a Lorentzian.[15] An additional analysis allows us to express the power spectrum FWHM as in (5.74). The analysis of *intensity fluctuations* will not be completed in the present section and the treatment is left to Section 5.13.4.

We start from the electric field $E(x,t)$ (x is the longitudinal direction) of the cavity dominant mode. We assume that the fluctuations of E can be described by an envelope $E_0(t)$, slowly varying with time, as[16]

$$E(x,t) \approx E_0(t)\exp(-\mathrm{j}kx + \mathrm{j}\omega t). \tag{5.43}$$

The cavity propagation constant is $\mathrm{j}k = \mathrm{j}\omega n_{\text{eff}}/c_0$, where n_{eff} is the (generally complex) cavity effective index and c_0 is the free space light velocity. The perturbed field $E(x,t)$ should satisfy the 1D wave equation:

$$\frac{\partial^2 E}{\partial x^2} - \frac{n_{\text{eff}}^2}{c_0^2}\frac{\partial^2 E}{\partial t^2} = 0, \tag{5.44}$$

where

$$n_{\text{eff}}^2 = (n_r - \mathrm{j}n_i)^2 = n_r^2 - n_i^2 - 2\mathrm{j}n_r n_i,$$

n_r and n_i being related to the field propagation constant β and attenuation $\overline{\alpha}$ as

$$-\mathrm{j}k = -\mathrm{j}\omega n_{\text{eff}}/c_0 = -\mathrm{j}\underbrace{\omega n_r/c_0}_{\beta} - \underbrace{\omega n_i/c_0}_{\overline{\alpha}}.$$

In a laser cavity, n_i accounts for the gain $g_c = \Gamma_{ov}g$, the mirror loss α_m, and the external loss α_{loss}; since $\alpha = 2\overline{\alpha} = \alpha_{loss} + \alpha_m - g_c$ we immediately obtain

$$n_i = \frac{c_0}{2\omega}\left(\alpha_{loss} + \alpha_m - g_c\right).$$

At and above threshold the gain and losses compensate, i.e., $n_i = 0$. However, carrier populations fluctuations induce a fluctuation δn_i in n_i, so that $n_i + \delta n_i \approx \delta n_i = (c_0/2\omega)\left(\alpha_{loss} + \alpha_m - g_c\right)$. Because of the Kramers–Kronig relations, δn_i will imply a fluctuation δn_r in the real part, such as

$$\alpha_H = \frac{\delta n_r}{\delta n_i}, \tag{5.45}$$

[15] A Lorentzian spectrum has a frequency behavior $(1+\omega^2\tau^2)^{-1}$ around the central frequency, where τ is a characteristic time.

[16] Remember that we mean slow with respect to the optical field period (i.e., with frequency components below the teraherz range).

where α_H is the Henry parameter. Assuming small variations we can therefore write

$$n_{\text{eff}}^2 = (n_r + \delta n_r - \mathrm{j}\delta n_i)^2 \approx n_r^2 - 2\mathrm{j}n_r\delta n_i (1 + \mathrm{j}\alpha_H). \tag{5.46}$$

We now show that a fluctuation in the effective index will cause a fluctuation in the perturbed field E. In fact, substituting (5.46) into (5.44) we obtain

$$\frac{\partial^2 E}{\partial x^2} - \frac{n_r^2 - 2\mathrm{j}n_r\delta n_i (1 + \mathrm{j}\alpha_H)}{c_0^2}\frac{\partial^2 E}{\partial t^2} = 0. \tag{5.47}$$

We express the derivatives of E through (5.43), neglecting the second-order time derivative of $E_0(t)$, as

$$\frac{\partial^2 E}{\partial x^2} = -k^2 E_0(t)\exp(-\mathrm{j}kx + \mathrm{j}\omega t)$$
$$\frac{\partial^2 E}{\partial t^2} \approx -\omega^2 E_0(t)\exp(-\mathrm{j}kx + \mathrm{j}\omega t) + 2\mathrm{j}\omega\frac{\mathrm{d}E_0(t)}{\mathrm{d}t}\exp(-\mathrm{j}kx + \mathrm{j}\omega t).$$

Substituting into (5.47) and simplifying the exponentials we obtain

$$\left[-k^2 + \frac{n_r^2\omega^2}{c_0^2}\right]E_0(t) - \frac{n_r^2 - 2\mathrm{j}n_r\delta n_i (1 + \mathrm{j}\alpha_H)}{c_0^2}2\mathrm{j}\omega\frac{\mathrm{d}E_0(t)}{\mathrm{d}t}$$
$$- \frac{2\mathrm{j}n_r\delta n_i (1 + \mathrm{j}\alpha_H)\,\omega^2}{c_0^2}E_0(t)$$
$$\approx -\frac{2\mathrm{j}\omega n_r^2}{c_0^2}\frac{\mathrm{d}E_0(t)}{\mathrm{d}t} - \frac{2\mathrm{j}\omega^2 n_r\delta n_i (1 + \mathrm{j}\alpha_H)}{c_0^2}E_0(t) = 0,$$

where we have assumed that $2\mathrm{j}n_r\delta n_i (1 + \mathrm{j}\alpha_H) \ll n_r^2$. Expressing

$$\delta n_i = \frac{c_0}{2\omega}(\alpha_c - g_c),$$

where $\alpha_c = \alpha_{loss} + \alpha_m$, the dynamic equation for the electric field envelope assumes the form

$$\frac{\mathrm{d}E_0(t)}{\mathrm{d}t} = -\frac{c}{2}(\alpha_c - g_c)(1 + \mathrm{j}\alpha_H)E_0(t), \tag{5.48}$$

where c is the phase velocity of the mode.

Equation (5.48) includes the fluctuating term $\alpha_c - g_c$, but does not account yet for the effect of spontaneous emission and the related fluctuations (the r.h.s. term is clearly connected to absorption and stimulated emission); we will introduce this missing term at a further stage, after having expressed E_0 in terms of a normalized intensity I and a phase ϕ:

$$E_0(t) = \sqrt{I(t)}\mathrm{e}^{\mathrm{j}\phi(t)}.$$

For convenience, the field amplitude is normalized so that $|E_0|^2 \equiv I = n_{ph}$, the photon number in the cavity (the photon average density is $N = n_{ph}/V$, where V

is the equivalent cavity volume for the photon mode). Substituting into (5.48), we obtain

$$\frac{\mathrm{d}E_0}{\mathrm{d}t} = \frac{1}{2\sqrt{I}}\frac{\mathrm{d}I}{\mathrm{d}t}\mathrm{e}^{\mathrm{j}\phi} + \mathrm{j}\sqrt{I}\frac{\mathrm{d}\phi}{\mathrm{d}t}\mathrm{e}^{\mathrm{j}\phi} = -\frac{c}{2}(\alpha_c - g_c)(1 + \mathrm{j}\alpha_H)\sqrt{I}\mathrm{e}^{\mathrm{j}\phi},$$

i.e., simplifying the exponentials and separating real and imaginary parts:

$$\frac{1}{2I(t)}\frac{\mathrm{d}I(t)}{\mathrm{d}t} = -\frac{c}{2}(\alpha_c - g_c) \tag{5.49}$$

$$\frac{\mathrm{d}\phi(t)}{\mathrm{d}t} = -\frac{c}{2}(\alpha_c - g_c)\alpha_H. \tag{5.50}$$

From the ratio of (5.49) and (5.50) we obtain an auxiliary equation which relates phase and amplitude fluctuations:

$$\frac{\mathrm{d}\phi(t)}{\mathrm{d}t} = \alpha_H \frac{1}{2I(t)}\frac{\mathrm{d}I(t)}{\mathrm{d}t}, \tag{5.51}$$

from which we immediately obtain the alternative definition of the Henry parameter:

$$\alpha_H = \frac{\dfrac{\mathrm{d}\phi(t)}{\mathrm{d}t}}{\dfrac{1}{2I(t)}\dfrac{\mathrm{d}I(t)}{\mathrm{d}t}} = \frac{4\pi\Delta f(t)}{\dfrac{1}{p_{out}(t)}\dfrac{\mathrm{d}p_{out}(t)}{\mathrm{d}t}}, \tag{5.52}$$

since $p_{out} \propto I$; Δf is the instantaneous frequency deviation or chirp. From (5.51), expressing differentials in terms of variations we have

$$\delta\phi' \approx \frac{\alpha_H}{2I}\delta I, \tag{5.53}$$

where we have denoted as $\delta\phi'$ the phase fluctuation proportional to the amplitude fluctuation δI through δn_r (it vanishes, in fact, if $\alpha_H = 0$).

Let us now introduce, in the dynamic model for the intensity, the effect of spontaneous emission, including in (5.49) the spontaneous emission rate $\overline{r}_o^{sp}$ (dimension, s^{-1}) and the related intensity fluctuation F_I:

$$\frac{\mathrm{d}I(t)}{\mathrm{d}t} = -c(\alpha_c - g_c)I(t) + \overline{r}_o^{sp} + F_I. \tag{5.54}$$

Before exploiting (5.54), we evaluate the effect of the addition, to the unperturbed field, of a *single photon with random phase*. The amplitude fluctuations δI can be considered as the collective result of this process (see note 17, p. 311), which, however, also leads to the phase fluctuation $\delta\phi''$. We call this fluctuation *direct*, to distinguish it from phase fluctuations related to the amplitude fluctuations in (5.53). Suppose that the unperturbed field E_0 has $\phi = 0$ and $|E_0| = \sqrt{I}$; we add a field $\delta E_0 = \sqrt{1}\exp(\mathrm{j}\theta)$ corresponding to one photon with random phase θ. The resulting total field is

$$E_0 + \delta E_0 = \sqrt{I} + \sqrt{1}\exp(\mathrm{j}\theta) = \sqrt{I} + \sqrt{1}\cos(\theta) + \mathrm{j}\sqrt{1}\sin(\theta), \tag{5.55}$$

but we can, alternatively, write

$$E_0 + \delta E_0 = \sqrt{I + \delta I}\exp\left(\mathrm{j}\delta\phi''\right) = \sqrt{I + \delta I}\cos(\delta\phi'') + \mathrm{j}\sqrt{I + \delta I}\sin(\delta\phi''). \tag{5.56}$$

Taking the magnitude squared of (5.55), we obtain

$$\begin{aligned}|E_0 + \delta E_0|^2 &= \left(\sqrt{I}\right)^2 + 2\sqrt{I}\sqrt{1}\cos(\theta) + \left(\sqrt{1}\right)^2\cos^2(\theta) + \left(\sqrt{1}\right)^2\sin^2(\theta)\\ &= I + 2\sqrt{I}\cos(\theta) + 1 = I + \delta I,\end{aligned}$$

i.e., the amplitude fluctuation δI is

$$\delta I = 2\sqrt{I}\cos(\theta) + 1. \tag{5.57}$$

Adding photons with random phase at times t_i can therefore be modeled as a photon generation process with generation rate

$$\left.\frac{\mathrm{d}I}{\mathrm{d}t}\right|_{em} = \sum_i 2\sqrt{I}\cos(\theta_i)\delta(t - t_i) + \sum_i \delta(t - t_i) = \left\langle \left.\frac{\mathrm{d}I}{\mathrm{d}t}\right|_{em} \right\rangle + F_I, \tag{5.58}$$

where the term in angular brackets is the (ensemble) average emission rate and F_I is the zero-average fluctuation. One has, with reference to an average generation rate G (for the moment to be determined):

$$\begin{aligned}\left\langle \left.\frac{\mathrm{d}I}{\mathrm{d}t}\right|_{em} \right\rangle &= \lim_{T\to\infty}\frac{1}{T}\int_0^T \sum_i^{GT} 2\sqrt{I}\cos(\theta_i)\delta(t - t_i)\,\mathrm{d}t\\ &+ \lim_{T\to\infty}\frac{1}{T}\int_0^T \sum_i^{GT} \delta(t - t_i)\,\mathrm{d}t = \lim_{T\to\infty}\frac{1}{T}\sum_i^{GT} 2\sqrt{I}\cos(\theta_i) + G = G,\end{aligned}$$

since θ_i is uniformly distributed. Indeed, G is the photon generation rate to be included in (5.54) besides stimulated emission; thus we identify $G = \overline{r}_o^{sp}$, the spontaneous emission rate.

We now come back to the direct phase fluctuation $\delta\phi''$. Comparing the imaginary parts in (5.55) and (5.56) we obtain, assuming small $\delta\phi''$:

$$\sin(\delta\phi'') \approx \delta\phi'' = \frac{1}{\sqrt{I + \delta I}}\sin(\theta) \approx \frac{1}{\sqrt{I}}\sin(\theta). \tag{5.59}$$

We can now express the *total* phase fluctuation $\delta\phi' + \delta\phi''$ arising from the addition of a single photon with random phase; $\delta\phi''$ was evaluated directly from (5.59), while $\delta\phi'$ is recovered through (5.53):[17]

$$\delta\phi = \delta\phi' + \delta\phi'' = \frac{\alpha_H}{2I}\left(2\sqrt{I}\cos(\theta) + 1\right) + \frac{1}{\sqrt{I}}\sin(\theta)$$
$$= \frac{\alpha_H}{2I} + \frac{1}{\sqrt{I}}\left[\sin(\theta) + \alpha_H\cos(\theta)\right].$$

We have expressed the phase fluctuation $\delta\phi$ in terms of the single photon random phase θ. We will now derive for $\delta\phi$ suitable statistical properties.

According to the previous identification $G \equiv \overline{r}_o^{sp}$, during a certain observation time t, a number of photons with random phase adds to the initial field, with a rate:

$$\overline{r}_o^{sp} = V_{ac}\beta_k R_o^{sp} \approx V_{ac}\beta_k n/\tau_n^{sp}, \tag{5.60}$$

where V_{ac} is the active region volume; β_k is the so-called *spontaneous emission factor* (expressing the relative amount of photons emitted by spontaneous emission into the lasing mode k);[18] $R_o^{sp} \approx n/\tau_n^{sp}$ is the electron spontaneous recombination rate per unit volume, where n is the cavity electron concentration; τ_n^{sp} the spontaneous lifetime. In what follows we will also exploit the electron spontaneous recombination rate (per unit volume) leading to spontaneous emission into the lasing mode k:

$$\overline{R}_o^{sp} = \beta_k R_o^{sp}. \tag{5.61}$$

The total number of photons emitted within the laser bandwidth at t is therefore $\overline{r}_o^{sp}t$. The total phase fluctuation $\delta\phi$ over the observation time t can be now expressed by adding all individual phase fluctuations:

$$\delta\phi = \sum_{i=1}^{\overline{r}_o^{sp}t}\delta\phi_i = \sum_{i=1}^{\overline{r}_o^{sp}t}\left\{\frac{\alpha_H}{2I} + \frac{1}{\sqrt{I}}\left[\sin(\theta_i) + \alpha_H\cos(\theta_i)\right]\right\}. \tag{5.62}$$

From (5.62), we can derive the statistical properties of $\delta\phi$, having made some assumptions on the statistics of θ_i. Assuming that the process $\delta\phi$ is (at least second-order) ergodic (i.e., time and ensemble averages coincide) and denoting the time average of $a(t)$ as $\langle a\rangle$ and the ensemble average as $\overline{a}$, we can now derive the average phase

[17] According to the present approach, intensity fluctuations are derived from a field analysis rather than from a photon number analysis. This implies that the equivalent emission rate considered is associated with spontaneous emission only (see, e.g., [82], Ch. 7). In fact, intensity or photon number fluctuations introduced this way also account for stimulated emission. Section 5.13 introduces a different approach to photon number fluctuations, in which δI also derives from fluctuations in the stimulated emission; as shown in Example 5.8, the two approaches are completely equivalent.

[18] In fact $\overline{r}_o^{sp}$ is the spontaneous recombination rate for the (single wavelength and frequency) photons in the lasing mode k, derived from the total spontaneous recombination rate through a proper weight β_k.

deviation $\langle \delta\phi \rangle$ and the quadratic mean of $\delta\phi$, $\left\langle (\delta\phi)^2 \right\rangle$. Since θ_i is uniformly distributed between 0 and 2π, $\langle \sin(\theta_i) \rangle = \langle \cos(\theta_i) \rangle = 0$; we then have:

$$\begin{aligned}
\langle \delta\phi \rangle &= \sum_{i=1}^{\overline{r}_o^{sp} t} \frac{\alpha_H}{2I} + \left\langle \sum_{i=1}^{\overline{r}_o^{sp} t} \frac{1}{\sqrt{I}} \left[\sin(\theta_i) + \alpha_H \cos(\theta_i) \right] \right\rangle \\
&= \frac{\alpha_H}{2I} \overline{r}_o^{sp} t + \sum_{i=1}^{\overline{r}_o^{sp} t} \frac{1}{\sqrt{I}} \left[\langle \sin(\theta_i) \rangle + \alpha_H \langle \cos(\theta_i) \rangle \right] = \frac{\alpha_H}{2I} \overline{r}_o^{sp} t,
\end{aligned} \tag{5.63}$$

i.e., the average phase deviation linearly increases with time. Concerning the quadratic mean, we obtain, taking into account that the phases of different photons are uncorrelated:

$$\begin{aligned}
\left\langle (\delta\phi)^2 \right\rangle &= \left\langle \left\{ \sum_{i=1}^{\overline{r}_o^{sp} t} \frac{\alpha_H}{2I} + \sum_{i=1}^{\overline{r}_o^{sp} t} \frac{1}{\sqrt{I}} \left[\sin(\theta_i) + \alpha_H \cos(\theta_i) \right] \right\} \right. \\
&\quad \times \left. \left\{ \sum_{j=1}^{\overline{r}_o^{sp} t} \frac{\alpha_H}{2I} + \sum_{j=1}^{\overline{r}_o^{sp} t} \frac{1}{\sqrt{I}} \left[\sin(\theta_j) + \alpha_H \cos(\theta_j) \right] \right\} \right\rangle \\
&= \left(\frac{\alpha_H}{2I} \overline{r}_o^{sp} t \right)^2 + \left\langle \sum_{i=1}^{\overline{r}_o^{sp} t} \frac{1}{I} \left[\sin^2(\theta_i) + \alpha_H^2 \cos^2(\theta_i) \right] \right\rangle \\
&= \left(\frac{\alpha_H}{2I} \overline{r}_o^{sp} t \right)^2 + \frac{1}{I} \left(\frac{1}{2} + \frac{1}{2} \alpha_H^2 \right) \overline{r}_o^{sp} \, |t| \,.
\end{aligned}$$

Note the absolute value $|t|$, implying that the quadratic mean should always be positive.

The average phase deviation $\langle \delta\phi \rangle$ in (5.63) increases linearly with time; this corresponds to a constant deviation in the average instantaneous frequency:

$$\delta f = \frac{1}{2\pi} \frac{\mathrm{d} \langle \delta\phi \rangle}{\mathrm{d}t} = \frac{1}{2\pi} \frac{\mathrm{d}}{\mathrm{d}t} \left(\frac{\alpha_H}{2I} \overline{r}_o^{sp} t \right) = \frac{\alpha_H}{4\pi I} \overline{r}_o^{sp}, \tag{5.64}$$

but does *not* imply linewidth broadening. On the other hand, the variance $\sigma_{\delta\phi}$ of the total phase fluctuations:

$$\sigma_{\delta\phi} = \left\langle (\delta\phi)^2 \right\rangle - \langle \delta\phi \rangle^2 = \frac{1 + \alpha_H^2}{2I} \overline{r}_o^{sp} \, |t| \tag{5.65}$$

also increases linearly with time. We will now show that such a behavior is compatible with a finite linewidth and a Lorentzian power spectrum.

To this purpose, we assume that $\delta\phi$ has a Gaussian probability distribution; since $\langle \delta\phi \rangle$ leads to a *constant frequency deviation* and has therefore no impact on linewidth broadening, we will neglect it and write the Gaussian probability distribution of $\delta\phi$ as

$$P(\delta\phi) = \frac{1}{\sqrt{2\pi\sigma_{\delta\phi}}} \exp\left[-\frac{(\delta\phi)^2}{2\sigma_{\delta\phi}} \right], \tag{5.66}$$

where $\sigma_{\delta\phi}$ depends on the observation time according to (5.65). Let us turn now to the electric field (without loss in generality, we suppress the space dependence):

$$E(t) = \sqrt{I(t)}e^{j\phi(t)}e^{j\omega t} = \sqrt{I(t)}e^{j\delta\phi(t)}e^{j\omega_0 t},$$

where ω_0 takes into account the constant frequency deviation from (5.64). To obtain the laser linewidth, we need to evaluate the *power spectrum* of E, which can be expressed as the Fourier transform of the field *autocorrelation function*:

$$\begin{aligned} R_E(\tau) &= \left\langle E(t+\tau)E^*(t)\right\rangle = \left\langle \sqrt{I(t+\tau)}e^{j\delta\phi(t+\tau)}e^{j\omega_0(t+\tau)}\sqrt{I(t)}e^{-j\delta\phi(t)}e^{-j\omega_0 t}\right\rangle \\ &\approx I e^{j\omega_0\tau}\left\langle e^{j[\delta\phi(t+\tau)-\delta\phi(t)]}\right\rangle = I e^{j\omega_0\tau}\left\langle e^{j\delta\phi(\tau)}\right\rangle, \end{aligned} \tag{5.67}$$

where we have neglected amplitude fluctuations. Since $\delta\phi(t)$ is stationary,[19] the statistical properties of $\delta\phi(t+\tau) - \delta\phi(t)$ are independent of t and therefore coincide with those of $\delta\phi(\tau) - \delta\phi(0) \equiv \delta\phi(\tau)$, assuming the phase in $t = 0$ as the reference phase.

The time average $\langle \exp(j\delta\phi(\tau))\rangle$ can be now evaluated (from ergodicity) as an *ensemble average*, exploiting (5.66):

$$\begin{aligned} \left\langle e^{j\delta\phi(\tau)}\right\rangle &= \overline{e^{j\delta\phi(\tau)}} = \int_{-\infty}^{\infty} f(\delta\phi)e^{j\delta\phi}\, d\delta\phi = \frac{1}{\sqrt{2\pi\sigma_{\delta\phi}}}\int_{-\infty}^{\infty} e^{-(\delta\phi)^2/2\sigma_{\delta\phi}}e^{j\delta\phi}\, d\delta\phi \\ &= \frac{1}{\sqrt{2\pi\sigma_{\delta\phi}}}\int_{-\infty}^{\infty} e^{-(\delta\phi)^2/2\sigma_{\delta\phi}}\cos(\delta\phi)\, d\delta\phi = e^{-\sigma_{\delta\phi}/2}, \end{aligned} \tag{5.68}$$

where the following integral has been used

$$\int_0^{\infty} e^{-a^2x^2}\cos(bx)\, dx = \frac{\sqrt{\pi}}{2a}\exp\left(-\frac{b^2}{4a^2}\right).$$

In conclusion, from (5.67), (5.68), and (5.65), the auto-correlation function is

$$R_E(\tau) = I e^{j\omega_0\tau}\exp\left(-\frac{1+\alpha_H^2}{4I}\overline{r}_o^{sp}\,|\tau|\right) = I e^{j\omega_0\tau}e^{-|\tau|/\tau_c}, \tag{5.69}$$

where we have introduced the *coherence time* τ_c:

$$\frac{1}{\tau_c} = \frac{1+\alpha_H^2}{4I}\overline{r}_o^{sp}. \tag{5.70}$$

[19] Remember that we have removed the average phase deviation $\langle\delta\phi\rangle$ from (5.63), which increases with time.

We can now evaluate the power spectrum of the field E by taking the Fourier transform of $R_E(\tau)$ from (5.69):

$$S_E(\omega) = \int_{-\infty}^{\infty} I e^{j\omega_0\tau} e^{-\frac{|\tau|}{\tau_c}} e^{-j\omega\tau}\, d\tau$$
$$= I \int_{-\infty}^{0} e^{\frac{\tau}{\tau_c}} e^{-j(\omega-\omega_0)\tau}\, d\tau + I \int_{0}^{\infty} e^{-\frac{\tau}{\tau_c}} e^{-j(\omega-\omega_0)\tau}\, d\tau$$
$$= I \frac{\tau_c + j(\omega-\omega_0)\tau_c^2}{1+(\omega-\omega_0)^2\tau_c^2} - I \frac{j(\omega-\omega_0)\tau_c^2 - \tau_c}{1+(\omega-\omega_0)^2\tau_c^2} = \frac{2\tau_c I}{1+(\omega-\omega_0)^2\tau_c^2},$$

which is a Lorentzian spectrum centered around ω_0 with full width at half maximum (FWHM):

$$\Delta\omega = \frac{2}{\tau_c}.$$

It follows that the laser linewidth can be finally expressed as

$$\Delta f = \frac{1+\alpha_H^2}{4\pi I} \overline{r}_o^{sp}. \tag{5.71}$$

A more suitable formulation can be derived from (5.71) by expressing the field intensity in terms of the laser output power, and by replacing $\overline{r}_o^{sp}$ with a more convenient parameter. Neglecting external losses vs. mirror losses, we have

$$P_{out} \approx \frac{hf n_{ph}}{\tau_{ph}} \equiv \frac{hf I}{\tau_{ph}} \rightarrow \frac{1}{I} \approx \frac{hf}{\tau_{ph} P_{out}},$$

from which (5.71) becomes

$$\Delta f = \frac{1+\alpha_H^2}{4\pi} \frac{hf}{\tau_{ph} P_{out}} \overline{r}_o^{sp}. \tag{5.72}$$

Note that, if the power is referred to a single facet (assuming equal facets), one has $P_{out} = 2P'_{out}$, and therefore the equation is modified as

$$\Delta f = \frac{1+\alpha_H^2}{8\pi} \frac{hf}{\tau_{ph} P'_{out}} \overline{r}_o^{sp}.$$

However, if the two facets are highly asymmetrical (5.72) can be directly applied.

A more suitable expression for the spontaneous emission scattering rate $\overline{r}_o^{sp}$ was developed in Section 2.4.5. From (2.38), we have

$$\overline{r}_o^{sp} = \frac{c_0}{n_r} g n_{sp}, \tag{5.73}$$

where n_{sp} is the nondimensional *spontaneous emission factor* defined in (2.39); we have $n_{sp} > 1$ but, typically, $n_{sp} \approx 2-3$. Exploiting (5.73), we finally obtain

$$\Delta f = \frac{1+\alpha_H^2}{4\pi} \frac{hf}{\tau_{ph} P_{out}} \frac{c_0}{n_r} g n_{sp}.$$

Further approximations are possible if we take into account that $g \approx g_{th}$ and that, at threshold, the gain compensates for the total loss:

$$\frac{c_0}{n_r} g \approx \frac{c_0}{n_r} g_{th} = \frac{1}{\tau_{ph}}.$$

We thus obtain:

$$\Delta f \approx \frac{1 + \alpha_H^2}{4\pi} \frac{hf n_{sp}}{\tau_{ph}^2 P_{out}},$$

which, for $n_{sp} \approx 1$, yields the simplified expression

$$\Delta f = \frac{1}{2\pi \tau_{ph}} \frac{hf}{2\tau_{ph} P_{out}} (1 + \alpha_H^2). \tag{5.74}$$

Notice that a different choice of the n_{sp} value leads to linewidths differing by a factor of 2 or 3 with respect to (5.74).

5.11 Laser dynamics and modulation response

A digital data stream can be transmitted through a fiber by switching the source on (ones) and off (zeros), i.e., through the *direct modulation* of the laser. In LEDs, direct modulation leads to low maximum speed or modulation bandwidth, limited to a few hundred Mbps. The situation is potentially better in lasers, due to the expected benefits from stimulated emission, leading to a shorter carrier lifetime vs. the spontaneous lifetime. However, on–off (or *large-signal*) laser modulation can be carried out in two ways: (a) both the zero and the one levels are above threshold; (b) only the one level is above threshold, while the zero level is below (see Fig. 5.33, left). The second alternative is inefficient from the standpoint of speed, since, before turn-on, the laser is dominated by spontaneous recombination; the turn-on time delay is, therefore, of the order of magnitude of the spontaneous lifetime. To model large-signal operation above threshold, let us consider a simplified situation in which the laser undergoes small-signal modulation with respect to a bias point above threshold (Fig. 5.33, right).

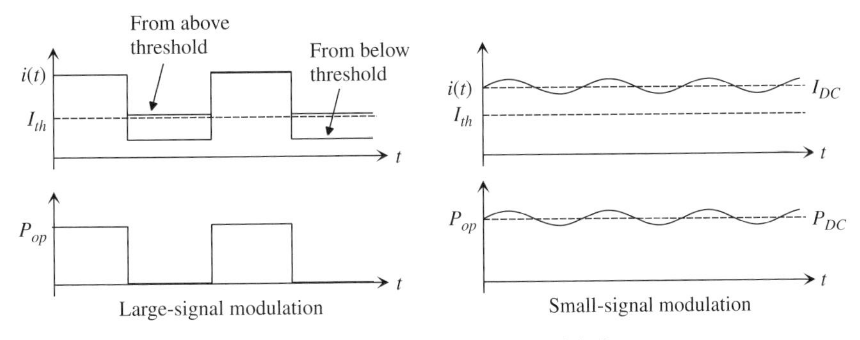

Figure 5.33 Large-signal (left) and small-signal (right) laser direct modulation.

We anticipate here some conclusions of the analysis. Laser dynamics is generally described by a dynamic system in the electron and photon densities, that is (at least) second order. Such a system is nonlinear, but can be linearized for small-signal analysis. Second-order systems, however, allow for an oscillatory behavior of the solution; we therefore expect that such a resonant behavior will be found both in the time-domain on–off switching and in the small-signal response.

Assuming a small-signal resonant behavior, i.e., a low-pass transfer function (between the input laser current and output power) described by two complex conjugate poles, the cutoff frequency will be decreasing (as the stimulated lifetime) with increasing photon density and increasing current. In general, this leads to a faster dynamics than for the LED, with record cutoff frequencies of the order of 30–40 GHz. The theoretical (intrinsic) maximum modulation frequency, however, is ultimately limited (as the stimulated lifetime) by gain compression. Moreover, laser amplitude modulation implies spurious phase and frequency modulation, denoted as *chirp*. Chirp leads to source broadening and limits the practical modulation speed to about 10 Gbps. Direct laser modulation cannot so far cover long-haul applications, for which external modulation is needed. A qualitative example of large-signal laser dynamic response to a current pulse (from below threshold) is shown in Fig. 5.34; the initial delay can be suppressed if the laser is driven by a pulse from above threshold.

The laser small-signal frequency response (see Section 5.12) is described by the normalized transfer function:

$$m(\omega) = \left|\frac{P_{out}(\omega)}{P_{out}(0)}\right| = \left|\frac{\widehat{N}_k(\omega)}{\widehat{N}_k(0)}\right| = \left|\frac{\omega_r^2}{\left(\omega_r^2 - \omega^2\right) + \mathrm{j}\omega\gamma}\right|, \tag{5.75}$$

where $\widehat{N}_k(\omega)$ is the small-signal component of the photon density relative to the dominant lasing mode, $P_{out}(\omega)$ is the small-signal laser output power, and the resonant frequency and damping factor are given (see (5.106), (5.107), (5.108)) by

$$\omega_r = \sqrt{\left.\frac{\mathrm{d}g}{\mathrm{d}n}\right|_{n_0} \frac{N_{k0}}{1 + \epsilon_c N_{k0}} \frac{c_0}{n_r} \frac{1}{\tau_{ph}}} \tag{5.76}$$

$$\gamma = K'\omega_r^2 + \frac{1}{\tau_n}, \quad K' = \frac{K}{(2\pi)^2} = \tau_{ph} + \frac{\epsilon_c n_r}{\mathrm{d}g/\mathrm{d}n|_{n_0} c_0}, \tag{5.77}$$

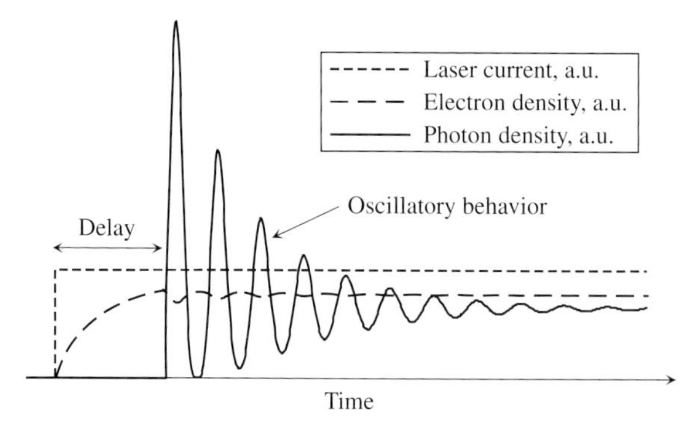

Figure 5.34 Qualitative behavior of the laser response to a current pulse (from subthreshold).

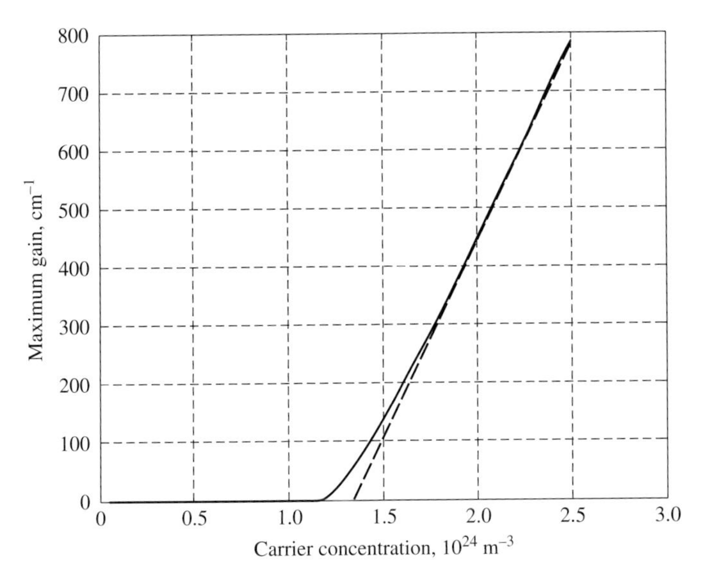

Figure 5.35 Behavior of the maximum gain vs. the electron density; the dotted line is the linear approximation assuming constant differential gain.

where τ_n is the total electron lifetime resulting from spontaneous and nonradiative recombination, and τ_{ph} is the photon lifetime (5.31); $n_r = n_{\text{eff}}$ is the lasing mode effective index, $N_{\lambda 0}$ is the DC photon density of the lasing mode. The parameter $\mathrm{d}g/\mathrm{d}n$ is the variation of gain with respect to the carrier density (*differential gain*); by plotting the maximum gain as a function of the electron density we can derive the linear approximation

$$g_{\max}(n) \approx a(n - n_{tr}),$$

where a is the differential gain and n_{tr} is the transparency charge density; see Fig. 5.35 for an example referring to bulk GaAs.

Since the DC photon density is proportional to the output power, and therefore to the DC bias current, the resonant frequency (and therefore the laser bandwidth) increases with the square root of bias current according to the law

$$f_r \propto \sqrt{I - I_{th}}.$$

However, for large current and photon densities the 3 dB cutoff frequency tends to saturate. Taking into account that the maximum 3 dB bandwidth is achieved when $2\omega_r = \gamma$ and is equal to the resonant frequency f_r, we find that the theoretical intrinsic maximum corresponds to the value (see (5.109)):

$$f_{3\text{dB,max}} = \frac{1}{\sqrt{2}\pi \left(\tau_{ph} + \dfrac{\epsilon_c n_{\text{eff}}}{\mathrm{d}g/\mathrm{d}n|_{n_0} c_0} \right)}.$$

The laser modulation bandwidth (see also (5.76)) is therefore positively affected by a large value of the differential gain, while it is limited by gain compression also for vanishingly small photon lifetime. Gain compression has, however, a positive influence on

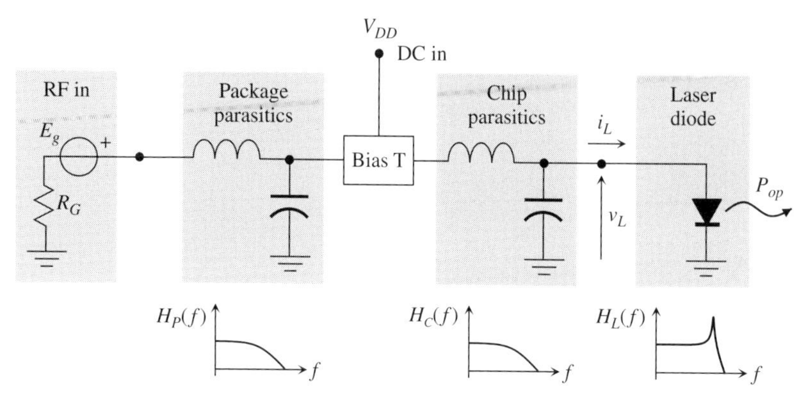

Figure 5.36 Intrinsic (laser) and extrinsic (chip, package and bias T) contributions to the laser response.

the response damping, as seen from (5.77); note that the damping enhancement provided by gain compression is constant for a given structure, i.e., it does not depend directly on laser bias.

Although $f_{3\mathrm{dB,max}}$ can, in theory, be as high as 100 GHz, in practice the maximum modulation bandwidths are in the range 20–40 GHz. Moreover, further effects tend to limit the laser speed under direct modulation, like transport effects (see e.g., [83], Section 5.6) and the effect of package and chip parasitics. Figure 5.36 shows the equivalent circuit of a junction laser, where the package and chip parasitics introduce low-pass filtering. External parasitics can, in practice, limit the frequency response of the laser. Careful RF design has to be implemented to allow the laser to approach the ideal intrinsic response.

As already recalled, the laser amplitude modulation also implies spurious phase (frequency or wavelength) modulation (*chirp*). Chirp ultimately arises because the laser current modulation changes the gain profile, but also (from the Kramers–Kronig relations) the real part of the refractive index. The relative variation $\Delta n'/\Delta g = (\Delta n_r/\Delta n)\,/\,(\Delta g/\Delta n)$, where n is the carrier population and n' is the real part of the material refractive index, is expressed again through the Henry parameter α_H, which we have already introduced as the line enhancement factor. Common values of this parameter are 3–6 in lasers. In Fig. 5.37 a typical plot of the gain vs. the photon energy is shown; note that the differential gain (proportional, with a constant Δn, to the absolute variation of the gain Δg) has a maximum for energies slightly above the maximum gain energy. Owing to the profile of the real refractive index variation (resulting from Kramers–Kronig) we obtain that the chirp parameter is typically positive in lasers, apart from high energies for which it may become negative.

The Henry parameter is related to the frequency deviation according to the following relation, already introduced in the discussion of the linewidth, see (5.45) and (5.52):

$$\alpha_H = \frac{\mathrm{d}n'}{\mathrm{d}n''} = -\frac{4\pi}{\lambda_0}\frac{\mathrm{d}n'/\mathrm{d}n}{\mathrm{d}g/\mathrm{d}n} = -\frac{4\pi}{\lambda_0}\frac{\mathrm{d}n'}{\mathrm{d}g} = -\frac{4\pi}{\lambda_0}\frac{\mathrm{d}n_{\mathrm{eff}}}{\mathrm{d}g_c} = 4\pi\frac{\Delta f(t)}{\dfrac{1}{p_{out}}\dfrac{\mathrm{d}p_{out}(t)}{\mathrm{d}t}}. \quad (5.78)$$

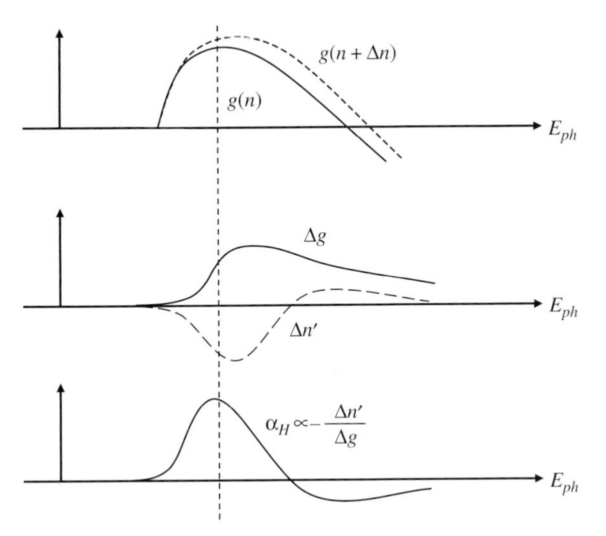

Figure 5.37 The Henry chirp parameter as a function of photon energy, as derived from the typical gain profile vs. energy and vs. the active layer electron concentration. The vertical line denotes the gain maximum, where α_H is typically positive.

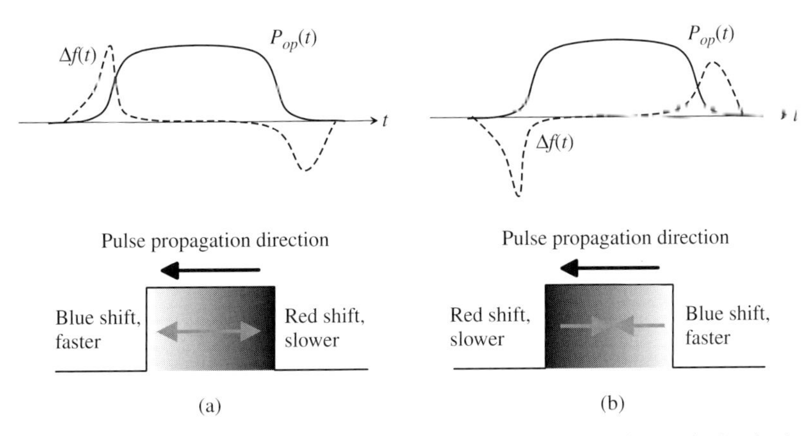

Figure 5.38 Effect of positive (a) and negative (b) chirp on the transmission of an optical pulse in a conventional dispersive fiber.

The meaning of positive and negative chirp parameter is further illustrated in Fig. 5.38. Since the frequency deviation is proportional (through α_H) to the relative amplitude variation with time, a *positive* α_H implies that, in the presence of a square amplitude pulse, the frequency deviation is positive (blue shift) on the rising pulse front and negative (red shift) on the settling pulse front; *negative* chirp leads to the opposite behavior. In optical fibers operating at $1.5\,\mu$m wavelength, the group refractive index n_g typically increases with λ; red shift therefore increases n_g and decreases the group velocity;

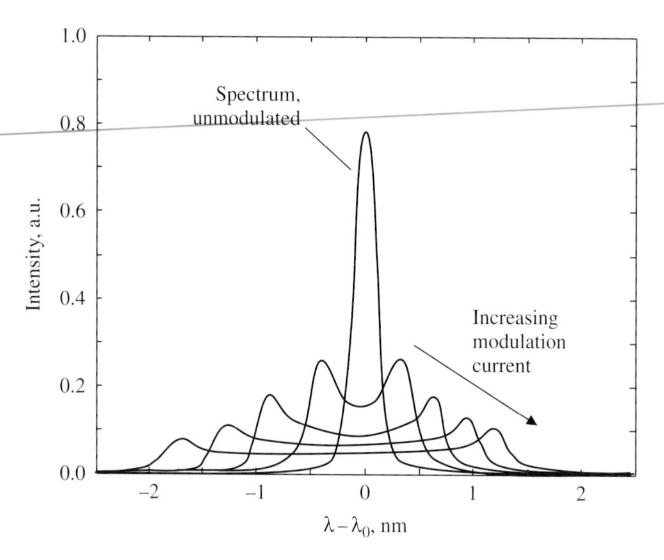

Figure 5.39 Measured spectrum of a 1.3 μm InGaAsP laser under sinusoidal modulation at 100 MHz. Increasing the modulation current leads to spectrum broadening, associated with chirp. Reused with permission from [84], Fig. 1. (©1984 American Institute of Physics).

blue shift, on the contrary, increases the group velocity.[20] If we consider the pulse as traveling in space, positive chirp leads to a fast (blue shift) pulse head and slow (red shift) pulse tail; this implies pulse broadening and negatively adds to the effect of fiber dispersion. Negative chirp, on the other hand, leads to a slow pulse head and to a fast pulse tail; the resulting pulse compression can compensate for fiber dispersion. In conclusion, a controlled amount of negative chirp can be often favorable to system design, while positive chirp is detrimental. Lasers are characterized by a somewhat large positive chirp, and therefore their direct modulation does not allow for high speed and/or long distance fiber links.

[20] Let n_{eff} be the effective refractive index in a single-mode fiber; the group velocity is

$$v_g = \frac{c_0}{n_g},$$

with the group refractive index defined as

$$n_g = n_{\text{eff}} + \omega \frac{\mathrm{d}n_{\text{eff}}}{\mathrm{d}\omega} = n_{\text{eff}} - \lambda \frac{\mathrm{d}n_{\text{eff}}}{\mathrm{d}\lambda}.$$

The wavelength variation of the group velocity with λ is often defined through the dispersion parameter D:

$$D = \frac{1}{c}\frac{\mathrm{d}n_g}{\mathrm{d}\lambda} = -\frac{\lambda}{c}\frac{\mathrm{d}^2 n_{\text{eff}}}{\mathrm{d}\lambda^2}.$$

Note that D includes the effect of both the material and modal dispersion. In conventional fibers, $D \approx 0$ around 1.3 μm and positive for longer wavelengths, implying increasing n_g and decreasing v_g vs. λ.

Finally, using (5.78) in small-signal conditions where $P_{out} \propto I_{DC} + i(t) \approx I_{DC}$, $\mathrm{d}p_{out}/\mathrm{d}t \propto \mathrm{d}i/\mathrm{d}t$, it follows that

$$\Delta f = \frac{\alpha_H}{4\pi}\Gamma_{ov} a \frac{c_0}{n_r}\Delta n = \frac{\alpha_H}{4\pi}\frac{1}{I_{DC}}\frac{\mathrm{d}i(t)}{\mathrm{d}t}. \qquad (5.79)$$

The chirp is therefore proportional to the injected charge variation and to the input current modulation amplitude $i(t)$. A larger amplitude leads to a larger chirp, as shown in Fig. 5.39 [84], and so does a larger modulation frequency.

The chirp introduced so far is denoted as *transient chirp*, since the frequency deviation vanishes in DC steady-state conditions. However, if we apply a current step to a laser, the final frequency will be slightly different with respect to the initial one, mainly due to the effect of gain compression; the resulting chirp is called *adiabatic chirp* and is an (unwanted) peculiarity of lasers under large-signal modulation.

5.12 Dynamic large-signal and small-signal laser modeling

In this section, we develop a simple, single-mode, laser dynamic model, which will be exploited to derive DC properties (already analyzed on a heuristic base) and a small-signal frequency-domain model, yielding the laser frequency response. The laser chirp can be also derived as a result of the small-signal analysis.

It is convenient to recall a number of definitions and parameters, which will be exploited in the following discussion:

- N_k is the photon density of the k-th mode (the dominant lasing mode); $N_P \equiv n_{ph}$ is the total photon number in the k-th mode.
- $n \approx p$ (n electron concentration) holds in the active region (quasi-neutrality); N is the total electron number in the active region.
- d is the thickness of laser active region.
- $g(\hbar\omega_k) \approx a(n - n_{tr})(1 + \epsilon_c N_k)^{-1}$ is the (maximum) material gain as a function of the electron density, accounting for gain compression.
- Γ_{ov} is the overlap integral.
- $n_r \equiv n_{\mathrm{eff}}$ is the effective refractive index of the lasing mode.
- c_0 is the speed of light in vacuo.
- β_k is the spontaneous emission factor for mode k; it describes the amount of the spontaneous emission spectrum falling within the linewidth of the lasing mode.
- $\tau_r \equiv \tau_n^{sp}$ is the spontaneous radiative lifetime.
- τ_{nr} is nonradiative electron lifetime (thermal and Auger); it typically depends on the electron concentration.
- τ_{ph} is photon lifetime due to losses outside the active region (α_{loss}) and mirror (or end) loss, see (5.31). The total lifetime can be split into the external loss (5.32) and mirror (5.33) contributions as $\tau_{ph}^{-1} = \tau_{loss}^{-1} + \tau_m^{-1}$.
- J is the injected current density.

The definition of the electron density n and of photon density N_k requires some care. In the present model n is an average value, with reference to the (almost constant) carrier density in the active region. Given N as the total electron number in the active region, we therefore simply define

$$n = N/V_{ac}, \tag{5.80}$$

where $V_{ac} = Ad$ is the active region volume. Concerning the photon density N_k, this can be interpreted as the *average density* of photons *in the active region*. Taking into account that the local photon density n_k is proportional to the energy of the optical field, we can assume (AR stands for active region, C for the whole cavity):

$$n_k = \frac{|E_{op}|^2 N_P}{\int_C |E_{op}|^2 \, dV}, \quad N_k = \frac{\int_{AR} n_k \, dV}{\int_{AR} dV} = \frac{\int_{AR} |E_{op}|^2 \, dV}{\int_C |E_{op}|^2 \, dV} \frac{N_P}{V_{ac}} = \frac{\Gamma_{ov}}{V_{ac}} N_P \equiv \frac{N_P}{V},$$

where the overlap integral is expressed as

$$\Gamma_{ov} = \frac{\int_{AR} |E_{op}|^2 \, dV}{\int_C |E_{op}|^2 \, dV},$$

and V is the modal volume of photon mode k:

$$V = V_{ac}/\Gamma_{ov}, \tag{5.81}$$

such as

$$N_k = N_P/V. \tag{5.82}$$

Let us now introduce two conservation equations (or rate equations), one for the charge (electron) density $n \approx p$, the other for the photon density N_k. The charge control equation includes electron generation due to current injection,

$$I = \frac{dQ_n}{dt} = \frac{d(Adqn)}{dt} = Adq\frac{dn}{dt} \rightarrow \left.\frac{dn}{dt}\right|_{gen} = \frac{I}{qdA} = \frac{J}{qd},$$

and all sources of electron recombination, i.e., recombination due to spontaneous and stimulated emission, and nonradiative recombination:

$$\frac{dn}{dt} = \underbrace{\frac{J}{qd}}_{\text{current injection}} \underbrace{\frac{n}{\tau_r}}_{\text{spontaneous recombination}} - \underbrace{g(\hbar\omega_k)N_k\frac{c_0}{n_r}}_{\text{stimulated recombination}} - \underbrace{\frac{n}{\tau_{nr}}}_{\text{nonradiative recombination}}. \tag{5.83}$$

To derive a similar rate equation for the photon density of the lasing mode k, N_k, we exploit the photon flux continuity equation (2.25). Since the photon density increases (because of stimulated emission) as $N_k = N_k(0)\exp(\Gamma_{ov} g x)$, the continuity equation for the photon flux $\Phi_{ph} = N_k c_0/n_r$ yields

$$\left.\frac{dN_k}{dt}\right|_{gain} = \frac{d\Phi_{ph}}{dx} = \Gamma_{ov} g N_k \frac{c_0}{n_r}. \tag{5.84}$$

The source term for the photon density associated with stimulated emission is, apart from the weight Γ_{ov}, the same as the stimulated recombination term in the electron density rate equation. The same remark holds for the spontaneous emission term (proportional to the nonradiative recombination term in the electron density rate equation through Γ_{ov}), but we further have to properly weight the contribution of this wideband spectrum over the narrowband laser emission through the parameter β_k:

$$\left.\frac{\mathrm{d}N_k}{\mathrm{d}t}\right|_{\mathrm{sp}} = \Gamma_{ov}\beta_k\frac{n}{\tau_r}. \tag{5.85}$$

Photon loss is suitably described through the already defined photon lifetimes τ_{ph}, see Section 5.6.5 and in particular (5.32) and (5.33), as

$$\left.\frac{\mathrm{d}N_k}{\mathrm{d}t}\right|_{\mathrm{loss}} = -\frac{N_k}{\tau_{ph}} = -\frac{N_k}{\tau_{loss}} - \frac{N_k}{\tau_m}. \tag{5.86}$$

Assembling (5.84), (5.85), and (5.86) we finally have

$$\frac{\mathrm{d}N_k}{\mathrm{d}t} = \underbrace{\Gamma_{ov}\beta_k\frac{n}{\tau_r}}_{\substack{\text{spontaneous}\\ \text{emission}}} + \underbrace{\Gamma_{ov}gN_k\frac{c_0}{n_r}}_{\substack{\text{stimulated}\\ \text{emission}}} - \underbrace{\frac{N_k}{\tau_{ph}}}_{\substack{\text{photon}\\ \text{total loss}}}. \tag{5.87}$$

Equations (5.83) and (5.87) yield the coupled large-signal (nonlinear) dynamic model:

$$\frac{\mathrm{d}N_k}{\mathrm{d}t} = \Gamma_{ov}\beta_k\frac{n}{\tau_r} + \Gamma_{ov}g(\hbar\omega_k)N_k\frac{c_0}{n_r} - \frac{N_k}{\tau_{ph}} \tag{5.88}$$

$$\frac{\mathrm{d}n}{\mathrm{d}t} = \frac{J}{qd} - \frac{n}{\tau_r} - g(\hbar\omega_k)N_k\frac{c_0}{n_r} - \frac{n}{\tau_{nr}}. \tag{5.89}$$

5.12.1 Steady-state (DC) solution

Assume that we neglect, at DC, the effect of spontaneous emission on the photon rate equation, and the effect of nonradiative recombination in the electron rate equation; from (5.88) and (5.89) we have

$$\frac{\mathrm{d}N_k}{\mathrm{d}t} \approx \Gamma_{ov}g(\hbar\omega_k)N_k\frac{c_0}{n_r} - \frac{N_k}{\tau_{ph}} = 0 \tag{5.90}$$

$$\frac{\mathrm{d}n}{\mathrm{d}t} \approx \frac{J}{qd} - g(\hbar\omega_k)N_k\frac{c_0}{n_r} - \frac{n}{\tau_r} = 0. \tag{5.91}$$

From (5.90) it follows that, in steady-state DC conditions:

$$\Gamma_{ov}g(\hbar\omega_k)\frac{c_0}{n_r} \approx \Gamma_{ov}g_{th}(\hbar\omega_k)\frac{c_0}{n_r} = \frac{1}{\tau_{ph}} = \alpha_{loss} + \frac{1}{L}\log R^{-1}, \tag{5.92}$$

which also yields the *threshold condition*. Assuming for the gain the approximation $g \approx a(n - n_{tr})$, where n_{tr} is the transparency charge density, one can further derive the threshold charge density as

$$\Gamma_{ov} g_{th}(\hbar\omega_k)\frac{c_0}{n_r} = \frac{1}{\tau_{ph}} \rightarrow n_{th} = n_{tr} + \frac{1}{\Gamma_{ov} a}\frac{n_r}{c_0}\frac{1}{\tau_{ph}}. \tag{5.93}$$

Substituting the threshold charge in (5.91) and taking into account (5.93), the threshold current is

$$J_{th} = \frac{qd}{\tau_r} n_{th} + qda(n_{th} - n_{tr})N_k \frac{c_0}{n_r} = \frac{qd}{\tau_r}\left(n_{th} + \frac{\tau_r N_k}{\Gamma_{ov}\tau_{ph}}\right).$$

Expressing the DC current from (5.91), and assuming (above threshold) $n \approx n_{th}$ and $J_{th} \approx qd/\tau_r n_{th}$, we finally obtain

$$J = qdg_{th}N_k\frac{c_0}{n_r} + \underbrace{\frac{qd}{\tau_r}n_{th}}_{J_{th}} \rightarrow J - J_{th} = qdg_{th}N_k\frac{c_0}{n_r} = \frac{qdN_k}{\Gamma_{ov}\tau_{ph}}.$$

The DC photon density can therefore be written as

$$N_k = \frac{\Gamma_{ov}\tau_{ph}}{qd}(J - J_{th}). \tag{5.94}$$

The output power P_k can be evaluated as the variation of the optical field energy with time:

$$W_k = \hbar\omega_k N_P = \hbar\omega_k V N_k = \frac{Ad}{\Gamma_{ov}}\hbar\omega_k N_k,$$

(see (5.81) and (5.82)), caused by the mirror loss. From the definition of the mirror loss lifetime (5.33) and from (5.86) we have

$$P_k = \left.\frac{dW_k}{dt}\right|_{\text{mirror}} = \frac{W_k}{\tau_m} = \frac{Ad}{\Gamma_{ov}}\hbar\omega_k \left.\frac{dN_k}{dt}\right|_{\text{mirror}} = \frac{Ad}{\Gamma_{ov}}\frac{\hbar\omega_k N_k}{\tau_m}, \tag{5.95}$$

i.e., taking into account (5.94),

$$P_k = \frac{\hbar\omega_k}{q}\frac{\tau_{ph}}{\tau_m}A(J - J_{th}) = \frac{\hbar\omega_k}{q}\frac{\tau_{ph}}{\tau_m}(I - I_{th}), \tag{5.96}$$

or, equivalently,

$$P_k = \frac{\hbar\omega_k}{q}\frac{\frac{1}{L}\log R^{-1}}{\alpha_{loss} + \frac{1}{L}\log R^{-1}}(I - I_{th}). \tag{5.97}$$

Equation (5.97) was already obtained by inspection; see (5.36). Note that if the mirror loss is close to the total loss the above expression reduces to

$$P_k \approx \frac{\hbar\omega_k}{q}(I - I_{th}) \approx \frac{E_g}{q}(I - I_{th})$$

already derived as an ideal, limiting approximation to the slope efficiency of the laser; see (5.37). The above expression is also equivalent to (5.30), already introduced in the analysis; see Example 5.7.

Example 5.7: Show that (5.96) is equivalent to (5.30), written here neglecting the output power at threshold:

$$P_k \equiv P_{out} \approx \frac{P_{out,th}}{Ad}\frac{\tau_{n,th}^{st}}{n_{th}q}(I - I_{th}) \equiv \frac{P_{k,th}}{Ad}\frac{\tau_{n,th}^{st}}{n_{th}q}(I - I_{th}).$$

In fact, from (5.95) evaluated at threshold we have

$$P_{k,th} = \frac{W_{k,th}}{\tau_m} = \frac{Ad}{\Gamma_{ov}}\frac{\hbar\omega_k N_{k,th}}{\tau_m};$$

thus

$$\begin{aligned} P_k &= \frac{P_{k,th}}{Ad}\frac{\tau_{n,th}^{st}}{n_{th}q}(I - I_{th}) = \frac{Ad}{\Gamma_{ov}}\frac{\hbar\omega_k N_{k,th}}{\tau_m}\frac{1}{Ad}\frac{\tau_{n,th}^{st}}{n_{th}q}(I - I_{th}) \\ &= \frac{\tau_{n,th}^{st}}{n_{th}\Gamma_{ov}}\frac{N_{k,th}}{\tau_m}\frac{\hbar\omega_k}{q}(I - I_{th}), \end{aligned}$$

but, comparing the photon and electron rate equations, we obtain at threshold:

$$\frac{\Gamma_{ov} n_{th}}{\tau_{n,th}^{st}} = \Gamma_{ov} g_{th} N_{k,th}\frac{c_0}{n_r} = \frac{N_{k,th}}{\tau_{ph}}. \tag{5.98}$$

Therefore, from (5.98),

$$P_k = \frac{\tau_{n,th}^{st}}{n_{th}\Gamma_{ov}}\frac{N_{k,th}}{\tau_m}\frac{\hbar\omega_k}{q}(I - I_{th}) = \frac{\hbar\omega_k}{q}\frac{\tau_{ph}}{\tau_m}(I - I_{th}),$$

which coincides with (5.96).

5.12.2 Small-signal model

In the dynamic equations (5.88) and (5.89), let us assume for all variables a small-signal perturbation superimposed on a DC component, as

$$\begin{aligned} &N_k(t) = N_{k0} + \widehat{N}_k(t), \quad n(t) = n_0 + \widehat{n}(t), \\ &J(t) = J_0 + \widehat{J}(t), \quad g = g_0 + \left.\frac{\partial g}{\partial n}\right|_0 \widehat{n} + \left.\frac{\partial g}{\partial N_k}\right|_0 \widehat{N}_k, \end{aligned}$$

or, in the frequency domain:

$$\widehat{N}_k(t) \to \widehat{N}_k(\omega), \quad \widehat{n}(t) \to \widehat{n}(\omega), \quad \widehat{J}(t) \to \widehat{J}(\omega).$$

The model for the gain linearization is obtained from (2.30), written here as

$$g\,(n, N_k) = \frac{g_F(n)}{1 + \epsilon_c N_k}, \tag{5.99}$$

where g_F is the gain at low photon density (quasi-Fermi distributions, i.e., neglecting gain compression). By assuming a as the differential gain for low photon density, i.e., $a = \partial g_F/\partial n$, we approximate

$$\begin{aligned} g\left(\widehat{n}, \widehat{N}_k\right) &\approx \frac{g_F(n_0)}{1 + \epsilon_c N_{k0}} + \left.\frac{\partial g}{\partial n}\right|_0 \widehat{n} + \left.\frac{\partial g}{\partial N_k}\right|_0 \widehat{N}_k \\ &= g_0 + \frac{a}{1 + \epsilon_c N_{k0}}\widehat{n} - \frac{\epsilon_c g_0}{1 + \epsilon_c N_{k0}}\widehat{N}_k. \end{aligned}$$

Substituting into (5.88) and (5.89) we have

$$\begin{aligned} \frac{\mathrm{d}\widehat{n}(t)}{\mathrm{d}t} = {} & \underbrace{\frac{J_0}{qd}}_{\mathrm{DC}} + \frac{\widehat{J}(t)}{qd} - \underbrace{\frac{n_0}{\tau_r}}_{\mathrm{DC}} - \frac{\widehat{n}(t)}{\tau_r} - \underbrace{g_0 N_{k0}\frac{c_0}{n_r}}_{\mathrm{DC}} - g_0 \widehat{N}_k(t)\frac{c_0}{n_r} \\ & - \frac{a N_{k0}}{1 + \epsilon_c N_{k0}}\frac{c_0}{n_r}\widehat{n}(t) + \frac{\epsilon_c g_0 N_{k0}}{1 + \epsilon_c N_{k0}}\frac{c_0}{n_r}\widehat{N}_k(t) - \underbrace{\frac{n_0}{\tau_{nr}}}_{\mathrm{DC}} - \frac{\widehat{n}(t)}{\tau_{nr}} \end{aligned}$$

$$\begin{aligned} \frac{\mathrm{d}\widehat{N}_k(t)}{\mathrm{d}t} = {} & \underbrace{\Gamma_{ov}\beta_k \frac{n_0}{\tau_r}}_{\mathrm{DC}} + \Gamma_{ov}\beta_k \frac{\widehat{n}(t)}{\tau_r} + \underbrace{\Gamma_{ov} g_0 N_{k0}\frac{c_0}{n_r}}_{\mathrm{DC}} + \Gamma_{ov} g_0 \widehat{N}_k(t)\frac{c_0}{n_r} \\ & + \frac{\Gamma_{ov} a N_{k0}}{1 + \epsilon_c N_{k0}}\frac{c_0}{n_r}\widehat{n}(t) - \frac{\Gamma_{ov}\epsilon_c g_0 N_{k0}}{1 + \epsilon_c N_{k0}}\frac{c_0}{n_r}\widehat{N}_k(t) - \underbrace{\frac{N_{k0}}{\tau_{ph}}}_{\mathrm{DC}} - \frac{\widehat{N}_k(t)}{\tau_{ph}}. \end{aligned}$$

Eliminating the DC part, transforming into the frequency domain and introducing the total lifetime $\tau_n^{-1} = \tau_r^{-1} + \tau_{nr}^{-1}$, we obtain

$$\begin{aligned} \mathrm{j}\omega\widehat{n}(\omega) = {} & \frac{\widehat{J}(\omega)}{qd} - \frac{\widehat{n}(\omega)}{\tau_r} - g_0\frac{c_0}{n_r}\widehat{N}_k(\omega) + \frac{\epsilon_c g_0 N_{k0}}{1 + \epsilon_c N_{k0}}\frac{c_0}{n_r}\widehat{N}_k(\omega) \\ & - \frac{a N_{k0}}{1 + \epsilon_c N_{k0}}\frac{c_0}{n_r}\widehat{n}(\omega) - \frac{\widehat{n}(\omega)}{\tau_{nr}} \\ = {} & \frac{\widehat{J}(\omega)}{qd} - \frac{g_0}{1 + \epsilon_c N_{k0}}\frac{c_0}{n_r}\widehat{N}_k(\omega) - \frac{a N_{k0}}{1 + \epsilon_c N_{k0}}\frac{c_0}{n_r}\widehat{n}(\omega) - \frac{\widehat{n}(\omega)}{\tau_n} \end{aligned} \tag{5.100}$$

$$
\begin{aligned}
\mathrm{j}\omega\widehat{N}_k(\omega) &= \Gamma_{ov}\beta_k\frac{\widehat{n}(\omega)}{\tau_r} + \Gamma_{ov}g_0\frac{c_0}{n_r}\widehat{N}_k(\omega) + \frac{\Gamma_{ov}aN_{k0}}{1+\epsilon_c N_{k0}}\frac{c_0}{n_r}\widehat{n}(\omega) \\
&\quad - \frac{\Gamma_{ov}\epsilon_c g_0 N_{k0}}{1+\epsilon_c N_{k0}}\frac{c_0}{n_r}\widehat{N}_k(\omega) - \frac{\widehat{N}_k(\omega)}{\tau_{ph}} \\
&\approx \frac{\Gamma_{ov}g_0}{1+\epsilon_c N_{k0}}\frac{c_0}{n_r}\widehat{N}_k(\omega) + \frac{\Gamma_{ov}aN_{k0}}{1+\epsilon_c N_{k0}}\frac{c_0}{n_r}\widehat{n}(\omega) - \frac{\widehat{N}_k(\omega)}{\tau_{ph}} \\
&\approx \frac{\Gamma_{ov}aN_{k0}}{1+\epsilon_c N_{k0}}\frac{c_0}{n_r}\widehat{n}(\omega) - \frac{\epsilon_c N_{k0}}{1+\epsilon_c N_{k0}}\frac{\widehat{N}_k(\omega)}{\tau_{ph}}
\end{aligned}
\tag{5.101}
$$

where we have neglected the contribution of spontaneous emission and taken into account that, from the threshold condition (5.92), we have

$$
\frac{\Gamma_{ov}g_0}{1+\epsilon_c N_{k0}}\frac{c_0}{n_r}\widehat{N}_k(\omega) - \frac{\widehat{N}_k(\omega)}{\tau_{ph}} = -\frac{\epsilon_c N_{k0}}{1+\epsilon_c N_{k0}}\frac{1}{\tau_{ph}}\widehat{N}_k(\omega).
$$

Defining for brevity $\delta_c = 1 + \epsilon_c N_{k0}$ ($\delta_c = 1$ if we neglect gain compression), we can now formulate the small-signal system in matrix form as

$$
\begin{pmatrix}
\mathrm{j}\omega + \dfrac{aN_{k0}}{\delta_c}\dfrac{c_0}{n_r} + \dfrac{1}{\tau_n} & \dfrac{g_0}{\delta_c}\dfrac{c_0}{n_r} \\
-\dfrac{\Gamma_{ov}aN_{k0}}{\delta_c}\dfrac{c_0}{n_r} & \mathrm{j}\omega + \dfrac{\epsilon_c N_{k0}}{\delta_c}\dfrac{1}{\tau_{ph}}
\end{pmatrix}
\begin{pmatrix}
\widehat{n}(\omega) \\ \widehat{N}_k(\omega)
\end{pmatrix}
=
\begin{pmatrix}
\dfrac{\widehat{J}(\omega)}{qd} \\ 0
\end{pmatrix}.
\tag{5.102}
$$

Solution of (5.102) yields:

$$
\widehat{n}(\omega) = \frac{\mathrm{j}\omega + \dfrac{\epsilon_c N_{k0}}{\delta_c}\dfrac{1}{\tau_{ph}}}{\left(\mathrm{j}\omega + \dfrac{\epsilon_c N_{k0}}{\delta_c}\dfrac{1}{\tau_{ph}}\right)\left(\mathrm{j}\omega + \dfrac{aN_{k0}}{\delta_c}\dfrac{c_0}{n_r} + \dfrac{1}{\tau_n}\right) + \dfrac{\Gamma_{ov}aN_{k0}g_0}{\delta_c^2}\left(\dfrac{c_0}{n_r}\right)^2}\frac{\widehat{J}(\omega)}{qd}
\tag{5.103}
$$

$$
\widehat{N}_k(\omega) = \frac{\dfrac{\Gamma_{ov}aN_{k0}}{\delta_c}\dfrac{c_0}{n_r}}{\left(\mathrm{j}\omega + \dfrac{\epsilon_c N_{k0}}{\delta_c}\dfrac{1}{\tau_{ph}}\right)\left(\mathrm{j}\omega + \dfrac{aN_{k0}}{\delta_c}\dfrac{c_0}{n_r} + \dfrac{1}{\tau_n}\right) + \dfrac{\Gamma_{ov}aN_{k0}g_0}{\delta_c^2}\left(\dfrac{c_0}{n_r}\right)^2}\frac{\widehat{J}(\omega)}{qd}.
\tag{5.104}
$$

From (5.104), exploiting the threshold condition (5.92), we finally obtain the damped resonant modulation response of the laser (corresponding to a Bode plot with two complex conjugate poles):

$$
\frac{\widehat{P}_{out}(\omega)}{P_{out}(0)} = \frac{\widehat{N}_k(\omega)}{\widehat{N}_k(0)} \equiv H(\omega) = \frac{\omega_r^2}{(\omega_r^2 - \omega^2) + \mathrm{j}\omega\gamma},
\tag{5.105}
$$

where the resonant frequency ω_r and the damping factor γ are expressed as

$$\omega_r^2 = \frac{aN_{k0}}{\delta_c}\frac{c_0}{n_r}\frac{1}{\tau_{ph}}\left(1+\frac{\epsilon_c n_r}{ac_0}\frac{1}{\tau_n}\right) \approx \frac{aN_{k0}}{\delta_c}\frac{c_0}{n_r}\frac{1}{\tau_{ph}}$$

$$\to \omega_r = 2\pi f_r = \sqrt{\frac{a}{\delta_c}\frac{c_0}{n_r}\frac{N_{k0}}{\tau_{ph}}} = \sqrt{\left.\frac{\mathrm{d}g}{\mathrm{d}n}\right|_{n_0}\frac{c_0}{n_r}\frac{N_{k0}}{\tau_{ph}}} \tag{5.106}$$

$$\gamma = \frac{aN_{k0}}{\delta_c}\frac{c_0}{n_r}\left(1+\frac{\epsilon_c n_r}{ac_0\tau_{ph}}\right)+\frac{1}{\tau_n} = Kf_r^2+\frac{1}{\tau_n} \approx Kf_r^2 \tag{5.107}$$

$$K = 4\pi^2\left(\tau_{ph}+\frac{\epsilon_c n_r}{ac_0}\right) \underset{\epsilon_c\to 0}{\to} 4\pi^2\tau_{ph}, \tag{5.108}$$

i.e., neglecting gain compression we have $\gamma \approx \tau_{ph}\omega_r^2$. The parameter $\mathrm{d}g/\mathrm{d}n|_{n_0}$ is the differential gain including the effect of gain compression. We thus see that improvements in the differential gain (obtained, e.g., in QW devices) lead to a higher cutoff frequency under direct modulation. Taking into account (5.94), we obtain from (5.106) that the resonant frequency increases with the DC current as

$$f_r = \frac{1}{2\pi}\sqrt{\left.\frac{\mathrm{d}g}{\mathrm{d}n}\right|_{n_0}\frac{c_0}{n_r}\frac{\Gamma_{ov}}{qAd}(I-I_{th})} \propto \sqrt{I-I_{th}}.$$

However, a limit to the increase of resonant frequency (and also to the laser cutoff frequency) with increasing current originates from gain saturation. In fact, according to the electrical definition of the 3 dB bandwidth (see Section 6.2.3), this is defined by the equation:

$$\left|\left(\omega_r^2-\omega_{3\mathrm{dB}}^2\right)+\mathrm{j}\omega_{3\mathrm{dB}}\gamma\right| = \sqrt{\left(\omega_r^2-\omega_{3\mathrm{dB}}^2\right)^2+\omega_{3\mathrm{dB}}^2\gamma^2} = \sqrt{2}\omega_r^2.$$

Solving for $\omega_{3\mathrm{dB}}$ we find that the maximum 3 dB cutoff frequency is obtained when the solution shows critical damping, i.e., when

$$2\omega_r^2 = \gamma^2 = K^2 f_r^4 \to f_r = \frac{2\sqrt{2}\pi}{K}.$$

Under such conditions we have $f_{3\mathrm{dB,max}} = f_r$ (the response is monotonically decreasing, with no resonant peak); therefore,

$$f_{3\mathrm{dB,max}} = \frac{2\sqrt{2}\pi}{K}. \tag{5.109}$$

The lower limit of K, assuming negligible photon lifetime, is

$$K_{\min} \approx K = 4\pi^2\frac{\epsilon_c n_r}{ac_0} = 4\pi^2\tau_{n\,\min}^{st},$$

where $\tau_{n\,\min}^{st}$ is the limit stimulated lifetime in the presence of gain compression; see (2.58). This yields the theoretical intrinsic limit to the laser modulation bandwidth:

$$f_{3\mathrm{dB,M}} = \frac{2\sqrt{2}\pi}{K} = \frac{\sqrt{2}}{2\pi\tau_{n\,\min}^{st}}.$$

For $\tau^{st}_{n\,\min} \approx 1$ ps the maximum bandwidth is of the order of 200 GHz; assuming a photon lifetime of 1 ps, such a value is still of the order of 100 GHz. However, practical values of the K parameter (around 0.2–0.6 ns, see, e.g., [11], Section 11.2.1) lead to maximum cutoff frequencies between 20 and 40 GHz.

5.12.3 Chirp analysis

A variation of the carrier density modulates the cavity gain, but also (due to the Kramers–Kronig relation) the (effective) complex cavity refractive index $n_{\text{eff}} = n_r - \mathrm{j}n_i$. Since the power of the cavity optical field propagating in the x direction can be written as

$$P_{out}(x) = P_{out}(0)\left|\exp\left(-\mathrm{j}\frac{2\pi}{\lambda_0}n_r x - \frac{2\pi}{\lambda_0}n_i x\right)\right|^2 = P_{out}(0)\exp\left(-\frac{4\pi}{\lambda_0}n_i x\right),$$

the imaginary part of the refractive index is related to the cavity gain as

$$g_c = \Gamma_{ov} g = -\frac{4\pi}{\lambda_0} n_i.$$

The Henry parameter can be thus written as

$$\alpha_H = \frac{\mathrm{d}n_r}{\mathrm{d}n_i} = -\frac{4\pi}{\lambda_0}\frac{1}{\Gamma_{ov}}\frac{\mathrm{d}n_r}{\mathrm{d}g} = -\frac{4\pi}{\lambda_0}\frac{1}{\Gamma_{ov} a}\frac{\mathrm{d}n_r}{\mathrm{d}n}, \tag{5.110}$$

where we have assumed for the differential gain $\mathrm{d}g/dn \approx a$, n being the electron density. Assume now that the laser emission wavelength and frequency satisfy the FP cavity resonance condition:

$$L = m\frac{\lambda_0}{2n_r} = m\frac{c_0}{2n_r}\frac{1}{f} \rightarrow \lambda_0 = \frac{2n_r L}{m} \rightarrow f = m\frac{c_0}{2n_r}\frac{1}{L}.$$

Taking the derivative of f vs. the carrier density n we have

$$\frac{\mathrm{d}f}{\mathrm{d}n} = -\frac{m}{2Ln_r}\frac{c_0}{n_r}\frac{\mathrm{d}n_r}{\mathrm{d}n} = -\frac{1}{\lambda_0}\frac{c_0}{n_r}\frac{\mathrm{d}n_r}{\mathrm{d}n} \rightarrow \Delta f \approx -\frac{1}{\lambda_0}\frac{c_0}{n_r}\frac{\mathrm{d}n_r}{\mathrm{d}n}\Delta n.$$

Expressing dn_r/dn from (5.110) we finally obtain

$$\Delta f = \frac{\alpha_H}{4\pi}\Gamma_{ov} a\frac{c_0}{n_r}\Delta n. \tag{5.111}$$

Typical values of α_H are, as already recalled, 3–6 in lasers. From (5.103) we see that (at least in small-signal conditions) the charge variation *increases* with the modulating current intensity and modulation frequency; therefore, a larger chirp is expected for a larger intensity modulation and modulation speed.

From system (5.102), second equation, and taking into account that the output power P_{out} is proportional to the charge density, i.e., $P_{out} \propto N_{k0}$, $\widehat{P}_{out}(\omega) \propto \widehat{N}_k(\omega)$, and setting $\delta_c \approx 1$, we obtain

$$\widehat{n}(\omega) = \frac{\mathrm{j}\omega\widehat{N}_k(\omega)}{\Gamma_{ov} a N_{k0}\dfrac{c_0}{n_r}} + \frac{\epsilon_c\widehat{N}_k(\omega)}{\tau_{ph}\Gamma_{ov} a\dfrac{c_0}{n_r}} = \frac{1}{\Gamma_{ov} a\dfrac{c_0}{n_r}}\frac{\mathrm{j}\omega\widehat{P}_{out}(\omega)}{P_{out}} + \frac{\epsilon_c\widehat{N}_k(\omega)}{\tau_{ph}\Gamma_{ov} a\dfrac{c_0}{n_r}},$$

i.e., identifying $\Delta n \equiv \widehat{n}(t)$, $\Delta N_k \equiv \widehat{N}_k(t)$,

$$\Delta n = \frac{1}{\Gamma_{ov} a \dfrac{c_0}{n_r}} \frac{1}{P_{out}} \frac{\mathrm{d}\widehat{p}_{out}(t)}{\mathrm{d}t} + \frac{\epsilon_c \Delta N_k}{\tau_{ph} \Gamma_{ov} a \dfrac{c_0}{n_r}}. \tag{5.112}$$

Equation (5.112) can be exploited to analyze chirp in two conditions.

In the presence of a fast-varying optical power, the first term dominates and leads to transient chirp. In fact, from (5.111) and (5.112), neglecting ΔN_k, the frequency chirp $\Delta f(t)$ can be expressed as

$$\Delta f(t) = \frac{\alpha_H}{4\pi} \Gamma_{ov} a \frac{c_0}{n_r} \Delta n = \frac{\alpha_H}{4\pi} \frac{1}{P_{out}} \frac{\mathrm{d}\widehat{p}_{out}(t)}{\mathrm{d}t}, \tag{5.113}$$

which corresponds again to the standard definition of the Henry parameter in small-signal conditions, see (5.78) with $\widehat{p}_{out} \equiv p_{out}$. Equations (5.113) or (5.78) can also be extended to the large-signal regime to define a time-domain (large-signal) chirp parameter; moreover, taking into account that the output optical power is related to the laser current $i(t)$ through the slope efficiency, a final expression for the chirp in terms of time-varying and DC laser currents can be derived as in (5.79).

Chirp in large-signal conditions includes a further effect: due to the dependence of the emission wavelength on the input current, the *high* and *low* digital levels typically correspond to slightly different frequencies. Applying, for instance, a current step leading the laser from the 0 to the 1 logical state, *transient* chirp occurs, as already described; however, the asymptotic frequency deviation is not zero because the 1 state emission frequency and the 0 state emission frequency are different. Such a residual, steady-state chirp is called the *adiabatic chirp*. Adiabatic chirp can be derived taking into account, from (5.112), that, even in steady-state DC conditions a variation of the bias current leads to a variation of the photon density ΔN_k and therefore (because of gain compression) to a decrease of gain that has to be compensated through an increase of the carrier density Δn. The resulting frequency deviation will be

$$\Delta f = \frac{\alpha_H}{4\pi} \frac{\epsilon_c}{\tau_{ph}} \Delta N_k, \tag{5.114}$$

and is proportional, as expected, to the gain compression factor ϵ_c.

5.13 Laser relative intensity noise

The analysis of the laser linewidth in Section 5.10.1 has already stressed that the (spontaneous and stimulated) emission of photons leads to both phase and amplitude fluctuations in the optical field, and therefore in the output power. Fluctuations in stimulated emission were associated with gain fluctuations, in turn caused by carrier population fluctuations. In the present section the issue is addressed anew starting from the large-signal dynamic system (5.88), (5.89). The fluctuation problem can be treated according to a well known mathematical formulation (called the *Langevin approach*), in which random fluctuations of the carrier and photon densities originated by random photon generation events (and the related carrier recombination events) appear as *forcing terms* in the dynamic system for the averages of such quantities.[21] Since fluctuations are small perturbations of the averages, these can be derived through a small-signal approach, which already led to system (5.102). The procedure consists of the following steps:

1. Identify the random process describing *photon density fluctuations* and evaluate its second-order statistical properties (the self-correlation function and power spectrum). This is called the Langevin source for the photon density fluctuations.
2. Do the same for the *carrier density fluctuations* and derive the statistical properties of the corresponding Langevin source (and of its correlation to the Langevin source for photons).
3. Apply the two random processes introduced above as the Langevin sources to the small-signal laser dynamic system and obtain, from its solution, the second-order statistical characterization of the average photon and carrier density fluctuations; such characterization is suitably derived in the frequency domain as the power (and correlation) spectrum.

For simplicity, the analysis will be carried out exploiting a small-signal model where gain compression is neglected ($\epsilon_c \approx 0$). Expressions including gain compression can be found in [83], Section 5.5.

5.13.1 Analysis of Langevin sources

Let us consider again the electron and photon density rate equations, where, for the sake of the noise analysis, we separate the contributions due to absorption and stimulated emission which were initially included in the net gain g (i.e., we separate the gain $\overline{g}$ and the absorption α so that the net gain is $g = \overline{g} - \alpha$):

$$\frac{\mathrm{d}n}{\mathrm{d}t} = \underbrace{\frac{J}{qd}}_{\substack{\text{current}\\ \text{injection}}} - \underbrace{\frac{n}{\tau_r}}_{\substack{\text{spontaneous}\\ \text{recombination}}} - \underbrace{\overline{g}N_k\frac{c_0}{n_r}}_{\substack{\text{stimulated}\\ \text{recombination}}} + \underbrace{\alpha N_k\frac{c_0}{n_r}}_{\substack{\text{generation due}\\ \text{to absorption}}} - \underbrace{\frac{n}{\tau_{nr}}}_{\substack{\text{nonradiative}\\ \text{recombination}}} \tag{5.115}$$

$$\frac{\mathrm{d}N_k}{\mathrm{d}t} = \underbrace{\Gamma_{ov}\beta_k\frac{n}{\tau_r}}_{\substack{\text{spontaneous}\\ \text{emission}}} + \underbrace{\Gamma_{ov}\overline{g}N_k\frac{c_0}{n_r}}_{\substack{\text{stimulated}\\ \text{emission}}} - \underbrace{\Gamma_{ov}\alpha N_k\frac{c_0}{n_r}}_{\text{absorption}} - \underbrace{\frac{N_k}{\tau_{ph}}}_{\substack{\text{photon}\\ \text{loss}}} . \tag{5.116}$$

The electron and photon populations can be readily interpreted as reservoirs, that are *filled* or *emptied* by a number of processes listed above. Spontaneous and stimulated

[21] For the Langevin analysis of the laser noise based on the photon and carrier rate equations, see, e.g., [83].

radiative processes contribute, at the same time, to emptying the electron reservoir and filling the photon reservoir.

Let us consider first a simplified case, in which two particle reservoirs exist with populations n_a and n_b (populations here are numbers, not densities). Each reservoir has a generation and a recombination rate (numbers per unit time), as defined by the following equations:

$$\frac{dn_a}{dt} = G_{aa} + G_{ab} - R_{aa} - R_{ab} \quad (5.117)$$

$$\frac{dn_b}{dt} = G_{bb} + G_{ba} - R_{bb} - R_{ba}. \quad (5.118)$$

The terms R_{ii}, G_{ii} describe the recombination and generation rates of each reservoir by itself, while $R_{ij} = G_{ji}$ defines the *particle transfer* $a \to b$ or $b \to a$; therefore, $G_{ab} = R_{ba}$ and $G_{ba} = R_{ab}$. Moreover, from DC steady-state we obtain

$$G_{aa} + G_{ab} - R_{aa} - R_{ab} = G_{aa} + R_{ba} - R_{aa} - R_{ab} = 0 \quad (5.119)$$

$$G_{bb} + G_{ba} - R_{bb} - R_{ba} = G_{bb} + R_{ab} - R_{bb} - R_{ba} = 0. \quad (5.120)$$

Let us now introduce a way to consider the random nature of the generation and recombination processes, in which particles appear or disappear, respectively, at random times t_i, leading to a ± 1 variation of the particle population. Consider, for simplicity, a single carrier population n affected by an average generation rate G; if generation events happened at regular intervals, $n(t)$ would be an increasing staircase function of time. Generation at random times, however, implies that $n(t)$ is affected by fluctuations $\delta n(t)$ with respect to the (statistical) average value $\overline{n}$, so that $n(t) = \overline{n} + \delta n(t)$. The fluctuation $\delta n(t)$ is a zero-average random process assuming positive or negative integer values, so that its time derivative can be expressed as a sum of Dirac delta functions having (with the same probability) weight $\alpha_i = \pm 1$. According to the Langevin approach, the effect of fluctuations on the evolution of n can be derived by adding to the *average* dynamic model $d\overline{n}/dt = G$ a forcing term $F_n(t)$ (the *Langevin source*) equal to the time derivative of $\delta n(t)$:

$$F_n(t) \equiv \frac{d\delta n(t)}{dt} = \sum_i \alpha_i \delta(t - t_i). \quad (5.121)$$

The overall dynamic equation for the total population n can be therefore written as

$$\frac{dn}{dt} = G + F_n(t),$$

where the Langevin random source gives rise to the fluctuation δn, while the average of the right-hand side G leads to the average value $\overline{n}$ (linearly increasing with time). The above equation can be linearized (if needed, e.g., if G depends on n) with respect to the average value, so that the response to the Langevin source (small with respect to the average forcing term) can be recovered through small-signal analysis. The second-order statistical properties of the zero-average Langevin source ($\langle F_n \rangle = 0$) can be derived from the autocorrelation function

$$R_{nn}(\tau) = \langle F_n(t)F_n(t-\tau)\rangle = \left\langle \sum_i \alpha_i \delta(t-t_i) \sum_j \alpha_j \delta(t-\tau-t_j) \right\rangle$$
$$= \left\langle \sum_i \delta(t-t_i)\delta(t-\tau-t_i) \right\rangle,$$

since generation or recombination events at different times are uncorrelated, and $\alpha_i^2 = 1$. Taking into account that $\delta(t-t_i)\delta(t-\tau-t_i) \neq 0$ only if $t-\tau = t_i$ and at the same time $t = t_i$, i.e., with $\tau = 0$, we can write

$$\delta(t-t_i)\delta(t-\tau-t_i) = \delta(\tau)\delta(t-t_i),$$

and therefore

$$\left\langle \sum_i \delta(t-t_i)\delta(t-\tau-t_i) \right\rangle = \delta(\tau)\left\langle \sum_i \delta(t-t_i) \right\rangle$$
$$= \delta(\tau) \lim_{T\to\infty} \frac{1}{T} \int_{-T/2}^{T/2} \sum_{i=1}^{GT} \delta(t-t_i)\, \mathrm{d}t = \delta(\tau) \lim_{T\to\infty} \frac{1}{T} \cdot GT = \delta(\tau)G,$$

where the summation is extended to all generation events occurring over T. Thus, the autocorrelation function of F_n is

$$R_{nn}(\tau) = G\delta(\tau), \tag{5.122}$$

and by Fourier transforming we immediately derive the white power spectrum of F_n, $S_{nn}(\omega)$, as

$$S_{nn}(\omega) = G.$$

The power spectrum of the fluctuations associated with the generation process is therefore equal to the *average value of the generation rate*. This result was expected, since F_n is a Poisson process (like shot noise); the above property is also known as Campbell's theorem (see e.g., [34]).

Let us now extend the Langevin source technique to system (5.117) and (5.118) by introducing the relevant Langevin sources associated with each generation and recombination process (denoted as k_{ij}, $k = r, g$, $i, j = a, b$, for a process with average rate K_{ij}, $K = R, G$) and the total Langevin sources F_a and F_b:

$$\frac{\mathrm{d}n_a}{\mathrm{d}t} = G_{aa} + g_{aa}(t) + R_{ba} + r_{ba}(t) - R_{aa} + r_{aa}(t) - R_{ab} - r_{ab}(t)$$
$$= G_{aa} + R_{ba} - R_{aa} - R_{ab} + F_a(t)$$
$$\frac{\mathrm{d}n_b}{\mathrm{d}t} = G_{bb} + g_{bb}(t) + R_{ab} + r_{ab}(t) - R_{bb} + r_{bb}(t) - R_{ba} - r_{ba}(t)$$
$$= G_{bb} + R_{ab} - R_{bb} - R_{ba} + F_b(t),$$

where

$$F_a(t) = g_{aa}(t) + r_{ba}(t) + r_{aa}(t) - r_{ab}(t)$$
$$F_b(t) = g_{bb}(t) + r_{ab}(t) + r_{bb}(t) - r_{ba}(t).$$

Note that contributions corresponding to particle transitions $a \to b$ have the same value in the total Langevin sources F_a and F_b, but opposite sign (while a particle disappears from reservoir a, it appears in reservoir b). Since all terms in F_a and F_b are uncorrelated, the auto- and mutual correlation functions of $F_a(t)$ and $F_b(t)$ can readily be shown to be expressed as

$$\begin{aligned}
R_{aa}(\tau) = \langle F_a(t)F_a(t-\tau)\rangle &= (G_{aa} + R_{ba} + R_{aa} + R_{ab})\,\delta(\tau) \\
&= 2\,(G_{aa} + R_{ba})\,\delta(\tau) \\
R_{bb}(\tau) = \langle F_b(t)F_b(t-\tau)\rangle &= (G_{bb} + R_{ab} + R_{bb} + R_{ba})\,\delta(\tau) \\
&= 2\,(G_{bb} + R_{ab})\,\delta(\tau) \\
R_{ab}(\tau) = \langle F_a(t)F_b(t-\tau)\rangle &= -\langle r_{ba}(t)r_{ba}(t-\tau)\rangle - \langle r_{ab}(t)r_{ab}(t-\tau)\rangle \\
&= -\,(R_{ba} + R_{ab})\,\delta(\tau),
\end{aligned}$$

where we have taken into account the model equation (5.122) along with (5.119) and (5.120), yielding $G_{aa} + R_{ba} = R_{aa} + R_{ab}$ and $G_{bb} + R_{ab} = R_{bb} + R_{ba}$. The corresponding power and correlation spectra are

$$\begin{aligned}
S_{aa}(\omega) &= 2\,(G_{aa} + R_{ba}) \\
S_{bb}(\omega) &= 2\,(G_{bb} + R_{ab}) \\
S_{ab}(\omega) &= -\,(R_{ba} + R_{ab})\,.
\end{aligned}$$

In other words, the *power spectrum* of the fluctuations in each reservoir is twice the total generation or recombination rate (equal in steady state) while the *correlation spectrum* is equal to (minus) the sum of the common transition rates (i.e., involving both reservoirs).

Let us now go back to the laser dynamic system (5.115), (5.116). In order to derive in a more straightforward way the relevant Langevin sources and their statistical properties we start, however, by expressing system (5.115), (5.116) in the total *photon* and *electron numbers*, denoted as N_P and N. If V is the modal volume of photon mode k, with density N_k, and $V_{ac} = A \cdot d$ is the active-region volume (by definition, $V_{ac} = \Gamma_{ov} V$, see (5.80) and (5.81)) we have $N_P = V N_k$, and $N = V_{ac} n$. Thus, multiplying (5.115) by V_{ac} and (5.116) by V, the dynamic system becomes

$$\begin{aligned}
\frac{\mathrm{d}V_{ac}n}{\mathrm{d}t} &= \frac{V_{ac}J}{qd} - \frac{V_{ac}n}{\tau_r} - \overline{g}V_{ac}N_k\frac{c_0}{n_r} + \alpha V_{ac}N_k\frac{c_0}{n_r} - \frac{V_{ac}n}{\tau_{nr}} \\
\frac{\mathrm{d}VN_k}{\mathrm{d}t} &= \Gamma_{ov}\beta_k\frac{Vn}{\tau_r} + \Gamma_{ov}\overline{g}VN_k\frac{c_0}{n_r} - \Gamma_{ov}\alpha VN_k\frac{c_0}{n_r} - \frac{VN_k}{\tau_{ph}};
\end{aligned}$$

i.e., substituting for N and N_P, and including the Langevin sources $F_N(t)$ and $F_P(t)$, we obtain

$$\frac{\mathrm{d}N}{\mathrm{d}t} = \frac{I}{q} - \frac{N}{\tau_r} - \Gamma_{ov}\overline{g}N_P\frac{c_0}{n_r} + \alpha N_P\frac{c_0}{n_r} - \frac{N}{\tau_{nr}} + F_N(t) \tag{5.123}$$

$$\frac{\mathrm{d}N_P}{\mathrm{d}t} = \beta_k\frac{N}{\tau_r} + \Gamma_{ov}\overline{g}N_P\frac{c_0}{n_r} - \Gamma_{ov}\alpha N_P\frac{c_0}{n_r} - \frac{N_P}{\tau_{ph}} + F_P(t). \tag{5.124}$$

Taking into account the average generation and recombination rates, we can immediately derive the auto- and mutual correlation of the Langevin sources:

$$\begin{aligned} R_{NN}(\tau) &= \langle F_N(t)F_N(t-\tau)\rangle \\ &= \left(\frac{I}{q} + \frac{N}{\tau_r} + \Gamma_{ov}\overline{g}N_P\frac{c_0}{n_r} + \Gamma_{ov}\alpha N_P\frac{c_0}{n_r} + \frac{N}{\tau_{nr}}\right)\delta(\tau) \\ &= \left[\frac{I}{q} + \beta_k\frac{N}{\tau_r} + (1-\beta_k)\frac{N}{\tau_r} + \Gamma_{ov}\overline{g}N_P\frac{c_0}{n_r} + \Gamma_{ov}\alpha N_P\frac{c_0}{n_r} + \frac{N}{\tau_{nr}}\right]\delta(\tau) \end{aligned} \tag{5.125}$$

$$\begin{aligned} R_{PP}(\tau) &= \langle F_P(t)F_P(t-\tau)\rangle \\ &= \left(\beta_k\frac{N}{\tau_r} + \Gamma_{ov}\overline{g}N_P\frac{c_0}{n_r} + \Gamma_{ov}\alpha N_P\frac{c_0}{n_r} + \frac{N_P}{\tau_{ph}}\right)\delta(\tau) \end{aligned} \tag{5.126}$$

$$\begin{aligned} R_{NP}(\tau) &= \langle F_N(t)F_P(t-\tau)\rangle \\ &= \left(\beta_k\frac{N}{\tau_r} + \Gamma_{ov}\overline{g}N_P\frac{c_0}{n_r} + \Gamma_{ov}\alpha N_P\frac{c_0}{n_r}\right)\delta(\tau). \end{aligned} \tag{5.127}$$

Note that the correlation function includes terms relevant to the cross rates, and in particular those relevant to stimulated emission and absorption (second and third terms) and the part of the spontaneous emission falling into the lasing mode (with weight β_k, first term). The expressions can be somewhat simplified taking into account the steady-state relations derived from (5.123) and (5.124):

$$\frac{I}{q} - \frac{N}{\tau_r} - \Gamma_{ov}\overline{g}N_P\frac{c_0}{n_r} + \alpha N_P\frac{c_0}{n_r} - \frac{N}{\tau_{nr}} = 0 \tag{5.128}$$

$$\beta_k\frac{N}{\tau_r} + \Gamma_{ov}\overline{g}N_P\frac{c_0}{n_r} - \Gamma_{ov}\alpha N_P\frac{c_0}{n_r} - \frac{N_P}{\tau_{ph}} = 0. \tag{5.129}$$

From (5.129) and (5.126) we have

$$\begin{aligned} R_{PP}(\tau) &= \left(\beta_k\frac{N}{\tau_r} + \Gamma_{ov}\overline{g}N_P\frac{c_0}{n_r} + \Gamma_{ov}\alpha N_P\frac{c_0}{n_r} + \frac{N_P}{\tau_{ph}}\right)\delta(\tau) \\ &= \left(\beta_k\frac{N}{\tau_r} + \Gamma_{ov}\overline{g}N_P\frac{c_0}{n_r} + \Gamma_{ov}\alpha N_P\frac{c_0}{n_r} + \beta_k\frac{N}{\tau_r}\right. \\ &\qquad \left. + \Gamma_{ov}\overline{g}N_P\frac{c_0}{n_r} - \Gamma_{ov}\alpha N_P\frac{c_0}{n_r}\right)\delta(\tau) \\ &= 2\left(\beta_k\frac{N}{\tau_r} + \Gamma_{ov}\overline{g}N_P\frac{c_0}{n_r}\right)\delta(\tau). \end{aligned}$$

However, from the relation $W_{em}^{st} = n_{ph} W_{em}^{sp}$ (2.20) which, in the current notation, reads

$$\Gamma_{ov} \overline{g} N_P \frac{c_0}{n_r} = \beta_k \frac{N}{\tau_r} N_P = \overline{r}_o^{sp} N_P, \tag{5.130}$$

where $\overline{r}_o^{sp}$ is the spontaneous emission rate into mode k (photons per second), one can write

$$\begin{aligned} R_{PP}(\tau) &= 2\left(\beta_k \frac{N}{\tau_r} + \Gamma_{ov} \overline{g} N_P \frac{c_0}{n_r}\right) \delta(\tau) \\ &= 2\overline{r}_o^{sp} N_P \left(1 + \frac{1}{N_P}\right) \delta(\tau) \approx 2\overline{r}_o^{sp} N_P \delta(\tau), \end{aligned} \tag{5.131}$$

assuming $N_P \gg 1$. The autocorrelation function of the carrier population can be treated in a similar way; exploiting (5.128), (5.130) and (5.125), we obtain

$$\begin{aligned} R_{NN}(\tau) &= 2\left(\frac{N}{\tau_r} + \Gamma_{ov} \overline{g} N_P \frac{c_0}{n_r} + \frac{N}{\tau_{nr}}\right) \delta(\tau) \\ &= 2\overline{r}_o^{sp} N_P \left(1 + \frac{1}{\beta_k N_P}\right) \delta(\tau) + 2\frac{N}{\tau_{nr}} \delta(\tau) \approx 2\overline{r}_o^{sp} N_P \delta(\tau) + 2\frac{N}{\tau_{nr}} \delta(\tau). \end{aligned}$$

With similar arguments, we have

$$\begin{aligned} R_{NP}(\tau) &= -\left(2\beta_k \frac{N}{\tau_r} + 2\Gamma_{ov} \overline{g} N_P \frac{c_0}{n_r} - \frac{N_P}{\tau_{ph}}\right) \delta(\tau) \\ &= -2\beta_k \frac{N}{\tau_r} N_P \left(1 + \frac{1}{N_P}\right) \delta(\tau) + \frac{N_P}{\tau_{ph}} \delta(\tau) \\ &= -2\overline{r}_o^{sp} N_P \left(1 + \frac{1}{N_P}\right) \delta(\tau) + \frac{N_P}{\tau_{ph}} \delta(\tau) \approx -2\overline{r}_o^{sp} N_P \delta(\tau) + \frac{N_P}{\tau_{ph}} \delta(\tau). \end{aligned}$$

We can now go back to (5.115) and (5.116) in the photon and electron densities, where we have reintroduced the net gain:

$$\frac{\mathrm{d}n}{\mathrm{d}t} = \frac{J}{qd} - \frac{n}{\tau_r} - g N_k \frac{c_0}{n_r} - \frac{n}{\tau_{nr}} + F_n(t) \tag{5.132}$$

$$\frac{\mathrm{d}N_k}{\mathrm{d}t} = \Gamma_{ov} \beta_k \frac{n}{\tau_r} + \Gamma_{ov} g N_k \frac{c_0}{n_r} - \frac{N_k}{\tau_{ph}} + F_k(t). \tag{5.133}$$

The relevant Langevin sources $F_n(t)$ and $F_k(t)$ are simply related to $F_N(t)$ and $F_P(t)$ defined as

$$F_n(t) = \frac{1}{V_{ac}} F_N(t), \quad F_k(t) = \frac{1}{V} F_P(t). \tag{5.134}$$

Taking into account that the auto- and mutual correlations scale as the products of the proportionality factors, we obtain

$$
\begin{aligned}
R_{nn}(\tau) &= \langle F_n(t)F_n(t-\tau)\rangle = \frac{1}{V_{ac}^2}\langle F_N(t)F_N(t-\tau)\rangle = \frac{1}{V_{ac}^2}R_{NN}(\tau)\\
&= \frac{2\overline{R}_o^{sp}N_k}{\Gamma_{ov}}\delta(\tau) + 2\frac{n}{V_{ac}\tau_{nr}}\delta(\tau)\\
R_{kk}(\tau) &= \langle F_k(t)F_k(t-\tau)\rangle = \frac{1}{V^2}\langle F_P(t)F_P(t-\tau)\rangle = \frac{1}{V^2}R_{PP}(\tau)\\
&= 2\Gamma_{ov}\overline{R}_o^{sp}N_k\delta(\tau)\\
R_{nk}(\tau) &= \langle F_n(t)F_k(t-\tau)\rangle = \frac{1}{VV_{ac}}\langle F_N(t)F_P(t-\tau)\rangle = \frac{1}{VV_{ac}}R_{NP}(\tau)\\
&= -2\overline{R}_o^{sp}N_k\delta(\tau) + \frac{N_k}{V_{ac}\tau_{ph}}\delta(\tau),
\end{aligned}
$$

where $\overline{R}_o^{sp} = \overline{r}_o^{sp}/V_{ac}$ is the recombination rate per unit volume leading to spontaneous emission into mode k (note that the relevant volume is the electrons' volume, not the photons'). From the above results we immediately obtain the white power spectra:

$$S_{nn}(\omega) = \frac{2\overline{R}_o^{sp}N_k}{\Gamma_{ov}} + 2\frac{n}{V_{ac}\tau_{nr}} \tag{5.135}$$

$$S_{kk}(\omega) = 2\Gamma_{ov}\overline{R}_o^{sp}N_k \tag{5.136}$$

$$S_{nk}(\omega) = -2\overline{R}_o^{sp}N_k + \frac{N_k}{V_{ac}\tau_{ph}}. \tag{5.137}$$

Example 5.8: Show that the autocorrelation function of the photon number fluctuation Langevin source (5.131), can also be directly obtained from the treatment in Section 5.10.1, provided that we consider, as the generation process, only spontaneous emission.

From the analysis in Section 5.10.1 we found that the emission of a single photon with random phase θ_i leads to a photon number fluctuation $\delta I = 2\sqrt{I}\cos(\theta_i) + 1 \approx 2\sqrt{I}\cos(\theta_i)$; see (5.57). Note that with this simplification the ensemble average of δI is zero. Using the template equation for the definition of the Langevin source (5.121), and with reference to (5.58), we can introduce a total Langevin source F_I as

$$\left.\frac{\mathrm{d}I}{\mathrm{d}t}\right|_{sp} = F_I \approx \sum_i 2\sqrt{I}\cos(\theta_i)\delta(t-t_i). \tag{5.138}$$

where the summation is extended to all spontaneous emission events occurring at random times t_i. Since photons are spontaneously emitted with an average rate $\overline{r}_o^{sp}$ per unit time, after a time T the total emission number will be $\overline{r}_o^{sp} T$. As θ_i is uniformly distributed, $\langle F_I \rangle = 0$. The autocorrelation function of F_I reduces to

$$
\begin{aligned}
R_{II}(\tau) &= \langle F_I(t) F_I(t-\tau) \rangle \\
&= \left\langle \left[\sum_i 2\sqrt{I} \cos(\theta_i) \delta(t - t_i) \right] \left[\sum_j 2\sqrt{I} \cos(\theta_j) \delta(t - \tau - t_j) \right] \right\rangle \\
&= \left\langle \sum_i 4I \cos^2(\theta_i) \delta(t - t_i) \delta(t - \tau - t_i) \right\rangle,
\end{aligned}
$$

since emission events occurring at different times are completely uncorrelated. Taking into account that $\delta(t - t_i)\delta(t - \tau - t_i)$ is different from zero only if $t - \tau = t_i$ and at the same time $t = t_i$, i.e., with $\tau = 0$, we have $\delta(t - t_i)\delta(t - \tau - t_i) = \delta(\tau)\delta(t - t_i)$, and therefore

$$
\begin{aligned}
R_{II}(\tau) &= \left\langle \sum_i 4I \cos^2(\theta_i) \delta(t - t_i) \delta(t - \tau - t_i) \right\rangle \\
&= 4I\delta(\tau) \left\langle \sum_i \cos^2(\theta_i) \delta(t - t_i) \right\rangle \approx 2\overline{r}_o^{sp} I \delta(\tau),
\end{aligned}
$$

since

$$
\begin{aligned}
\left\langle \sum_i \cos^2(\theta_i) \delta(t - t_i) \right\rangle &= \lim_{T \to \infty} \frac{1}{T} \int_{-T/2}^{T/2} \sum_{i=1}^{\overline{r}_o^{sp} T} \cos^2(\theta_i) \delta(t - t_i) \, dt \\
&= \lim_{T \to \infty} \frac{1}{T} \sum_{i=1}^{\overline{r}_o^{sp} T} \int_{-T/2}^{T/2} \cos^2(\theta_i) \delta(t - t_i) \, dt \\
&= \lim_{T \to \infty} \frac{1}{T} \cdot \overline{r}_o^{sp} T \cdot \frac{1}{2} = \frac{\overline{r}_o^{sp}}{2},
\end{aligned}
$$

where we have averaged $\cos^2(\theta_i)$ to $1/2$. From the above expression for $R_{II}(\tau)$ and identifying $I \equiv N_P$, we obtain the result $R_{II}(\tau) = 2\overline{r}_o^{sp} N_P \delta(\tau)$; see (5.131). We can therefore conclude that the Langevin source for photon number fluctuations can be derived by considering as "equivalent" fluctuation events the spontaneous emissions only.

5.13.2 Carrier and photon population fluctuations

According to the Langevin approach to fluctuation analysis, we start from the dynamic system with random Langevin sources (5.132), (5.133) and evaluate the small-signal response to F_n and F_k (defined in (5.134) and with power and correlation spectra (5.135), (5.136) and (5.137)), setting to zero all deterministic forcing terms (such as the small-signal current source). This is equivalent to rewriting the small-signal dynamic system (5.102) with *zero impressed current density* but *including the Langevin sources*; the notation is also suitably modified by introducing the fluctuations $\delta n(\omega)$ and $\delta N_k(\omega)$, and gain compression has been neglected ($\epsilon_c = 0$, $\delta_c = 1$):

$$\begin{pmatrix} \mathrm{j}\omega + aN_{k0}\dfrac{c_0}{n_r} + \dfrac{1}{\tau_n} & g_0\dfrac{c_0}{n_r} \\ -\Gamma_{ov}aN_{k0}\dfrac{c_0}{n_r} & \mathrm{j}\omega \end{pmatrix} \begin{pmatrix} \delta n(\omega) \\ \delta N_k(\omega) \end{pmatrix} = \begin{pmatrix} F_n(\omega) \\ F_k(\omega) \end{pmatrix}. \tag{5.139}$$

Solving, we obtain

$$\delta n(\omega) = \frac{\mathrm{j}\omega F_n(\omega) - g_0\dfrac{c_0}{n_r}F_k(\omega)}{\mathrm{j}\omega\left(\mathrm{j}\omega + aN_{k0}\dfrac{c_0}{n_r} + \dfrac{1}{\tau_n}\right) + g_0\dfrac{c_0}{n_r}\Gamma_{ov}aN_{k0}\dfrac{c_0}{n_r}}$$
$$= \frac{\mathrm{j}\omega F_n(\omega) - \dfrac{1}{\Gamma_{ov}\tau_{ph}}F_k(\omega)}{\left(\omega_r^2 - \omega^2\right) + \mathrm{j}\omega\gamma} \tag{5.140}$$

$$\delta N_k(\omega) = \frac{\Gamma_{ov}aN_{k0}\dfrac{c_0}{n_r}F_n(\omega) + \left(\mathrm{j}\omega + aN_{k0}\dfrac{c_0}{n_r} + \dfrac{1}{\tau_n}\right)F_k(\omega)}{\mathrm{j}\omega\left(\mathrm{j}\omega + aN_{k0}\dfrac{c_0}{n_r} + \dfrac{1}{\tau_n}\right) + g_0\dfrac{c_0}{n_r}\Gamma_{ov}aN_{k0}\dfrac{c_0}{n_r}}$$
$$= \frac{\omega_r^2\Gamma_{ov}\tau_{ph}F_n(\omega) + (\mathrm{j}\omega + \gamma)F_k(\omega)}{\left(\omega_r^2 - \omega^2\right) + \mathrm{j}\omega\gamma}, \tag{5.141}$$

with ω_r and γ defined in (5.106) and (5.107), which have also been exploited in the simplifications, together with the DC condition (5.92) yielding $\Gamma_{ov}g_0c_0/n_r \approx \tau_{ph}^{-1}$.

We can now formally derive the power and correlation spectra of δn and δN_k through a spectral ensemble average as follows:

$$S_{\delta n\delta n} = \left\langle \delta n(\omega)\delta n^*(\omega')\right\rangle$$
$$= \frac{\omega^2\left\langle F_n(\omega)F_n^*(\omega')\right\rangle + \dfrac{\left\langle F_k(\omega)F_k^*(\omega')\right\rangle}{\left(\Gamma_{ov}\tau_{ph}\right)^2} + 2\,\mathrm{Re}\left(-\dfrac{\mathrm{j}\omega\left\langle F_n(\omega)F_k^*(\omega')\right\rangle}{\Gamma_{ov}\tau_{ph}}\right)}{\left(\omega_r^2 - \omega^2\right)^2 + (\omega\gamma)^2}$$
$$= \frac{\omega^2\left(\dfrac{2\overline{R}_o^{sp}N_{k0}}{\Gamma_{ov}} + \dfrac{2n}{V_{ac}\tau_{nr}}\right) + \dfrac{2\overline{R}_o^{sp}N_{k0}}{\Gamma_{ov}\tau_{ph}^2}}{\left(\omega_r^2 - \omega^2\right)^2 + (\omega\gamma)^2}, \tag{5.142}$$

$$
\begin{aligned}
S_{\delta N_k \delta N_k} &= \left\langle \delta N_k(\omega)\delta N_k^*(\omega')\right\rangle \\
&= \frac{\left(\omega_r^2\Gamma_{ov}\tau_{ph}\right)^2\left\langle F_n(\omega)F_n^*(\omega')\right\rangle + 2\,\mathrm{Re}\left[(\mathrm{j}\omega+\gamma)\,\omega_r^2\Gamma_{ov}\tau_{ph}\left\langle F_k(\omega)F_n^*(\omega')\right\rangle\right]}{\left(\omega_r^2-\omega^2\right)^2+(\omega\gamma)^2} \\
&\quad + \frac{\left(\omega^2+\gamma^2\right)\left\langle F_k(\omega)F_k^*(\omega')\right\rangle}{\left(\omega_r^2-\omega^2\right)^2+(\omega\gamma)^2} \\
&= \frac{\left(\omega_r^2\Gamma_{ov}\tau_{ph}\right)^2\left(\dfrac{2\overline{R}_o^{sp}N_{k0}}{\Gamma_{ov}}+\dfrac{2n}{V_{ac}\tau_{nr}}\right) - 4\gamma\omega_r^2\Gamma_{ov}\tau_{ph}\overline{R}_o^{sp}N_{k0} + 2\gamma\omega_r^2\dfrac{N_{k0}}{V}}{\left(\omega_r^2-\omega^2\right)^2+(\omega\gamma)^2} \\
&\quad + \frac{2\left(\omega^2+\gamma^2\right)\Gamma_{ov}\overline{R}_o^{sp}N_{k0}}{\left(\omega_r^2-\omega^2\right)^2+(\omega\gamma)^2} \\
&= \frac{2\overline{R}_o^{sp}N_{k0}\Gamma_{ov}\left(\dfrac{1}{\tau_n^2}+\omega^2\right) + 2\gamma\omega_r^2\dfrac{N_{k0}}{V} + 2\dfrac{\left(\omega_r^2\Gamma_{ov}\tau_{ph}\right)^2 n}{V_{ac}\tau_{nr}}}{\left(\omega_r^2-\omega^2\right)^2+(\omega\gamma)^2},
\end{aligned}
\tag{5.143}
$$

$$
\begin{aligned}
S_{\delta n \delta N_k} &= \left\langle \delta n(\omega)\delta N_k^*(\omega')\right\rangle \\
&= \left\langle \frac{\mathrm{j}\omega F_n(\omega) - \dfrac{1}{\Gamma_{ov}\tau_{ph}}F_k(\omega)}{\left(\omega_r^2-\omega^2\right)+\mathrm{j}\omega\gamma}\,\frac{\omega_r^2\Gamma_{ov}\tau_{ph}F_n^*(\omega) - (-\mathrm{j}\omega+\gamma)\,F_k^*(\omega)}{\left(\omega_r^2-\omega^2\right)-\mathrm{j}\omega\gamma}\right\rangle \\
&= \frac{\mathrm{j}\omega\omega_r^2\Gamma_{ov}\tau_{ph}\left(\dfrac{2\overline{R}_o^{sp}N_k}{\Gamma_{ov}}+\dfrac{2n}{V_{ac}\tau_{nr}}\right)}{\left(\omega_r^2-\omega^2\right)^2+(\omega\gamma)^2} \\
&\quad + \frac{2\left(-\mathrm{j}\omega+\gamma\right)\left(\mathrm{j}\omega+\dfrac{1}{\tau_{ph}}\right)\left(\overline{R}_o^{sp}N_k - \dfrac{N_k}{2V_{ac}\tau_{ph}}\right) + 2\omega_r^2\overline{R}_o^{sp}N_k}{\left(\omega_r^2-\omega^2\right)^2+(\omega\gamma)^2}.
\end{aligned}
\tag{5.144}
$$

Both the power and correlation spectra have low-pass resonant behavior with the same resonant frequency as the small-signal resonant frequency ω_r.

5.13.3 Output power fluctuations

On the basis of the above discussion, the photon density in the cavity can be modeled as a random process with average N_{k0} and fluctuation δN_k:

$$N_k(t) = N_{k0} + \delta N_k(t),$$

where the fluctuation has the Fourier representation (5.141) as a function of the Langevin sources F_n and F_k. The cavity output power depends on the photon flux transmitted by the mirrors. Since photons disappear from the cavity because of the mirror loss with a lifetime τ_m (mirror loss lifetime), the average transmitted power is related to the average photon number as

$$P_{out} = \hbar\omega \frac{V N_{k0}}{\tau_m}.$$

However, the fluctuation in the instantaneous output power, denoted for brevity as $\delta P(t)$, does not depend only on the cavity photon density fluctuations δN_k. In fact, the exterior of the cavity can be considered as another reservoir fed by an average "photon generation rate" $V N_{k0}/\tau_m$. Since photons escape the cavity through mirrors at random times, mirror transmission is a cause of fluctuations in the number of transmitted photons, which again can be modeled through a shot noise approach. Let us define such power fluctuations in terms of the random process $F_{pn}(t)$, with autocorrelation $R_{pn}(\tau)$ (pn stands for *partition noise* due to mirrors). Expressing the random processes in terms of *particle number fluctuations* we have, from the Campbell theorem,

$$\left(\frac{1}{\hbar\omega}\right)^2 R_{pn}(\tau) = \left(\frac{1}{\hbar\omega}\right)^2 \left\langle F_{pn}(t) F_{pn}(t-\tau)\right\rangle = \frac{V N_{k0}}{\tau_m}\delta(\tau),$$

i.e.,

$$R_{pn}(\tau) = (\hbar\omega)^2 \frac{V N_{k0}}{\tau_m}\delta(\tau) = \hbar\omega P_{out}\delta(\tau), \qquad S_{pn}(\omega) = \hbar\omega P_{out}.$$

We can finally express the power fluctuations $\delta P(\omega)$ as

$$\begin{aligned}
\delta P(\omega) &= \frac{\hbar\omega V}{\tau_m}\delta N_k(\omega) + F_{pn}(\omega) \\
&= \frac{\hbar\omega V}{\tau_m} \frac{\omega_r^2 \Gamma_{ov}\tau_{ph} F_n(\omega) + (\mathrm{j}\omega + \gamma) F_k(\omega)}{\left(\omega_r^2 - \omega^2\right) + \mathrm{j}\omega\gamma} + F_{pn}(\omega).
\end{aligned}$$

Thus, the power spectrum of the power fluctuations, $S_P(\omega)$, will be

$$\begin{aligned}
S_P(\omega) &= \left\langle \delta P(\omega)\delta P^*(\omega')\right\rangle \\
&= \left(\frac{\hbar\omega V}{\tau_m}\right)^2 S_{\delta N_k \delta N_k}(\omega) + \frac{\hbar\omega V}{\tau_m}\frac{(\mathrm{j}\omega + \gamma)\left\langle F_k(\omega) F_{pn}^*(\omega')\right\rangle}{\left(\omega_r^2 - \omega^2\right) + \mathrm{j}\omega\gamma} \\
&+ \frac{\hbar\omega V}{\tau_m}\frac{(-\mathrm{j}\omega + \gamma)\left\langle F_{pn}(\omega) F_k^*(\omega')\right\rangle}{\left(\omega_r^2 - \omega^2\right) - \mathrm{j}\omega\gamma} + S_{pn}(\omega),
\end{aligned} \tag{5.145}$$

where we have taken into account that photon fluctuations due to mirror partition noise are uncorrelated with population fluctuations. Concerning the evaluation of $\left\langle F_k(\omega) F_{pn}^*(\omega')\right\rangle$, the process $F_k = F_P/V$, see (5.134), is the Langevin force for photon density fluctuations. The only part physically correlated with F_{pn} is the fluctuation related to mirror losses. Reducing F_k and F_{pn} to particle number fluctuations we conclude that the correlated part is indeed the same process, with average value $V N_{k0}/\tau_m$

(the photon disappearance rate due to mirror loss). Applying the Campbell theorem we therefore obtain

$$\left\langle F_P(\omega)\frac{F^*_{pn}(\omega')}{\hbar\omega}\right\rangle = \frac{V}{\hbar\omega}\left\langle F_k(\omega)F^*_{pn}(\omega')\right\rangle = -\frac{VN_{k0}}{\tau_m}$$
$$\rightarrow \left\langle F_k(\omega)F^*_{pn}(\omega')\right\rangle = -\frac{P_{out}}{V}. \tag{5.146}$$

The minus sign in the correlation function takes into account that a positive fluctuation on the cavity photon number implies that a photon is *not* transmitted – i.e., a negative fluctuation in the output power. Substituting into (5.145) $S_{\delta N_k \delta N_k}(\omega)$ from (5.144) and taking into account (5.146), we find

$$\begin{aligned} S_P(\omega) &= \left(\frac{\hbar\omega V}{\tau_m}\right)^2 \frac{2\overline{R}^{sp}_o N_{k0}\Gamma_{ov}\left(\dfrac{1}{\tau_n^2}+\omega^2\right) + 2\gamma\omega_r^2\dfrac{N_{k0}}{V} + 2\dfrac{\left(\omega_r^2\Gamma_{ov}\tau_{ph}\right)^2 n}{V_{ac}\tau_{nr}}}{\left(\omega_r^2-\omega^2\right)^2+(\omega\gamma)^2} \\ &\quad - 2\frac{\hbar\omega V}{\tau_m}\frac{\gamma\omega_r^2}{\left(\omega_r^2-\omega^2\right)^2+(\omega\gamma)^2}\frac{P_{out}}{V} + \hbar\omega P_{out} \\ &= \hbar\omega P_{out}\left[\frac{\dfrac{2V\overline{R}^{sp}_o\Gamma_{ov}}{\tau_m}\left(\dfrac{1}{\tau_n^2}+\omega^2\right)}{\left(\omega_r^2-\omega^2\right)^2+(\omega\gamma)^2}+1\right] + \frac{1}{\tau_{nr}}\left(\frac{\hbar\omega}{\tau_m}\right)^2\frac{2\Gamma_{ov}\omega_r^4\tau_{ph}^2 n}{\left(\omega_r^2-\omega^2\right)^2+(\omega\gamma)^2}. \end{aligned} \tag{5.147}$$

Equation (5.147) can be reformulated, introducing from (5.105) the magnitude (squared) of the small-signal normalized modulation response:

$$|H(\omega)|^2 = \frac{\omega_r^4}{\left(\omega_r^2-\omega^2\right)^2+(\omega\gamma)^2}. \tag{5.148}$$

We obtain:

$$\begin{aligned} S_P(\omega) &= \hbar\omega P_{out}\left[\frac{2V_{ac}\overline{R}^{sp}_o}{\omega_r^2\tau_m}\left(\frac{\omega^2}{\omega_r^2}+\frac{1}{\omega_r^2\tau_n^2}\right)|H(\omega)|^2+1\right] \\ &\quad + \frac{2(\hbar\omega)^2 V_{ac}\tau_{ph}^2 n}{\tau_m^2\tau_{nr}}|H(\omega)|^2 \\ &= \hbar\omega P_{out}\left[\left(a_1\omega^2+a_2\right)|H(\omega)|^2+1\right] + a_3|H(\omega)|^2. \end{aligned} \tag{5.149}$$

The coefficients a_1 and a_2 depend on P_{out} through the resonance frequency ω_r and damping factor γ. Fluctuations in the output power globally follow a low-pass frequency behavior similar to the small-signal response, with a magnitude proportional to the output power. However, the term $+1$ in square brackets implies that also when $H \rightarrow 0$ the power spectrum of power fluctuations cannot be lower than $\hbar\omega P_{out}$; this can be interpreted as the *shot noise limit* for $S_P(\omega)$ (in fact, $(\hbar\omega)^{-2} S_P(\omega) = P_{out}/\hbar\omega$ from the Campbell theorem is the shot noise limit).

5.13.4 Relative intensity noise

Suppose that a digital data stream is transmitted by the laser on–off direct modulation; the laser signal is then detected by a photodetector. In the ON state the laser power is P_{out} and the signal to noise ratio (SNR) at the detector output (assuming the detector is noiseless) is given by

$$\mathrm{SNR} = \frac{\langle i_S^2 \rangle}{\langle i_N^2 \rangle} = \frac{(\Re P_{out})^2}{\langle (\Re \delta P(t))^2 \rangle} = \frac{(P_{out})^2}{\langle \delta P(t)^2 \rangle},$$

where $\Re$ is the detector responsivity. The quadratic average of the power fluctuations can be recovered by integrating the power spectrum over the detector bandwidth; assuming that a narrowband filter was used with center frequency ω and bandwidth $\Delta\omega$ one has

$$\mathrm{SNR} \approx \frac{P_{out}^2}{S_P(\omega)\Delta\omega}.$$

The above relation can be expressed in terms of the so called relative intensity noise parameter (RIN), defined as

$$\mathrm{RIN} = \frac{\langle \delta P(t)^2 \rangle}{P_{out}^2} = \frac{1}{\mathrm{SNR}};$$

equivalently, for a narrowband filtered detector, we obtain the RIN per unit bandwidth (of the modulating signal):

$$\frac{\mathrm{RIN}}{\Delta f} = \frac{S_P(f)}{P_{out}^2}. \tag{5.150}$$

Example 5.9: Suppose for a digital link we require BER $= 1 \times 10^{-9}$. What is the minimum RIN for 1 Gbps or 10 Gbps transmission?

Taking into account that

$$\mathrm{BER} \approx \frac{\exp(-Q^2/2)}{Q\sqrt{2\pi}}, \quad \mathrm{SNR} = 4Q^2,$$

for BER $= 1 \times 10^{-9}$ one has $Q \approx 6$ and SNR ≈ 21.6 dB. Thus RIN $= -21.6$ dB and for 1 Gbps transmission

$$\frac{\mathrm{RIN}}{\Delta f} \approx 10 \log_{10}\left(\frac{1}{10^9 \cdot 4 \cdot 36}\right) = -111.6\,\mathrm{dB/Hz},$$

whereas for 10 Gbps transmission the RIN per unit bandwidth will be -121.6 dB/Hz.

In (5.150), we defined the power spectrum for positive frequencies only; for a bilateral power spectrum a factor 2 should be introduced. From (5.149), giving the bilateral spectrum $S_P(\omega)$, we therefore obtain

$$\frac{\mathrm{RIN}}{\Delta f} = \frac{2\hbar\omega}{P_{out}}\left[\frac{2V_{ac}\overline{R}_o^{sp}}{\omega_r^2\tau_m}\left(\frac{\omega^2}{\omega_r^2}+\frac{1}{\omega_r^2\tau_n^2}\right)|H(\omega)|^2+1\right] + \frac{1}{P_{out}^2}\frac{4(\hbar\omega)^2 V_{ac}\tau_{ph}^2 n}{\tau_m^2\tau_{nr}}|H(\omega)|^2, \quad (5.151)$$

where $2\hbar\omega/P_{out}$ is the RIN shot noise limit. From (5.106) and neglecting gain compression, we have

$$\omega_r = \sqrt{a\frac{c_0}{n_r}\frac{N_{k0}}{\tau_{ph}}} = \sqrt{\frac{a}{V}\frac{c_0}{n_r}\frac{\tau_m}{\tau_{ph}}\frac{VN_{k0}}{\tau_m}} = k\sqrt{P_{out}},$$

where k is a proportionality factor; therefore

$$\frac{\mathrm{RIN}}{\Delta f} \approx \frac{4(\hbar\omega)V_{ac}\overline{R}_o^{sp}}{P_{out}^3 k^4\tau_m}\left(\omega^2+\frac{1}{\tau_n^2}\right)|H(\omega)|^2 + \frac{4(\hbar\omega)^2 V_{ac}\tau_{ph}^2 n}{P_{out}^2\tau_m^2\tau_{nr}}|H(\omega)|^2 + \frac{2(\hbar\omega)}{P_{out}}$$

i.e., the low-frequency value of RIN decreases vs. P_{out}, at low power, by 30 dB per decade. At high power, on the other hand, the RIN decreases vs. P_{out} by 10 dB per decade, ultimately reaching the shot noise limit. It can readily be shown that the RIN increases from a low-frequency plateau, peaks at the resonance frequency ω_r and then decreases to the shot noise limit at very high frequency. If we assume that the shot noise contribution is negligible at low frequency, the peak and low-frequency RIN are proportional, and the peak RIN decreases with increasing power. Since the resonant value is typically much larger than the low-frequency value, bandwidth limitations are required at the receiver to limit the effect on the laser noise.[22]

Example 5.10: Evaluate the RIN and the small-signal response for a laser with the following parameters:

$$\begin{array}{llll}
\tau_m = 3\,\mathrm{ps} & V_{ac} = 300\times 2\times 0.01\ (\mu\mathrm{m})^3 & & \Gamma_{ov} = 0.03 \\
\tau_{nr} = 0.1\,\mathrm{ns} & \lambda = 1.55\,\mu\mathrm{m} & & \gamma = K\omega_r^2 + \dfrac{1}{\tau_n} \\
\tau_{ph} \approx 3\,\mathrm{ps} & \overline{R}_o^{sp} = 10^{23}\,\mathrm{cm^{-3}/s} & & K = 0.4\,\mathrm{ns} \\
\tau_n = 3\,\mathrm{ns} & \omega_r = 2\pi\cdot 10\times 10^9\sqrt{\dfrac{P}{25\,\mathrm{mW}}}\ \mathrm{rad/s}, & & n = 10^{19}\,\mathrm{cm^{-3}}
\end{array}$$

using (5.151) and (5.148). Exploit for the damping factor the relationship including gain compression (K factor) in evaluating the small-signal response.

[22] In the above treatment we have neglected the effect of the noisy laser diode driving current, see [83] Section 5.5.

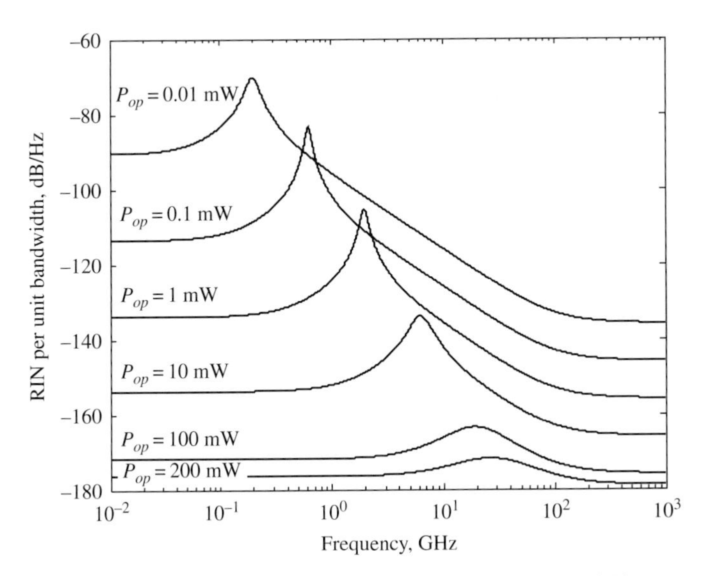

Figure 5.40 RIN per unit bandwidth response, parameters as in Example 5.10.

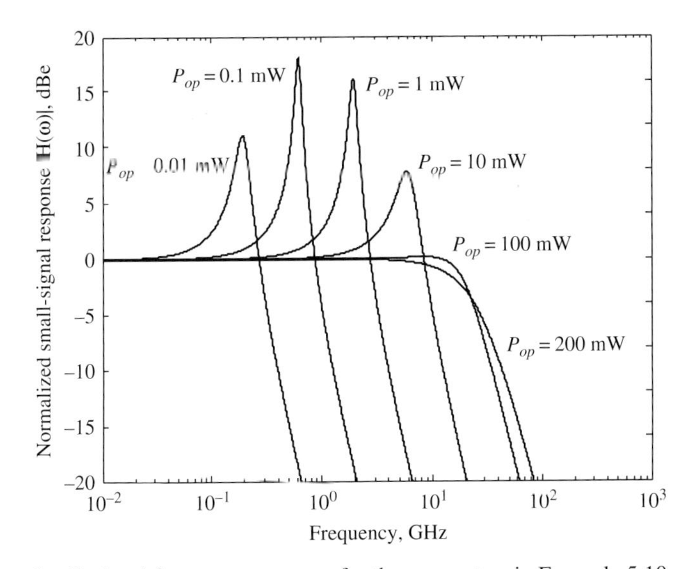

Figure 5.41 Small-signal frequency response for the parameters in Example 5.10.

The resulting RIN response is plotted in Fig. 5.40; the RIN decreases with increasing power, initially at about 30 dB per decade of power decrease; the resonant peak is clearly visible and for very large frequency the RIN decreases down to the shot noise limit. The small-signal response and the RIN response are closely related as far as the peak is concerned, see Fig. 5.41; notice that the resonant frequency increases with the laser power, but also the damping factor. For the largest power we notice a decrease in the small-signal bandwidth. The theoretical maximum 3 dB bandwidth is $2\sqrt{2}\pi/K = 2\sqrt{2}\pi/(0.4 \times 10^{-9}) = 22\,\text{GHz}$.

5.13.5 Phase noise and linewidth from the Langevin approach

Phase noise and the resulting laser linewidth can be recovered through an extension of the Langevin approach. From (5.111) we have, replacing finite differences with differentials:

$$\mathrm{d}f = \frac{\alpha_H}{4\pi}\Gamma_{ov} a \frac{c_0}{n_r}\,\mathrm{d}n. \tag{5.152}$$

Interpreting variations as fluctuations ($\mathrm{d}f \to \delta f$) and taking into account that frequency fluctuations are proportional to the derivative of phase fluctuations $\delta\phi$, we can obtain from (5.152) a rate equation for $\delta\phi$:

$$\frac{\mathrm{d}\delta\phi(t)}{\mathrm{d}t} = 2\pi\delta f(t) = \frac{\alpha_H}{2}\Gamma_{ov} a \frac{c_0}{n_r}\delta n(t) + F_\phi(t).$$

where the first term in the r.h.s., proportional to $\delta\phi$, is related to the modulation of the cavity effective index, while the Langevin force $F_\phi(t)$ expresses single-photon fluctuations. In the frequency domain we have

$$\delta f(\omega) = \frac{\alpha_H}{4\pi}\Gamma_{ov} a \frac{c_0}{n_r}\delta n(\omega) + \frac{1}{2\pi}F_\phi(\omega). \tag{5.153}$$

We now analyze the Langevin force F_ϕ. From the linewidth analysis, we already found that photon fluctuations lead to a direct phase fluctuation $\delta\phi''$ and to an indirect contribution $\delta\phi'$ related to amplitude fluctuations via the Henry parameter. We do not consider the latter contribution $\delta\phi'$, because this is included, see (5.153), in the term proportional to $\delta n(t)$. We already found that (see (5.59) with a slightly different notation)

$$\delta\phi'' \approx \frac{1}{\sqrt{I}}\sin(\theta)$$

where θ is a uniformly distributed random angle and I is the photon number. The Langevin term corresponding to a set of *spontaneous photon emissions* (stimulated emissions are coherent and do not cause direct phase fluctuations) is therefore

$$F_\phi = \sum_i \frac{1}{\sqrt{I}}\sin(\theta_i)\delta(t - t_i).$$

The autocorrelation function of F_ϕ will be

$$\begin{aligned} R_{\phi\phi}(\tau) &= \left\langle F_\phi(t)F_\phi(t-\tau)\right\rangle \\ &= \left\langle \sum_i \frac{1}{\sqrt{I}}\sin(\theta_i)\delta(t-t_i)\sum_j \frac{1}{\sqrt{I}}\sin(\theta_j)\delta(t-\tau-t_j)\right\rangle \\ &= \frac{1}{I}\left\langle \sum_i \sin^2(\theta_i)\delta(t-t_i)\delta(t-\tau-t_i)\right\rangle, \end{aligned}$$

since generation or recombination events at different times are completely uncorrelated. As already discussed, we have $\delta(t-t_i)\delta(t-\tau-t_i) = \delta(\tau)\delta(t-t_i)$, and therefore

$$
\begin{aligned}
R_{\phi\phi}(\tau) &= \frac{1}{I}\left\langle \sum_i \sin^2(\theta_i)\delta(t-t_i)\right\rangle \delta(\tau) \\
&= \frac{1}{I}\delta(\tau) \lim_{T\to\infty}\frac{1}{T}\int_{-T/2}^{T/2}\sum_{i=1}^{\overline{r}_o^{sp}T}\left\langle \sin^2(\theta_i)\right\rangle \delta(t-t_i)\,\mathrm{d}t \\
&= \frac{1}{2I}\delta(\tau)\lim_{T\to\infty}\frac{1}{T}\cdot\overline{r}_o^{sp}T = \frac{V_{ac}\overline{R}_o^{sp}}{2I}\delta(\tau) = \frac{V_{ac}\overline{R}_o^{sp}}{2N_{k0}V}\delta(\tau) = \frac{\Gamma_{ov}\overline{R}_o^{sp}}{2N_{k0}}\delta(\tau),
\end{aligned}
$$

where $\overline{R}_o^{sp} = \overline{r}_o^{sp}/V_{ac}$ is again the recombination rate per unit volume leading to spontaneous emission into mode k. Now, field fluctuations providing only magnitude variations are coherent, and therefore yield zero-phase variation; conversely, field fluctuations providing only phase variations are in quadrature, and therefore lead to zero-magnitude fluctuation; thus, the phase and amplitude (or photon number) fluctuations are uncorrelated. Similarly, carrier number fluctuations are uncorrelated with direct phase fluctuations related to the Langevin force F_ϕ. Taking into account such comments, we have for the power spectrum of the frequency fluctuations simply

$$
\begin{aligned}
S_{\delta f\delta f}(\omega) &= \left(\frac{\alpha_H}{4\pi}\Gamma_{ov}\frac{c_0}{n_r}a\right)^2 S_{\delta n\delta n}(\omega) + \left(\frac{1}{2\pi}\right)^2 S_{\phi\phi}(\omega) \\
&= \left(\frac{\alpha_H}{4\pi}\Gamma_{ov}\frac{c_0}{n_r}a\right)^2 \left[\frac{2\overline{R}_o^{sp}N_{k0}}{\omega_r^2\Gamma_{ov}}\left(\frac{\omega^2}{\omega_r^2}+\frac{1}{\omega_r^2\tau_{ph}^2}\right)\right. \\
&\quad \left. +\frac{\omega^2}{\omega_r^4}\left(2\frac{n}{V_{ac}\tau_{nr}}\right)\right]|H(\omega)|^2 + \left(\frac{1}{2\pi}\right)^2\frac{\Gamma_{ov}\overline{R}_o^{sp}}{2N_{k0}}.
\end{aligned}
$$

Note that the low-frequency value of the frequency fluctuation spectrum is

$$
S_{\delta f\delta f}(0) = \left(\frac{1}{2\pi}\right)^2\frac{\Gamma_{ov}\overline{R}_o^{sp}}{2N_{k0}}\left(1+\alpha_H^2\right) = \frac{1}{2\pi}\Delta f,
$$

where Δf is the laser linewidth; in fact from (5.72) we have[23]

[23] The result coincides with (5.71); in fact,

$$
\Delta f = \frac{1+\alpha_H^2}{4\pi}\frac{\Gamma_{ov}\overline{R}_o^{sp}}{N_{k0}} = \frac{1+\alpha_H^2}{4\pi}\frac{\Gamma_{ov}\overline{r}_o^{sp}}{V_{ac}N_{k0}} = \frac{1+\alpha_H^2}{4\pi}\frac{\overline{r}_o^{sp}}{VN_{k0}} = \frac{1+\alpha_H^2}{4\pi I}\overline{r}_o^{sp},
$$

where $VN_{k0} = I$ is the field intensity normalized to the photon number. Notice that the Schawlow–Townes linewidth ($\alpha_H = 0$) can be also expressed as

$$
\Delta f_{ST} = \frac{1}{4\pi I}\overline{r}_o^{sp}.
$$

$$\Delta f = \frac{1+\alpha_H^2}{4\pi}\frac{hf}{\tau_{ph}P_{out}}\overline{r}_o^{sp} = \frac{1+\alpha_H^2}{4\pi}\frac{hf}{\tau_{ph}\dfrac{hfVN_{k0}}{\tau_{ph}}}V_{ac}\overline{R}_o^{sp}$$

$$= \frac{1+\alpha_H^2}{4\pi}\frac{V_{ac}\overline{R}_o^{sp}}{VN_{k0}} = \frac{1+\alpha_H^2}{4\pi}\frac{\Gamma_{ov}\overline{R}_o^{sp}}{N_{k0}}. \quad (5.154)$$

Moreover, defining the so-called Schawlow–Townes linewidth $\Delta f_{ST} = \Delta f(\alpha_H = 0)$ (i.e., the linewidth in the absence of the broadening effect related to α_H), we obtain

$$S_{\delta f \delta f}(\omega) = \frac{\Delta f_{ST}}{2\pi}\left(1 + \alpha_H^2 |H(\omega)|^2\right). \quad (5.155)$$

But, how does the frequency behavior of the frequency fluctuations spectrum (and in particular the resonant peaks) impact on the laser spectral output shape and linewidth? We start from the expression of the optical field power spectrum:

$$S_E(\omega) = \int_{-\infty}^{\infty} R_E(\tau)\exp(-\mathrm{j}\omega\tau)\,\mathrm{d}\tau.$$

The autocorrelation $R_E(\tau)$ can be expressed, neglecting intensity noise, as in (5.67); combining (5.67) and (5.68), in the hypothesis that $\delta\phi$ is gaussian, we obtain that $R_E(\tau)$ can be expressed as a function of the process variance $\sigma_{\delta\phi}$:

$$R_E(\tau) = I\mathrm{e}^{\mathrm{j}\omega_0\tau}\mathrm{e}^{-\sigma_{\delta\phi}/2}. \quad (5.156)$$

We can generally write the variance $\sigma_{\delta\phi}$ as follows:

$$\sigma_{\delta\phi}(\tau) = \left\langle \delta\phi^2 \right\rangle = \langle [\delta\phi(t+\tau) - \delta\phi(t)]\,[\delta\phi(t+\tau) - \delta\phi(t)] \rangle$$

$$= \left\langle \delta\phi^2(t+\tau) \right\rangle + \left\langle \delta\phi^2(t) \right\rangle - 2\,\langle \delta\phi(t+\tau)\delta\phi(t) \rangle$$

$$= 2R_{\delta\phi\delta\phi}(0) - 2R_{\delta\phi\delta\phi}(\tau),$$

where $R_{\delta\phi\delta\phi}(\tau)$ is the autocorrelation function of $\delta\phi$. The related power spectrum can be derived from the power spectrum of frequency fluctuations (5.155) as $S_{\delta\phi\delta\phi} = \omega^{-2}S_{\delta f \delta f}$. Taking into account that

$$R_{\delta\phi\delta\phi}(\tau) = \frac{1}{2\pi}\int_{-\infty}^{\infty} S_{\delta\phi\delta\phi}(\omega)\exp(\mathrm{j}\omega\tau)\,\mathrm{d}\omega,$$

we immediately have

$$\sigma_{\delta\phi}(\tau) = \frac{1}{\pi}\int_{-\infty}^{\infty} S_{\delta\phi\delta\phi}(\omega)\,\mathrm{d}\omega - \frac{1}{\pi}\int_{-\infty}^{\infty} S_{\delta\phi\delta\phi}(\omega)\exp(\mathrm{j}\omega\tau)\,\mathrm{d}\omega$$

$$= \frac{1}{\pi}\int_{-\infty}^{\infty} S_{\delta\phi\delta\phi}(\omega)\,(1 - \cos\omega\tau)\,\mathrm{d}\omega = \frac{1}{\pi}\int_{-\infty}^{\infty} \frac{1-\cos\omega\tau}{\omega^2} S_{\delta f \delta f}(\omega)\,\mathrm{d}\omega$$

$$= 2\Delta f_{ST}\int_{-\infty}^{\infty} \frac{1+\alpha_H^2\,|H(\omega)|^2}{\omega^2}(1 - \cos\omega\tau)\,\mathrm{d}\omega$$

where we have taken into account that the power spectrum is a real even function of the angular frequency. The integral

$$A = \int_{-\infty}^{\infty} \frac{1+\alpha_H^2 |H(\omega)|^2}{\omega^2}(1-\cos\omega\tau)\, \mathrm{d}\omega$$
$$= \int_{-\infty}^{\infty} \left[\frac{1}{\omega^2} + \frac{1}{\omega^2}\frac{\alpha_H^2\omega_r^4}{\left(\omega_r^2-\omega^2\right)^2+(\omega\gamma)^2}\right](1-\cos\omega\tau)\, \mathrm{d}\omega \tag{5.157}$$

can be solved by contour integration in the complex plane. As shown in Example 5.11, we obtain

$$\sigma_{\delta\phi}(\tau) = 2\Delta f_{ST} A = 2\pi\Delta f_{ST}\left\{\left(1+\alpha_H^2\right)|\tau| + \frac{\alpha_H^2\left(\omega_r^2-\gamma^2\right)}{2\sigma\omega_r^2}\right.$$
$$\left. - \frac{\alpha_H^2 \mathrm{e}^{-\sigma|\tau|}}{2\omega_r^2}\left[\frac{\omega_r^2-\gamma^2}{\sigma}\cos\omega_R|\tau| + \frac{3\omega_r^2-\gamma^2}{\omega_0}\sin\omega_R|\tau|\right]\right\}, \tag{5.158}$$

where

$$\omega_R = \sqrt{\omega_r^2-\sigma^2}, \quad \sigma = \frac{1}{2}\gamma.$$

The variance therefore exhibits a first term increasing with $|\tau|$, which has already been derived in the simplified analysis (see Section 5.10.1, Eq. (5.65)). The Langevin technique further suggests the presence of a damped oscillatory component (note that the constant term is such as to lead to $\sigma_{\delta\phi}(0) = 0$).

Example 5.11: Evaluate the integral A in (5.157) by contour integration.

Owing to the parity of the integrand with respect to ω, we can conveniently express A as

$$A = \underbrace{\lim_{a\to 0}\int_{-\infty}^{\infty}\frac{1-\mathrm{e}^{\mathrm{j}\omega\tau}}{\omega^2+a^2}\mathrm{d}\omega}_{A_1} + \underbrace{\lim_{a\to 0}\int_{-\infty}^{\infty}\frac{\alpha_H^2\omega_r^4}{\omega^2+a^2}\frac{1-\mathrm{e}^{\mathrm{j}\omega\tau}}{\left(\omega_r^2-\omega^2\right)^2+(\omega\gamma)^2}\mathrm{d}\omega}_{A_2}.$$

From (5.157) we immediately note that A is an even function of τ. We can confine the discussion to the case $\tau \geq 0$ and extend to $\tau < 0$, imposing parity. Applying Cauchy's theorem, for the first term one has (we close the integration path in the upper complex ω plane):

$$A_1 = \lim_{a\to 0}\int_{-\infty}^{\infty}\frac{1-\mathrm{e}^{\mathrm{j}\omega\tau}}{(\omega+\mathrm{j}a)(\omega-\mathrm{j}a)}\mathrm{d}\omega = 2\pi\mathrm{j}\lim_{a\to 0}\frac{1-\mathrm{e}^{a\tau}}{2\mathrm{j}a} = \pi\tau, \quad \tau \geq 0$$

i.e., $A_1 = \pi|\tau|$. For the second term, we have ($\tau \geq 0$) three poles in the upper complex ω plane; on applying Cauchy's theorem,

$$A_2 = \lim_{a\to 0}\int_{-\infty}^{\infty}\frac{\alpha_H^2\omega_r^4\left(1-\mathrm{e}^{\mathrm{j}\omega\tau}\right)}{P(\omega)}\mathrm{d}\omega,$$

where

$$P(\omega) = (\omega + ja)(\omega - ja)(\omega - \omega_R - j\sigma) \times (\omega + \omega_R - j\sigma)(\omega - \omega_R + j\sigma)(\omega + \omega_R + j\sigma).$$

Thus, decomposing, we obtain for $\tau \geq 0$:

$$\begin{aligned}
\frac{A_2}{2\pi j} &= \lim_{a\to 0} \frac{1}{2ja} \frac{\alpha_H^2 \omega_r^4 (1 - e^{-a\tau})}{(ja - \omega_R - j\sigma)(ja + \omega_R - j\sigma)(ja - \omega_R + j\sigma)(ja + \omega_R + j\sigma)} \\
&+ \lim_{a\to 0} \frac{1}{(\omega_R + j\sigma + ja)(\omega_R + j\sigma - ja)} \\
&\times \frac{\alpha_H^2 \omega_r^4 (1 - e^{j\omega_R \tau - \sigma\tau})}{(\omega_R + j\sigma + \omega_R - j\sigma)(\omega_R + j\sigma - \omega_R + j\sigma)(\omega_R + j\sigma + \omega_R + j\sigma)} \\
&+ \lim_{a\to 0} \frac{1}{(-\omega_R + j\sigma + ja)(-\omega_R + j\sigma - ja)} \\
&\times \frac{\alpha_H^2 \omega_r^4 (1 - e^{-j\omega_R \tau - \sigma\tau})}{(-\omega_R + j\sigma - \omega_R - j\sigma)(-\omega_R + j\sigma - \omega_R + j\sigma)(-\omega_R + j\sigma + \omega_R + j\sigma)}.
\end{aligned}$$

That is,

$$\begin{aligned}
A_2 &= \pi \frac{\alpha_H^2 \omega_r^4 |\tau|}{\left(\omega_R + \sigma^2\right)^2} + \pi \frac{1}{(\omega_R + j\sigma)^3} \frac{\alpha_H^2 \omega_r^4}{4\omega_R \sigma} \left(1 - e^{j\omega_R \tau - \sigma|\tau|}\right) \\
&+ \pi \frac{1}{(\omega_R - j\sigma)^3} \frac{\alpha_H^2 \omega_r^4}{4\omega_R \sigma} \left(1 - e^{-j\omega_R |\tau| - \sigma|\tau|}\right) \\
&= \pi \alpha_H^2 |\tau| + \frac{\pi \alpha_H^2}{2\sigma \omega_r^2} \left(\omega_r^2 - \gamma^2\right) \\
&- \frac{\pi \alpha_H^2 e^{-\sigma\tau}}{2\omega_r^2} \left(\frac{\omega_r^2 - \gamma^2}{\sigma} \cos \omega_R |\tau| + \frac{3\omega_r^2 - \gamma^2}{\omega_R} \sin \omega_R |\tau| \right).
\end{aligned}$$

Thus, extending to $\tau < 0$ for parity,

$$\begin{aligned}
A &= \pi \left(1 + \alpha_H^2\right) |\tau| + \frac{\pi \alpha_H^2}{2\sigma \omega_r^2} \left(\omega_r^2 - \gamma^2\right) \\
&- \frac{\pi \alpha_H^2 e^{-\sigma\tau}}{2\omega_r^2} \left(\frac{\omega_r^2 - \gamma^2}{\sigma} \cos \omega_R |\tau| + \frac{3\omega_r^2 - \gamma^2}{\omega_R} \sin \omega_R |\tau| \right).
\end{aligned}$$

From $\sigma_{\delta\phi}(\tau)$, we can finally derive the normalized power spectrum of the optical field by taking the Fourier transform of (5.156):

$$\frac{S_E(\omega)}{I} = \int_{-\infty}^{\infty} e^{-\frac{\sigma_{\delta\phi}(\tau)}{2}} e^{-j(\omega - \omega_0)\tau} \, d\tau.$$

Neglecting the oscillatory terms in (5.158) and considering only the first term proportional to $|\tau|$, we have again

$$\sigma_{\delta\phi}(\tau) \approx 2\pi \Delta f_{ST} \left(1 + \alpha_H^2\right) |\tau|,$$

which corresponds to (5.65). Therefore, the spectrum is Lorentzian around ω_0:

$$\frac{S_E(\omega)}{I} = \int_{-\infty}^{\infty} \mathrm{e}^{-\pi \Delta f_{ST}\left(1+\alpha_H^2\right)|\tau|} \mathrm{e}^{-\mathrm{j}(\omega-\omega_0)\tau} \, \mathrm{d}\tau = \frac{2\tau_c}{1+(\omega-\omega_0)^2\tau_c^2},$$

where the *coherence time* is defined as

$$\tau_c = \frac{1}{\pi \Delta f_{ST} \left(1+\alpha_H^2\right)},$$

coinciding with (5.70). The FWHM linewidth is

$$\Delta f = \frac{1}{\pi \tau_c} = \Delta f_{ST} \left(1+\alpha_H^2\right),$$

as already derived in Section 5.10.1.

The effect of the oscillatory component of $\sigma_{\delta\phi}(\tau)$ cannot be readily expressed in closed form, since the resulting integral does not admit a representation in terms of elementary functions.

Considering, however, that the kernel of the integrand includes the exponential of an oscillating function, we can assume that the effect on the resulting power spectrum S_E is somewhat similar to the spectrum of a frequency-modulated signal with modulation frequency ω_R. Qualitatively, the power spectrum will have a Lorentzian shape around ω_0 plus sidebands corresponding to $\omega_0 \pm k\omega_R$ with k integer. However, the sideband amplitude is typically very small, so that the linewidth is, in practical cases, unaffected, as shown in Fig. 5.42 [81]; note that the resonance peaks decrease with increasing laser power, while the displacement of the peaks increases, consistent with the fact that ω_R increases with increasing output power.

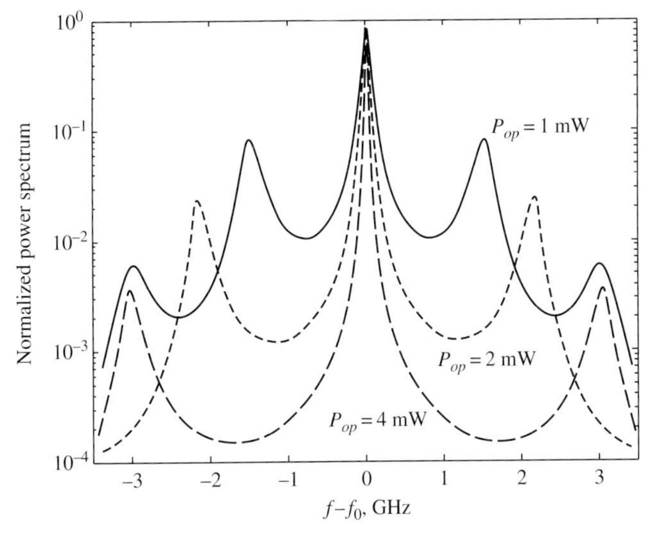

Figure 5.42 Emission spectrum of a laser as a function of emitted power, showing the central Lorentzian peak and the secondary resonances with spacing equal to the laser resonant frequency. Adapted after [81], Fig. 5 (©1983 IEEE).

5.14 Questions and problems

5.14.1 Questions

1. Explain why in a pn homostructure LED with a p surface layer the n substrate layer should by highly doped, while in a heterostructure PpN LED we do not need the substrate to be more doped than the surface layer.
2. Explain why heterostructure LEDs exploit PnN or PpN structures, but not pNn or pPn.
3. Sketch a vertical emission (Burrus) LED structure.
4. Sketch a lateral emission LED structure.
5. Explain the difference between a conventional and superradiant LED.
6. What limits the modulation bandwidth of a LED? What is the maximum theoretical value for non-superradiant LEDs?
7. Justify the LED emission spectrum linewidth (order of magnitude $k_B T$).
8. The LED linewidth can be:
   (a) around 50 GHz;
   (b) around 200 Å at 1 μm emission wavelength;
   (c) around E_g.
9. Explain the high-current saturation of the output LED power.
10. Explain why the LED current–power characteristic is sublinear at low current.
11. The external LED quantum efficiency is lower than the internal one. Why?
12. Suppose that the external quantum efficiency of a LED is unity. Justify the fact that $P_{op} \approx I$ (each in its own units) for materials with a gap around 1 eV.
13. A heterostructure laser exploits a PnN or PpN junction, exactly as a heterostructure LED. Which of those remarks is true?
   (a) Both in lasers and in LEDs the double heterojunction confines carriers *and* photons.
   (b) In a laser the double heterostructure is for carrier confinement, in a LED for photon confinement.
   (c) The laser double heterojunction is for photon and carrier confinement, in a LED it is for carrier confinement only.
14. Explain why the modulation small-signal response of a laser is faster than the small-signal response of a LED.
15. A laser is switched on from below the threshold. Explain why the response is not significantly faster than the LED turn-on response.
16. Define the *threshold* and *transparency* condition in a laser.
17. The spontaneous emission radiative lifetime:
   (a) increases with increasing carrier concentration;
   (b) decreases with increasing carrier concentration;
   (c) decreases with increasing carrier concentration, but ultimately saturates to values less than 1 ns.
18. The stimulated emission lifetime:
   (a) increases with increasing carrier concentration;
   (b) increases with increasing photon concentration;

(c) keeping the carrier concentration constant, decreases with increasing photon concentration.

19. Define the superposition integral Γ_{ov} and explain its physical meaning.
20. Describe a Fabry–Perot and a DFB laser.
21. What is the difference between a DFB and a DBR laser?
22. Explain why the laser spectral purity is much better than the LED spectral purity.
23. Define the order of magnitude of the spectral purity (in Å or GHz at 1 μm emission wavelength) of the Fabry–Perot, DFB, DBR lasers.
24. Discuss the temperature stability of a Fabry–Perot and DBR laser.
25. Explain the advantages of a stripe Fabry–Perot laser vs. a simple one.
26. Discuss the behavior of the threshold current of a laser as a function of the active region thickness d.
27. What is a GRINSCH laser?
28. Discuss tunable lasers. What are the applications of such devices?
29. Describe the structure of a vertical cavity laser (VCSEL), and in particular the mirror realization.
30. Sketch the behavior of the power–current characteristic of a laser.
31. Both in a laser and in a LED the number of photons generated per unit time is proportional to the input current. Explain why, with the same current, the maximum optical power of the laser (vs. the photon energy) is much larger than that of the LED.
32. Discuss the small-signal response of a laser above threshold.
33. Explain the role of gain compression in limiting the maximum modulation bandwidth of a semiconductor laser.
34. Sketch the turn-on response of a laser from below and from above threshold.
35. Explain why the amplitude modulation of a laser leads to a spurious frequency modulation (chirp). What is the Henry chirp parameter?
36. Does positive and negative chirp affect signal transmission on a dispersive fiber in the same way? Explain.
37. Discuss the laser linewidth and explain why this is different from the cavity linewidth.
38. Why is the linewidth of a gas laser better than the linewidth of a semiconductor laser? Explain the role of the chirp parameter α_H (also called linewidth enhancement factor) in this context.
39. What are the physical origins of laser noise? Comment on the interpretation of laser noise in terms of spontaneous emission.
40. What is the power spectrum of laser noise around the carrier wavelength?

5.14.2 Problems

1. A heterojunction LED has an active layer of $In_{0.53}Ga_{0.47}As$ (assume $E_g = 0.8\,eV$). Excess electrons and holes are injected with a concentration $n' \approx p' = 10^{15}\ cm^{-3}$. Calculate the photon generation rate in the system, assuming $E_g = 0.8$, $m_e^* = 0.042m_0$, and $m_h^* = 0.4m_0$. Use $n_r = 3$ and $E_p = 20\,eV$ for the dipole matrix

element energy. Assuming a heterojunction LED with an active region thickness $d = 0.1\ \mu m$ and an area $A = 1\ mm^2$, approximately evaluate the generated power neglecting reflections and nonradiative effects.

2. A homojunction GaAs LED is made with a pn^+ structure (p side on surface) with dopings $N_D = 10^{17}\ cm^{-3}$, $N_A = 10^{16}\ cm^{-3}$. The carrier mobilities are $\mu_n = 3000\ cm^2\ V^{-1}s^{-1}$ and $\mu_h = 500\ cm^2\ V^{-1}\ s^{-1}$, the electron and hole radiative lifetimes are $\tau_{n,r} = \tau_{h,r} = 10$ ns, while nonradiative lifetimes are $\tau_{n,nr} = \tau_{h,nr} = 0.1\ \mu s$. For GaAs assume $n_i = 2.1 \times 10^6\ cm^{-3}$ and operating temperature $T = 300$ K.
   (a) Evaluate the injection efficiency of the device, the radiative efficiency, and the total efficiency, assuming a surface power reflection coefficient of 40%.
   (b) Suppose the LED must transmit a binary sequence of 0s and 1s with powers of 1 and 100 μW. Estimate the total current needed, assuming the total external efficiency evaluated above, and bias voltage corresponding to the low and high levels. Suppose that the area is $A = 1\ mm^2$.
   (c) What is the maximum bit rate that the LED can support in direct modulation?
3. A heterojunction AlGaAs/GaAs/AlGaAs LED is biased at a current of 1 mA; the active region thickness is $d = 1\ \mu m$.
   (a) Estimate the radiative lifetime in the GaAs layer, knowing that the LED bandwidth is 100 MHz (neglect nonradiative recombination). What is the bandwidth for a current of 10 mA? What is the limiting modulation bandwidth of the device (assume a spontaneous radiative lifetime of 0.5 ns)?
   (b) Assuming for GaAs $E_g = 1.42$ eV, estimate the output optical power with 1 mA bias and a device efficiency of 30%. Estimate the total charge stored in the GaAs layer and the injected electron density. The LED area is $A = 0.1\ mm^2$.
4. We want to design a heterojunction InGaAsP/InGaAs/InP LED for 1.3 μm emission, with an area $A = 0.01\ mm^2$ and an output optical power $P_{out} = 10$ mW. The maximum injected charge density in the active InGaAsP layer is $n = 5 \times 10^{17}\ cm^{-3}$. Evaluate the thickness of the active region assuming a radiative lifetime $\tau_{n,r} = 3$ ns and an external device efficiency of 40%.
5. A double PnN heterojunction laser has an active region of InGaAsP with $E_g = 0.8$ eV and thickness $d = 0.2\ \mu m$. Assuming as effective masses $m_n^* = 0.04 m_0$, $m_h^* = 0.35 m_0$, evaluate the injected electron density $n \approx p$ needed to achieve the population inversion condition at 300 K (hint: use the Joyce–Dixon approximation). Assuming a radiative lifetime $\tau_r = 2$ ns, evaluate the corresponding current for a junction effective area $A = 3 \times 300\ \mu m^2$. What might the laser application be?
6. A GaAs Fabry–Perot laser has a cavity length of $L = 125\ \mu m$. Assuming $E_g = 1.42$ eV, $n_{\text{eff}} = 3.3$, evaluate the number of longitudinal cavity modes within the LED bandwidth (i.e., approximately from E_g to $E_g + 2k_BT$, $T = 300$ K).
7. A Fabry–Perot cavity has a cavity loss $\alpha_{loss} = 15\ cm^{-1}$ and a mirror power reflectivity of 35%. Evaluate the photon lifetime for a cavity length $L_1 = 100\ \mu m$. What should be the Bragg mirror reflectivity to obtain the same lifetime in a VCSEL with $L_2 = 2\ \mu m$? Assume an effective refractive index $n_{\text{eff}} = 3.3$.
8. An AlGaAs/GaAs/AlGaAs laser has a cavity length $L = 100\ \mu m$ and a cavity width $w = 5\ \mu m$. The mirror reflectivity is $R = 0.3$, the cladding

loss $\alpha_{loss} = 20\,\text{cm}^{-1}$, and the overlap integral $\Gamma_{ov} = 0.1$. Assume 1.55 μm emission.

(a) Estimate the injected electron density needed at threshold, assuming a differential gain $a = 8.0 \times 10^{-16}$ cm^2 and a transparency electron density $n_{tr} = 1 \times 10^{18}\,\text{cm}^{-3}$.

(b) Assuming a total radiative lifetime $\tau_r = 1.5$ ns, evaluate the threshold current for an active region thickness $d = 0.3$ μm.

(c) Estimate the power–current characteristic of the laser above threshold and compute the output power for $I = 10I_{th}$.

9. A laser with 1.3 μm emission has a cavity length $L = 150$ μm and a cavity width $w = 5$ μm. The threshold current is $I_{th} = 1$ mA and the bias current is $I = 15$ mA. Assume a mirror reflectivity $R = 0.4$, cladding loss $\alpha_{loss} = 10\,\text{cm}^{-1}$, overlap integral $\Gamma_{ov} = 0.1$, and differential gain $a = 20 \times 10^{-20}$ m^2. The cavity refractive index is $n_{\text{eff}} = 3.1$.

(a) Evaluate the laser output power at the given bias and the cavity photon density N. The active region thickness is $d = 0.25$ μm.

(b) Assuming an average below-threshold spontaneous lifetime $\tau_n^{sp} = 4$ ns, evaluate the time required to reach the threshold starting from zero bias and with a final value $I = 15$ mA.

(c) Evaluate the modulation bandwidth of the laser at the given bias and plot the frequency response. Assume the total nonradiative carrier lifetime $\tau_n = 100$ ns. Neglect gain compression.

(d) Assuming for the Henry chirp parameter $\alpha_{H} = 4.5$, evaluate the laser linewidth at the given bias.

10. A laser emitting at 1.55 μm has a threshold current $I_{th} = 1$ mA. The photon lifetime is $\tau_{ph} = 1$ ps. Assuming that the mirror loss is ten times α_{loss}, derive the current–power characteristic. Evaluate the output power for 10 mA current excitation.

11. In a laser, assume a small-signal resonance frequency of 2 GHz, a K factor of 0.3 ns, and a carrier lifetime of 5 ns. Estimate the peak RIN, taking into account that the low-frequency RIN is −110 dB/Hz. (Hint: assume that the optical power is low.)

12. Consider an AlGaAs/GaAs laser (0.8 μm emission) with active region thickness $d = 0.1$ μm. The GaAs index is $n_f = 3.66$, the AlGaAs index is $n_s = 3.52$. Estimate the overlap integral.

13. A DBR laser is made with a cleaved cavity with length $L = 100$ μm and end (power) reflectivity $R = 0.5$ coupled to two gratings of length $L_g = 150$ μm and coupling coefficient such as $\kappa L_g = 4$. Suppose that the emission wavelength is $\lambda_0 = 1.55$ μm. Design the grating periodicity Λ to obtain the desired emission and evaluate the end loss corresponding to the cleaved cavity modes around the selected mode. Assume a cavity refractive index $n_{\text{eff}} = 3.3$.

6 Modulators

6.1 Light modulation and modulator choices

LEDs and lasers biased at a constant current emit CW (continuous wave) light with constant average optical power. To transmit information we need to *modulate* the light source by changing its amplitude and/or phase. For simplicity, we will consider only intensity modulation (IM), where digital or analog information is associated with the light's instantaneous power. In digital IM, binary symbols are associated with intervals of high (ON state) or low (OFF state) optical power; the power ratio between the ON and OFF states is the *extinction* or *contrast ratio* (ER or CR), while the maximum bit rate (maximum bandwidth) of the modulation process is the *modulation speed* (or *modulation bandwidth*).

Light modulation can be *internal* (*direct*) or *external* (*indirect*). Direct modulation is based on the modulation of the instantaneous bias point of the light source (LED or laser); the solution is compact, since no additional external device is required, but has worse extinction ratio and a higher chirp (i.e., spurious frequency modulation associated with the intensity modulation) than indirect modulation schemes. The modulation bandwidth is very limited (below 1 Gbps) in LEDs, much wider (up to 10 Gbps in the field) for lasers.

External (indirect) modulation requires an additional (sometimes large and separately packaged) voltage-driven device, the *light modulator*, acting as a light switch; see Fig. 6.1. The advantages are a higher extinction ratio, lower chirp (ideally zero in some devices), and higher bit rate (above 40 Gbps).

Several solutions exist for the external modulation; the devices of choice today are the electrooptic modulator (EOM) and the electroabsorption modulator (EAM). Both EAMs and EOMs are able to achieve high-speed operation; commercially available devices exist for 40 Gbps systems. Among less conventional modulator choices we mention the current-driven variable-gain semiconductor optical amplifiers (SOAs); the resulting amplitude modulator has large chirp and the contrast ratio is dominated by the gain control dynamics, but, the modulator is active, i.e., it can provide power amplification. In what follows, however, attention will be focused on conventional EOM and EAM modulator approaches.

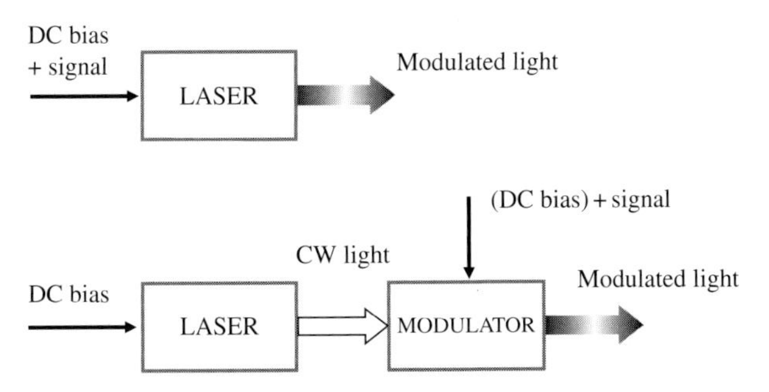

Figure 6.1 Internal (direct) vs. external (indirect) modulation of light.

Electrooptic modulators exploit the modulation of material's refractive index induced by an electric field.[1] The electric field is induced by a voltage applied to a semiconductor junction in reverse bias, or to an electrode system deposited on a dielectric having electrooptic properties. Although several possible applications of this principle in specific devices exist, the most popular today is the Mach–Zehnder (MZ) interferometric modulator, in which IM is carried out through the constructive or destructive interference of two phase-modulated beams. EOMs can be implemented with several electrooptic materials: piezoelectric (perovskites), semiconductor (GaAs, InP, SiGe), and more recently polymers and Si. Compared with EAMs, EOMs generally have lower chirp, higher optical saturation power, and wider optical bandwidth (up to 100 nm); the same device can thus modulate different WDM channels. The optical insertion loss is typically low, since the absorption of the material is negligible (not always, however, in semiconductor modulators). Perovskite-based modulators exhibit interaction lengths up to 1–2 cm and are therefore large, packaged devices, with little perspective for integration; the on–off voltage is typically high (5 V for high-speed devices), thus requiring a complex high-speed, wideband amplifier as a driver. Semiconductor EOMs are smaller (down to 500 μm) and can be integrated (with some difficulty) with the laser source in the so-called EOL (electrooptic integrated laser).

Electroabsorption modulators are based on the modulation of the material absorption by an electric field, typically induced by a voltage applied to a semiconductor *pin* junction in reverse bias. In the ON state, absorption is low and light travels almost unaffected through the device; in the OFF state, absorption is high and the EAM operates as a photodiode with a long active region, so that the unabsorbed light escaping from the device is negligible. Absorption modulation can occur in bulk semiconductors through the Franz–Keldysh effect (FKE) or in a quantum well (QW) via the quantum confined Stark effect (QCSE). Owing to the greater strength of the QCSE, QW-based modulators have better extinction ratio than bulk modulators. EAMs generally have larger chirp than EOMs; moreover, while in the latter chirp is dominated by the device geometry, in EAMs (as in lasers) it is uniquely determined by the material properties. The EAM optical bandwidth is moderately narrow for FKE-based devices (typical values

[1] The refractive index variation can also be induced by injected charge; this effect was exploited in the design of Si-based electrooptic modulators, see Section 6.6.2.

are around 10 nm, but record 40 nm bandwidths have also been demonstrated in EAMs integrated with tunable lasers, see [85]) and very narrow, single-channel, for QCSE-based devices. QCSE-based EAMs are small (interaction length below 500 μm) and potentially have lower on–off voltages (e.g., 2 V or less). Integration with the source is possible and highly convenient in QCSE-based EAMs, where the optical bandwidth is closely matched to the specific source; this device, also referred to as the EAL or electroabsorption integrated laser, is gaining increasing popularity due to its superior performance vs. direct modulation, which is obtained while preserving the source's compactness.

From a physical standpoint, both EAMs and EOMs exploit the modulation of the material refractive index induced by the E-field; in EOMs the modulation of the *real* part n_r is used, whereas EAMs are based on the modulation of the *imaginary* part n_i. However, n_r and n_i are related via to the Kramers–Kronig relations, and so is their variation in the presence of an applied field. Such an interdependence is the basic cause of chirp in EAMs (as in lasers), since in this device intensity modulation implies a certain amount of phase modulation. In EOMs, on the other hand, the absorption coefficient is so low (both in the perovskite and in the semiconductor case) that its variation has no appreciable influence on the device performance; in such devices chirp originates from the device geometry and dissymmetry.

6.2 Modulator parameters

Modulators (electrooptic or electroabsorption) are characterized by a number of system parameters; some of these can be specific to the *digital* or *analog* nature of the IM. In *digital IM*, the optical carrier is modulated by a baseband bit sequence whose frequency bandwidth ideally extends from DC (or a few hundred kHz) to a maximum frequency B related to the baseband bit rate R_b; for simplicity we will assume $B \approx R_b$. In *analog IM*, on the other hand, an analog, often narrowband, signal modulates the optical carrier.

Analog IM is in principle compatible with a modulating system having narrow modulation bandwidth but high central frequency; digital IM requires instead a modulating system whose bandwidth ranges approximately from DC to R_b.[2] Moreover, analog modulation should ideally be a linear process; since practical modulators are not linear, analog operation typically occurs under small-signal conditions. Modulator *linearity* and *distortion* are important parameters in analog IM, less so in digital IM.

6.2.1 Electrooptic (static) response

The electrooptic response, also called the transmission characteristics or switching curve, is the ratio of the output and input light intensity or power vs. the electrical

[2] In practice, however, special-purpose, narrowband analog modulators are uncommon, and wideband devices are typically exploited both for digital and analog modulation.

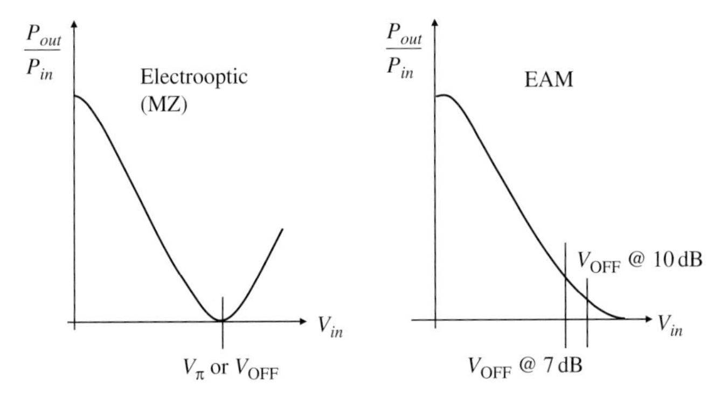

Figure 6.2 Static (DC) input–output characteristics of electrooptic (left) and electroabsorption (right) modulator.

input voltage in static conditions, i.e., for a DC or (sometimes) low-frequency voltage input:

$$T(V_{in}) = \frac{P_{out}(V_{in})}{P_{in}}.$$

Typical EOM and EAM responses are shown in Fig. 6.2. In EOMs the response is periodic vs. the input voltage, while in EAMs it is ideally monotonic (however, in real devices the response often has a broad minimum, and then increases for large values of V_{in}). A number of characteristic parameters can be derived from $T(V_{in})$.

The *off-state voltage*, V_{OFF} (often called V_π with reference to EOM operation, see Section 6.4.2) is the voltage required to turn off the input light at the output with a specific extinction or contrast ratio ER > 1 (see later); i.e., in natural and log units:

$$T(V_{ON})/T(V_{OFF}) = \text{ER}$$
$$T(V_{ON})|_{dB} - T(V_{OFF})|_{dB} = \text{ER}|_{dB} = 10\log_{10}\text{ER}.$$

The *on-state voltage* V_{ON} (often $V_{ON} \approx 0$) corresponds to maximum transmission. In practice $T(V_{ON}) < 1$, since the modulator is affected by a residual modulator loss in the ON state called the *optical insertion loss* L_{op}:

$$L_{op} = \frac{P_{out}(V_{ON})}{P_{in}} \equiv T(V_{ON}).$$

Finally $V_{SW} = V_{OFF} - V_{ON} \approx V_{OFF}$ is the on–off voltage, also called the *switching voltage*.

The *extinction ratio* or *contrast ratio* ER > 1 or $\text{ER}|_{dB} > 0$ (also referred to as CR) is the ratio of the ON state output light intensity to the OFF state output light intensity. It describes the ability of the modulator to switch the input light down to a specific level. Due to the response periodicity, for the EOM the extinction ratio is uniquely defined as

$$\text{ER}|_{dB} = T(0)|_{dB} - T(V_\pi)|_{dB}\,.$$

In EAMs, the extinction ratio increases with V_{OFF}; therefore, V_{OFF} refers to a specific ER (e.g., we define V_{OFF}@7 dB, meaning the OFF voltage that yields an extinction ratio of 7 dB).

6.2.2 Dynamic response

The dynamic large-signal response is the electrooptic modulator response when the input signal is time-varying and has arbitrary amplitude. Since the static modulator response is not strictly linear, see Fig. 6.2, and modulator structures typically include *memory effects*, the modulator large-signal dynamic response should generally be modeled by a *dynamic nonlinear* system; this implies that the response changes according to both the speed and the amplitude of the input driving signal. In digital applications, the large-signal response is often represented in terms of the output eye diagram with different input signal speeds. In most practical cases, the dynamic response has a low-pass frequency behavior, i.e., the modulator response deteriorates with increasing input bit rate. In order to rigorously define the frequency-domain response, we can conveniently investigate the modulator behavior in linear, small-signal operation.

6.2.3 Small-signal frequency response

To define the small-signal modulator frequency response, assume a (small-signal) sinusoidal input voltage at angular frequency ω (the modulation frequency) superimposed to a bias voltage $V_{in,DC}$:

$$v_{in}(t) = V_{in,DC} + \widehat{v}_{in}(t) = V_{in,DC} + \mathrm{Re}\left(\widehat{V}_{in} \exp(\mathrm{j}\omega t)\right),$$

where $\widehat{V}_{in}$ is the phasor associated with the small-signal input voltage. Since the system operates, according to the small-signal assumption, in linearity, the optical instantaneous output power will have a DC and a harmonic modulation component at ω, as

$$P_{out}(t) = P_{out,DC} + \widehat{p}_{out}(t) = P_{out,DC} + \mathrm{Re}\left(\widehat{P}_{out} \exp(\mathrm{j}\omega t)\right).$$

$\widehat{P}_{out}$ is a complex phasor, whose phase expresses the phase difference between the input voltage and the output (small-signal) modulation component of the optical power; see Fig. 6.3. The complex transfer function $M(\omega) = \widehat{P}_{out}/\widehat{V}_{in}$ can be assumed as the small-signal frequency response of the modulator, often normalized (considering only the magnitude) to the low-frequency or DC value as $m(\omega) = |M(\omega)/M(0)|$.

The limit value $M(0)$ can be derived from linearizing the DC electrooptic response. In fact, in quasi-static conditions (DC, low frequency) where memory effects are negligible, we have ($v_{in}(t) = V_{in,DC} + \widehat{v}_{in}(t)$)

$$T(v_{in}(t)) \approx T\left(V_{in,DC}\right) + \left.\frac{\mathrm{d}T}{\mathrm{d}v_{in}}\right|_{DC} \widehat{v}_{in}(t) = \frac{P_{out,DC}}{P_{in}} + \frac{\widehat{p}_{out}(t)}{P_{in}}.$$

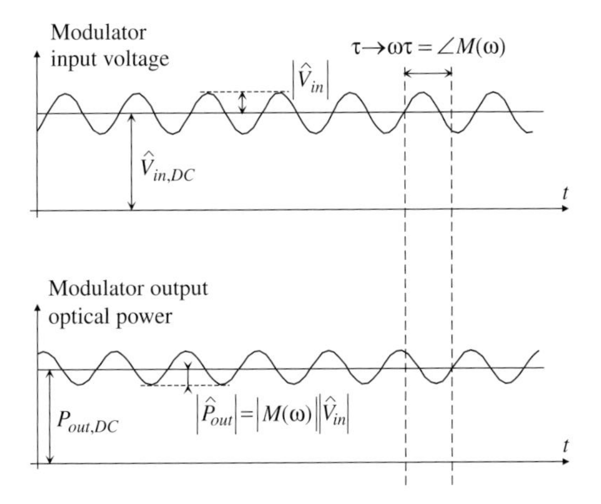

Figure 6.3 Definition of the small-signal frequency domain modulator response.

Assuming harmonic input and output, the signal part can be written as

$$\frac{\widehat{p}_{out}(t)}{P_{in}} = \mathrm{Re}\left(\left.\frac{\mathrm{d}T}{\mathrm{d}v_{in}}\right|_{DC} \widehat{V}_{in} \exp(\mathrm{j}\omega t)\right) = \mathrm{Re}\left(\frac{\widehat{P}_{out}}{P_{in}} \exp(\mathrm{j}\omega t)\right),$$

and, in terms of phasors, as

$$\frac{\widehat{P}_{out}}{P_{in}} = \left.\frac{\mathrm{d}T}{\mathrm{d}v_{in}}\right|_{DC} \widehat{V}_{in} \rightarrow \frac{\widehat{P}_{out}}{\widehat{V}_{in}} = P_{in} \left.\frac{\mathrm{d}T}{\mathrm{d}v_{in}}\right|_{DC} = M(0). \tag{6.1}$$

The quasi-static relation (6.1) can be extended to the dynamic case, as already mentioned, by the transfer function

$$M(\omega) = \frac{\widehat{P}_{out}(\omega)}{\widehat{V}_{in}(\omega)} = |M(\omega)| \exp(\mathrm{j}\angle M(\omega)).$$

See Fig. 6.3 for the meaning of $\angle M$. The modulation frequency response $m(\omega)$ is usually defined (and measured) with respect to the low-frequency (or DC) value; typically, only the amplitude of the response is considered:

$$m(\omega) = \left|\frac{M(\omega)}{M(0)}\right|.$$

The modulation response $m(\omega)$ is often described in log units, according to two different approaches, termed the *optical* and the *electrical* definition. We start from the definition of $m(\omega)$; in natural units we have, assuming a frequency-independent input voltage $\widehat{V}_{in}$,

$$m(\omega) = \frac{\left|\widehat{P}_{out}(\omega)/\widehat{P}_{out}(0)\right|}{\left|\widehat{V}_{in}(\omega)/\widehat{V}_{in}(0)\right|} = \frac{\left|\widehat{P}_{out}(\omega)\right|}{\left|\widehat{P}_{out}(0)\right|}.$$

Since the modulation response is formulated as a power ratio, direct application of log units leads to the so-called *optical* definition of the response:

$$m_{op}(\omega)\big|_{\mathrm{dB}} = 10\log_{10}[m(\omega)] = \left|\widehat{P}_{out}(\omega)\right|_{\mathrm{dBm}} - \left|\widehat{P}_{out}(0)\right|_{\mathrm{dBm}}. \tag{6.2}$$

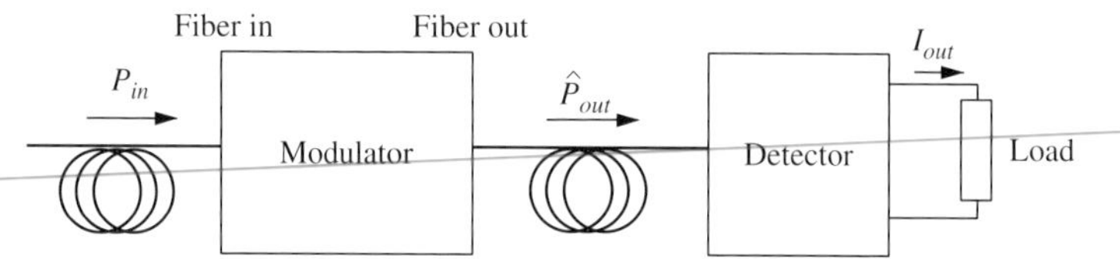

Figure 6.4 On the optical and electrical definition of the modulator small-signal response.

Alternatively, we measure the output power through a photodetector of responsivity $\Re$ (see Fig. 6.4). Taking into account that the detector phasor current is $I_{out} = \Re \left| \widehat{P}_{out} \right|$, we can express the modulator small-signal response as

$$m(\omega) = \frac{\left| \widehat{P}_{out}(\omega) \right|}{\left| \widehat{P}_{out}(0) \right|} = \frac{\Re \left| \widehat{P}_{out}(\omega) \right|}{\Re \left| \widehat{P}_{out}(0) \right|} = \frac{I_{out}(\omega)}{I_{out}(0)}.$$

We now define the modulation response in log units (according to the so-called *electrical* definition) with reference to the electrical power delivered to a load:

$$\begin{aligned} m_{el}(\omega)|_{\mathrm{dB}} &= 20 \log_{10}[m(\omega)] = 20 \log_{10}\left[\frac{I_{out}(\omega)}{I_{out}(0)}\right] \\ &= 2 \left| \widehat{P}_{out}(\omega) \right|_{\mathrm{dBm}} - 2 \left| \widehat{P}_{out}(0) \right|_{\mathrm{dBm}} = 2\, m_{op}(\omega)\big|_{\mathrm{dB}}. \end{aligned} \tag{6.3}$$

Thus, the electrical and optical definitions are related as $m_{el}(\omega)|_{\mathrm{dB}} = 2\, m_{op}(\omega)\big|_{\mathrm{dB}}$. According to another common notation, the optical definition is expressed in dBo (optical dB) and the electrical definition in dBe (electrical dB). One has, therefore,

$$m_{el,op}(\omega)\big|_{\mathrm{dB}} \equiv m(\omega)|_{\mathrm{dBe,dBo}},$$

respectively; it follows that

$$m(\omega)|_{\mathrm{dBe}} = 2\, m(\omega)|_{\mathrm{dBo}}.$$

6.2.4 Optical and electrical modulation bandwidth

From $m(\omega)$ the bandwidth can be derived, according to the optical (electrical) definition of the response, as the modulation frequency $f_{3\mathrm{dB,op}}$ ($f_{3\mathrm{dB,el}}$) at which the optical (electrical) response has a 3 dB decay with respect to the low-frequency value:

$$m_{op}(f_{3\mathrm{dB,op}})\big|_{\mathrm{dB}} = 10 \log_{10}\left[m(f_{3\mathrm{dB,op}})\right] = -3\,\mathrm{dB} \tag{6.4}$$

$$m_{el}(f_{3\mathrm{dB,el}})\big|_{\mathrm{dB}} = 20 \log_{10}\left[m(f_{3\mathrm{dB,el}})\right] = -3\,\mathrm{dB}. \tag{6.5}$$

From the definition we see that an optical 3 dB bandwidth corresponds to an electrical 6 dB bandwidth, while an electrical 3 dB bandwidth corresponds to an optical 1.5 dB bandwidth. Due to the low-pass modulator behavior, the optical 3 dB bandwidth is wider than the electrical 3 dB bandwidth.

6.2.5 Chirp

The chirp, or instantaneous frequency deviation, is the spurious frequency modulation Δf of the intensity-modulated output light with respect to the input light. Usually, the frequency deviation is expressed through the help of the Henry parameter α_H, see (5.40) and (5.52). In large-signal conditions the Henry parameter generally is a function of time, while it is constant (and bias point dependent) in small-signal operation. The EOM chirp depends only on device geometry, while in EAMs chirp depends, as in lasers, on material parameters. However, while digitally modulated lasers exhibit both transient and adiabatic chirp (Section 5.12.3), EAMs have only transient chirp; moreover, the EAM Henry parameter is typically smaller than in lasers.

6.2.6 Optical bandwidth

The optical bandwidth (not to be confused with the optical definition of the modulation bandwidth) is the photon wavelength range on which the modulator operates within specifications.

6.2.7 Electrical or RF input matching

The RF input matching is defined in terms of the modulator input reflection coefficient, seen from the electrical input port:

$$\Gamma_{in} = \frac{Z_{in} - Z_G}{Z_{in} + Z_G},$$

where Z_{in} is the modulator input impedance and Z_G is the generator impedance (here real, often $50\,\Omega$). From the power reflection coefficient $|\Gamma_{in}|^2$ one can also define the RF input insertion loss, i.e., the power loss resulting from the input reflection $1 - |\Gamma_{in}|^2$. Typical design values for the power input reflection coefficient are below -10 dB or $|\Gamma_{in}| \approx \sqrt{0.1} = 0.32$. Note that input matching can be a difficult problem in modulators, owing to their broadband behavior.

6.2.8 Linearity and distortion

Analog modulators can be characterized in terms of signal distortion caused by the nonlinearity of the electrooptic response $T(v_{in})$. The response can be expanded in power series of v_{in} around a bias point and a number of figures of merit can be defined, related to the generation of *harmonics* in the presence of a sinusoidal single-tone input signal (harmonic distortion), or to the generation of *intermodulation products* in the presence of two or more closely spaced sinusoidal input signals (intermodulation distortion).[3]

[3] Given two input tones f_1 and f_2, the intermodulation products (IMPs) are output frequencies $mf_1 \pm nf_2$ where m and n are integers. Third-order IMPs with a minus sign ($2f_1 - f_2$ and $2f_2 - f_1$) are particularly important since they fall close to the signal bandwidth, and are the main cause of nonlinear distortion.

The intermodulation distortion also defines the upper limit of the spurious-free dynamic range (SFDR).[4]

6.3 Electrooptic modulators

Electrooptic modulators are based on the modulation of the material refractive index from the electric field, induced by an applied input voltage. A simplified setup is shown in Fig. 6.5, consisting of a capacitor, whose dielectric is a slab of electrooptic material. In static conditions, the input voltage applied to the capacitor is $V_{in} = V_G$, and the related electric field has magnitude $E = V_G/h$, where h is the distance between the metal plates. The induced index variation $\Delta n \propto E \propto V_G$ linearly modulates the phase of the optical wave crossing the capacitor, with a phase variation $\Delta\phi \propto V_G$. Amplitude modulation can be obtained from phase modulation by making two phase-modulated beams interfere constructively or destructively. This is the principle of the *Mach–Zehnder modulator*; other solutions include *polarization-based modulators*, where the applied field changes the field polarization and an output polarizer blocks the optical field in the OFF state, and *directional coupler modulators*, where the applied field changes the state of an optical directional coupler from transmission (ON state) to coupling (OFF state).

Several electrooptic materials are available for implementing electrooptic modulators: ferroelectric crystals having piezoelectric properties, such as lithium niobate ($LiNbO_3$, often denoted as LN, probably the most important material today), lithium tantalate, barium titanate, in bulk or thin film; semiconductors (GaAs and InP); and finally electrooptic polymers. Silicon-based modulators have been proposed for exploiting the refractive index change resulting from charge injection.

Semiconductors exhibit a weaker electrooptic effect than ferroelectric crystals, but the related structures are able to enhance the induced electric field due to junction effects (i.e., the applied voltage acts across a very short distance, as in the intrinsic region of a *pin* structure). Moreover, QW and MQW structures can be exploited to tailor the material response. In the next sections, we will focus on the implementation of the phase modulation part of the device in different materials; the Mach–Zehnder amplitude modulator structure will be discussed in Section. 6.4.

Since the phase modulation structure in Fig. 6.5 is, from the electrical standpoint, a capacitor, the modulation bandwidth is limited by the RC cutoff (the resistance is the generator resistance R_G). Concentrated structures achieve in practice, with realistic driving voltages, speeds of only a few Gbps; high-speed modulators are based on a distributed, traveling-wave approach (see Section 6.5).

[4] The SFDR is the input power or voltage interval in which the system operates with a signal-to-noise ratio (SNR) compatible with the system specifications (lower limit), and with acceptable distortion (upper limit). Several definitions of the upper limit are available (e.g., the input power at which the intermodulation level at the system output equals the noise output level).

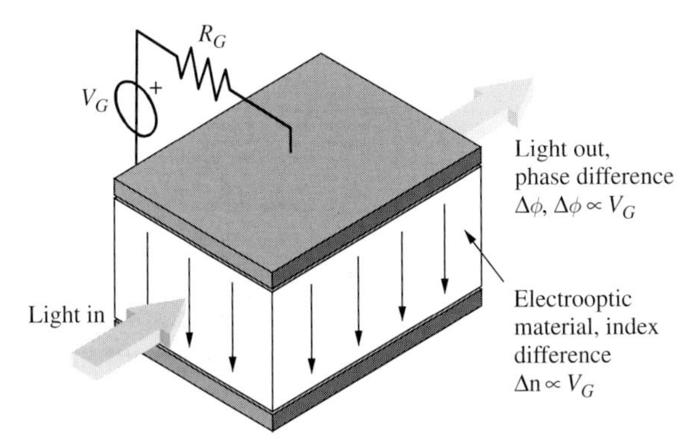

Figure 6.5 The principle of phase modulation through an applied voltage in an electrooptic modulator.

6.3.1 Lithium niobate electrooptic modulators

Lithium niobate ($LiNbO_3$) is an anisotropic uniaxial crystal (for the properties of lithium niobate and other electrooptic materials, see, e.g., [55], Table 9.2). The material permittivity is a tensor (whose rectangular components can be assembled in a 3×3 matrix), which becomes diagonal when the reference system corresponds to the crystal principal axes. Since the crystal is uniaxial, one of the principal axes (called the *optical* or *extraordinary* axis) is fixed, while the two other axes (called the *ordinary* axes) can be chosen arbitrarily in the plane orthogonal to the optical (extraordinary) axis. In such a reference frame (let x and y be the ordinary axes, z the extraordinary axis), the dielectric permittivity is a diagonal matrix:

$$\boldsymbol{\epsilon} = \epsilon_0 \boldsymbol{\epsilon}_r = \epsilon_0 \begin{pmatrix} \epsilon_x & 0 & 0 \\ 0 & \epsilon_x & 0 \\ 0 & 0 & \epsilon_z \end{pmatrix},$$

where

$$\epsilon_x = n_o^2, \quad \epsilon_z = n_e^2.$$

The two refractive indices n_o and n_e are the *ordinary* and *extraordinary* indices, respectively. Due to material dispersion, the RF or microwave values and the optical values are quite different. At RF:

$$\epsilon_{xm} = n_{om}^2 = 43 \rightarrow n_{om} = \sqrt{43} = 6.56$$
$$\epsilon_{zm} = n_{em}^2 = 28 \rightarrow n_{em} = \sqrt{28} = 5.29.$$

The optical values, on the other hand, are much lower:

$$n_{oo} = 2.35, \quad n_{eo} = 2.25.$$

A further effect to be considered is the strong frequency dispersion that LN exhibits (due to its piezoelectric nature) from DC to RF:

$$\epsilon_{xDC} = n_{oDC}^2 = 85 \rightarrow n_{oDC} = \sqrt{85} = 9.22 \gg n_{om} = 6.56$$
$$\epsilon_{zDC} = n_{eDC}^2 = 29 \rightarrow n_{eDC} = \sqrt{29} = 5.38 \approx n_{em} = 5.29.$$

At low frequency piezoelectricity increases the dielectric response, but at RF and above the associated mechanical effects are under cutoff, leading to a marked dispersion of the dielectric parameters. This may cause an anomalous modulator response at low-frequency (e.g., in the MHz range).

Due to the electrooptic effect, the elements of the (relative) permittivity matrix $\boldsymbol{\epsilon}_r$, ϵ_{ij}, are a function of the applied field components $\mathcal{E}_k$. Such a function can be conveniently expressed by expanding the variation of $1/\epsilon_{ij}$ in power series as

$$\Delta\left(\frac{1}{\epsilon_{ij}}\right) = \left(\frac{1}{\epsilon_{ij}(\underline{\mathcal{E}})} - \frac{1}{\epsilon_{ij}(0)}\right) = \sum_{k=1}^{3} r_{ijk}\mathcal{E}_k + \sum_{k,l=1}^{3} s_{ijkl}\mathcal{E}_k\mathcal{E}_l + \cdots$$

where $\underline{\mathcal{E}} = \mathcal{E}_i\widehat{u}_i + \mathcal{E}_j\widehat{u}_j + \mathcal{E}_k\widehat{u}_k$ is the applied electric field expanded into rectangular components (with unit vectors $\widehat{u}_i$, $\widehat{u}_j$, $\widehat{u}_k$), r_{ijk} are the elements of the linear electrooptic tensor $\boldsymbol{r}$ (dimension $3 \times 3 \times 3$, 27 components), and s_{ijkl} are the elements of the quadratic electrooptic tensor $\boldsymbol{s}$ (dimension $3 \times 3 \times 3 \times 3$, 81 components). Since in LN the linear electrooptic effect (also called the *Pockels effect*) dominates over the quadratic effect (also called the *Kerr effect*), we will focus on the former and write

$$\Delta\left(\frac{1}{\epsilon_{ij}}\right) = \Delta\left(\frac{1}{n_{ij}^2}\right) = \sum_{k=1}^{3} r_{ijk}\mathcal{E}_k \rightarrow \Delta n_{ij} \approx -\frac{n_{ij}^3}{2}\sum_{k=1}^{3} r_{ijk}\mathcal{E}_k, \tag{6.6}$$

where we have associated, with each element of the relative permittivity matrix, a refractive index such as $\epsilon_{ij} = n_{ij}^2$, and have further assumed that the variation is small, so that it can be obtained by applying derivation rules. The linear electrooptic tensor elements satisfy reciprocity, independently of the applied field; this implies $r_{ijk} = r_{jik}$, so that only 18 elements out of 27 are independent. To simplify the notation, the *contracted index representation* is introduced: the 18 independent elements are assembled into a 3×6 matrix by contracting the first two indices ij into a unique index i according to the rules:

$$\begin{array}{ll} ij = 11 \rightarrow i = 1 & ij = (23, 32) \rightarrow i = 4 \\ ij = 22 \rightarrow i = 2 & ij = (31, 13) \rightarrow i = 5 \\ ij = 33 \rightarrow i = 3 & ij = (21, 12) \rightarrow i = 6. \end{array}$$

In the contracted index notation, the variation of the refractive index $n_{ij} \rightarrow n_i$ can be expressed as

$$\Delta n_i = n_i(\underline{\mathcal{E}}) - n_i(0) = -\frac{n_i^3}{2}\sum_{k=1}^{3} r_{ik}\mathcal{E}_k$$

or, expanding,

$$\Delta n_i = -\frac{n_i^3}{2}\left(r_{i1}\mathcal{E}_1 + r_{i2}\mathcal{E}_2 + r_{i3}\mathcal{E}_3\right), \quad i = 1\ldots6. \tag{6.7}$$

Let us now consider the LN case in detail. We express the permittivity in the principal axis reference system and associate the indices $(1, 2, 3)$ with the rectangular axes (x, y, z), where z is the optical or extraordinary axis, and x, y are the ordinary axes. Due to LN crystal symmetry, most of the elements in $\boldsymbol{r}$ are zero, some are equal; the only nonzero elements have values:[5]

$$r_{13} = r_{23} = 9 \text{ pm/V}$$
$$r_{33} = r_{51} = r_{42} = 30 \text{ pm/V}$$
$$r_{22} = -r_{12} = -r_{61} = 6.6 \text{ pm/V}.$$

In general, an applied field will change all the components of the permittivity matrix, thus leading to a change of the principal axes. We confine the analysis to the case relevant to applications and assume that the applied field is directed along the optical axis, i.e., that $\mathcal{E}_z \equiv \mathcal{E}_3 \neq 0$ while both $\mathcal{E}_x = \mathcal{E}_1 = 0$ and $\mathcal{E}_y = \mathcal{E}_2 = 0$. In this case, considering only the nonzero elements of $\boldsymbol{r}$, (6.7) yields (for clarity, we partly revert to the expanded index notation in rectangular coordinates):

$$\Delta n_{xx} = \Delta n_{yy} = \Delta n_o = -\frac{n_o^3}{2} r_{13}\mathcal{E}_z$$
$$\Delta n_{zz} = \Delta n_e = -\frac{n_e^3}{2} r_{33}\mathcal{E}_z$$
$$\Delta n_{yz} = \Delta n_{xz} = \Delta n_{xy} = 0.$$

With such a choice of the applied electric field, the off-diagonal elements of the permittivity matrix are always zero, and the principal axes do not vary with respect to the zero-field case. Moreover, the largest element r_{33} is exploited, leading to a variation of the extraordinary index Δn_e; the ordinary index Δn_o exhibits a smaller variation.

The applied field configuration described above can be implemented in LN modulators by properly arranging the optical waveguide and electrode settings. Optical waveguides can be realized in a LN substrate by diffusion of titanium (Ti) at 1000°C for 4–10 hours. The Ti diffusion slightly increases the LN refractive index, yielding a waveguide with approximate Gaussian refractive index profile. Typical diffusion widths are of the order of 5–10 μm, leading to a comparable optical mode spot size in the waveguide plane. Concerning the waveguide orientation with respect to the crystal axes, two solutions can be implemented:

- In the *Z-cut configuration*, the optical axis is orthogonal to the crystal surface and the waveguide axis is parallel to the x or y axes (remember that such ordinary axes are arbitrary); see Fig. 6.6(left). The optical mode polarization is transverse magnetic

[5] The reported values are at $\lambda = 0.633$ μm, see [55], Table 9.2.

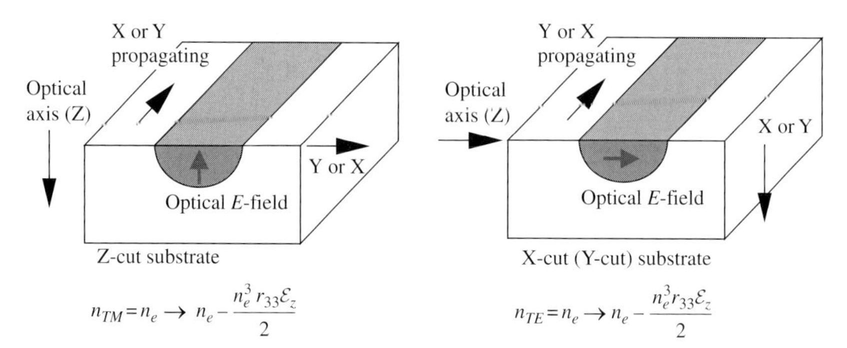

Figure 6.6 Z-cut (left) and X-cut (right) configuration for a diffused waveguide in a lithium niobate substrate.

(TM); i.e., the electric field polarization is along the optical axis. In such conditions, the modal refractive index is $n_{TM} \approx n_e$. Application of a RF electric field along the z axis (parallel to the optical field) leads to a variation of the modal refractive index:

$$\Delta n_{TM} = -\frac{n_e^3 r_{33} \mathcal{E}_z}{2}.$$

- In the *X-cut configuration*, the optical axis is parallel to the crystal surface and orthogonal to the waveguide axis; see Fig. 6.6(right). The waveguide axis runs along x (or y), while the crystal surface is orthogonal to y (or x), respectively. The optical mode polarization is transverse electric (TE); i.e., the electric field polarization is again along the optical axis, this time horizontal. In such conditions, the modal refractive index is $n_{TE} \approx n_e$. Application of a RF electric field along the z axis (parallel to the optical field and to the crystal surface) leads to a variation of the modal refractive index:

$$\Delta n_{TE} = -\frac{n_e^3 r_{33} \mathcal{E}_z}{2}, \tag{6.8}$$

the same as in the Z-cut case. The Y-cut configuration is of course equivalent to the X-cut one.

The Z-cut and X-cut configurations appear to be altogether equivalent from the standpoint of the optical waveguide; however, the RF electrode setting needed to create a properly oriented electric field is different. In fact, in Z cut substrates the optical field must interact with a *perpendicular* RF field, so that both the coplanar waveguide (all electrodes on the crystal surface) and the microstrip (one electrode on the surface, a ground plane on the crystal bottom) configurations are possible. On the other hand, in X-cut substrates the interaction is with the *parallel* component of the microwave field, so that only a coplanar electrode configuration is feasible. The RF field lines and optical waveguide position are sketched in Fig. 6.7 for the Z-cut and X-cut configurations, respectively. In the Z-cut case, the optical field must interact with the vertical electric field, and therefore the waveguide position is below the RF electrode. In this way, the field strength is maximum, but at the same time a dielectric buffer must be inserted between the optical waveguide and the metal electrode to avoid large optical losses in

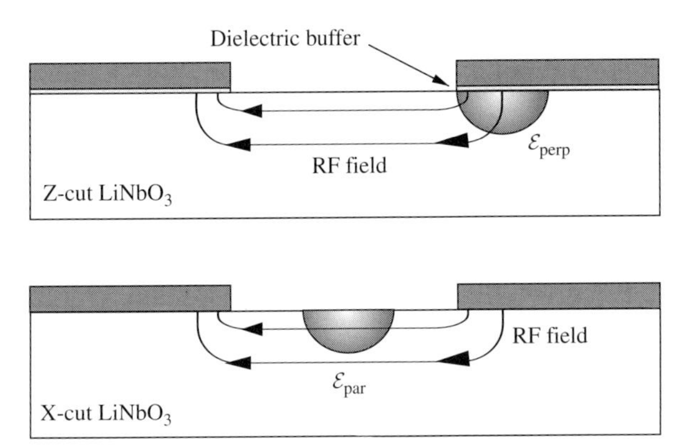

Figure 6.7 RF electrode configuration for the Z-cut substrate (above) and the X-cut substrate (below).

the TM mode. As discussed in Section 6.6, introduction of a dielectric buffer is also needed to improve the synchronous coupling between the RF and optical mode in a distributed modulator. In the X-cut case, the waveguide has to interact with the horizontal field, so that it must be located between the electrodes. No dielectric buffer is needed in principle to avoid losses, but the field strength is somewhat lower than in the Z-cut case. Notice that both the X-cut and Z-cut configurations are polarization dependent, i.e., the interaction requires TE or TM polarized fields. From a system standpoint, this implies that the polarization of the input optical signal or carrier has to be established through a polarization-maintaining (PM) fiber.

The Z-cut configuration has approximately a 30% advantage over the X-cut one in terms of interaction strength; however, Z-cut operation is affected by problems connected with static induced charge: due to the piezoelectric nature of the material, a static charge is induced on the free surfaces orthogonal to the optical axis. In the Z-cut configuration, a layer of induced charge slowly builds up on the top surface, where electrodes are located; the induced static field is superimposed on the applied bias, leading to long-term fluctuations in the bias point that may compromise the device operation and reliability. In the X-cut configuration, on the other hand, charge buildup occurs on the two lateral surfaces (the LN edges), which are contacted by the device package and can be suitably metallized so as to short induced fields. In the X-cut configuration, therefore, charge build up is ineffective and the bias point does not experience slow drifts. Suppression of the bias drift can be obtained in the Z-cut configuration by depositing a resistive layer, called the *charge bleed* layer, on the top surface; the layer redistributes the excess induced charge, while its resistance is large enough not to compromise RF operation and biasing. Charge bleed layers can be implemented, for example, through polysilicon sputtering.

A last point to be considered is the fact that both the RF and optical field intensity are nonuniform over the interaction region. To relate the external applied voltage to the variation of the modal refractive index, we introduce the *overlap integral* Γ_{mo} between the microwave and optical fields. Consider, the wave equation for the modal electric field (assume, for example, the TE mode; the same treatment applies for the TM mode), with propagation constant (along z) equal to β_{TE}:

$$\nabla_t^2 E_{op} + \left[n^2 k_0^2 - \beta_{TE}^2\right] E_{op} = 0. \tag{6.9}$$

A perturbation $(n + \Delta n)^2 \approx n^2 + 2n\Delta n$ leads to a perturbation in the propagation constant $(\beta_{TE} + \Delta\beta_{TE})^2 \approx \beta_{TE}^2 + 2\beta_{TE}\Delta\beta_{TE}$. Taking into account (6.9) and assuming that the perturbed field E' is $E' \approx E_{op}$, we obtain

$$\nabla_t^2 E' + \left[\left(n^2 + 2n\Delta n\right) k_0^2 - \left(\beta_{TE}^2 + 2\beta_{TE}\Delta\beta_{TE}\right)\right] E'$$
$$\approx \left[2n\Delta n k_0^2 - 2\beta_{TE}\Delta\beta_{TE}\right] E_{op} = 0.$$

Multiplying both sides by E_{op}^* and integrating over the waveguide cross section we find, taking into account that $\beta_{TE} = n_{TE}k_0$ and that $n \approx n_e$ (n_e and Δn_e are real),

$$\iint \left|E_{op}(\underline{r})\right|^2 n_e \Delta n_e \, \mathrm{d}S = n_{TE}\Delta n_{TE} \iint \left|E_{op}(\underline{r})\right|^2 \, \mathrm{d}S,$$

from which, since $n_{TE} \approx n_e$,

$$\Delta n_{TE} \approx \frac{\displaystyle\iint \left|E_{op}(\underline{r})\right|^2 \Delta n_e \, \mathrm{d}S}{\displaystyle\iint \left|E_{op}(\underline{r})\right|^2 \, \mathrm{d}S}.$$

Denoting the RF field as $\mathcal{E}_z$ and taking into account (6.8), we have

$$\Delta n_{TE} = -\frac{n_e^3 r_{33}}{2} \frac{\displaystyle\iint \left|E_{op}(\underline{r})\right|^2 \mathcal{E}_z(\underline{r}) \, \mathrm{d}S}{\displaystyle\iint \left|E_{op}(\underline{r})\right|^2 \, \mathrm{d}S} = -\frac{n_e^3 r_{33} V_A}{2G} \frac{\displaystyle\iint \left|E_{op}(\underline{r})\right|^2 e_z(\underline{r}) \, \mathrm{d}S}{\displaystyle\iint \left|E_{op}(\underline{r})\right|^2 \, \mathrm{d}S},$$

where

$$e_z(\underline{r}) = \frac{G}{V_A}\mathcal{E}_z(\underline{r})$$

is the normalized electric field (normalization is with respect to a uniform field induced by a voltage V_A over a distance G); V_A is the applied voltage, G is the gap between the coplanar electrodes, see Fig. 6.8. The variation of the extraordinary index corresponds to a variation of the modal (TM or TE) index, so that

$$\Delta n_{TE} = -\frac{n_e^3 r_{33} V_A}{2G} \Gamma_{mo}, \tag{6.10}$$

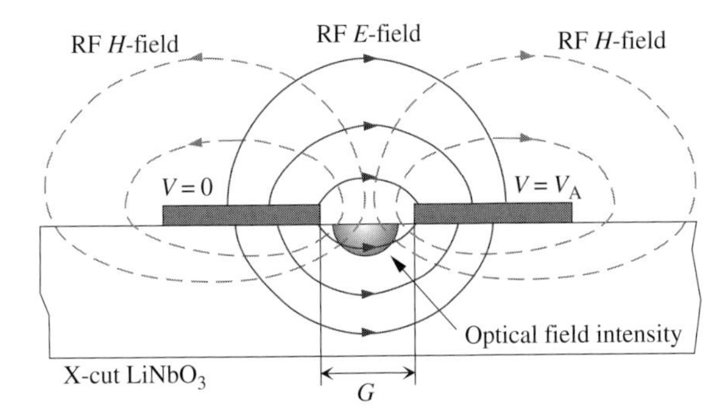

Figure 6.8 Microwave field lines and optical mode pattern in the cross section of a X-cut LN modulator with coplanar electrodes. The electrode distance is the gap G.

where the overlap integral is[6]

$$\Gamma_{mo} = \frac{G}{V_A} \frac{\iint \left|\mathcal{E}_{op}(\underline{r})\right|^2 \mathcal{E}_z(\underline{r})\, \mathrm{d}S}{\iint \left|\mathcal{E}_{op}(\underline{r})\right|^2 \mathrm{d}S}.$$

This is equivalent to postulating a uniform equivalent RF field $\mathcal{E}_z = -\Gamma_{mo} V_A / G$. The overlap integral is equal to unity if the field is uniform (as in a parallel-plate configuration, in which the gap coincides with the distance between the two plates). For the TM field, the principle is the same but the relevant overlap integral may be different due to the different optical field pattern.

A concluding remark concerns the RF propagation characteristics of a coplanar electrode configuration deposited on LN substrate. Such a coplanar waveguide supports a quasi-TEM propagation mode, in which the electric and magnetic fields lie almost entirely in the transverse plane. The refractive index in the transmission line cross section (corresponding to the plane determined by the ordinary and extraordinary axes both in the Z-cut and in the X-cut configurations) is anisotropic, but the quasi-TEM mode effective microwave propagation index n_m is given simply by[7]

$$n_m \approx \sqrt{\frac{1 + \sqrt{\epsilon_{xm}\epsilon_{zm}}}{2}} = \sqrt{\frac{1 + n_{om}n_{em}}{2}} = \sqrt{\frac{1 + 6.56 \cdot 5.29}{2}} = 4.22. \qquad (6.11)$$

The RF index n_m is therefore much larger than the optical refractive index $n_{TE/TM} \equiv n_o \approx 2.2$. Since the RF phase velocity is about twice as large as the optical phase velocity, n_m must be suitably decreased in order to achieve synchronous coupling in distributed modulators; see Section 6.6.

[6] The definition of the overlap integral is consistent with (5.18), introducing the same parameter for a dielectric slab waveguide, where Δn (in the laser case, the variation is related to the material gain) is assumed as uniform in the active region.

[7] The parameters of a coplanar waveguide on a semi-infinite dielectric substrate, such that two principal axes are parallel and perpendicular to the dielectric surface, respectively, with permittivities ϵ_1 and ϵ_2, can be shown to coincide with the parameters of the same waveguide on an equivalent isotropic substrate with $\epsilon = \sqrt{\epsilon_1\epsilon_2}$, see [86].

6.3.2 Semiconductor electrooptic modulators

GaAs- and InP-based Mach–Zehnder modulators have been developed both in discrete form and integrated with a source. For the sake of brevity, let us confine the treatment to GaAs. GaAs is isotropic, with microwave relative permittivity $\epsilon_m \approx 13$, microwave refractive index $n_m \approx \sqrt{13} = 3.61$, and optical refractive index $n_o \approx 3.4$. Due to crystal symmetry, most of the elements in the linear electrooptic tensor of GaAs are again zero, many are equal, and the only nonzero elements have values (at $\lambda = 0.8\,\mu\text{m}$)

$$r_{41} = r_{52} = r_{63} = 1.2 \ \text{pm/V}.$$

By inspection of (6.6), only the nondiagonal elements of the permittivity matrix change when applying an electric field; i.e., the material becomes anisotropic. For simplicity we refer here to two practically exploited electrode configurations, shown in Fig. 6.9. In configuration (a), the semiconductor is grown along the z direction, while the optical TE field is polarized along the y' direction; the applied electric field is vertical (along z). In configuration (b), the growth is again along z, and the polarization and the applied field are both along y'. In the primed reference system, we have $z' = z$, while the other axes are rotated by $\pi/4$ with respect to the unprimed axes x, y (defined along the crystal directions). In both cases, the guiding structure includes a double AlGaAs-GaAs heterostructure providing vertical photon confinement. Doped junctions can be exploited to enhance the effect of the applied voltage.

In an anisotropic material, the identification of the principal axes can be carried out through the help of the *index ellipsoid*, defined by the equation

$$\sum_{i,j=1}^{3} \frac{x_i x_j}{n_{ij}^2} = \frac{x_1^2}{n_1^2} + \frac{x_2^2}{n_2^2} + \frac{x_3^2}{n_3^2} + \frac{2x_1x_2}{n_6^2} + \frac{2x_1x_3}{n_5^2} + \frac{2x_2x_3}{n_4^2} = 1,$$

where we have accounted for the material reciprocity and used the contracted index notation. For GaAs we generally have

$$\Delta\left(\frac{1}{n_1^2}\right) = \Delta\left(\frac{1}{n_2^2}\right) = \Delta\left(\frac{1}{n_3^2}\right) = 0$$

$$\Delta\left(\frac{1}{n_4^2}\right) = r_{41}\mathcal{E}_x, \quad \Delta\left(\frac{1}{n_5^2}\right) = r_{52}\mathcal{E}_y = r_{41}\mathcal{E}_y, \quad \Delta\left(\frac{1}{n_6^2}\right) = r_{63}\mathcal{E}_z = r_{41}\mathcal{E}_z,$$

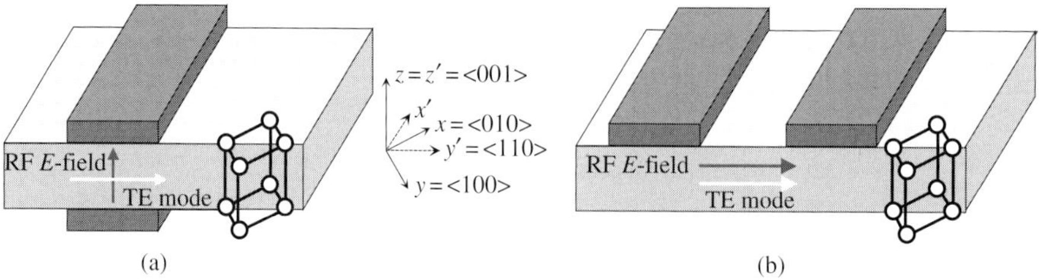

Figure 6.9 Electrode configurations for GaAs-based electrooptic modulators: (a) with microstrip electrodes, (b) with coplanar electrodes.

so that, in the presence of an applied field, the index ellipsoid becomes

$$\left(\frac{1}{n_o^2}\right)x^2 + \left(\frac{1}{n_o^2}\right)y^2 + \left(\frac{1}{n_o^2}\right)z^2 + 2r_{41}\mathcal{E}_z xy + 2r_{41}\mathcal{E}_y xz + 2r_{41}\mathcal{E}_x yz = 1.$$

Consider now the case in Fig. 6.9(a); the applied electric field is vertical (along z) so that the index ellipsoid becomes

$$\left(\frac{1}{n_o^2}\right)x^2 + \left(\frac{1}{n_o^2}\right)y^2 + \left(\frac{1}{n_o^2}\right)z^2 + 2r_{41}\mathcal{E}_z xy = 1.$$

The already introduced primed coordinate system is related to the unprimed one through a $\pi/4$ rotation in the xy plane, as

$$x' = \frac{x-y}{\sqrt{2}}, \quad y' = \frac{x+y}{\sqrt{2}}, \quad z' = z,$$

i.e.

$$x = \frac{x'+y'}{\sqrt{2}}, \quad y = \frac{x'-y'}{\sqrt{2}}.$$

The new y' axis is now parallel to the optical field (cf. Fig. 6.9(a)). Substituting, we have

$$\frac{x'^2 + y'^2 + 2x'y'}{2n_o^2} + \frac{x'^2 + y'^2 - 2x'y'}{2n_o^2} + \frac{z'^2}{n_o^2} + 2r_{41}\mathcal{E}_z\frac{x'^2 - y'^2}{2}$$
$$= \left(\frac{1}{n_o^2} + r_{41}\mathcal{E}_z\right)x'^2 + \left(\frac{1}{n_o^2} - r_{41}\mathcal{E}_z\right)y'^2 + \left(\frac{1}{n_o^2}\right)z'^2 = 1.$$

Therefore, the new (diagonal) refractive indices in the principal axes reference frame will be

$$n'_{xx} \approx n_o - \frac{n_o^2 r_{41}\mathcal{E}_z}{2} = n_o - \Delta n_V$$
$$n'_{yy} \approx n_o + \frac{n_o^2 r_{41}\mathcal{E}_z}{2} = n_o + \Delta n_V$$
$$n'_{zz} = n_o.$$

The TE mode experiences a modal index variation of $-\Delta n_V$ (along y'); in the same conditions, the TM mode (vertical polarization) is unaffected. The refractive index also varies along x', so that the optical waveguide could also be defined orthogonal to that direction. The resulting modulator is therefore *polarization dependent*, and similar to a LN Mach–Zehnder modulator.

The case in Fig. 6.9(b), is more complex, since an electric field parallel to the TE field makes the material biaxial, with principal axes having nonzero projection on both the TE and the TM modes. The resulting mode coupling leads to mode conversion: a TE input mode is converted after a certain length into a TM mode with opposite phase, and vice versa. In a Mach–Zehnder configuration, under applied voltage V_π, the TE components in each arm are converted to TM, with opposite sign at the output (destructive interference), while the TM components are converted into TE components with

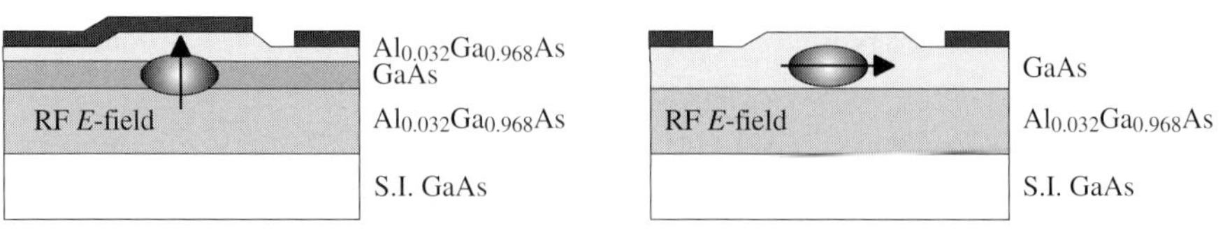

Figure 6.10 Electrode setting for a polarization-dependent modulator with vertical RF field (left) and polarization-independent modulator with horizontal RF field (right). The waveguide is a double heterostructure GaAs-AlGaAs; the ridge width is typically a few μm. Adapted from [88], Fig. 4.

opposite sign (destructive interference again); moreover, it can be shown that the effective variation of modal index is $2\Delta n_V$, i.e., the configuration is twice as effective as the one with vertical electric field [87]. The resulting modulator is therefore *polarization independent*. A more detailed representation of the two structures is shown in Fig. 6.10 [88].

To compare, at a material level, the relative merits of LN and of GaAs, we should consider that the material figure of merit is $n_0^3 r$, rather than r alone. From this standpoint, in LN $n_0^3 r \approx 2.2^3 \cdot 30 = 320$ pm/V, while in GaAs $n_0^3 r \approx 3.4^3 \cdot 1.3 = 51$ pm/V. However, the electrooptic potential of GaAs can be enhanced in several ways. First, QW or MQW structures with excitonic effects at room temperature can lead to an increase of the r coefficient with respect to bulk. Secondly, junction effects (which cannot of course be implemented in LN) allow enhancement of the applied field with the same applied voltage. To make a comparison, imagine applying the same voltage on a coplanar electrode pair with separation $G = 10\,\mu$m (about the minimum allowed by the optical spot size in Ti-diffused waveguides) and on a *pin* junction with intrinsic layer of thickness $d = 0.5\,\mu$m; a factor of 20 in favor of GaAs arises, so that GaAs turns out to be 3–4 times better than LN. In practice, this advantage is shown by the comparative lengths of LN-based modulators (about 10 mm) and of GaAs- or InP-based modulators (about 1 mm in bulk modulators and even less in QW structures). However, since the realization of semiconductor-based modulators is fraught with many technological difficulties, this solution is less significant than the LN-based one for applications.

More recently, all-silicon modulators were developed exploiting the refractive index variation associated with charge injection; some further details and structure examples are provided in Section 6.6.2.

6.3.3 Polymer modulators

Polymer-based electrooptic modulators have promising properties, although material degradation in the long term is still a practical problem in system applications. Polymers can be made electrooptic by means of *chromophore molecules*, characterized by large dipole moments and optical nonlinearity. Chromophores can be activated in a polymer through a process called high-temperature poling, whereby chromophores are aligned by a strong applied electric field in a heated polymer. The randomly aligned dipole moments keep their alignments after cooling and result in a polymer having macroscopic electrooptic properties that can be expressed in terms of a linear electrooptic

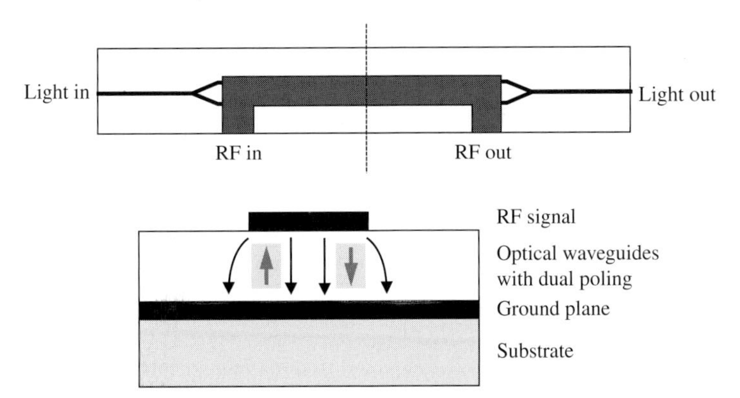

Figure 6.11 Cross section of microstrip-based polymer modulator; the layout is a conventional Mach–Zehnder interferometer. After [89], Fig. 1 (©1999 IEEE).

tensor. Typically, the largest coefficient is for TM polarization (i.e., the applied electric field is vertical, like the poling field, and parallel to the optical field); the situation is somewhat similar to LN modulators in the Z-cut configuration and microstrip electrodes. The corresponding r_{33} can be as large as 20–60 pm/V, and the refractive index is around 1.6. Owing to the microstrip configuration (see Fig. 6.11 [89]), and to the low refractive index, the optical and microwave refractive indices are similar, leading to very efficient synchronous coupling in traveling-wave modulators. From a system standpoint, polymer modulators are comparable to LN modulators concerning bandwidth and driving voltages, but their lifetime and practical use is still impaired by long-term polymer degradation. The material figure of merit is $n_0^3 r \approx 1.6^3 \cdot 20 = 81$ pm/V, somewhat similar to GaAs.

6.4 The Mach–Zehnder electrooptic modulator

The Mach–Zehnder amplitude modulator is based on the constructive or destructive interference of two phase-modulated beams generated by splitting the input optical beam into two parts, each running in a separate optical waveguide; see Fig. 6.12. An electric field is applied to each arm of the modulator through a set of coplanar or microstrip electrodes; the electrode cross section design depends on the material and on the optical field polarization. The applied RF field causes the two optical beams to be phase modulated before recombination in the output optical combiner. At zero applied voltage and RF field, the two beams reach the combiner with equal phase and recombine constructively. When the applied voltage is equal to V_π, the two recombining optical fields have a phase difference of π, and in the output arm of the combiner a higher-order, below threshold mode is radiated, leading to zero output optical power.

The field profiles in the combiner are shown qualitatively in Fig. 6.12. Due to the conversion of phase modulation into amplitude modulation through beam interference, we can call this class of modulators *interferometric modulators*.

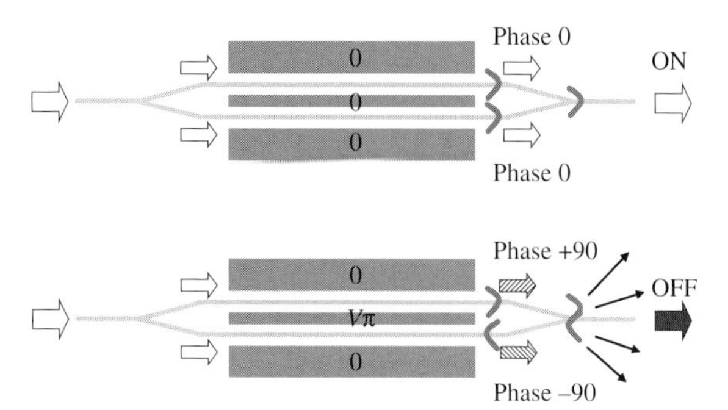

Figure 6.12 Layout of a coplanar Mach–Zehnder modulator: ON state (above), OFF state (below). In the OFF state the optical fields are radiated away out of the combiner.

6.4.1 The lumped Mach–Zehnder modulator

The basic building block of the Mach–Zehnder interferometric modulator (Fig. 6.13) is the phase modulator section. We assume that the electrooptic material is LN; for semiconductors, the treatment is similar. Suppose that the optical waveguide lies between a ground plane and an electrode at potential V_A (this is the case for the upper modulation section in a Mach–Zehnder modulator; see Fig. 6.12); the electric field will be directed toward the negative z axis (consider, e.g., an X-cut case, Fig. 6.6). The phase difference induced by V_A over an interaction region of length L, with respect to the case in which $V_A = 0$, can be expressed as[8]

$$\Delta\phi = \int_0^L \left[k\left(V_A\right) - k\left(0\right)\right] \, \mathrm{d}x = \frac{2\pi}{\lambda_0}\int_0^L \Delta n_o \, \mathrm{d}x = \frac{\pi n_e^3 r_{33} \Gamma_{mo}}{\lambda_0} \frac{L}{G} V_A, \qquad (6.12)$$

where n_o is the TE or TM modal index. The phase variation $\Delta\phi$ is therefore a linear function of the applied voltage. In the symmetric Mach–Zehnder configuration, the voltages applied to the two phase modulation sections (upper and lower, see Fig. 6.12) induce, in the upper and lower waveguides, RF fields with opposite direction. Since the dominant electrooptic effect is linear, the modal index variation will also be equal in magnitude in the upper and lower modulation sections, but with opposite sign. If we assume that the central electrode in Fig. 6.12 is at V_A, while the upper and lower strips are grounded, the phase variations of the upper and lower arms will add, so that the phase difference between the optical fields at the end of the phase modulation section will be

$$\left|\Delta\phi_U - \Delta\phi_L\right| = \frac{2\pi n_e^3 r_{33} \Gamma_{mo}}{\lambda_0} \frac{L}{G} V_A.$$

In the symmetric case, $\Delta\phi_U = -\Delta\phi_L$. When $\left|\Delta\phi_U - \Delta\phi_L\right| = \pi$, the optical fields in the combiner interfere destructively and the modulator is in the OFF state. The voltage

[8] According to our notation, a harmonic wave propagating in the positive x direction has a phase term $\exp(-\mathrm{j}kx + \mathrm{j}\omega t)$. Thus, the additional phase rotation induced by the modulator is $\exp(-\mathrm{j}\Delta\phi)$ rather than $\exp(\mathrm{j}\Delta\phi)$, see (6.12).

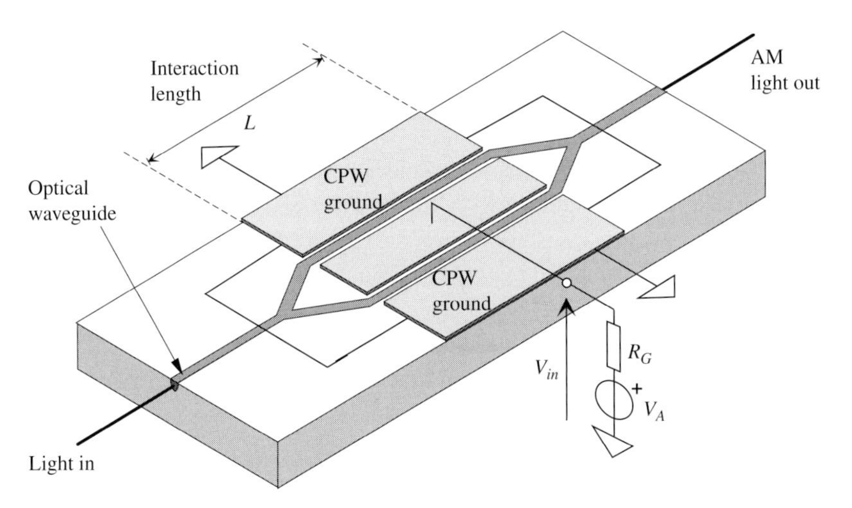

Figure 6.13 Coplanar lumped Mach–Zehnder modulator.

needed to bring the modulator into the OFF state from $V_A = 0$ (the ON state), i.e., the modulator V_π, can be derived from the condition:

$$|\Delta\phi_U - \Delta\phi_L| = \frac{2\pi n_e^3 r_{33}\Gamma_{mo}}{\lambda_0}\frac{L}{G}V_\pi = \pi \rightarrow V_\pi = \frac{\lambda_0}{2n_e^3 r_{33}\Gamma_{mo}}\frac{G}{L}.$$

Note that the phase variation for each arm is $\pi/2$ at $V_A = V_\pi$.

6.4.2 Static electrooptic response

The DC electrooptic response of the modulator can be derived as follows. Suppose that the input splitter is matched at port 1 (the input) and at the two outputs (ports 2 and 3); suppose that the two output ports are isolated. Neglecting losses, the scattering matrix of the splitter can be expressed by inspection as

$$\mathbf{S}_{sp} = \begin{pmatrix} 0 & \sqrt{\alpha}\,\mathrm{e}^{\mathrm{j}\phi_{sp}} & \sqrt{1-\alpha}\,\mathrm{e}^{\mathrm{j}\phi_{sp}} \\ \sqrt{\alpha}\,\mathrm{e}^{\mathrm{j}\phi_{sp}} & 0 & 0 \\ \sqrt{1-\alpha}\,\mathrm{e}^{\mathrm{j}\phi_{sp}} & 0 & 0 \end{pmatrix}.$$

The parameter α accounts for the power asymmetry of an asymmetrical splitter; if $\alpha = 1/2$ the splitter is symmetrical. Note that $\mathbf{S}_{sp}$ satisfies the lossless condition $\mathbf{S}_{sp}\mathbf{S}_{sp}^\dagger = 1$. Similarly, the scattering matrix of the combiner with input ports 2 and 3 and output port 1 is $\mathbf{S}_c = \mathbf{S}_{sp}$. Assume that the incident optical power wave into port 1 is a_1; the transmitted power waves at ports 2 and 3 will be

$$b_2 = \sqrt{\alpha}\mathrm{e}^{\mathrm{j}\phi_{sp}}a_1, \quad b_3 = \sqrt{1-\alpha}\mathrm{e}^{\mathrm{j}\phi_{sp}}a_1.$$

Assuming that the upper and lower arms are perfectly matched, the power waves at the combiner inputs, a_2' (upper arm) and a_3' (lower arm), are

$$a_2' = \sqrt{\alpha}\mathrm{e}^{\mathrm{j}\phi_{sp}}a_1\mathrm{e}^{-\mathrm{j}k_oL-\mathrm{j}\Delta\phi_U}, \quad a_3' = \sqrt{1-\alpha}\mathrm{e}^{\mathrm{j}\phi_{sp}}a_1\mathrm{e}^{-\mathrm{j}k_oL-\mathrm{j}\Delta\phi_L},$$

and therefore the combiner output b_1' will be

$$b_1' = e^{2j\phi_{sp}} e^{-jk_o L} \left[\alpha e^{-j\Delta\phi_U} + (1-\alpha) e^{-j\Delta\phi_L} \right] a_1. \tag{6.13}$$

If the input and output optical waveguides have the same impedance, the ratio between the input and output optical power can be expressed as

$$T(V_{in}) = \frac{P_{out}}{P_{in}} = \left| \frac{b_1'}{a_1} \right|^2 = \eta \left\{ 1 + 2\alpha (1-\alpha) \left[\cos(\Delta\phi_U - \Delta\phi_L) - 1 \right] \right\}, \tag{6.14}$$

where the additional factor η accounts for the optical insertion loss of the modulator.

Let us now confine the treatment to a modulator with a symmetrical splitter, but with possibly asymmetrical upper and lower arms. The case is practically important since it corresponds to the ordinary (non-dual-drive) Z-cut modulators, due to the different overlap integrals.

From (6.14) we have, for $\alpha = 1/2$,

$$T(V_{in}) = \frac{\eta}{2} \left[1 + \cos(\Delta\phi_U - \Delta\phi_L) \right].$$

Define now as $V_{\pi U}$ and $V_{\pi L}$ the voltages (applied to the upper and lower arms, respectively) for which $\Delta\phi_U = \pi$, $\Delta\phi_L = -\pi$. We have

$$\Delta\phi_U = \pi \frac{v_{inU}}{V_{\pi U}}, \quad \Delta\phi_L = -\pi \frac{v_{inL}}{V_{\pi L}}. \tag{6.15}$$

The two voltages $V_{\pi U,L}$ are defined by the relation:

$$\left| \Delta\phi_{U,L} \right| = \frac{\pi n_e^3 r_{33} \Gamma_{moU,L}}{\lambda_0} \frac{L}{G} V_{\pi U,L} = \pi \rightarrow V_{\pi U,L} = \frac{G}{L} \frac{\lambda_0}{n_e^3 r_{33} \Gamma_{moU,L}}. \tag{6.16}$$

Assume now that $v_{inU} = v_{inL} = v_{in}$; we have

$$T(v_{in}) = \frac{\eta}{2} \left\{ 1 + \cos \left[\pi \left(\frac{1}{V_{\pi U}} + \frac{1}{V_{\pi L}} \right) v_{in} \right] \right\} = \frac{\eta}{2} \left\{ 1 + \cos \left(\pi \frac{v_{in}}{V_\pi} \right) \right\}, \tag{6.17}$$

where

$$V_\pi = \frac{V_{\pi U} V_{\pi L}}{V_{\pi U} + V_{\pi L}}$$

is the total modulator on–off voltage for which $T(v_{in}) = 0$. The modulator static transfer curve is therefore a raised cosine, as shown in Fig. 6.14.

6.4.3 Lumped modulator dynamic response

Suppose now that the generator open-circuit voltage $v_A(t)$ is time-varying; to evaluate the modulator response we express the input voltage $v_{in}(t)$, see Fig. 6.13, as a function of $v_A(t)$, taking into account the input equivalent circuit including the generator resistance R_G and the modulator input capacitance C (the circuit considered is the same as in Fig. 6.42, in which the gray box is removed, the stray inductances are not considered, $R_L \rightarrow \infty$, and C_{pad} is the modulator capacitance). The effect of other elements, such

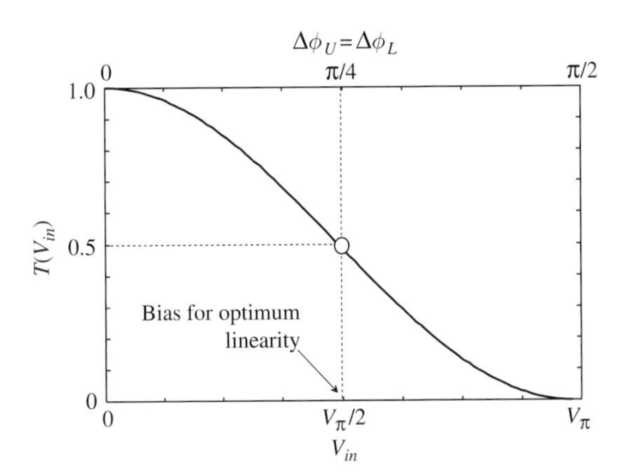

Figure 6.14 DC electrooptic response of a symmetric Mach–Zehnder modulator. The $V_{in} = V_\pi/2$ bias corresponds to the inflexion point in the transfer curve and therefore to the maximum (local) linearity.

as a modulator output load resistance and parasitic stray inductances and capacitances, can easily be accounted for at circuit level. In general, we can conveniently work in the frequency domain and relate the modulator input voltage phasor $V_{in}(\omega)$ to the generator voltage phasor $V_A(\omega)$ through a low-pass transfer function $H(\omega)$ as

$$V_{in}(\omega) = H(\omega) V_A(\omega),$$

where $H(0) = 1$. Assuming that (6.15) holds for time varying input voltages, we can associate with the upper and lower phase modulation section phase delays a proper phasor $\Delta\phi_\alpha(\omega)$, $\alpha = U, L$. Applying the same input voltage to the upper and lower phase modulator sections, we therefore obtain

$$\Delta\phi_U(\omega) = \pi \frac{V_{in}(\omega)}{V_{\pi U}} = \pi H(\omega) \frac{V_A(\omega)}{V_{\pi U}}$$
$$\Delta\phi_L(\omega) = -\pi \frac{V_{in}(\omega)}{V_{\pi L}} = -\pi H(\omega) \frac{V_A(\omega)}{V_{\pi L}}.$$

In dynamic conditions, the relation between the phase delay and the input voltage is *linear* but *dispersive* (with low-pass behavior). Taking into account that the relation between the phase delay and the output power (6.17) is *nonlinear* but *memoriless*, we can finally express the dynamic modulator response as

$$\frac{p_{out}(t)}{P_{in}} = \frac{\eta}{2}\left\{1 + \cos\left(\pi \frac{v_{in}(t)}{V_\pi}\right)\right\} = \frac{\eta}{2}\left\{1 + \cos\left(\pi \frac{\mathcal{F}^{-1}\left[H(\omega)V_A(\omega)\right]}{V_\pi}\right)\right\},$$

where $\mathcal{F}$ is the Fourier transform and $\mathcal{F}^{-1}$ the inverse transform.[9]

[9] From a system standpoint, the modulator can be decomposed into a dispersive linear system (the phase modulation sections) in cascade with a nonlinear memoriless system (the combiner). Such a decomposition can be conveniently exploited to generate system-level CAD models for electrooptic modulators, see [90].

To express the overall frequency response of the modulator, we assume small-signal operation with respect to a DC working point, i.e.,

$$v_A(t) = V_{A,DC} + \widehat{v}_A(t), \quad v_{in}(t) = V_{in,DC} + \widehat{v}_{in}(t),$$

where $\widehat{v}_A(t)$, $\widehat{v}_{in}(t)$ are small-signal components. Linearizing around $V_{in,DC}$ we obtain

$$\begin{aligned}\frac{p_{out}(t)}{P_{in}} &= \frac{\eta}{2}\left\{1+\cos\left(\pi\frac{V_{in,DC}+\widehat{v}_{in}(t)}{V_\pi}\right)\right\} \\ &\approx \underbrace{\frac{\eta}{2}\left\{1+\cos\left(\pi\frac{V_{in,DC}}{V_\pi}\right)\right\}}_{P_{out,DC}/P_{in}} \underbrace{- \eta\frac{\pi}{2V_\pi}\sin\left(\pi\frac{V_{in,DC}}{V_\pi}\right)\widehat{v}_{in}(t).}_{\widehat{p}_{out}(t)/P_{in}}\end{aligned}$$

To maximize the modulator linearity (e.g., in analog applications), the bias point can be chosen at $V_\pi/2$; in this case, one has

$$\frac{p_{out}(t)}{P_{in}} \approx \frac{\eta}{2} - \eta\frac{\pi}{2}\frac{\widehat{v}_{in}(t)}{V_\pi},$$

i.e., the modulator bias is at half optical power (assuming no losses), and

$$\frac{\widehat{p}_{out}(t)}{P_{in}} = \eta\frac{\pi}{2}\frac{\widehat{v}_{in}(t)}{V_\pi} \to \frac{\widehat{P}_{out}(\omega)}{P_{in}} = \eta\frac{\pi}{2}\frac{V_{in}(\omega)}{V_\pi} = \eta\frac{\pi}{2}H(\omega)\frac{V_A(\omega)}{V_\pi}.$$

The normalized modulator frequency response is, therefore

$$m(\omega) = \frac{\left|\widehat{P}_{out}(\omega)\right|}{\left|\widehat{P}_{out}(0)\right|} = |H(\omega)|,$$

since $H(0) = 1$, and *coincides with the frequency response of the phase modulation section*:

$$\frac{\Delta\phi_\alpha(\omega)}{\Delta\phi_\alpha(0)} = H(\omega), \quad \alpha = U, L.$$

This is by no means surprising, since the phase modulation section is the only dispersive block in the modulator.

6.4.4 Efficiency–bandwidth trade-off in lumped MZ modulators

To increase the modulator *efficiency* (i.e., the modulator's ability to be operated by a generator with a reduced available power and open-circuit voltage), V_π should be decreased. The product $V_\pi L$ is, however, constant for a given geometry, see (6.16), and can be minimized by increasing the overlap integral, reducing the gap G, and improving the material properties; once $V_\pi L$ has been minimized, V_π can be decreased only by increasing the modulator length L. Unfortunately, increasing L reduces the lumped modulator bandwidth, since it increases the input capacitance. For the sake of definiteness, let us consider a coplanar modulator and denote as C_0 the per-unit-length capacitance of the central strip with respect to the two ground planes; R_G is the generator internal resistance. The modulator input voltage induced by a generator voltage $V_A(\omega)$ will be

$$V_{in}(\omega) = V_A(\omega)\frac{1}{1 + \mathrm{j}\omega R_G C_{in}} = H(\omega) V_A(\omega),$$

where $C_{in} = C_0 L$ is the total input capacitance of the device. The modulator response is then

$$m(\omega) = |H(\omega)| = \frac{1}{\sqrt{1 + \omega^2 R_G^2 C_{in}^2}}. \tag{6.18}$$

From the definition of the 3 dB electrical bandwidth:

$$m_{op}(f_{3\mathrm{dB,el}})\big|_{\mathrm{dB}} = 20\log_{10}\left[m(f_{3\mathrm{dB,el}})\right] = 20\log_{10}\frac{1}{\sqrt{1 + \omega_{3\mathrm{dB,el}}^2 R_G^2 C_{in}^2}} = -3\ \mathrm{dB}$$

we obtain:

$$2\pi f_{3\mathrm{dB,el}} R_G C_{in} = 1 \rightarrow f_{3\mathrm{dB,el}} = \frac{1}{2\pi R_G C_{in}} \propto \frac{1}{L}.$$

Following the definition of the optical bandwidth, we find instead:

$$f_{3\mathrm{dB,op}} = \frac{\sqrt{3}}{2\pi R_G C_{in}} \propto \frac{1}{L}.$$

Independently of the definition considered, it is clear that *increasing the modulator length decreases the on–off voltage* but also *decreases the modulator bandwidth*. As shown in Example 6.1, with realistic input voltages (of the order of 5 V), the bandwidth of a lumped modulator is unsuited to applications with speed above 1 Gbps.

Example 6.1: Evaluate the product $f_{3\mathrm{dB,op}}L$ for a X-cut LN modulator with "typical" parameters. Assume as a goal $V_\pi \approx 5$ V, with $r_{33} = 30$ pm/V, $n_e = n_{eo} = 2.25$, $\lambda_0 = 1.55\ \mu$m, $\Gamma_{mo} = 0.5$, $R_G = 50\ \Omega$, and suppose that the gap is kept to a minimum value to allow the optical waveguide to be defined between the central electrode and ground plane ($G = 10\ \mu$m). Furthermore, assume that the central electrode width is $W = 20\ \mu$m to ensure decoupling between the two optical waveguides.

We have

$$V_\pi L = \frac{\lambda_0 G}{2n_e^3 r_{33}\Gamma_{mo}} = \frac{1.55\times 10^{-6}\cdot 10\times 10^{-6}}{2\cdot 2.25^3\cdot 30\times 10^{-12}\cdot 0.5} = 4.5\times 10^{-2}\ \mathrm{V\,m} = 4.5\ \mathrm{V\,cm}.$$

To evaluate the bandwidth, we need to compute the capacitance of the central strip with respect to the two lateral coplanar ground planes. The capacitance per unit length can be expressed, from (3.29) and (3.31), as

$$C_0 = \frac{\epsilon_{\mathrm{eff}}}{30\pi c_0}\frac{K(k)}{K(k')},$$

where c_0 is the velocity of light in vacuo, K is the elliptic integral of the first kind, $k = W/(W + 2G) = 0.5$, ϵ_{eff} is the coplanar effective permittivity (3.32):

$$\epsilon_{\text{eff}} = n_m^2 = \frac{1 + \sqrt{\epsilon_{xm}\epsilon_{zm}}}{2} = 17.64.$$

Taking into account the approximation (3.33), we have $k' = \sqrt{1 - 0.5^2} = 0.87$ and

$$\frac{K(k)}{K(k')} \approx \frac{1}{\pi} \log\left(2\frac{1+\sqrt{k'}}{1-\sqrt{k'}}\right) = \frac{1}{\pi} \log\left(2\frac{1+\sqrt{0.87}}{1-\sqrt{0.87}}\right) = 1.29;$$

i.e.,

$$C_0 = \frac{\epsilon_{\text{eff}}}{30\pi c_0} \frac{K(k)}{K(k')} = \frac{17.64 \cdot 1.29}{30\pi \cdot 3 \times 10^8} = 8.05 \text{ pF/cm}.$$

The resulting bandwidth–length product will be

$$f_{3\text{dB,op}} L = \frac{\sqrt{3}}{2\pi R_G C_0} = \frac{\sqrt{3}}{2\pi \cdot 50 \cdot 8.05 \times 10^{-12}} = 690 \text{ MHz cm}.$$

We can therefore conclude that a modulator with suitable V_π (requiring a length of the order of 1 cm) has a bandwidth of the order of 1 GHz (take into account that the capacitance can be somewhat reduced by means of thick electrodes and the use of low-permittivity dielectric layers).

6.5 The traveling-wave Mach–Zehnder modulator

As discussed in Section 6.4.4, the bandwidth of a LN concentrated modulator having realistic on–off voltage (≈ 5 V) is of the order of 1 GHz cm. Reducing the modulator length leads to an increase of the bandwidth, but also of the switching voltage V_π. The efficiency–bandwidth trade-off, however, can be substantially improved by making use of a *distributed* rather than of a *concentrated* (lumped) modulator structure, as shown in Fig. 6.15. In the distributed or *traveling-wave* modulator the RF structure works as a transmission line rather than as a concentrated capacitor. The RF signal is fed from the RF input through a coaxial connector (shown in idealized form), and propagates on the coplanar waveguide (CPW) through the interaction length, inducing an electric field that travels together with the optical signal. The optical and RF fields therefore undergo synchronous interaction (as long as their phase velocities are the same). The coplanar transmission line is terminated by an RF load, which may be connected (internally or externally) to an output coaxial connector. Note that an optical fiber is also pigtailed at the optical input and output of the modulator.

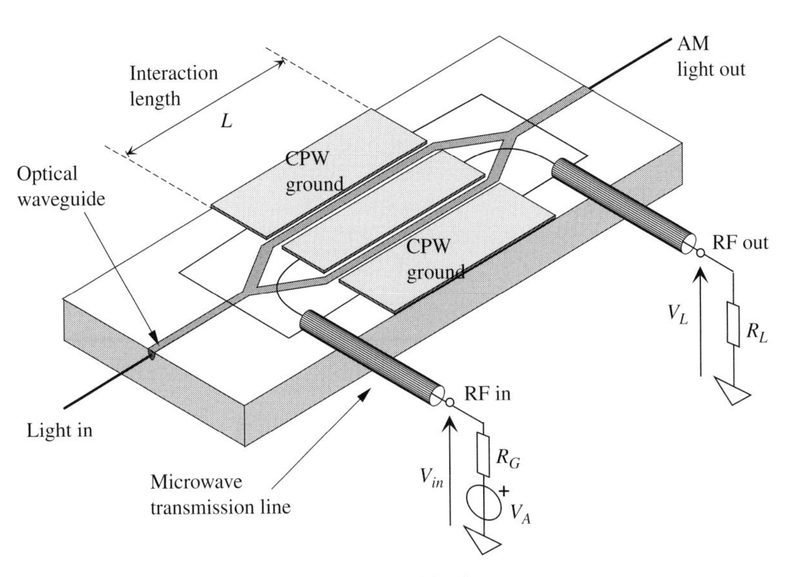

Figure 6.15 Distributed Mach–Zehnder modulator on LN substrate.

6.5.1 Mach–Zehnder traveling-wave modulator dynamic response

In Section 6.4.3 we have shown that, since the optical combiners and dividers are memoriless components, the modulator small-signal frequency response coincides with the frequency response of the phase modulation section. We now analyze the phase modulation frequency response of the traveling-wave modulator on the basis of the distributed interaction between the RF and the optical fields. The RF waveguide is a quasi-TEM transmission line, whose voltage can be expressed as a superposition of forward and backward (reflected) waves:

$$v_m(z, t) = V^+ e^{j\omega(t - z/v_m) - \alpha_m z} + V^- e^{j\omega(t + z/v_m) + \alpha_m z} \tag{6.19}$$

where v_m is the RF phase velocity, α_m is the RF attenuation, and V^+ and V^- are complex constants to be determined from the generator and load conditions. The RF voltage $v_m(z, t)$ in (6.19) is expressed as a complex signal; the corresponding real signal can be obtained as $\text{Re}(v_m)$. Suppose now that an optical wave is traveling in the forward direction (positive z) with group velocity v_o,[10] entering the line ($z = 0$) at time t_0. At $t_1 > t_0$ the optical wavefront is at $z = (t_1 - t_0)v_o$, i.e., $t_1 = t_0 + z/v_o$. The time t_2 at which the optical wavefront reaches the end of the interaction region will therefore be $t_2 = t_0 + L/v_o$, where L is the length of the interaction region see Fig. 6.16.

While traveling along the line, the optical wave experiences, due to the effect of the RF field, a local variation of the refractive index n; for the sake of generality we will assume that n is complex, $n = n_r - jn_i$. Let us call such a variation $\Delta n = \Delta n_r - j\Delta n_i$,

[10] For a monochromatic laser beam entering the modulation region, v_o should rather be interpreted as the phase velocity; however, as soon as phase modulation takes place, the optical signal becomes a narrowband modulated signal and the wavefront propagation velocity should be the group velocity. See also [91].

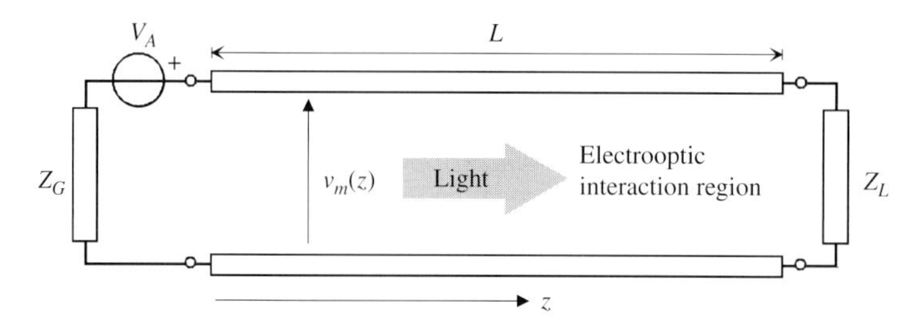

Figure 6.16 Transmission line model of the phase modulation section of an electrooptic modulator.

and assume that Δn depends linearly on the local voltage $v_m(z, t)$:

$$\Delta n(z, t) = a v_m(z, t), \tag{6.20}$$

where a is a proper constant. The total amplitude and phase variation of the optical field can be recovered by integrating Δn as *seen* by the optical field while traveling along the transmission line, i.e., $\Delta n(z, t(z))$, where

$$t(z) = t_0 + z/v_o = t_2 - L/v_o + z/v_o.$$

Defining E as the optical field at the output of the interaction region, we have

$$\begin{aligned} E(t_2) &= E(t_2, \Delta n = 0) \exp\left[-\mathrm{j}k_o \int_0^L \Delta n \left(z, t_2 - \frac{L}{v_o} + \frac{z}{v_o}\right) \mathrm{d}z\right] \\ &= E(t_2, \Delta n = 0) \exp\left[-\mathrm{j}k_o \int_0^L \Delta n_r \left(z, t_2 - \frac{L}{v_o} + \frac{z}{v_o}\right) \mathrm{d}z \right. \\ &\qquad \left. - \int_0^L \Delta\bar{\alpha}_o \left(z, t_2 - \frac{L}{v_o} + \frac{z}{v_o}\right) \mathrm{d}z\right], \end{aligned}$$

where $\Delta\bar{\alpha}_o = k_0 \Delta n_i$ is the variation in the optical field *attenuation*. The first integral describes a cumulative *phase modulation*, the second one a cumulative *amplitude modulation*.

Introducing (6.20) and (6.19), neglecting optical losses and their modulation ($\Delta n = \Delta n_r$), we express the phase variation as[11]

$$\begin{aligned} \Delta\phi &= k_0 \int_0^L \Delta n \left(z, t_2 - \frac{L}{v_o} + \frac{z}{v_o}\right) \mathrm{d}z = k_0 a \int_0^L v_m \left(z, t_2 - \frac{L}{v_o} + \frac{z}{v_o}\right) \mathrm{d}z \\ &= k_0 a \mathrm{e}^{\mathrm{j}\omega\left(t_2 - \frac{L}{v_o}\right)} \left[V^+ \frac{\mathrm{e}^{\mathrm{j}\omega\left(\frac{L}{v_o} - \frac{L}{v_m}\right) - \alpha_m L} - 1}{\mathrm{j}\omega\left(\frac{1}{v_o} - \frac{1}{v_m}\right) - \alpha_m} + V^- \frac{\mathrm{e}^{\mathrm{j}\omega\left(\frac{L}{v_o} + \frac{L}{v_m}\right) + \alpha_m L} - 1}{\mathrm{j}\omega\left(\frac{1}{v_o} + \frac{1}{v_m}\right) + \alpha_m} \right]. \end{aligned} \tag{6.21}$$

[11] The phase variation $\Delta\phi = \Delta\phi(t, \omega)$ is expressed as a complex time-domain signal, with reference time $t = t_2$. In principle, it could be associated with a phasor $\Delta\Phi$ such as $\Delta\phi = \Delta\Phi \exp(\mathrm{j}\omega t)$.

The coefficients V^+ and V^- can be now derived by imposing generator ($z = 0$) and load ($z = L$) boundary conditions:

$$v_m(0, t) = V_A \exp(\mathrm{j}\omega t) - i_m(0, t) Z_G \tag{6.22}$$

$$v_m(L, t) = i_m(L, t) Z_L, \tag{6.23}$$

where Z_G (Z_L) are the generator (load) impedance, respectively (typically real), and V_A is the open-circuit generator voltage. To this end, the line voltage and current can be written, taking into account (6.19) and (3.19), (3.20), as

$$\begin{aligned} v_m &= V^+ \exp(-\gamma_m z + \mathrm{j}\omega t) + V^- \exp(\gamma_m z + \mathrm{j}\omega t) \\ i_m &= \frac{V^+}{Z_0} \exp(-\gamma_m z + \mathrm{j}\omega t) - \frac{V^-}{Z_0} \exp(\gamma_m z + \mathrm{j}\omega t), \end{aligned}$$

where γ_m is the RF complex propagation constant,

$$\gamma_m = \alpha_m + \mathrm{j}\beta_m = \alpha_m + \mathrm{j}\omega / v_m,$$

and Z_0 is the line characteristic impedance. Imposing (6.22) and (6.23) leads to the expressions

$$V^+ = \frac{1}{1 - \Gamma_L \Gamma_G \exp(-2\gamma_m L)} \frac{Z_0}{Z_0 + Z_G} V_A \tag{6.24}$$

$$V^- = \frac{\Gamma_L \exp(-2\gamma_m L)}{1 - \Gamma_L \Gamma_G \exp(-2\gamma_m L)} \frac{Z_0}{Z_0 + Z_G} V_A, \tag{6.25}$$

where the load and source reflection coefficients have been introduced:

$$\Gamma_L = \frac{Z_0 - Z_L}{Z_0 + Z_L}, \quad \Gamma_G = \frac{Z_0 - Z_G}{Z_0 + Z_G}.$$

We finally take into account that in the Mach–Zehnder LN amplitude modulator $n_r = n_e$ and that, from (6.10) and (6.20),

$$k_0 a = \frac{\pi n_r^3 r_{33}}{\lambda_o} \frac{\Gamma_{mo} L}{G} = \frac{\pi n_e^3 r_{33}}{\lambda_o} \frac{\Gamma_{mo} L}{G}.$$

Substituting V^+ and V^- from (6.24) and (6.25) into (6.21) and defining

$$\beta_o = \omega / v_o$$

(note that $\beta_o \neq 2\pi/\lambda_o$!), the phase variation of the optical field at the end of the phase modulation section and at $t = t_2$ can finally be expressed as

$$\begin{aligned} \Delta\phi(t_2, \omega) = {} & \frac{\pi n_e^3 r_{33}}{\lambda_o} \frac{\Gamma_{mo} L V_A}{G} \frac{Z_0}{Z_0 + Z_G} \frac{\mathrm{e}^{\mathrm{j}\omega t_2} \mathrm{e}^{-\gamma_m L}}{1 - \Gamma_L \Gamma_G \mathrm{e}^{-2\gamma_m L}} \\ & \times \left[\frac{1 - \mathrm{e}^{-\mathrm{j}(\beta_o - \beta_m)L + \alpha_m L}}{\mathrm{j}(\beta_m - \beta_o)L + \alpha_m L} - \Gamma_L \frac{1 - \mathrm{e}^{-\mathrm{j}(\beta_o + \beta_m)L - \alpha_m L}}{\mathrm{j}(\beta_m + \beta_o)L + \alpha_m L} \right], \end{aligned} \tag{6.26}$$

or, in a more compact form and assuming $t = t_2$ (the time at which the wavefront reaches the end of the modulating region) as the reference time,

$$\Delta\phi(\omega) = \frac{\pi n_e^3 r_{33}}{\lambda_o} \frac{\Gamma_{mo} L}{G} \frac{Z_0 e^{-\gamma_m L}}{Z_0 + Z_G} \frac{F(u_+) + \Gamma_L F(u_-)}{1 - \Gamma_L \Gamma_G e^{-2\gamma_m L}} V_A e^{j\omega t} = H(\omega) V_A e^{j\omega t}, \tag{6.27}$$

where

$$F(u) = \frac{1 - \exp(u)}{u} \tag{6.28}$$

$$u_\pm(\omega) = j(\pm\beta_m - \beta_o)L \pm \alpha_m L = \pm\alpha_m L + j\frac{\omega}{c_0}(\pm n_m - n_o)L. \tag{6.29}$$

An alternative, equivalent form for $H(\omega)$ is

$$H(\omega) = \frac{\pi n_e^3 r_{33} \Gamma_{mo}}{\lambda_o} \frac{L}{G} \frac{Z_{in}}{Z_{in} + Z_G} \frac{(Z_L + Z_0)F(u_+) + (Z_L - Z_0)F(u_-)}{(Z_L + Z_0)e^{\gamma_m L} + (Z_L - Z_0)e^{-\gamma_m L}}, \tag{6.30}$$

where Z_{in} is the RF line input impedance:

$$Z_{in} = Z_0 \frac{Z_L + Z_0 \tanh(\gamma_m L)}{Z_0 + Z_L \tanh(\gamma_m L)}.$$

The modulation response of the distributed modulator can be finally expressed as

$$\begin{aligned} m(\omega) &= \left|\frac{\Delta\phi(\omega)}{\Delta\phi(0)}\right| = \left|\frac{H(\omega)}{H(0)}\right| \\ &= \frac{R_L + R_G}{R_L} \left|\frac{Z_{in}}{Z_{in} + Z_G}\right| \left|\frac{(Z_L + Z_0)F(u_+) + (Z_L - Z_0)F(u_-)}{(Z_L + Z_0)e^{\gamma_m L} + (Z_L - Z_0)e^{-\gamma_m L}}\right|, \end{aligned} \tag{6.31}$$

where we have assumed (neglecting DC ohmic losses) $Z_{in}(0) \approx Z_L(0) = R_L$, $Z_G(0) = R_G$, $F(u(0)) \approx 1$, $\gamma_m(0)L \approx 0$ and therefore

$$\Delta\phi(0) \approx \frac{\pi n_e^3 r_{33} \Gamma_{mo} V_A}{\lambda_o} \frac{L}{G} \frac{R_L}{R_L + R_G}.$$

Note that the DC value coincides with the result already found, see (6.12), assuming that the load is open ($R_L \to \infty$). In particular, the distributed frequency response coincides with the lumped-parameter response if we neglect RF losses in the line. In fact, assume $R_L \to \infty$ and make $L \to 0$; in this case $F(u) \approx 1$ and

$$m(\omega) \approx \left|\frac{Z_{in}}{Z_{in} + R_G}\right|,$$

but

$$Z_{in} \underset{Z_L = R_L \to \infty}{\to} \frac{Z_0}{\tanh(\gamma_m L)} \approx \frac{Z_0 v_m}{j\omega L} = \frac{1}{j\omega L} \underbrace{\sqrt{\frac{\mathcal{L}}{\mathcal{C}}}}_{Z_0} \underbrace{\frac{1}{\sqrt{\mathcal{L}\mathcal{C}}}}_{v_m} = \frac{1}{j\omega C_{in}}$$

where $\mathcal{L}$ and $\mathcal{C}$ are the inductance and capacitance per unit length of the line, C_{in} is the input capacitance of the line, and we have approximated $\tanh(\gamma_m L) = \tanh(j\omega L/v_m) \approx j\omega L/v_m$. Therefore we obtain

$$m(\omega) = \frac{1}{\sqrt{1+\omega^2 R_G^2 C_{in}^2}},$$

the result we already found for the lumped-parameter modulator, see (6.18).

6.5.2 Analysis of the TW Mach–Zehnder modulator response

The frequency response of the distributed Mach–Zehnder modulator is dominated by the following main factors:

1. The velocity mismatch between the optical and RF propagating waves, related to the index mismatch $\Delta n_{mo} = n_m - n_o$. This parameter has an influence on u_+ and therefore on $F(u_+)$.
2. The impedance mismatch between the load, the transmission line, and the generator.
3. The RF losses of the modulator line.

In order to single out the effect of each factor, consider first the modulation response $m_1(\omega)$ of a lossless modulator ($\alpha_m = 0$) with a matched load and generator ($Z_L = Z_G = Z_0$). From (6.31) we obtain

$$m_1(\omega) = |F(u_+)| = \left|\frac{\sin(U)}{U}\right|, \qquad U = \frac{\omega}{2c_0}(n_m - n_o)L = \frac{\pi f \Delta n_{mo} L}{c_0},$$

and the frequency response is dominated by the velocity mismatch between the RF and optical signals. With zero velocity mismatch, $m_1(\omega) = 1$, independent of frequency, and the bandwidth is ideally infinite. For finite velocity mismatch, the behavior is shown in Fig. 6.17 for $L = 11$ mm; the relative mismatch is defined as $\Delta_{\%} = \left[(n_m - n_o)/n_o\right] \times 100$; in LN we always have $n_m > n_o$.

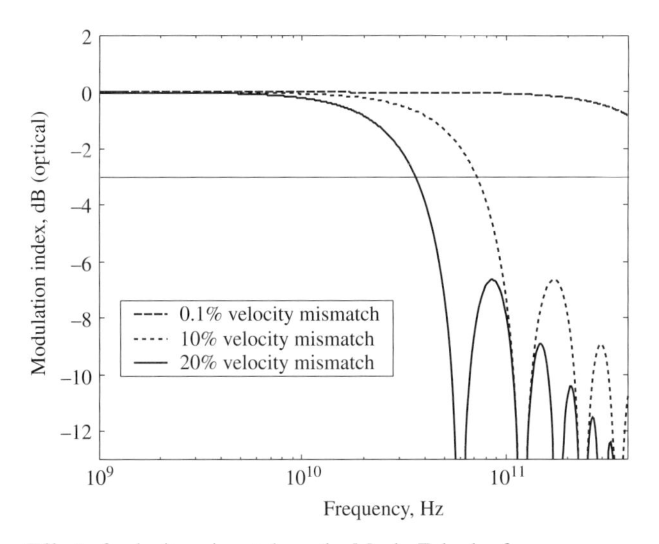

Figure 6.17 Effect of velocity mismatch on the Mach–Zehnder frequency response.

The velocity mismatch-dependent modulator electrical and optical bandwidth can be evaluated by imposing (6.4) and (6.5) to $m_1(U)$; we obtain $U_{3\text{dB,op}} = 1.89$, $U_{3\text{dB,el}} = 1.39$, i.e.,

$$f_{3\text{dB,op}}L = \frac{1.89c_0}{\pi\,\Delta n_{mo}} = \frac{0.180}{\Delta n_{mo}}\ \text{GHz m}$$

$$f_{3\text{dB,el}}L = \frac{1.39c_0}{\pi\,\Delta n_{mo}} = \frac{0.133}{\Delta n_{mo}}\ \text{GHz m.}$$

Taking into account that $L \approx 1$ cm to obtain $V_\pi \approx 5$ V, we have that the velocity mismatch for 10 Gbps operation (optical bandwidth) is $\Delta n_{mo} = 1.8$. For a coplanar modulator on a LN niobate one has $\Delta n_{mo} = n_m - n_o = 4.22 - 2.2 \approx 2$. Thus, even without accounting for losses, a straightforward modulator without any structure optimization aimed at improving synchronous coupling can cover 2.5 Gbps applications but can hardly satisfy 10 Gbps specifications.

RF losses cause a decrease of the bandwidth and the filling of response zeros, see Fig. 6.18. Since the RF signal line cross section is very small (e.g., the width and thickness are of the order of 10 μm), and skin effect losses increase anyway as $\sqrt{f}$, conductor losses can be a major limitation to the high-speed modulator operation.

Assuming synchronous coupling and input and output impedance matching, the RF loss-limited response $m_2(\omega)$ is, from (6.31),

$$m_2(\omega) = \exp(-W)\left|\frac{\sinh(W)}{W}\right|, \qquad W = \frac{\alpha_m(f)L}{2} = \frac{\alpha_m(f_0)L}{2}\sqrt{\frac{f}{f_0}}.$$

The loss-dependent modulator electrical and optical bandwidth can be evaluated by imposing (6.4) and (6.5) on $m_2(W)$;

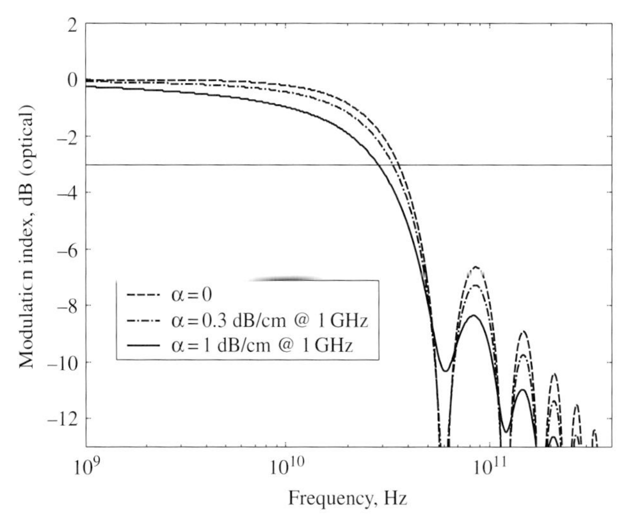

Figure 6.18 Effect of losses on the distributed Mach–Zehnder modulator frequency response. Losses are introduced in the case where the velocity mismatch is 5%.

we obtain $W_{3\text{dB,op}} = 0.794$ and $W_{3\text{dB,el}} = 0.368$, i.e.,

$$\frac{\alpha_m(f_{3\text{dB,op}})L}{2} = 0.794 \rightarrow \alpha_m(f_{3\text{dB,op}})L = 1.588 \rightarrow \alpha_m(f_{3\text{dB,op}})L\big|_{\text{dB}} = 6.897\,\text{dB}$$

$$\frac{\alpha_m(f_{3\text{dB,el}})L}{2} = 0.368 \rightarrow \alpha_m(f_{3\text{dB,el}})L = 0.736 \rightarrow \alpha_m(f_{3\text{dB,el}})L\big|_{\text{dB}} = 6.393\,\text{dB}.$$

The total line attenuation at the frequency corresponding to the bandwidth is therefore 6.9 dB for the optical bandwidth and 6.4 dB for the electrical 3 dB bandwidth. As an example, in order to allow for 40 Gbps operation, the (loss-limited) 3 dB optical bandwidth of a 1 cm long modulator should satisfy the constraint

$$\alpha_m(40\,\text{GHz})L|_{\text{dB}} = 6.897\,\text{dB} \rightarrow \alpha_m(40\,\text{GHz})|_{\text{dB}} = 6.897\,\text{dB/cm};$$

i.e., the 1 GHz loss should be

$$\alpha_m(1\,\text{GHz})\sqrt{\frac{40\,\text{GHz}}{1\,\text{GHz}}} = 6.897\,\text{dB/cm} \rightarrow \alpha_m(1\,\text{GHz}) = \frac{6.897}{\sqrt{40}} = 1.1\,\text{dB/cm}.$$

This is a very low value, considering the small line cross section of modulator lines.

Finally, assume that no losses are present and that the velocities are matched; the effect of impedance mismatch can be better understood by separating the phase delay into two contributions:

$$\Delta\phi(\omega) = \Delta\phi_+(\omega) + \Delta\phi_-(\omega),$$

where $\Delta\phi_+(\omega)$ is the phase modulation due to *codirectional coupling*, i.e., the coupling between the RF forward traveling wave and the optical mode, while $\Delta\phi_-(\omega)$ is the phase modulation due to *contradirectional coupling*, i.e., the coupling between the RF backward or reflected wave and the forward-traveling optical wave. From (6.27) we obtain

$$\Delta\phi_+(\omega) = \frac{\pi n_e^3 r_{33}}{\lambda_o}\frac{\Gamma_{mo}L}{G}\frac{Z_0}{Z_0+Z_G}\frac{e^{-\gamma_m L}}{1-\Gamma_L\Gamma_G e^{-2\gamma_m L}}F(u_+)V_A e^{j\omega t}$$

$$\Delta\phi_-(\omega) = \frac{\pi n_e^3 r_{33}}{\lambda_o}\frac{\Gamma_{mo}L}{G}\frac{Z_0}{Z_0+Z_G}\frac{e^{-\gamma_m L}}{1-\Gamma_L\Gamma_G e^{-2\gamma_m L}}\Gamma_L F(u_-)V_A e^{j\omega t}.$$

Contradirectional coupling is never synchronous, and typically leads to a rapidly decreasing $\sin x/x$ behavior with increasing frequency. Neglecting losses and velocity mismatch, we obtain from (6.27) the impedance mismatch-limited response $m_3(\omega)$:

$$m_3(\omega) = \left|\frac{\Delta\phi(\omega)}{\Delta\phi(0)}\right| = \left|\frac{(1-\Gamma_L\Gamma_G)\left[1+\Gamma_L F(U_-)\right]}{1-\Gamma_L\Gamma_G\exp(-2j\beta_m L)}\right|, \quad U_- = -j\frac{2\omega}{c_0}n_o L.$$

Impedance mismatch leads to ripples in the frequency response; see Fig. 6.19.

Complete mismatch is achieved with an open-circuit or short-circuit modulator. The open-circuit modulator has a narrowband low-pass response, somewhat similar to the response of a lumped modulator, while the short-circuit modulator has zero efficiency at DC (due to the cancellation of co- and counterpropagating responses) and exhibits a resonant peak at higher frequency; see Fig. 6.20. Shorted resonant modulators can sometimes be used for narrowband operation, since they increase their efficiency

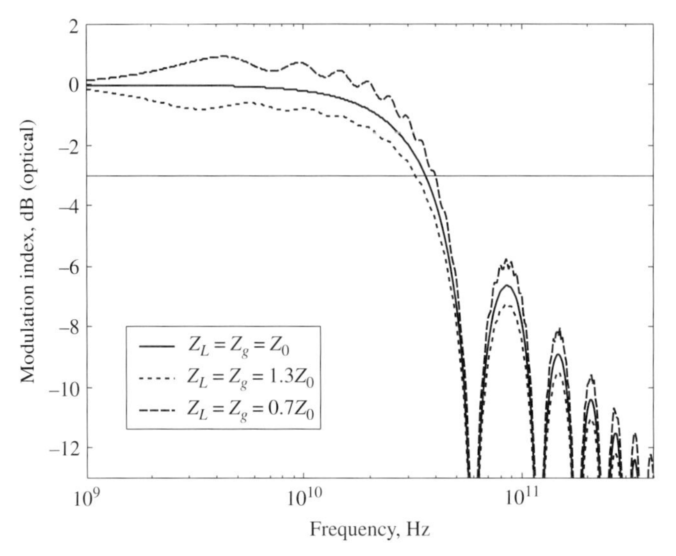

Figure 6.19 Effect of impedance mismatch on the modulator response; the zero impedance mismatch case is lossless with a velocity mismatch of 20%.

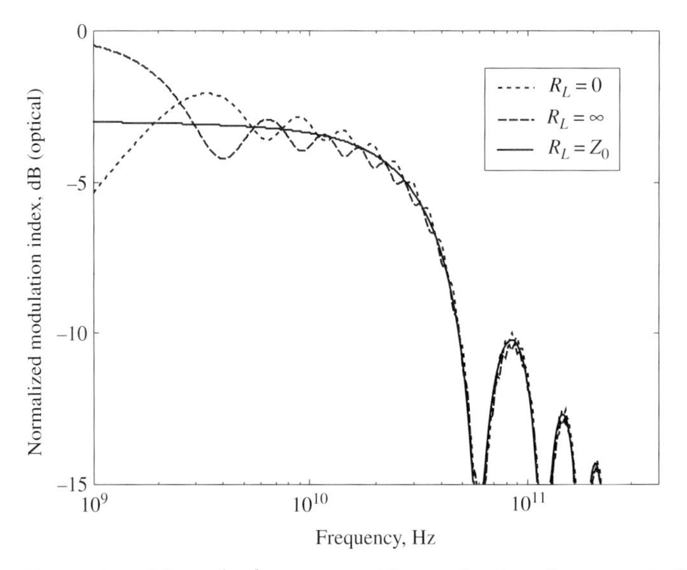

Figure 6.20 Resonant modulators in short or open. The starting impedance matched structure is lossless and the velocity mismatch is 20%.

with respect to velocity-matched modulators on a narrow band. Other examples of narrowband modulators, which could be suited to narrowband transmission of an analog signal over an optical fiber, are the *phase reversal* modulators [92].

Phase reversal modulators can be realized by periodically reversing the sign of the applied electric field. The optical signal therefore alternately undergoes variations of the refractive index with opposite sign. This is easily obtained with a Z-cut substrate,

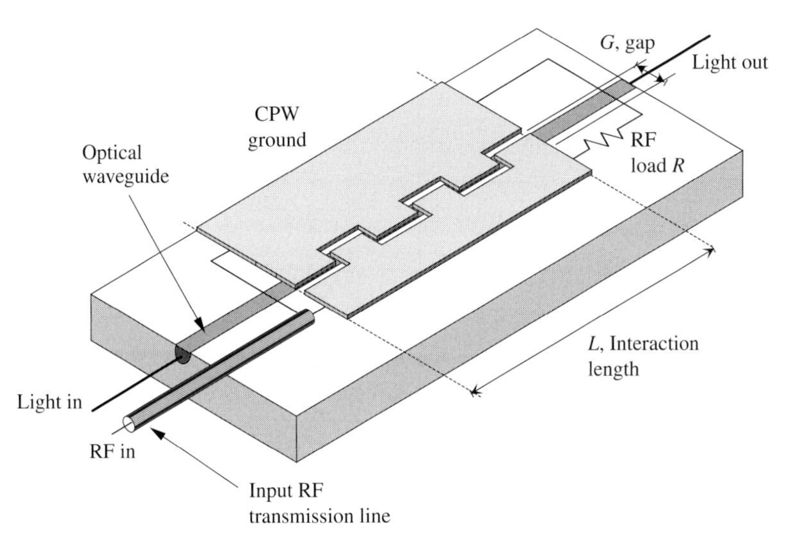

Figure 6.21 Phase modulation section of a phase reversal modulator on Z-cut LN. ACPW stands for asymmetric ω planar waveguide.

as shown in Fig. 6.21 (for the sake of simplicity, only the phase modulation section is shown).

The operation of the phase reversal device can be explained as follows. At low frequency the phase delay of the RF signal is very low, so that, integrating phase variations of opposite sign over the modulator length, the net result is zero (Fig. 6.22(a)). With increasing frequency the RF signal finally undergoes a large phase delay, assuming, at a certain frequency f_0, alternate positive and negative sign over the modulator length; at that frequency the phase variation of the RF wave and the phase variation in the periodic structure compensate, leading to an overall phase shift of the optical signal (Fig. 6.22(b)). Thus, the response of the modulator is zero at low frequency, while it exhibits a resonant peak around f_0; see Fig. 6.23 for a simplified example with only two reversal sections. It may be noted that the response peak occurs at the same frequency at which a zero response occurs in the absence of phase reversal. The resulting device is clearly suited to the transmission of narrowband analog signals only.

6.5.3 The Mach–Zehnder modulator chirp

The instantaneous frequency deviation of Mach–Zehnder modulators is dominated by geometrical factors rather than by material parameters. Let us consider again from (6.13) the modulator output optical wave

$$\begin{aligned} b_1' &= \exp(2\mathrm{j}\phi_{sp}) \exp(-\mathrm{j}k_o L) \left[\alpha \exp(-\mathrm{j}\Delta\phi_U(t)) + (1-\alpha)\exp(-\mathrm{j}\Delta\phi_L(t))\right] a_1 \\ &= K\left[\alpha \exp(-\mathrm{j}\Delta\phi_U(t)) + (1-\alpha)\exp(-\mathrm{j}\Delta\phi_L(t))\right], \end{aligned}$$

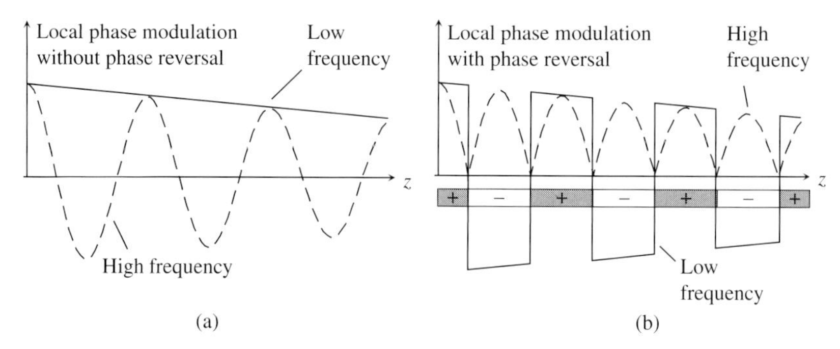

Figure 6.22 Behavior of the induced phase delay in a conventional (a) and phase reversal modulator (b) at low (continuous line) and high (dashed line) frequency.

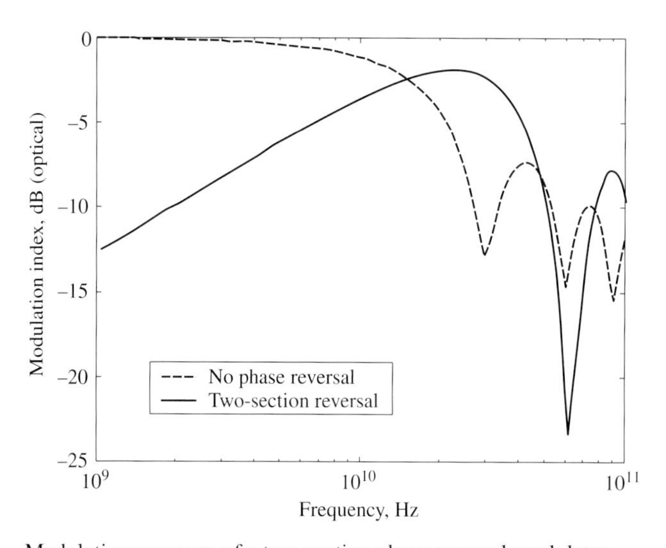

Figure 6.23 Modulation response of a two-section phase reversal modulator.

where K is a time-independent constant. The time-varying phase of the output signal will therefore be[12]

$$\varphi(t) = -\tan^{-1}\left[\frac{\mathrm{Im}(b_1'/K)}{\mathrm{Re}(b_1'/K)}\right] = \tan^{-1}\left[\frac{\alpha\sin(\Delta\phi_U(t)) + (1-\alpha)\sin(\Delta\phi_L(t))}{\alpha\cos(\Delta\phi_U(t)) + (1-\alpha)\cos(\Delta\phi_L(t))}\right].$$

The instantaneous frequency deviation can be now directly evaluated as

$$\Delta f(t) = \frac{1}{2\pi}\frac{\mathrm{d}\varphi(t)}{\mathrm{d}t} = \frac{1}{2\pi}\frac{\mathrm{d}}{\mathrm{d}t}\tan^{-1}\left[\frac{\alpha\sin(\Delta\phi_U(t)) + (1-\alpha)\sin(\Delta\phi_L(t))}{\alpha\cos(\Delta\phi_U(t)) + (1-\alpha)\cos(\Delta\phi_L(t))}\right].$$

[12] We refer here to a phase term $\exp(\mathrm{j}\varphi)$.

For simplicity, let us consider a symmetrical splitter and combiner ($\alpha = 1/2$); we have

$$\Delta f(t) = \frac{1}{2\pi}\frac{\mathrm{d}}{\mathrm{d}t}\tan^{-1}\left[\frac{\sin\Delta\phi_U + \sin\Delta\phi_L}{\cos\Delta\phi_U + \cos\Delta\phi_L}\right]$$
$$= \frac{1}{2\pi}\frac{\mathrm{d}}{\mathrm{d}t}\tan^{-1}\left[\frac{2\sin\dfrac{\Delta\phi_U + \Delta\phi_L}{2}\cos\dfrac{\Delta\phi_U - \Delta\phi_L}{2}}{2\cos\dfrac{\Delta\phi_U + \Delta\phi_L}{2}\cos\dfrac{\Delta\phi_U - \Delta\phi_L}{2}}\right]$$
$$= \frac{1}{2\pi}\frac{\mathrm{d}}{\mathrm{d}t}\tan^{-1}\left[\tan\frac{\Delta\phi_U + \Delta\phi_L}{2}\right]$$
$$= \frac{1}{4\pi}\frac{\mathrm{d}(\Delta\phi_U + \Delta\phi_L)}{\mathrm{d}t}\frac{1}{4}\left[\frac{1}{V_{\pi U}}\frac{\mathrm{d}v_{inU}}{\mathrm{d}t} - \frac{1}{V_{\pi L}}\frac{\mathrm{d}v_{inL}}{\mathrm{d}t}\right], \quad (6.32)$$

where we have expressed $\Delta\phi_U$ and $\Delta\phi_L$ through (6.15). The Henry chirp parameter can now be conveniently introduced according to (5.52). In fact, for the amplitude modulation we have

$$P_{out} = \frac{1}{2}\eta\left[1 + \cos\left(\Delta\phi_U(t) - \Delta\phi_L(t)\right)\right],$$

and therefore

$$\frac{1}{P_{out}}\frac{\mathrm{d}P_{out}}{\mathrm{d}t} = \pi\frac{\frac{1}{2}\eta\sin\left(\Delta\phi_U - \Delta\phi_L\right)}{\frac{1}{2}\eta\left[1 + \cos\left(\Delta\phi_U - \Delta\phi_L\right)\right]}\left[\frac{1}{V_{\pi U}}\frac{\mathrm{d}v_{inU}}{\mathrm{d}t} + \frac{1}{V_{\pi L}}\frac{\mathrm{d}v_{inL}}{\mathrm{d}t}\right]$$
$$= \tan\left(\frac{\Delta\phi_U(t) - \Delta\phi_L(t)}{2}\right)\left[\frac{1}{V_{\pi U}}\frac{\mathrm{d}v_{inU}}{\mathrm{d}t} + \frac{1}{V_{\pi L}}\frac{\mathrm{d}v_{inL}}{\mathrm{d}t}\right]. \quad (6.33)$$

Thus from (6.32) and (6.33) the Henry chirp parameter α_H becomes

$$\alpha_H = \frac{4\pi\Delta f}{\dfrac{1}{P_{out}}\dfrac{\mathrm{d}P_{out}}{\mathrm{d}t}} = \frac{\dfrac{1}{V_{\pi U}}\dfrac{\mathrm{d}v_{inU}}{\mathrm{d}t} - \dfrac{1}{V_{\pi L}}\dfrac{\mathrm{d}v_{inL}}{\mathrm{d}t}}{\dfrac{1}{V_{\pi U}}\dfrac{\mathrm{d}v_{inU}}{\mathrm{d}t} + \dfrac{1}{V_{\pi L}}\dfrac{\mathrm{d}v_{inL}}{\mathrm{d}t}}\cot\left(\frac{\Delta\phi_U(t) - \Delta\phi_L(t)}{2}\right). \quad (6.34)$$

In general, α_H is a function of time, which may assume positive or negative values according to the modulator parameters and the driving voltages. If we restrict the analysis to the case $v_{inU} = v_{inL} = v_{in}$ we have from (6.34)

$$\alpha_H(t) = \frac{V_{\pi L} - V_{\pi U}}{V_{\pi L} + V_{\pi U}}\cot\left[\frac{\Delta\phi_U(t) - \Delta\phi_L(t)}{2}\right] = \frac{V_{\pi L} - V_{\pi U}}{V_{\pi L} + V_{\pi U}}\cot\left[\frac{\pi}{2}\frac{v_{in}(t)}{V_\pi}\right],$$

where again

$$V_\pi = \frac{V_{\pi U}V_{\pi L}}{V_{\pi U} + V_{\pi L}}.$$

In small-signal conditions we have $v_{in}(t) = V_{in} + \widehat{v}_{in}(t) \approx V_{in}$ and therefore the Henry parameter becomes a constant:

$$\alpha_H = \frac{V_{\pi L} - V_{\pi U}}{V_{\pi L} + V_{\pi U}}\cot\left[\frac{\pi}{2}\frac{V_{in}}{V_\pi}\right].$$

If we bias the modulator at $V_{in} = V_\pi/2$, then cot $= 1$ and chirp can be made positive or negative (see Fig. 5.38) by proper design of the modulator arms. If the two arms are unbalanced, the amount of chirp can be varied somewhat by changing the bias point with respect to $V_{in} = V_\pi/2$. If, on the other hand, the modulator is completely symmetrical (as in X-cut Mach–Zehnder modulators on LN) the chirp is identically zero, independent of the bias voltage. Z-cut modulators have nonzero chirp, which can conveniently be designed to be negative to (partly) compensate for fiber dispersion.

On the other hand, variable-chirp modulators can be obtained in principle by exploiting two separate driving voltages for the upper and lower arms. The relevant structure, called a dual-drive modulator, will be discussed in Section 6.6.1. Dual-drive modulators typically have a symmetrical structure ($V_{\pi U} = V_{\pi L} = 2V_\pi$) but the upper and lower phase modulation sections are controlled by different voltages. Suppose we work under small-signal conditions and apply a symmetrical bias to the two modulator arms;[13] we have

$$v_{inU,L}(t) = V_{in} + \widehat{v}_{inU,L}(t), \quad \widehat{v}_{inU}(t) = V_U f(t), \quad \widehat{v}_{inL}(t) = V_L f(t),$$

where $f(t)$ is a function of time, so that, setting $V_{\pi L} = V_{\pi U}$ and $\cot\left[\left(\Delta\phi_U - \Delta\phi_L\right)/2\right] = \cot\left(\pi V_{in}/2V_\pi\right)$ in (6.34), we obtain

$$\alpha_H = \frac{V_U - V_L}{V_U + V_L} \cot\left[\frac{\pi}{2}\frac{V_{in}}{V_\pi}\right].$$

Working at $V_{in} = V_\pi/2$ and properly adjusting the magnitude of the RF driving voltages (V_U and V_L), the chirp parameter can be changed from positive to negative.

6.6 High-speed electrooptic modulator design

Achieving high-speed electrooptic modulators for 10 Gbps or 40 Gbps requires the use of traveling-wave structures. However, a large (more than 100%) velocity mismatch naturally exists between the optical and RF velocities in a coplanar LN modulator (see (6.11) and the following discussion), so that the RF structure has to be properly optimized in order to reduce the microwave refractive index and improve the velocity matching with the optical signal. Common recipes for LN are the use of low-permittivity (e.g. SiO_2) thin ($<1.5\,\mu m$) buffer layers interposed between the coplanar electrodes and the substrate, the increase of the electrode thickness through electroplating to store more RF energy in air rather than in the high-permittivity LN substrate, and the use of ridge optical waveguides to optimize the overlap integral.

[13] Note that in a X-cut modulator the applied voltage in the center conductor induces opposite fields in the upper and lower waveguides; in a dual drive Z-cut modulator, see Fig. 6.25(d), opposite driving voltages have to be imposed to obtain opposite fields. In order to use the same notation as in the single-drive case, the driving voltage in the lower arm has been assumed with a minus sign, so that equal external driving voltages can again induce an opposite phase delay.

For semiconductor modulators, the situation is reversed, since for GaAs the effective permittivity of a coplanar waveguide is $\approx (1+\epsilon_r)/2 = 7$, corresponding to an RF index $n_m = \sqrt{7} = 2.65$. On the other hand, the optical refractive index is $n_o \approx 3.4$. To achieve synchronous coupling, the RF index should be increased (i.e., the RF wave made slower) through the use of proper RF slow-wave structures. A similar issue was discussed in the context of velocity-matched traveling-wave photodetectors, see Section 4.10.3.

Reducing the RF line losses is also a must, since line losses are a major limiting factor to operation beyond 40 GHz. Losses can be partly decreased through electrode thickening and shaping, but, due to the skin effect, improvements are somewhat limited, and record attenuations are of the order of 0.3 dB/cm at 1 GHz [93]. Finally, the switching voltage should be kept low (typically 5 V or less) by reducing the line gap and sometimes placing the waveguide in a ridge or other special structure to increase the interaction with the RF field. In semiconductor modulators the use of junctions allows enhancement of the applied field in the active region; typical switching voltages are, however, similar to those of LN-based devices. Reducing the switching voltage is important in high-speed operation, since it makes the design of the driver stage less critical.

Finally, 50 Ω impedance matching would be highly desirable, since this would simplify the design of driver stages (wideband matching is virtually impossible). This goal is less difficult to achieve in LN traveling-wave modulators, whose typical line impedances are of the order of 40–50 Ω, while in semiconductor modulators very often the impedance level is significantly lower.

To investigate together the velocity matching and loss requirements for high-speed operation, we consider again the frequency response of a distributed modulator for the impedance-matched case, and we consider *together* losses and velocity mismatch. From (6.31), the response can be written in the form

$$m(\omega) = \exp(-u'_+/2) \left| \frac{\sinh(u_+/2)}{u_+/2} \right|,$$

where

$$u_+ = u'_+ + \mathrm{j}u''_+ = \alpha_m L + \mathrm{j}\frac{\omega(n_m - n_o)L}{c_0}.$$

We now plot the 3 dB bandwidth (electrical in this case) as a function of the normalized velocity mismatch $\Delta n_{\text{eff}} L$ and of the total loss at 1 GHz, $\alpha_{0m} L$ (assuming that $\alpha_m \propto \sqrt{f}$ due to the skin effect). The result, shown in Fig. 6.24, indicates that two regions exist, the first (horizontal lines) is limited by velocity mismatch, the second (vertical lines) by losses. With the present technology, 40 Gbps modulators can be manufactured but require total RF losses less than 1 dB over the interaction region and velocity mismatch of the order of 0.4 over an interaction length of 1 cm. Increasing speed appears far more demanding in terms of attenuation requirements than in terms of velocity matching.

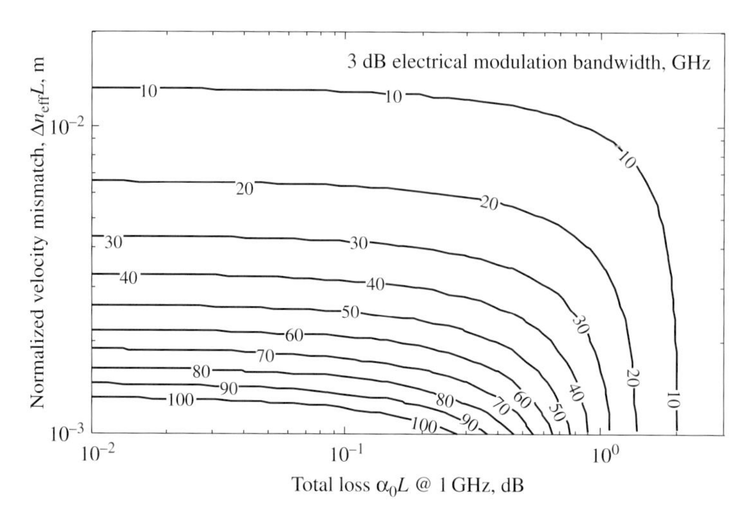

Figure 6.24 Requirements on total line loss and velocity matching in order to achieve a given modulation bandwidth.

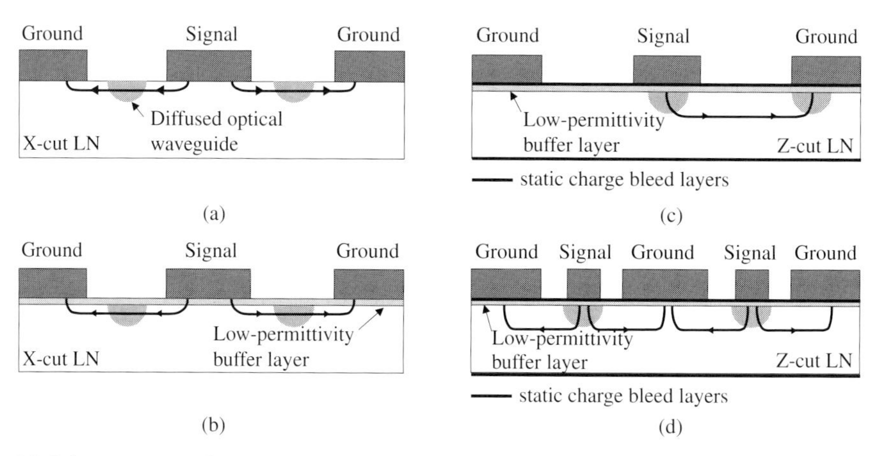

Figure 6.25 Modulator cross sections: (a) low-speed unbuffered X-cut LN modulator; (b) high-speed buffered X-cut modulator; (c) Z-cut buffered modulator with charge bleed layers; (d) dual drive Z-cut modulator. Adapted from [94], Fig. 3.

6.6.1 Lithium niobate modulators

Due to the easier technology and the lack of piezoelectrically induced static charge problems, X-cut modulators have been the initial choice for LN. Figure 6.25 shows a few solutions to the interaction region, implemented through coplanar RF distributed electrodes. Solution (a) is an unbuffered X-cut modulator which, due to large velocity mismatch between the RF and optical modes, has a better modulation bandwidth than the lumped modulator but cannot achieve 10 Gbps operation. Reduction of the velocity mismatch is possible through the use of a low-permittivity dielectric buffer (silicon oxide but also polymers have been exploited to this end) and of metal electrodes thickened through electroplating. Solution (c) is a Z-cut modulator, characterized by better efficiency, nonzero chirp, and the need to include a conductive buffer to handle the static charge accumulation problem. Finally, (d) is an example of dual-drive solution, in which each phase arm is driven by a different voltage level. The implementation shown is in Z-cut technology and is symmetrical; the dual-drive Z-cut configuration has a moderate margin in terms of modulation efficiency.

Example 6.2: Suppose (as a rule of thumb) that a single-drive Z-cut modulator (Fig. 6.25(c)) has a 30% advantage over an X-cut single-drive modulator of similar technology (Fig. 6.25(a)) in terms of on–off voltage, because of the superior overlap integral of the optical waveguide located below the signal conductor. What will be the advantage for a Z-cut dual-drive (Fig. 6.25(d))?

In the single-drive Z-cut we have $V_{\pi U} \neq V_{\pi L}$; denoting by $V_{\pi U}$ the voltage required for π phase shift in the waveguide located under the center (signal) conductor and by $V_{\pi L}$ the voltage required for π phase shift in the waveguide located under the ground plane, we have that the switching voltage for the Z-cut is

$$V_{\mathrm{OFF,Z}} = \frac{V_{\pi U} V_{\pi L}}{V_{\pi U} + V_{\pi L}} \approx 0.7 V_{\mathrm{OFF,X}}$$

where $V_{\mathrm{OFF,X}}$ is the switching voltage for the X-cut. We assume approximately that $V_{\pi U} < 2V_{\mathrm{OFF,X}} < V_{\pi L}$ (the modulating action is typically more effective for the central conductor) and that, in particular,

$$V_{\pi U} \approx 2\alpha V_{\mathrm{OFF,X}}, \quad V_{\pi L} \approx \frac{2V_{\mathrm{OFF,X}}}{\alpha},$$

with $\alpha < 1$. Substituting we have

$$V_{\mathrm{OFF,Z}} = \frac{\alpha}{1+\alpha^2} 2V_{\mathrm{OFF,X}} \approx 0.7 V_{\mathrm{OFF,X}} \rightarrow \alpha = 0.41.$$

In the dual-drive configuration, on the other hand, we have $V_{\pi U} = V_{\pi L} = 2\alpha V_{\mathrm{OFF,X}}$; thus the related on-off voltage V_{OFF}^{DD} will be

$$V_{\mathrm{OFF}}^{DD} = \alpha V_{\mathrm{OFF,X}} \approx 0.41 V_{\mathrm{OFF,X}}.$$

Therefore the Z-cut dual-drive has a 60% advantage over the single-drive X-cut configuration, or about half of the on–off voltage.

An example of realistic modulator cross section for a X-cut buffered configuration is shown in Fig. 6.26; the thick Au electrode technology can be based on electroplating through mask, or on electroplating followed by etching (see, e.g., [94], Section II C).

X-cut modulators can very well cover 10 Gbps applications and are able to reach the 40 Gbps goal with modulation voltages of the order of 5 V; Z-cut modulators have superior performances in terms of efficiency but may have inferior chirp; moreover, the static charge buildup issue requires additional complexities, such as the deposition of charge bleed layers. Both in X-cut and in Z-cut configurations, a number of solutions

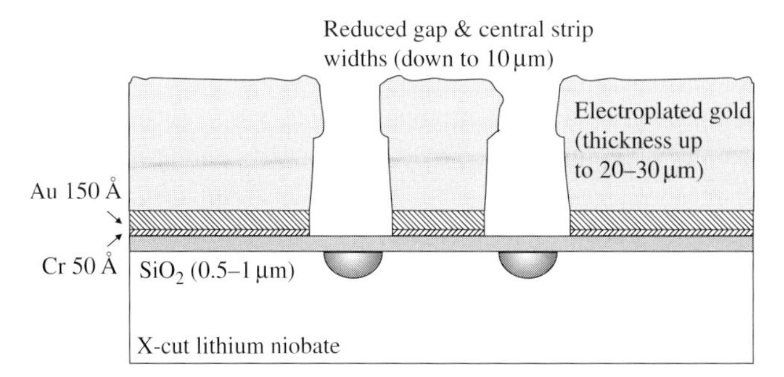

Figure 6.26 Typical cross section of a high-speed low-permittivity dielectric buffered X-cut modulator.

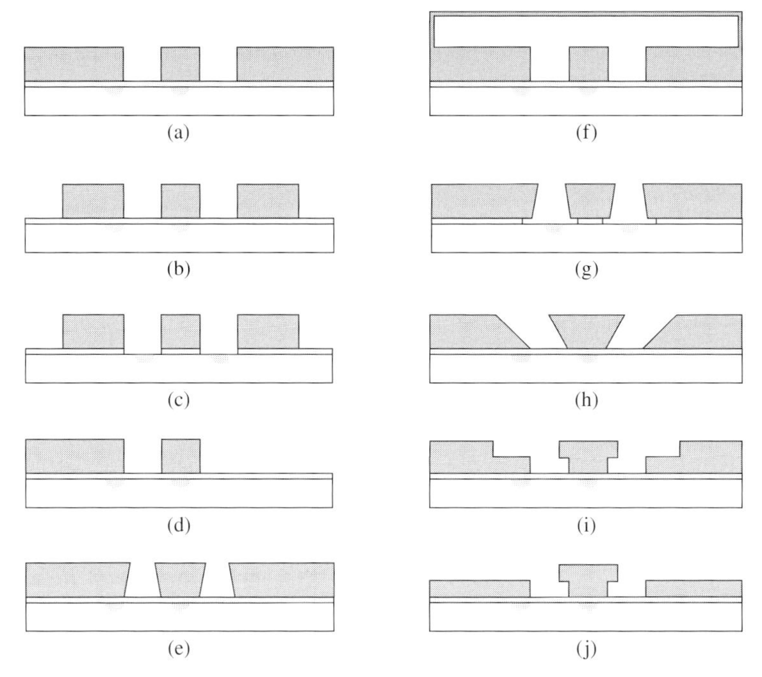

Figure 6.27 Some LN modulator structures: (a) coplanar waveguide, Z-cut, with dielectric buffer; (b) same, but with finite grounds; (c) same, but X-cut with dielectric etching; (d) asymmetric coplanar waveguide, Z-cut; (e) coplanar waveguide with slanted (overcut) electrodes; (f) coplanar with an upper shield; (g) with slanted electrodes and dielectric underetching; (h) with slanted electrodes (minimum loss structure); (i) and (j), same as (h), but with a two-metal layer design.

have been tried, as shown in Fig. 6.27, to reduce the modulation voltage and decrease the line effective index (at the same time increasing the line characteristic impedance).

The best LN modulator performances have probably been achieved through Z-cut modulators where the optical waveguides are located in ridge structures. An example of this design approach is shown in Fig. 6.28 [95]; the characteristic impedance is 50 Ω, the microwave effective index 2.2, and the length–bandwidth and length–on-off voltage products are $B_{3\mathrm{dB},el}L = 90\,\mathrm{GHz}$ cm and $V_\pi L = 10$ V cm, respectively. For 5 V operation a 2 cm modulator is needed whose electrical bandwidth exceeds 45 GHz.

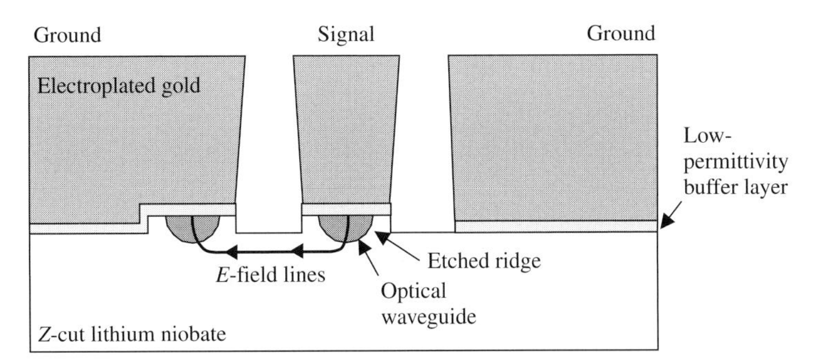

Figure 6.28 Example of Z-cut modulator structure with ridge electrodes. Adapted from [95], Fig. 1.

The RF attenuation presented (0.3 dB/cm at 1 GHz) is one of the lowest achieved in this kind of structure.

In conclusion, the LN modulator is well qualified as a high-speed device. On this mature technology are based many commercially available products that can meet the requirements of long-haul links up to 40 Gbps. However, further progress toward speeds in excess of 40 Gbps appears to be difficult without introducing entirely new elements into the design.

6.6.2 Compound semiconductor, polymer, and silicon modulators

Mach–Zehnder *semiconductor* (GaAs or InP-based) interferometric modulators have been studied extensively and have been developed to the market stage. Many semiconductor modulators exploit a TE optical mode with vertical RF applied field; the active region is a thin, undoped, bulk GaAs layer. Moreover, most high-speed devices are traveling-wave, with slow-wave electrodes aimed at improving the velocity matching to the optical signal [96]. An example is the Mach–Zehnder GaAs modulator developed by GEC-Marconi [97], see Fig. 6.29. The RF signal travels on two coplanar strips (one acting as the ground plane) loaded with T-shaped electrodes (T-rails). T-rails have the purpose of capacitively loading the transmission line, thus decreasing the phase velocity, through a set of lumped capacitors; moreover, T-rails apply the RF signal to the active region where the optical waveguide is located. The connection between the striplines and the center T-rails is made through a series of airbridges. In the active region, a floating internal backplane obtained by means of an n-doped layer allows the two ridge optical waveguides, located immediately below the T-rails (see Fig. 6.29(below)), to interact with two RF electric fields having opposite directions, thus creating the same push–pull effect as in a symmetric LN modulator. The splitter and combiner are realized through MMI (multimode interference) combiners and dividers, which are more compact and integrable than conventional adiabatic transitions. The modulator figures of merit are $B_{3\text{dB},el}L = 53\,\text{GHz cm}$ and $V_\pi L = 7.5\ \text{V cm}$, respectively, comparable to those of the best LN modulators; 40 Gbps operation is also achieved with driving voltages of about 5 V.

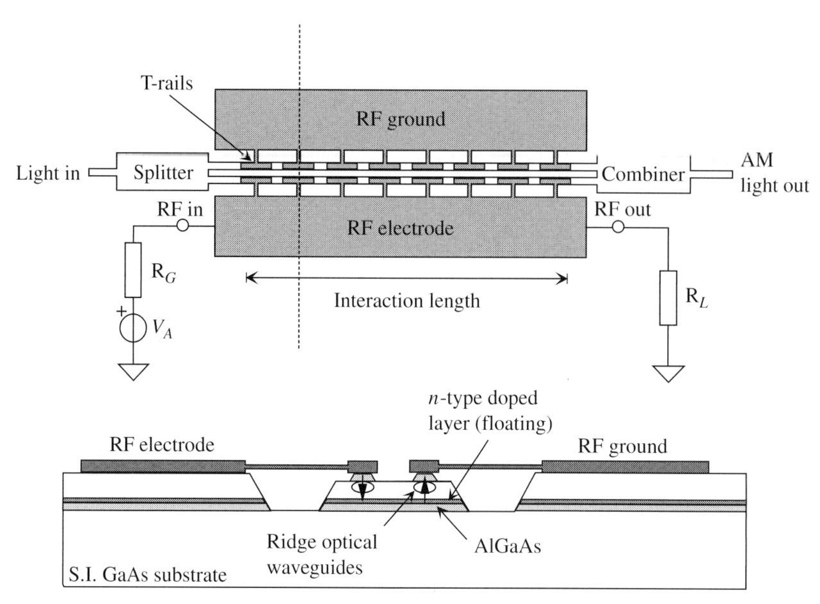

Figure 6.29 GEC-Marconi Mach–Zehnder GaAs-based intensity modulator. Adapted from [96], Fig. 2 and Fig. 3.

Integration with the laser source is a possible advantage of the semiconductor modulator when compared to the LN solution. A few examples of lasers with integrated modulators have been proposed so far; see, e.g., Fig. 6.30 for a lumped 10 Gbps Mach–Zehnder InP modulator integrated with a DFB laser [98]. The modulator waveguide is placed in the central part of a PpN structure with a MQW; the modulator exhibits negative chirp.

With respect to other integrated solutions, the Mach–Zehnder modulator certainly has some performance advantage (i.e., in terms of chirp) but also has the disadvantage of large size when compared to the source. On the other hand, integration allows the modulator efficiency to be improved through the use of MQW structures, which exploit exciton resonances and therefore have a narrow optical bandwidth, that must be tuned to a specific source.

Example 6.3: Evaluate the Henry parameter in the modulator shown in Fig. 6.30. The intensity has $\Delta P_{out} = 0.9 - 0.1 = 0.8$ in arbitrary units with an average power $P_{out} = 0.1$ and the increasing pulse has a risetime of $\Delta t \approx 500$ ps. The frequency deviation (maximum) is $\Delta f = -5$ GHz.

From the definition of α_H we have

$$\alpha_H = 2\frac{2\pi\Delta f}{\dfrac{1}{P_{out}}\dfrac{dp_{out}}{dt}} \approx 2\frac{2\pi\Delta f}{\dfrac{1}{P_{out}}\dfrac{\Delta P_{out}}{\Delta t}} = -2\frac{2\pi \cdot 5\times 10^9}{\dfrac{1}{0.5}\dfrac{0.8}{50\times 10^{-12}}} = -1.96.$$

The same parameter is somewhat smaller for the decreasing pulse wavefront. The modulator shown is actually dual-drive, with an extinction ratio of the order of $10\log_{10}(0.9/0.1) \approx 9.5$ dB.

Polymer-based modulators appear (from the standpoint of material properties) as a promising substitute to LN-based devices; performances are comparable to the best LN modulators (electrical bandwidth of 40 GHz with on–off voltage less than 4 V; optical insertion loss of 6 dB, zero chirp), with slightly inferior optical power-handling capabilities. As already mentioned, long-term stability and the need for hermetic packages is a major problem in polymer-based devices.

During the last few years, the prospect of monolithically integrating an optical transceiver (apart from the laser source) on a Si substrate has fostered research in the area of *Si-based modulators*. In Si, the linear electrooptic effect is zero because of crystal symmetries, and the only way to modulate the material refractive index is through the so-called plasma dispersion effect, related to the density of free carriers in the semiconductor. Such an effect, besides being rather weak (and therefore requiring long interaction lengths in a MZ configuration), can be implemented in a forward-bias *pn* junction, in a reverse-bias *pin* junction, or in a MOS junction exploiting an inversion channel [99]. The main problems with some Si-based configurations are the

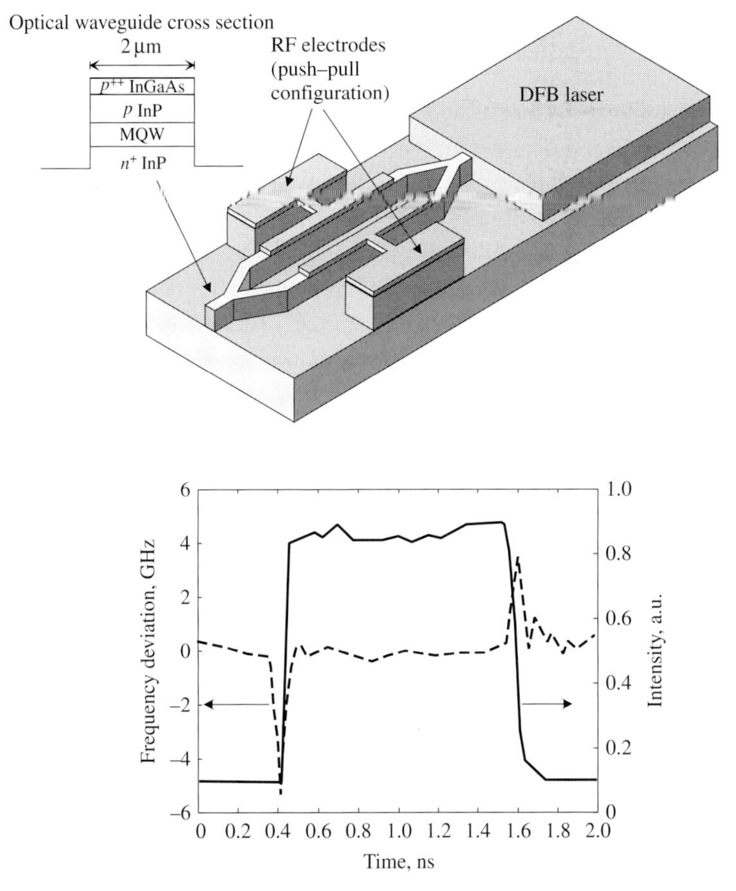

Figure 6.30 Integrated laser-MZ InP amplitude modulator: device structure (above), device chirp (below) in a 10 Gbps push–pull configuration. After [98], Fig. 1 and Fig. 2 (©1998 IEEE).

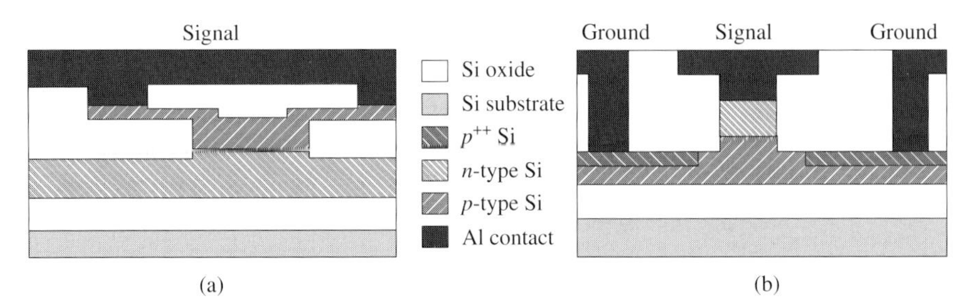

Figure 6.31 Simplified cross sections of the phase modulation section in Si-based MZ modulators: (a) 10 Gbps MOS modulator [100]; (b) 40 Gbps *pn* modulator [101].

switching times, often disappointing due to the limits posed by the minority carrier lifetime.

A 10 Gbps modulator was implemented by INTEL researchers in the MOS technology [100]; see Fig 6.31(a). The structure is made of n-type Si with an upper rib of p-type Si obtained through epitaxial lateral overgrowth; between the two Si regions there is a thin insulating layer. Positive bias applied to the p-type Si leads to charge accumulation in the interface, with a change of the refractive index. The structure operates as the phase shift block in a MZ modulator (note that no push–pull operation is strictly possible).

More recently, a 40 Gbps modulator with 30 GHz electrical bandwidth was obtained, again by INTEL researchers [101]. The phase modulator section makes use of a reverse-bias *pn* junction, see Fig 6.31(b); the reduced capacitance allows for larger speed. The interaction length is 1 mm, and the driving voltage exploited in the device measurement was 6 V peak-to-peak superimposed on a 3 V bias; the switching voltage was estimated as 4 V cm. To reduce the driving voltage without increasing the interaction length, alternative approaches, such as the use of photonic crystal waveguides, were also proposed, with a much shorter interaction length (80 μm) [102]. Other solutions for Si-based modulators exploited SiGe-Si QW and MQW structures (see [99] and references therein).

6.7 Electroabsorption modulator physics

Electroabsorption modulators (EAMs) exploit the variation in the semiconductor optical absorption caused by an applied electric field. EAMs can be based on two physical mechanisms: the *Franz–Keldysh effect*, FKE, or photon-assisted tunneling, typical of bulk semiconductors; or the *quantum confined Stark effect* (QCSE), occurring in quantum wells. The QCSE also implies the modulation of excitonic absorption. The FKE is weaker, but the resulting optical bandwidth is wider (though not as wide as in EOMs); on the other hand, the QCSE is stronger but optically narrowband. In practice, QCSE-based modulators are able to modulate a *single channel* and, therefore, more conveniently appear as devices integrated with the laser source.

6.7.1 The Franz–Keldysh effect (FKE)

Consider the band structure of a bulk, direct-bandgap semiconductor in the absence of an applied electric field. To promote a transition from the conduction to the valence band, a photon is needed with an energy larger than the gap E_g. An applied field, see Fig. 6.32, causes the bands to bend; now, a photon with $E_{ph} < E_g$ can *virtually* bring the electron very close to the conduction band edge; further motion in the real space, made possible by tunneling through a thin and possibly low potential barrier, allows the electron to be promoted to the conduction band.

From a macroscopic standpoint, the application of an external field allows the material to absorb photons having energy lower than the absorption edge $\approx E_g$. Below the absorption edge, an energy interval exists where $\alpha \approx 0$ without applied field, but $\alpha \gg 0$ in the presence of an applied field; in such a wavelength range, light switching can be carried out by applying to the absorption region of the device a voltage large enough to induce an electric field leading to a significant variation of the absorption.

The absorption profile resulting from the FKE in the presence of an applied field is shown qualitatively in Fig. 6.33. The ripples in the absorption profile are such as to have constant area below the α curve with respect to the zero-field case. The FKE is typically rather weak, and requires electric fields of the order of $\mathcal{E} > 10^5$ V/cm to obtain significant variations of α; it is optically rather wideband and polarization insensitive.

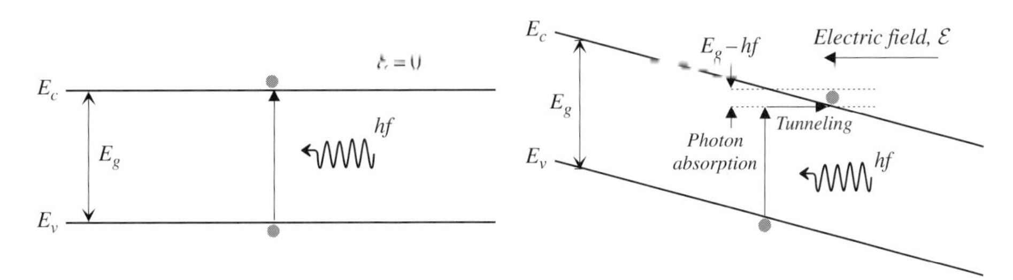

Figure 6.32 Franz–Keldysh effect: semiconductor bandstructure without (left) and with (right) applied electric field and tunneling-assisted photon absorption.

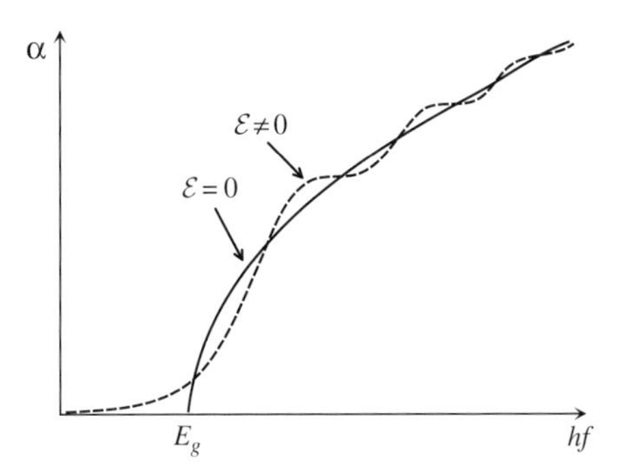

Figure 6.33 Absorption profile with and without applied field due to the FK effect.

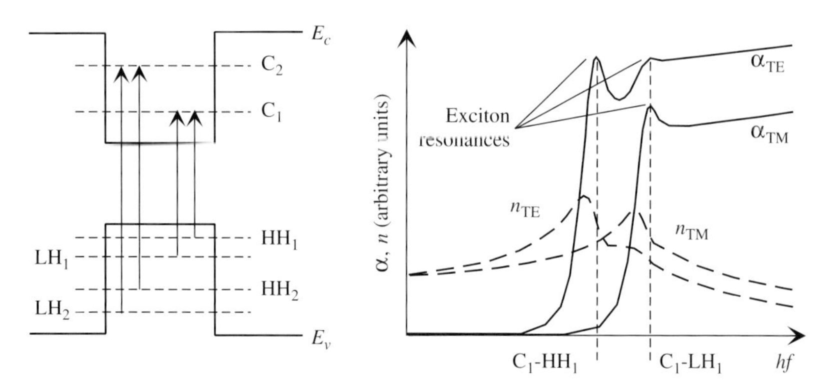

Figure 6.34 Bandstructure in a quantum well (left) and resulting absorption and refractive index profiles for TE and TM mode (right). Exciton resonances modify the absorption profile with peaks corresponding to JDOS steps. Note the QW dichroism (i.e., polarization sensitivity); the relative interaction strength for the TE mode is $(1/4, 3/4)$ for the LH and HH transitions, while for the TM mode this is $(1, 0)$ for the LH and HH transitions, respectively.

6.7.2 The quantum confined Stark effect (QCSE)

Quantum wells exhibit a band structure with several subbands in the conduction and valence bands. Let us denote the conduction band levels as C_1, $C_2 \ldots$ and the valence band levels (which split into heavy and light hole levels) as HH_1, $HH_2 \ldots$ and LH_1, $LH_2 \ldots$. Selection rules only allow some of the transitions between levels to occur, and, moreover, make the transition interaction strengths generally different for TE or TM polarized photons. The QW absorption profile is inherited from the staircase-shaped joint density of states (JDOS), see Section 2.4.3, although only the steps corresponding to allowed transitions are considered, see Fig. 6.34; the weight of each step will be different for different polarizations.

The effect of *excitons* (see Section 2.4.3), which can be detected in quantum wells at room temperature, play a significant role in the QW optical properties. Excitons are bound states involving one electron and one hole, and their binding energy is larger in a QW than in the bulk; they interact with photons, leading to absorption peaks immediately below each step of the JDOS; see Fig. 6.34.

Due to the combined effect of polarization-sensitive selection rules, QW show, even in the absence of an electric field, a marked dichroism, i.e., a different absorption and refractive index profile for each polarization; see Fig. 6.34(right) for the qualitative behavior. Since the interaction strength of the C_1-HH_1 transition is zero for TM polarization, the first step of the JDOS is missing from the absorption response to TM polarization, and so is the associated exciton peak.

Application of an external field causes the QW bands to bend, exactly as in the bulk case; band bending leads to three effects:

- The energy levels involved in the transitions become closer due to the shift in the QW potential profile; see Fig. 6.35 for an example referring to the C_1-H_1 transition (H_1 stands for HH_1 or LH_1 depending on the subband ordering, which may change

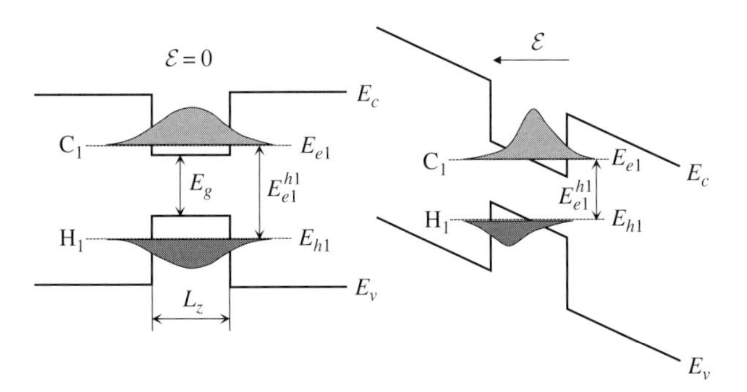

Figure 6.35 Quantum confined Stark effect in a quantum well: bandstructure without (left) and with (right) applied electric field. A qualitative picture of the envelope wavefunction is shown for the C_1 and H_1 states.

in strained QWs). As a consequence, the absorption edge is red-shifted toward lower energies.

- The exciton levels undergo a certain amount of energy shift.
- The envelope wavefunctions of the valence and conduction band states change, leading to a modulation of the transition interaction strength. For very large fields we expect the interaction strength to decrease, due to the decreasing overlap between the two wavefunctions.

We can express the photon energy as

$$E_{ph} = \hbar\omega = E_{h_1}^{e_1} - E_{ex},$$

where $E_{h_1}^{e_1}$ is the energy difference between the C_1 and H_1 levels, and E_{ex} is the exciton binding energy (of the order of 10 meV). The intersubband transition energy can be decomposed as

$$E_{h_1}^{e_1} = E_g + E_{e_1} - E_{h_1},$$

where E_{e_1} is the level of C_1 above the conduction band edge, E_{h_1} is the level of H_1 below the valence band edge E_v (E_{h_1} is therefore negative). It can be shown that the transition energy decreases with the applied field, according to a quadratic law:

$$\Delta E_{h_1}^{e_1} = E_{h_1}^{e_1}(F) - E_{h_1}^{e_1}(0) \propto -\frac{C\left(m_e^* + m_h^*\right) q^2 \mathcal{E}^2 L^4}{\hbar^2},$$

where $\mathcal{E}$ is the applied electric field, L is the (effective) QW thickness, and $C = 2.19 \times 10^{-3}$ is a dimensionless constant, see [11], Section 13.4.6. An example of the energy shift for a 9.5 nm GaAs/AlGaAs QW is shown in Fig. 6.36. For applied fields of the order of 100 kV/cm, corresponding to a potential difference $\Delta V = 100 \times 10^3 \cdot 10^2 \cdot 9.5 \times 10^{-9} \approx 100$ mV across the QW, the shift is of the order of 10 meV.

A second contribution to the absorption edge shift is caused by the change in the exciton binding energy; see Fig. 6.37 for an example referring to a GaAs QW [104]. For increasing applied field, the binding energy decreases slightly, i.e., excitons get closer to the state energy levels; this partly counteracts the shift of the energy levels,

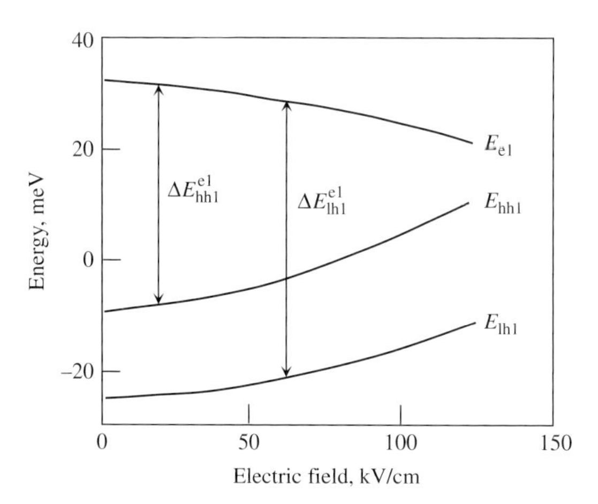

Figure 6.36 Subband transition energy level shift as a function of the applied electrical field. Adapted from [103], Fig. 1.

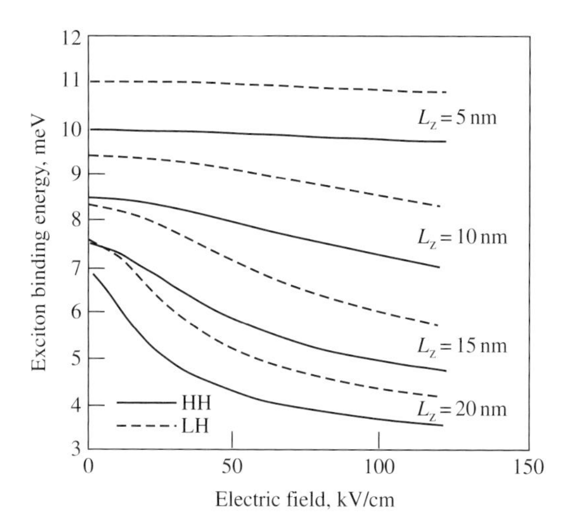

Figure 6.37 Shift in the heavy and light hole exciton binding energy as a function of the applied field in a GaAs/AlGaAs MQW for different well thicknesses. Adapted from [104], Fig. 1 (a) and (b).

but, for small QW thicknesses, the variation is about one order of magnitude less than for ΔE^{e1}_{h1}.

Let us consider now the total effect resulting from the application of an electric field. Since typical QW structures are *PiN* junctions, the applied voltage causes the junction to be in reverse bias (exactly as in a *pin* photodiode). The resulting almost uniform field in the intrinsic region decreases the intersubband transition energy and the magnitude of the exciton binding energy (due to decreased Coulomb interaction between electron and hole). As the overall effect, the absorption spectrum is red-shifted to longer wavelengths. With increasing applied field, we also observe a quenching of the peak absorption at the exciton resonance due to decreased electron–hole overlap. Variations in the absorption profile are finally influenced by the different interaction strength of

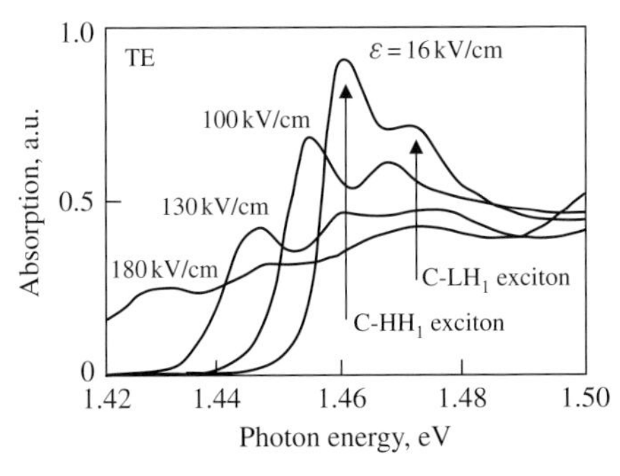

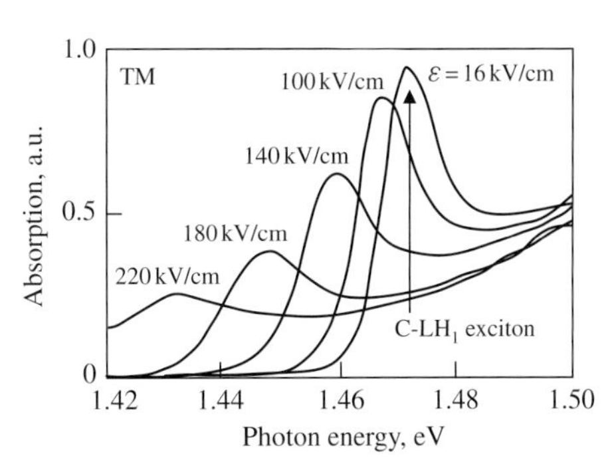

Figure 6.38 Absorption spectra (arbitrary units) for GaAs/$Al_{0.3}Ga_{0.7}As$ QW for TE (left) and TM (right) polarization. Adapted with permission from [105], Fig. 2, ((©1985 American Institute of Physics).

the transitions between the heavy and light hole valence band levels and the electron conduction band levels. We recall that the predicted relative oscillator strengths are

- for TE polarization, 3/4 (C_1-HH_1), 1/4 (C_1-LH_1),
- for TM polarization, 0 (C_1-HH_1), 1 (C_1-LH_1).

An example of normalized absorption spectra (derived as the material transmission in log scale) is shown in Fig. 6.38 [105]; the dichroism also exists in the presence of an applied field, and the effect of exciton absorption resonant peaks and their quenching with increasing field is clearly visible. Due to the exciton peaks, large variations of the absorption can be obtained at comparatively low fields, but only for a very narrow range of optical wavelengths.

Dichroism may be a limitation in EAMs, at least in discrete (nonintegrated) devices, since it requires the incoming light to be polarized in a specific TE or TM state. Polarization-insensitive devices can be developed by removing dichroism through the use of stressed QWs.

In the presence of applied stress, the QW bandstructure (as the bulk bandstructure) modifies in two ways: compressive (tensile) stress leads to a decrease (increase) of the lattice constant, and therefore to an increase (decrease) of E_g; in the bulk, stress causes heavy and light holes to lose their degeneracy, while in a quantum well it changes the relative alignments of the heavy and light hole levels. As a result, we have the situation shown in Fig. 6.39; tensile strain can lead to an alignment of the LH_1 and HH_1 energies, by which the polarization anisotropy is suppressed or, at least, reduced. Introduction of tensile strain in the wells limits the total thickness of the MQW (to avoid misfit dislocations); however, this effect can be compensated by using a compressively-strained barrier.

A final effect concerns the dependence of the absorption profile on the carrier density. Ideally, EAMs operate with almost zero carrier density: the conduction band is supposed to be empty, and the electric field is uniform and unperturbed by space-charge effects. However, photogenerated carriers recombine or are swept out of the undoped (absorption) region with an average lifetime τ_t, which can be considered as a transit

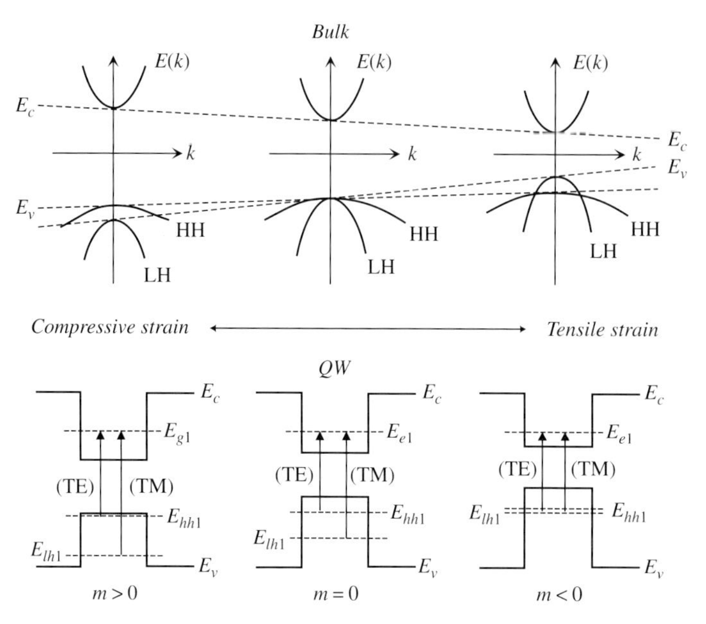

Figure 6.39 Effect of strain on the QW bandstructure: $m > 0$ corresponds to compressive strain, $m = 0$ to no strain, $m < 0$ to tensile strain. The bulk dispersion relation is shown above, the QW band alignment below.

or disappearance time. Taking into account that photocarriers are generated by photon absorption, and have lifetime τ_t, we obtain from the photon flux continuity equation in DC steady state:

$$\frac{n}{\tau_t} = -\frac{\mathrm{d}}{\mathrm{d}x}\left[\frac{P_{op}(0)}{A\hbar\omega}\exp(-\alpha x)\right] = \alpha\frac{P_{op}}{A\hbar\omega},$$

where P_{op} is the incident optical power and A is the device cross section. The resulting carrier density is therefore

$$n = \alpha\frac{P_{op}\tau_t}{A\hbar\omega}.$$

With increase in the optical power, the carrier density in the absorption region increases. Beyond a certain level, optical power saturation may occur due to two effects: first, generated photocarriers screen the applied external RF field, thus leading to a decrease of absorption; second, emission begins to play an important role due to band filling, thus counteracting the effect of absorption and reducing the effective absorption coefficient. As a result of both effects, absorption will decrease with increasing optical power.

The mechanisms that limit the disappearance time τ_t are, in bulk modulators, charge trapping at heterojunctions (e.g., InGaAsP/InP), and in QWs, the accumulation of holes in the well (due to their higher mass). Bandgap engineering is needed to lower the relevant band offset and reduce τ_t. Note that the same problem occurs in photodiodes (see Section 4.9.2), where, however, the input optical power is low enough to make power saturation effects often less significant.[14]

6.8 Electroabsorption modulator structures and parameters

Early electroabsorption modulators exploited *pin* vertical structures similar to vertical photodiodes, with short active (absorption) MQW region (1 μm), and large diameter (50–100 μm) [106]; an example of the structure cross section is shown in Fig. 6.40(a). In this vertical design, due to the small thickness of the absorbing stack (similar to the absorption region of a vertical photodetector), the FKE is too weak to cause significant contrast ratio, and the exploited physical mechanism was the QCSE.

To overcome these limitations, practical EAM structures are based on waveguide approaches (similar to those exploited in waveguide photodetectors, see Section 4.10.1); that is, the active region is an optical waveguide with applied electric field orthogonal to the optical mode propagation. Such structures allow the absorption region to be long enough in the OFF state, without compromising the driving voltage. An example is shown in Fig. 6.40(b); the *pin* structure is designed to form a slab waveguide, with typical thickness of the intrinsic absorption region $d = 0.1-0.5\,\mu\text{m}$. The total length L of the absorption region can be comparatively large ($L = 50-300\,\mu\text{m}$). Due to the small value of d, the applied field can be large even with a low reverse bias; typical switching voltages can be as low as $1.5-2V$. Both the FKE and the QCSE can be employed as the absorption modulation mechanism.

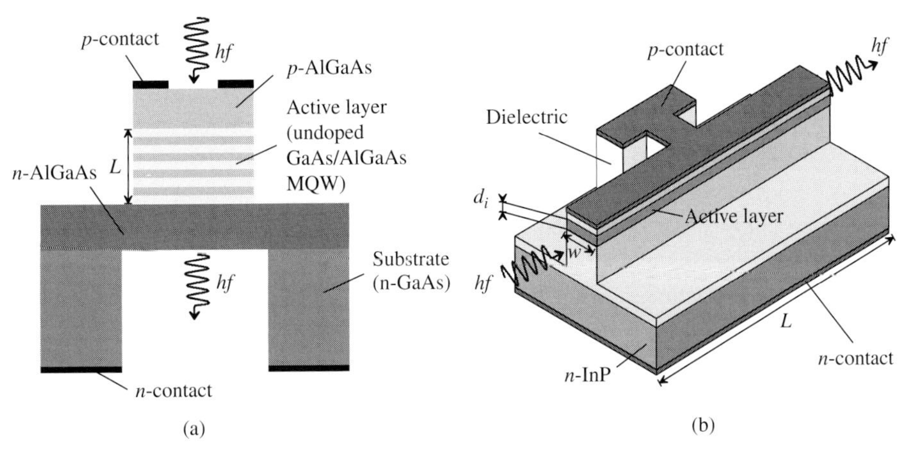

Figure 6.40 (a) Vertical QW electroabsorption modulator (adapted from [106], Fig. 1); (b) waveguide concentrated EAM.

[14] Photodiodes operate in the receiver stage, where the optical power is low, whereas EAMs are directly connected to the optical source. Photodiode saturation is a major concern, however, in applications such as the optical generation of microwaves or millimeter-waves, where the optical input power should be maximized to increase the output RF power.

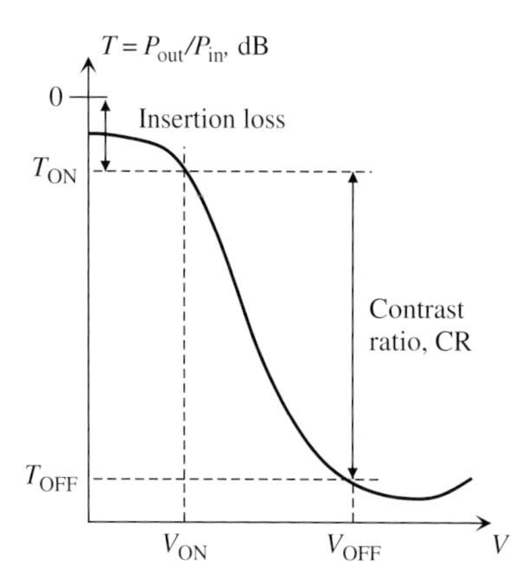

Figure 6.41 Transmission curve for an EAM.

6.8.1 EAM static response

The transmission characteristics of an EAM can be derived from the optical mode absorption $\alpha_o = \Gamma_{ov}\alpha$, where α is the material absorption in the active region and Γ_{ov} is the *overlap integral*, see Section. 5.5.1 and (5.18).[15] The RF field is almost uniform in the active region, inducing a uniform variation of α and negligible variation in the cladding. Assuming an input and output insertion loss $L_{op} < 1$, the transmission characteristics over a uniform device of length L and input voltage V_{in} can be expressed as[16]

$$T(V_{in}) = \frac{P_{out}}{P_{in}} = L_{op} \exp\left(-\int_0^L \Gamma_{ov}\alpha(V_{in})\,\mathrm{d}x\right) = L_{op} \exp\left[-\Gamma_{ov}\alpha(V_{in})L\right].$$

The resulting transmission curve is shown in Fig. 6.41; the ON and OFF state losses are $T_{\mathrm{ON,OFF}} = T(V_{\mathrm{ON,OFF}})$ and the extinction or contrast ratio is ER$= T_{\mathrm{ON}}/T_{\mathrm{OFF}}$ or $\mathrm{ER}|_{\mathrm{dB}} = T_{\mathrm{ON}}|_{\mathrm{dB}} - T_{\mathrm{OFF}}|_{\mathrm{dB}}$. Note that in QCSE-based devices the response typically increases again at large applied voltages, due to the weaker overlap between the initial and final states involved in the absorption process.

From the standpoint of material and structure optimization, a trade-off exists between the extinction ratio and the ON state residual transmission loss. In fact, we have

$$\begin{aligned}\mathrm{ER}|_{\mathrm{dB}} &= 10\log_{10}\left[\frac{T(V_{\mathrm{ON}})}{T(V_{\mathrm{OFF}})}\right] = 10\log_{10}\left[\frac{L_{op}\exp\left(-\Gamma_{ov}\alpha(V_{\mathrm{ON}})L\right)}{L_{op}\exp\left(-\Gamma_{ov}\alpha(V_{\mathrm{OFF}})L\right)}\right] \\ &= 4.343\left[\alpha(V_{\mathrm{OFF}}) - \alpha(V_{\mathrm{ON}})\right]\Gamma_{ov}L.\end{aligned}$$

[15] The overlap integral takes into account here the fact that the optical field also extends outside of the active region, where the absorption variation occurs; it is therefore similar in all respects to the parameter exploited in laser analysis. We assume here that the electric field is uniform in the active region, contrary to what happens in LN-based EOMs, where the overlap integral takes into account the nonuniform RF field distribution.

[16] Note that L_{op} in log units is a positive number, i.e., $L_{op}|_{\mathrm{dB}} = -10\log_{10} L_{op}$.

Defining

$$\Delta\alpha = \alpha(V_{\mathrm{OFF}}) - \alpha(V_{\mathrm{ON}}), \quad \alpha_0 = \alpha(V_{\mathrm{ON}}), \quad K_{pl} = T(V_{\mathrm{ON}})/L_{op} < 1,$$

where K_{pl} ($K_{pl}|_{\mathrm{dB}} = -10\log_{10} K_{pl}$) is the transmission loss in the ON state, neglecting the insertion loss, we have

$$\begin{aligned}
\mathrm{ER}|_{\mathrm{dB}} &= 4.343\,[\alpha(V_{\mathrm{OFF}}) - \alpha(V_{\mathrm{ON}})]\,\Gamma_{ov}L = \frac{\Delta\alpha}{\alpha_0} \times 4.343\alpha_0\Gamma_{ov}L \\
&= -\frac{\Delta\alpha}{\alpha_0} \times 10\log_{10}\{\exp[-\Gamma_{ov}\alpha(V_{\mathrm{ON}})L]\} \\
&= -\frac{\Delta\alpha}{\alpha_0} \times 10\log_{10}\left[\frac{T(V_{\mathrm{ON}})}{L_{op}}\right] = \frac{\Delta\alpha}{\alpha_0} K_{pl}\big|_{\mathrm{dB}}.
\end{aligned}$$

Apparently, large contrast or extinction ratios can be achieved by increasing the length L of the EAM; however, this also increases losses in the ON state. To increase the ER without compromising the ON state losses, the parameter to be optimized is $\Delta\alpha/\alpha_0$, i.e., the relative variation of the absorption from the ON to the OFF state. Typical ON transmission losses are in the range 10–20 dB/mm, while in practical cases $\Delta\alpha/\alpha_0 \approx$3–10.

Example 6.4: Suppose the ON loss is 10 dB/mm and $\Delta\alpha/\alpha_0 = 5$, with an additional insertion loss L_{op} of 1.5 dB. What extinction ratio can we achieve, and what is the device length, assuming we accept an overall ON insertion loss of 3 dB?

We have $K_{pl}|_{\mathrm{dB}} = 3 - 1.5 = 1.5$ dB, i.e.,

$$\mathrm{ER}|_{\mathrm{dB}} = \frac{\Delta\alpha}{\alpha_0} K_{pl}\big|_{\mathrm{dB}} = 5 \cdot 1.5 = 7.5\ \mathrm{dB}.$$

Moreover, the ON loss excluding L_{op} is 1.5 dB, which is achieved over a length $L =$ 150 μm.

The ON state transmission loss K_{pl} mainly consists of absorption in the waveguide in the ON state (equal to $\Gamma_{ov}\alpha_o$, where α_o is the active region absorption) but also of free-carrier absorption in the cladding layers (α_{fc}):

$$\alpha_0 \approx \Gamma_{ov}\alpha_o(V_{\mathrm{ON}}) + (1 - \Gamma_{ov})\,\alpha_{fc}.$$

Absorption in the widegap cladding layers is due to the free carrier effect, and increases with increasing doping (a particularly critical effect in p-doped layers). The residual absorption in the active region, on the other hand, can be reduced by increasing the difference between the operating wavelength and the absorption peak (or absorption edge) of the material, i.e., by *detuning* the operating wavelength with respect to the emission wavelength of the material; see Section 6.8.2. Large detuning implies low loss in the ON state, but also poor extinction ratio.

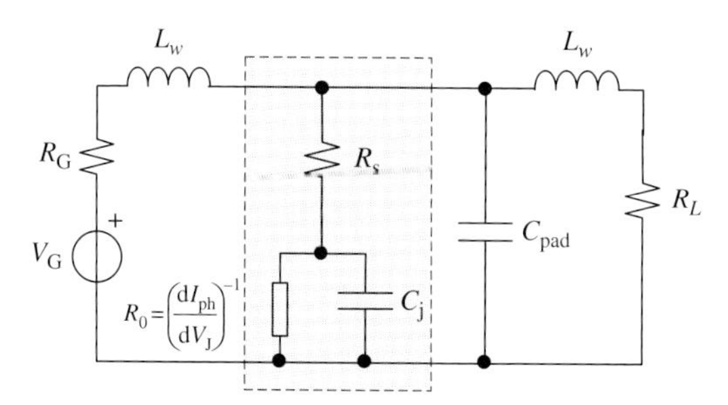

Figure 6.42 Equivalent circuit of a lumped EAM connected to a generator and a load resistance. The box denotes the intrinsic device.

6.8.2 Lumped EAM dynamic response

The dynamic response of lumped EAMs can be evaluated by considering the electric equivalent circuit, shown in Fig. 6.42. From the electrical standpoint, the EAM is a reverse-bias *pin* junction or heterojunction. The intrinsic equivalent circuit of the EAM includes the series parasitic resistance R_s and the junction capacitance C_j:

$$C_j = \frac{\epsilon_r \epsilon_0 A}{d} = \frac{\epsilon_r \epsilon_0 w L}{d},$$

where A is the junction area, w is the junction width, L is the active region length, d is the active layer thickness, and ϵ_r is the permittivity of the narrowgap intrinsic active region. The *photoresistance* R_0 originates from the photocurrent I_{ph} generated by the EAM.[17] In general, the absorption depends on the voltage applied to the junction; we therefore have

$$I_{ph} = I_{ph}(V_j),$$

where V_j is the voltage applied to C_j (across the active region), i.e., the driving voltage of the average field that modulates the absorption. The presence of the photocurrent can lead to fairly complex effects, since a large photocurrent is able to change the device bias point by counteracting the effect of the external generator; this is indeed a further cause of optical power saturation. Assume, however, that the bias effect of the photocurrent is negligible; we have from (4.29),

$$I_{ph} = -\frac{\eta_Q q}{\hbar\omega} P_{in} \Gamma_{ov} \alpha(V_j) \int_0^L \mathrm{e}^{-\Gamma_{ov}\alpha(V_j)x}\, \mathrm{d}x = \eta_Q \frac{q P_{in}}{\hbar\omega}\left[1 - \mathrm{e}^{-\Gamma_{ov}\alpha(V_j)L}\right].$$

In small-signal operation, $v_j(t) = V_j + \widehat{v}_j(t)$ and

$$I_{ph} = I_{ph}(V_j + \widehat{v}_j(t)) \approx I_{ph}(V_j) + \left.\frac{\mathrm{d}I_{ph}}{\mathrm{d}v_j}\right|_{V_j} \widehat{v}_j(t) = I_{ph}(V_j) + \frac{1}{R_0}\widehat{v}_j(t),$$

i.e., the small-signal photocurrent can be interpreted as the current in a conductance $1/R_0$ driven by the small-signal voltage $\widehat{v}_j(t)$. R_0 is large (and therefore negligible) in

[17] The EAM photocurrent is large in the OFF state, when the output power is low, and low in the ON state; it can be exploited, when characterizing the device, as a complementary monitor of the output power.

the OFF state, small in the ON state. The equivalent circuit in Fig. 6.42 finally includes some extrinsic, parasitic elements (L_w, the parasitic connector inductance, and C_{pad}, the external parasitic capacitance), the generator and the load resistance.

The evaluation of the dynamic EAM response cannot generally be performed in closed form, since the equivalent circuit includes, in principle, nonlinear elements such as the photocurrent conductance (and also, to a smaller extent, the junction capacitance). We therefore confine the analysis to the small-signal case, and assume that the generator voltage has a DC and signal component: $v_G(t) = V_G + \widehat{v}_G(t)$; as a consequence, the junction voltage can be decomposed as $v_j(t) = V_j + \widehat{v}_j(t)$, where

$$V_j = V_G \frac{R_L}{R_G + R_L}.$$

We associate the phasors $V_G(\omega)$ and $V_j(\omega)$, respectively, with $\widehat{v}_G(t)$ and $\widehat{v}_j(t)$; neglecting the parasitic resistances, the photoresistance and the parasitic inductances, and assuming $C = C_j + C_{\text{pad}}$, we obtain

$$V_j(\omega) = \frac{R_{eq}}{R_G} \frac{1}{1 + \mathrm{j}\omega C R_{eq}} V_G(\omega) = H(\omega) V_G(\omega), \tag{6.35}$$

where

$$R_{eq} = \frac{R_L R_G}{R_G + R_L}.$$

By linearizing the transfer curve around the DC bias point V_j we obtain

$$T(v_{in}(t)) = \frac{P_{out} + \widehat{p}_{out}(t)}{P_{in}} = T(V_j + \widehat{v}_j(t)) \approx \underbrace{T(V_j)}_{P_{out}/P_{in}} + \underbrace{\left.\frac{\mathrm{d}T}{\mathrm{d}v_j}\right|_{V_j} \widehat{v}_j(t)}_{\widehat{p}_{out}(t)/P_{in}}.$$

Assuming a sinusoidal input voltage, i.e., $\widehat{p}_{out}(t) = \mathrm{Re}\left[\widehat{P}_{out} \exp(\mathrm{j}\omega t)\right]$, we therefore have

$$\frac{\widehat{p}_{out}(t)}{P_{in}} = \mathrm{Re}\left[\frac{\widehat{P}_{out}(\omega)}{P_{in}} \exp(\mathrm{j}\omega t)\right] = \left.\frac{\mathrm{d}T}{\mathrm{d}v_j}\right|_{V_j} \widehat{v}_j(t) = \mathrm{Re}\left[H(\omega) V_G(\omega) \exp(\mathrm{j}\omega t)\right],$$

where $H(\omega)$ is defined in (6.35), i.e.,

$$\frac{\widehat{P}_{out}(\omega)}{P_{in}} = H(\omega) V_G(\omega). \tag{6.36}$$

From (6.36) the modulator frequency response results as

$$M(\omega) = \frac{\widehat{P}_{out}(\omega)}{P_{in}} = \frac{R_{eq}}{R_G} \frac{1}{1 + \mathrm{j}\omega C R_{eq}} \widehat{V}_G,$$

with normalized response

$$m(\omega) = \left|\frac{M(\omega)}{M(0)}\right| = \frac{1}{\sqrt{1 + \omega^2 C^2 R_{eq}^2}}.$$

Finally, from the definition of the electrical and optical modulator bandwidth we immediately have

$$f_{3\text{dB,el}} = \frac{1}{2\pi R_{eq} C}, \qquad f_{3\text{dB,op}} = \frac{\sqrt{3}}{2\pi R_{eq} C}.$$

With decreasing R_L, R_{eq} decreases, thus leading to a wider bandwidth. At the same time, however, the DC efficiency decreases (it is 50% for a matched resistor, 100% for an open load); the same effect occurs in LN modulators if we connect them to a matched load. Due to the rather small areas and low capacitance, lumped EAMs can readily cover 10 Gbps applications; for higher speed, distributed, traveling-wave modulators can be again introduced, see Section 6.9.

The design of EAMs has to face similar trade-offs as the EOM design. From the absorbing material viewpoint, we should maximize $\Delta\alpha/\alpha_0$ and increase the sensitivity of the absorption to the electric field, and therefore to the applied voltage, $\Delta\alpha/\Delta V$. The parameter $\Delta\alpha/\alpha_0$ depends on the difference between the EAM operating wavelength and the material absorption edge. Such a difference is often expressed through the *detuning energy*, i.e., the difference between the energy of the photons emitted by the source E_{ph} and the absorption edge energy of the EAM material, E_{EAM}. For $E_{ph} > E_{\text{EAM}}$, the zero-field absorption α_0 is large and therefore $\Delta\alpha/\alpha_0$ is small; the detuning $\Delta E = E_{\text{EAM}} - E_{ph}$ should therefore be positive. For very large detuning, however, $\Delta\alpha$ vanishes and therefore $\Delta\alpha/\alpha_0$ is also small; $\Delta\alpha/\alpha_0$ typically has a maximum for intermediate values of detuning.

Optimization of the optical and RF structures required, on the other hand, several trade-offs. Reducing the intrinsic layer thickness d_i increases the electric field with the same driving voltage, thus improving the switching voltage, but at the same time increases the EAM capacitance (thus reducing the bandwidth) and leads to poorer coupling with the optical fiber. Increasing the length L again improves the switching voltage and the extinction ratio, at the expense of the ON state loss and of the bandwidth (due to the increased capacitance).

6.8.3 EAM chirp

While in interferometric electrooptic modulators chirp is dominated by the geometry, the EAM chirp depends on material properties. Due to the Kramers–Kronig relations, varying the imaginary part of the refractive index (i.e., losses), leads to a variation of the real part, thus introducing spurious phase modulation. EAM chirp is typically lower than for directly modulated lasers, and can in principle be reduced and properly tailored (e.g., made negative) by selecting the bias point. Assuming a complex refractive index $n = n_r - \mathrm{j}n_i$ and taking into account that the complex propagation constant of the optical wave is

$$\gamma_o = \bar{\alpha}_o + \mathrm{j}\beta_o = \frac{2\pi}{\lambda_0}n_i + \mathrm{j}\frac{2\pi}{\lambda_0}n_r \rightarrow n_i = \bar{\alpha}_o \left(\frac{2\pi}{\lambda_0}\right)^{-1} = \frac{\alpha}{2}\left(\frac{2\pi}{\lambda_0}\right)^{-1},$$

where $\bar{\alpha}_o$ is the optical field *attenuation* and α is the *absorption*, the chirp parameter α_H can be expressed from (5.40) in the form

$$\alpha_H = \frac{\Delta n_r}{\Delta n_i} = \frac{4\pi}{\lambda_0}\frac{\Delta n_r}{\Delta \alpha}. \tag{6.37}$$

The output optical field and power can be now written as

$$E_{out} = E_0 \mathrm{e}^{-\mathrm{j}\frac{2\pi}{\lambda_0}\Delta n_r(t)L} \mathrm{e}^{-\Delta\bar{\alpha}_o(t)L}, \quad P_{out} = P_0 \mathrm{e}^{-\Delta\alpha(t)L},$$

where E_0 is the field in the absence of refractive index variation and P_0 the related power. Differentiating the field phase $\phi = -2\pi\Delta n_r(t)L/\lambda_0$ and the output power vs. time we obtain

$$\Delta f = \frac{1}{2\pi}\frac{\Delta\phi}{\Delta t} = -\frac{\Delta n_r(t)}{\Delta t}\frac{L}{\lambda_0} \rightarrow \frac{\Delta n_r}{\Delta t} = -\frac{\lambda_0 \Delta f}{L},$$

$$\frac{\Delta P_{out}}{\Delta t} = -\frac{\Delta\alpha}{\Delta t} L P_{out} \rightarrow \frac{\Delta\alpha}{\Delta t} = -\frac{\Delta P_{out}}{\Delta t}\frac{1}{L P_{out}},$$

i.e., substituting into (6.37),

$$\alpha_H = \frac{4\pi}{\lambda_0}\frac{\Delta n_r}{\Delta\alpha} = \frac{4\pi}{\lambda_0}\frac{\Delta n_r/\Delta t}{\Delta\alpha/\Delta t} = \frac{4\pi}{\lambda_0}\frac{\lambda_0\Delta f}{L} \times \frac{1}{\dfrac{1}{L P_{out}}\dfrac{\Delta P_{out}}{\Delta t}} \approx \frac{4\pi\Delta f}{\dfrac{1}{P_{out}}\dfrac{\mathrm{d}p_{out}}{\mathrm{d}t}},$$

which corresponds to the system-level definition of α_H already introduced in the discussion of laser chirp, see (5.78). The Henry parameter α_H is therefore related to the material characteristics and depends on the applied voltage (in small-signal conditions, on the DC bias) and on the detuning energy. An example of chirp behavior in an InGaAsP/InGaAsP QW is shown in Fig. 6.43 [107].

The parametric plot of the variation in the real and imaginary parts of the complex refractive index (the parameter is the applied bias) yields a curve whose slope is the chirp parameter α_H. Before the maximum the chirp parameter is positive, it is zero in the maximum, and then becomes negative. On the lower horizontal axis the absolute transmission over $L = 100\,\mu\text{m}$ is also shown. On the right, the chirp parameter is plotted as a function of the reverse applied voltage. For low bias voltage the chirp is large and positive, then it becomes zero, and finally negative; negative chirp is a desirable condition for transmission over a dispersive fiber, since it allows for compensating dispersion, see p. 320. The input bias voltage can be adjusted so as to make chirp small or negative; however, this corresponds to rather large ON-state attenuations. On the whole, chirp in EAMs can be substantially better than in directly modulated lasers, but LN modulators are certainly superior from this standpoint.

6.9 The distributed electroabsorption modulator

To estimate the bandwidth expected from a lumped parameter EAM, let us consider a structure with typical dimensions (e.g., $w \approx 2\,\mu\text{m}$, $L = 200\,\mu\text{m}$, $d = 0.3\,\mu\text{m}$).

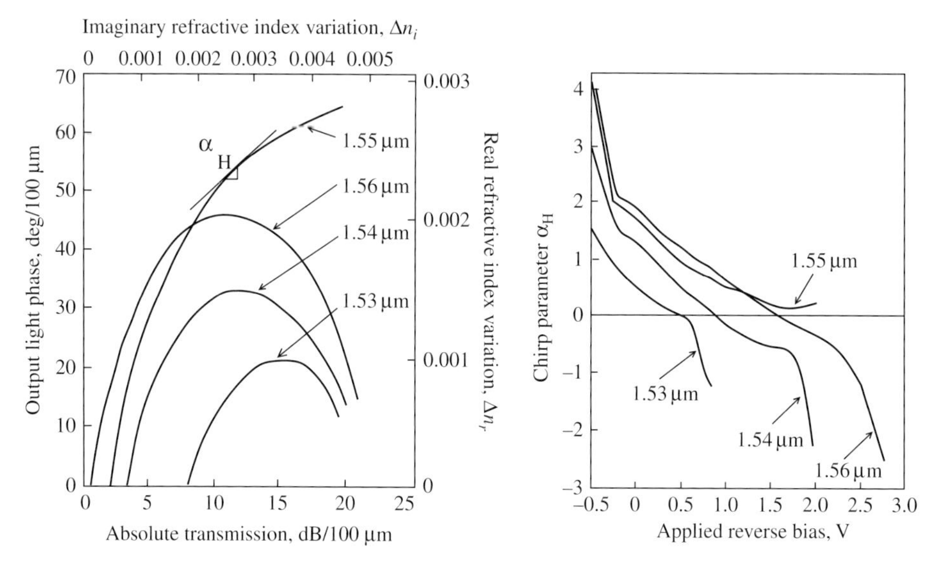

Figure 6.43 Chirp characteristics of an InGaAsP QW. Variation of the real and imaginary parts of the refractive index at different wavelengths (left). Chirp parameter as a function of the applied reverse bias (right). After [107], Fig. 1 and Fig. 2 (©1994 IEEE).

Assuming $\epsilon_r \approx 13$, the junction capacitance is

$$C_j = \epsilon_r \epsilon_0 \frac{wL}{d} = 13 \cdot 8.86 \times 10^{-12} \cdot \frac{2 \times 10^{-6} \cdot 200 \times 10^{-6}}{0.3 \times 10^{-6}} = 0.15 \text{ pF}$$

which leads, for a 50 Ω load and generator, to a 3 dB electrical bandwidth of

$$f_{3\text{dB,el}} = \frac{1}{2\pi R_{eq} C} = \frac{1}{2\pi \cdot 25 \cdot 0.15 \times 10^{-12}} = 42.5\,\text{GHz},$$

more than one order of magnitude larger than the bandwidth of lumped LN EOMs. Lumped EAMs are thus able to cover applications at least up to 10 Gbps. However, increasing the speed up to 40 Gbps and beyond may make the design critical, thus suggesting in this case also the use of *traveling-wave*, *distributed* structures, in which the RF electrodes are designed as quasi-TEM transmission lines, running parallel to the optical waveguide and supporting a modulating microwave signal which co-propagates with the optical signal. Also in this case, as in traveling-wave electrooptic modulators, no RC limitation to the bandwidth exists, and the device length can be increased without sacrificing speed. An example of distributed EAM structure is shown in Fig. 6.44. The structure of the active region is similar to that in the waveguide modulator, see Fig. 6.40(b), only the RF structure acts as a transmission line.

The modulation response of distributed EAMs can be derived, in small-signal conditions, following the same approach as for the distributed electrooptic modulator analysis; see Section 6.5. Since the RF transmission line is nonlinear (due to the voltage dependence of the junction capacitance and the photoresistance), we cannot readily separate the system into a linear part with memory and a memoriless nonlinear part, as

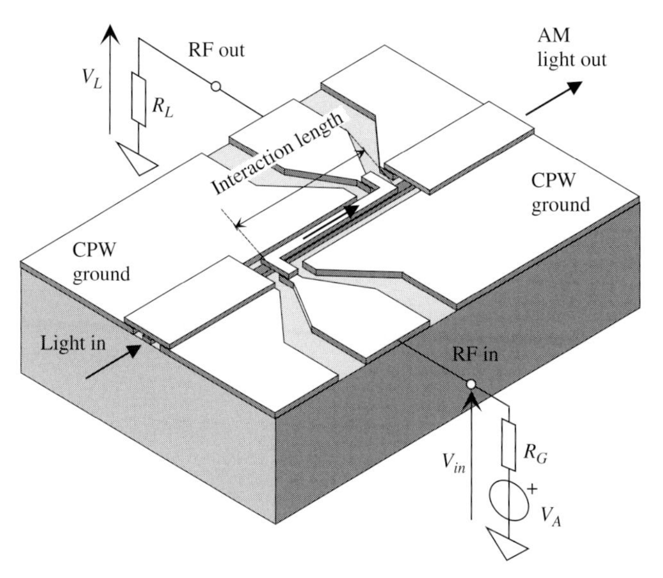

Figure 6.44 Coplanar waveguide (CPW) distributed EAM.

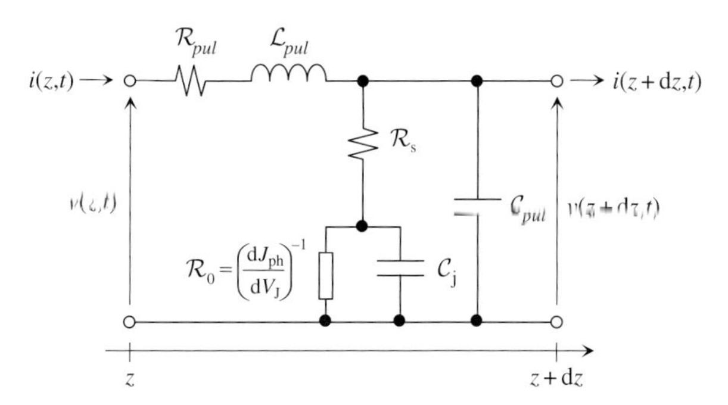

Figure 6.45 Equivalent circuit of an EAM transmission line section of infinitesimal length dz. The line is loaded with the EAM *pin* active cell.

in LN modulators, and a closed-form dynamic analysis can be only carried out in the small-signal case.

The RF voltage on the transmission line is expressed as

$$v_m(z, t) = V^+ e^{j\omega(t - z/v_m) - \alpha_m z} + V^- e^{j\omega(t + z/v_m) + \alpha_m z},$$

where the RF or microwave phase velocity v_m can be derived from the pseudo-lumped equivalent circuit of an infinitely short transmission line cell, shown in Fig. 6.45.

The per-unit-length line parameters can be associated with the following physical mechanisms:

- $\mathcal{R}_{pul}$ is the skin-effect series resistance of the metal layers (Ω/m).
- $\mathcal{L}_{pul}$ is the conductor per unit length inductance (H/m).
- $\mathcal{R}_s$ is the *pin* region series resistance, Ω·m.

- $\mathcal{C}_j(V_j)$ is the junction capacitance, F/m, which weakly depends on the junction applied voltage V_j.
- $\mathcal{C}_{pul}$ is the external parasitic capacitance, F/m.
- $\mathcal{R}_o = \mathcal{R}_o(V_j)$ is the distributed photoresistance, $\Omega\cdot$ m, depending on the DC junction voltage.

Defining the *per-unit-length series impedance* and *parallel admittance* of the EAM RF line:

$$\mathcal{Z}(\omega) = \mathcal{R}_{pul}(\omega) + \mathrm{j}\omega\mathcal{L}_{pul}(\omega)$$

$$\mathcal{Y}(\omega) = \mathrm{j}\omega\mathcal{C}_{pul} + \frac{1}{\mathcal{R}_s + \left\{\left[\mathcal{R}_o(V_j)\right]^{-1} + \mathrm{j}\omega\mathcal{C}_j(V_j)\right\}^{-1}},$$

the characteristic impedance and complex propagation constant are derived, from transmission line theory (3.14), as

$$Z_0 = \sqrt{\frac{\mathcal{Z}(\omega)}{\mathcal{Y}(\omega)}}, \quad \gamma_m = \alpha_m + \mathrm{j}\frac{\omega}{v_m} = \sqrt{\mathcal{Z}(\omega)\mathcal{Y}(\omega)}.$$

Traveling along the EAM line, the optical signal experiences a local variation of the complex refractive index $\Delta n = \Delta n_r - \mathrm{j}\Delta n_i$. Neglecting Δn_i (i.e., chirp), we have

$$\Delta n(z,t) = -\mathrm{j}\Delta n_i(z,t) \approx a v_m(z,t), \tag{6.38}$$

where we have again introduced the linear dependence on the local RF voltage through the parameter a (now imaginary), as already done in (6.20) in the discussion of the frequency response of the distributed EOM. We can now relate Δn_i to $\Delta\alpha$ (the variation in absorption), since

$$k_o\Delta n_i(z,t) = \mathrm{j}k_o a v_m(z,t) = \frac{1}{2}\Delta\alpha(z,t) \approx \frac{1}{2}\left.\frac{\mathrm{d}\alpha}{\mathrm{d}v_m}\right|_{V_0} v_m(z,t). \tag{6.39}$$

Thus, from (6.38) and (6.39) the parameter a is given by

$$a = \frac{1}{2\mathrm{j}k_o}\left.\frac{\mathrm{d}\alpha}{\mathrm{d}v_m}\right|_{V_0}, \tag{6.40}$$

where V_0 is the DC bias point, supposed to be uniform along the line (a condition which can be obtained by DC decoupling the load and biasing the input through a low-impedance bias T); the local bias junction voltage V_j can be derived from the DC equivalent circuit.

The total amplitude variation experienced by the optical field at the modulator output, $E(t_2)$, will be derived by considering again, as in the EOM distributed analysis, the voltage seen by the optical wave traveling through the interaction region (we neglect, as already stressed, the phase variation):

$$
\begin{aligned}
E(t_2) &= E(t_2, \Delta n = 0) \exp\left[-\mathrm{j}k_o \int_0^L \Delta n\left(z, t_2 - \frac{L}{v_o} + \frac{z}{v_o}\right) \mathrm{d}z\right] \\
&\approx E(t_2, \Delta n = 0) \exp\left[-\frac{1}{2}\left.\frac{\mathrm{d}\alpha}{\mathrm{d}v_m}\right|_{V_0} \int_0^L v_m\left(z, t_2 - \frac{L}{v_o} + \frac{z}{v_o}\right) \mathrm{d}z\right],
\end{aligned}
$$

since $-\mathrm{j}k_o\Delta n = -k_o\Delta n_i(z, t)$ from (6.38).

Deriving the optical power from the magnitude squared of the optical field, and taking into account that, in the absence of small-signal applied voltage, the input power undergoes a total absorption $\alpha(V_0)$, we obtain

$$
\begin{aligned}
\frac{p_{out}(t)}{P_{in}} &= \mathrm{e}^{-\alpha(V_0)L} \exp\left[-\left.\frac{\mathrm{d}\alpha}{\mathrm{d}v_m}\right|_{V_0} \int_0^L v_m\left(z, t_2 - \frac{L}{v_o} + \frac{z}{v_o}\right) \mathrm{d}z\right] \\
&\approx \mathrm{e}^{-\alpha(V_0)L} \left[1 - \left.\frac{\mathrm{d}\alpha}{\mathrm{d}v_m}\right|_{V_0} \int_0^L v_m\left(z, t_2 - \frac{L}{v_o} + \frac{z}{v_o}\right) \mathrm{d}z\right] \\
&= \frac{P_{out} + \widehat{p}_{out}(t)}{P_{in}},
\end{aligned}
$$

where a further small-signal linearization of the exponential $\exp[\epsilon] \approx [1 - \epsilon]$ for small ϵ has been carried out.

Using (6.21) to evaluate the integral and comparing with (6.26), we obtain

$$
\begin{aligned}
&-\left.\frac{\mathrm{d}\alpha}{\mathrm{d}v_m}\right|_{V_0} \int_0^L v_m\left(z, t_2 - \frac{L}{v_o} + \frac{z}{v_o}\right) \mathrm{d}z \\
&\quad = -\left.\frac{\mathrm{d}\alpha}{\mathrm{d}v_m}\right| \times \frac{Z_0 V_G}{Z_0 + Z_G} \frac{\mathrm{e}^{\mathrm{j}\omega t_2}\mathrm{e}^{-\gamma_m L}}{1 - \Gamma_L\Gamma_G\mathrm{e}^{-2\gamma_m L}} \\
&\quad \times \left[\frac{1 - \mathrm{e}^{-j(\beta_o - \beta_m)L + \alpha_m L}}{\mathrm{j}(\beta_m - \beta_o)L + \alpha_m L} - \Gamma_L \frac{1 - \mathrm{e}^{-\mathrm{j}(\beta_o + \beta_m)L - \alpha_m L}}{\mathrm{j}(\beta_m + \beta_o)L + \alpha_m L}\right].
\end{aligned}
$$

Assuming a sinusoidal variation of $\widehat{p}_{out}(t) = \widehat{P}_{out}(\omega)\exp(\mathrm{j}\omega t)$, one finally has the frequency response

$$
M(\omega) = \frac{\widehat{P}_{out}(\omega)}{P_{in}} = -\mathrm{e}^{-\alpha(V_0)L} \left.\frac{\mathrm{d}\alpha}{\mathrm{d}v_m}\right|_{V_0} \frac{Z_0\mathrm{e}^{-\gamma_m L}}{Z_0 + Z_G} \frac{F(u_+) + \Gamma_L F(u_-)}{1 - \Gamma_L\Gamma_G\mathrm{e}^{-2\gamma_m L}}
$$

where $F(u)$ and $u_\pm$ are given by (6.28) and (6.29), respectively. The normalized small-signal modulation index can be finally evaluated as

$$
m(\omega) = \left|\frac{\widehat{P}_{out}(\omega)}{\widehat{P}_{out}(0)}\right| = \left|\frac{M(\omega)}{M(0)}\right|.
$$

The small-signal frequency response of a distributed EAM is formally equivalent to the response of a traveling-wave electrooptic modulator, as can be seen by evaluating $|H(\omega)/H(0)|$ from (6.27); therefore, the alternative response from (6.31) can also be exploited. In practice, however, many differences arise. First, the high optical residual insertion loss limits actual EAM lengths to a few hundred μm; besides, the RF propagation loss is much higher than in dielectric-based electrooptic modulators due to the heavily doped layers (typically 10 dB/cm at 1 GHz against 0.3 dB/cm in LN). Finally, the velocity mismatch is fairly large, but the structure is so short that this is often no important limitation to the bandwidth; due to the small thickness of the intrinsic layer, lines with characteristic impedance much lower than 50 Ω are typically obtained. Despite such limitations, traveling-wave modulators have shown significant advantages in terms of bandwidth over lumped EAMs.

6.10 Electroabsorption modulator examples

An example of a lumped-parameter, high-speed EAM is shown in Fig. 6.46 [108]. The guiding structure is based on an intrinsic InGaAs/InAlAs MQW, sandwiched between a p-type InAlAs layer and an n-type layer of the same material. The substrate is n-type InP. The length of the modulation region of this QCSE-based modulator is varied between 50 and 150 μm, leading to the normalized transfer characteristics shown in Fig. 6.47 (left). The contrast ratio increases from 10 dB at 3 V reverse bias to over 30 dB, but at the same time also the ON-state insertion loss increases with the length. Note that the transmission curve grows again after the minimum, due to the weaker QCSE when the driving voltage becomes large, caused by the decrease in overlap between the valence and conduction band QW wavefunctions. Figure 6.47 (right) shows the frequency response normalized with respect to the low-frequency value; for the shortest device the electrical bandwidth is in excess of 30 GHz.

Figure 6.48 reports a traveling-wave EAM developed by the University of California at Santa Barbara (UCSB) [109], [110]. The bulk InGaAsP absorption layer operates through FKE, and the EAM parameters are: $L = 200\,\mu\mathrm{m}$, $w = 3\,\mu\mathrm{m}$, $d \approx 0.35\,\mu\mathrm{m}$; the optical saturation power is ≈ 45 mW@ -1.3 V applied reverse bias, and the matching impedance is 26 Ω. Figure 6.49 shows the propagation characteristics of

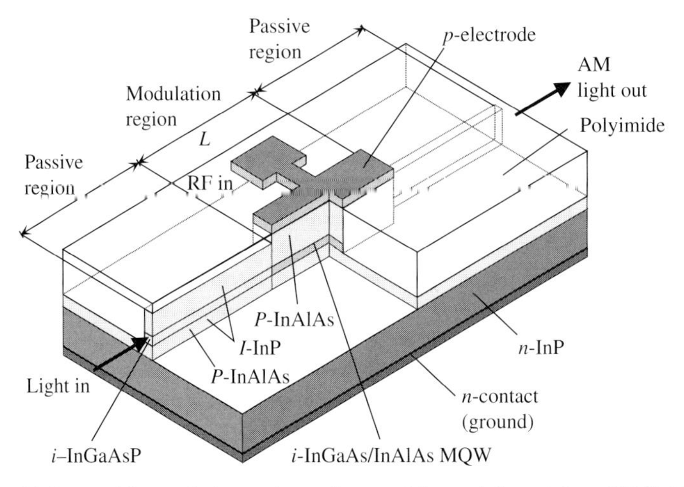

Figure 6.46 High-speed lumped electroabsorption modulator. Adapted from [108], Fig. 1 (©1996 IEEE).

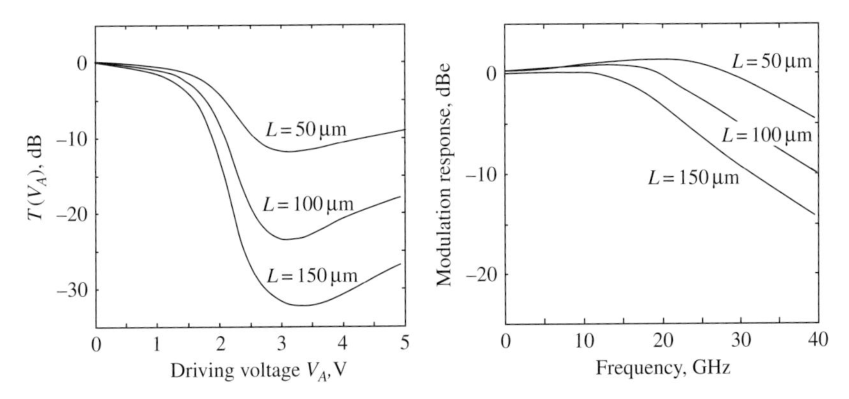

Figure 6.47 Static transmission curve (left) and normalized frequency response (right) of lumped EAM. After [108], Fig. 3 and (adapted) Fig. 4 (©1996 IEEE).

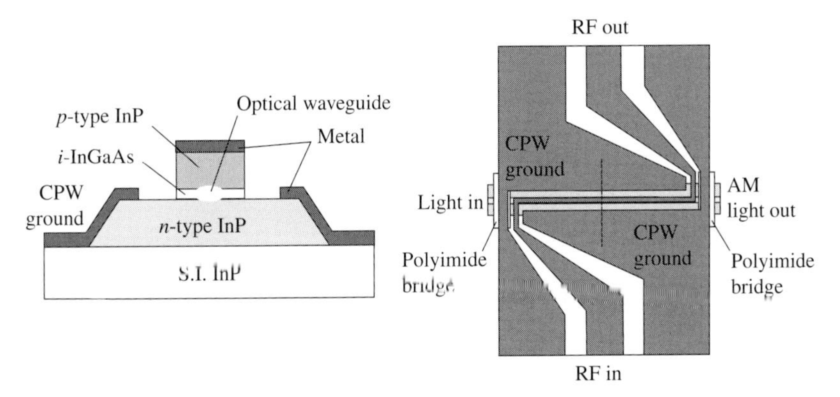

Figure 6.48 FKE-based (bulk) traveling-wave EAM. Cross section of the active region (left); Layout (right). After [110], Fig. 1 and Fig. 2 (©2001 IEEE).

the RF line; the refractive index exhibits the *slow-wave effect*, typical of transmission lines including semiconductors. Owing to such an effect, the RF index n_m exhibits strong low-frequency dispersion and is much larger than the refractive index of the material (about 7 at high frequency against $\sqrt{13} = 3.6$).

A simplified explanation of the slow-wave effect, which makes the synchronous coupling between the optical and RF waveguide more critical, is as follows. In a PiN structure, the electric field is completely confined in the intrinsic layer, due to the charge screening effect taking place in the doped surrounding layers. However, the magnetic field permeates the doped layers, unless the frequency is so high that the skin effect prevents the EM fields from penetrating doped layers altogether, see Fig. 6.50. While the line capacitance is the capacitance of the intrinsic layer only, the line inductance is (approximately) the inductance of the line without semiconductors (i.e., where only the metal electrodes are present). However, the inductance of such a line is equal to the inductance in air $\mathcal{L}_0$, and is related to the in-air capacitance $\mathcal{C}_0$ as $\mathcal{L}_0 = 1/\left(c_0^2\mathcal{C}_0\right)$, see

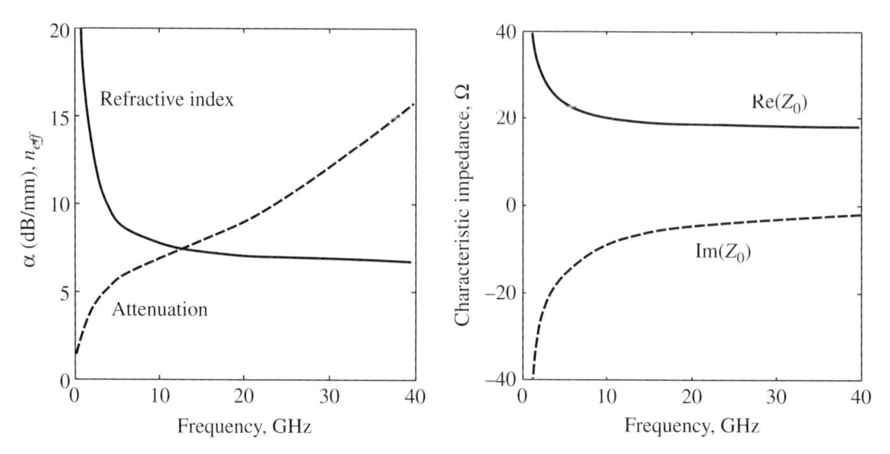

Figure 6.49 Propagation characteristics of the EAM as a distributed structure: Attenuation and effective refractive index (left); Real and imaginary part of the characteristic impedance (right). From [109], Fig. 2 and Fig. 3 (©2001 IEEE).

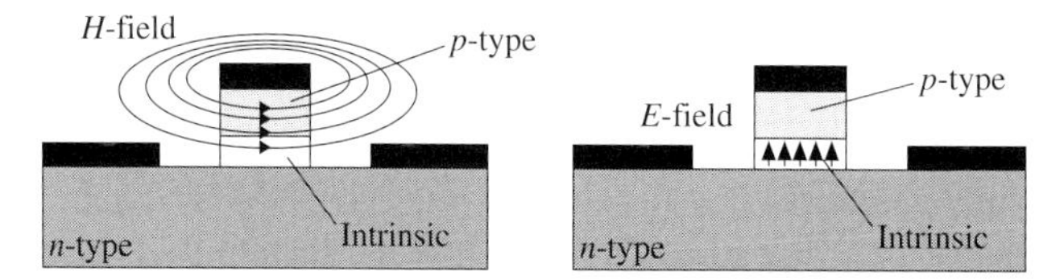

Figure 6.50 Slow-wave effect in semiconductor transmission lines: magnetic (left) and electric (right) field lines; the E-field concentration in the intrinsic layer leads to high capacitance and effective permittivity.

(3.27). Since $\mathcal{C}_0 \ll \mathcal{C}$, where $\mathcal{C}$ is the intrinsic layer capacitance per unit length, one has

$$v_f = \frac{1}{\sqrt{\mathcal{LC}}} = \frac{1}{\sqrt{\mathcal{L}_0\mathcal{C}}} = c_0\sqrt{\frac{\mathcal{C}_0}{\mathcal{C}}} \rightarrow n_m = \sqrt{\frac{\mathcal{C}}{\mathcal{C}_0}} \gg \sqrt{\epsilon_r},$$

as the structures in air and with dielectrics differ not only because of the different dielectric constant but also because of the different geometry.

Apart from the slow-wave effect, the EAM line losses are very high and the impedance level low, see Fig. 6.49. Note the large imaginary part of the characteristic impedance, typical of a RC line behavior (i.e., a line where the dominant parameters are the series resistance and the parallel capacitance, see Section 3.2.1).

By injecting the RF signal so as to make it co-propagate or counter-propagate with respect to the optical signal, the distributed operation can be experimentally detected. In fact, measured data show that for the co-propagating TW-EAM the optical bandwidth is around 35 GHz, while in the counter-propagating case it is reduced to 20 GHz. The expected large reduction in the bandwidth confirms that the modulator operates as a traveling-wave structure. The switching voltage with 10 dB extinction is 3 V; the smaller extinction associated with a larger length with respect to the MQW structure can be associated with the weaker FKE; however, the optical bandwidth is wider.

6.10.1 Integrated EAMs (EALs)

MQW EAMs typically have to be tailored to a specific laser source, due to the very narrow optical bandwidth. Since directly modulated lasers have poor chirp characteristics (α_H is always positive and somewhat large, e.g., 3–5), integration of a DFB laser with an EAM appears as an interesting solution to increase the potential of the modulated source for high-speed, long-distance transmission. Integrated lasers and EAMs (called EALs or EMLs) are a commercial solution available today for transmission speed up to 10 Gbps. Moreover, EAM are similar in size to the source, in contrast to electrooptic semiconductor modulators, which are typically much larger than the source.

The development of EALs has to overcome a number of difficulties. Excellent isolation must be provided both electronically and optically between the source and the modulator; moreover, the two epitaxial layers should not be exactly the same, but some detuning is needed to allow the EAM to operate properly. Figure 6.51 shows some solutions for integration. In the butt-joint approach, the EAM is obtained by epitaxial regrowth, and the structures can be independently optimized; however, the morphology of the interface between the two is critical. In the selective area growth (SAG) process the epitaxial growth takes place in both structures, but silicon dioxide layers are deposited parallel to the DFB area; see Fig. 6.52(a). The presence of such layers induces a slightly different composition (e.g., a larger In fraction) and thickness of the epitaxial layers, which leads to a different E_g and therefore to detuning. An example is shown in Fig. 6.52(b), where the SAG region (DFB) and the Field region (EAM) clearly exhibit detuning (namely, the absorption edge of the unbiased EAM is at slightly higher energy and lower wavelength than the DFB emission; note that the photoluminescence peak energy is slightly lower than the absorption edge and shifts with the absorption edge). Finally, the two structures can be derived from an identical layer, operating in direct or reverse bias; in this case the growth is easier but a strong compromise must be made between the two devices. An example of a 10 Gbps EAL developed by Avago Technologies is shown in Fig. 6.53; the EAM is isolated both optically and electrically from the laser, which operates in CW conditions. The structure chirp can be made positive or negative by varying the bias voltage; the resulting chirp typically is one order of magnitude less than in a directly modulated laser [111].

Figure 6.54 finally shows a schematic picture of the traveling-wave EAL from NTT [112]. In the traveling-wave mode the device exhibits a bandwidth larger than 40 GHz, which reduces to 10 GHz in the lumped configuration; see Fig. 6.55. Very-high speed

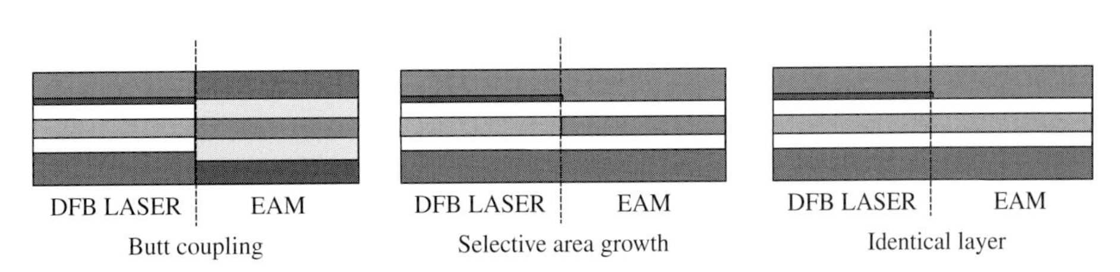

Figure 6.51 Fabrication techniques for integrated EAM-DFB lasers.

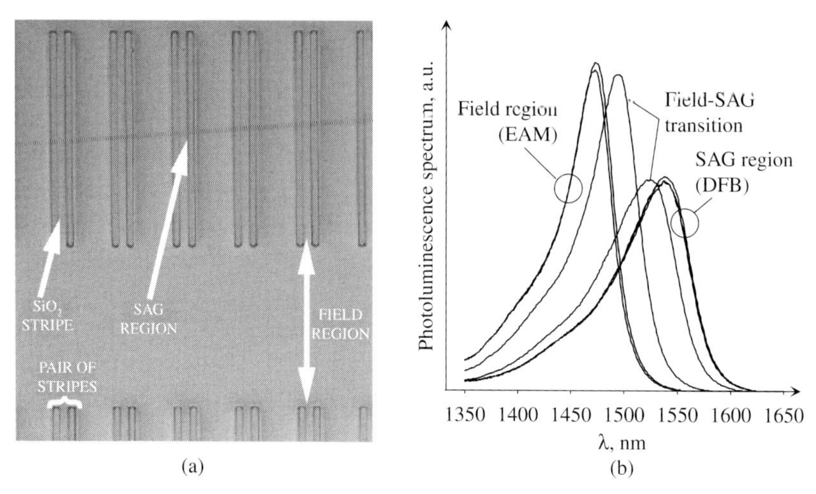

Figure 6.52 SAG technique: (a) lateral silicon dioxide deposition; (b) photoluminescence spectra from the SAG region (DFB) and the field region (EAM) revealing the relative detuning. Courtesy of Avago Technologies – Fiber Optic Product Division.

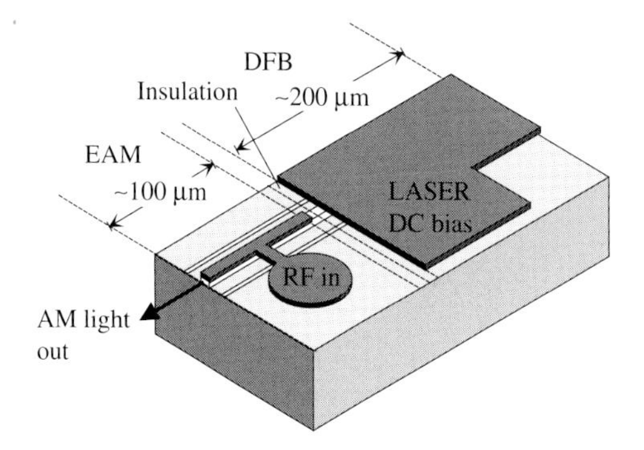

Figure 6.53 Schematic structure of lumped 10 Gbps EAL [111].

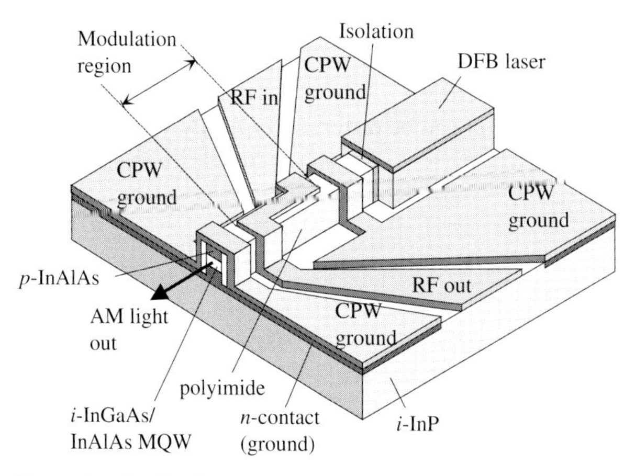

Figure 6.54 Example of a distributed electroabsorption modulator coupled to a DFB laser. From [112], Fig. 1 (©2001 IEEE).

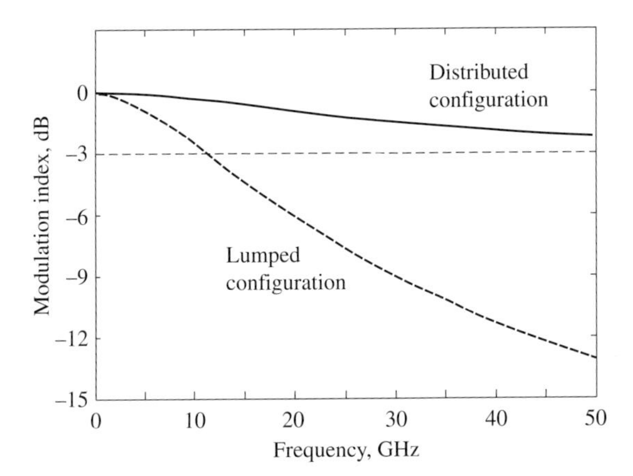

Figure 6.55 Frequency response of the integrated EAL: lumped-parameter and distributed-electrode setting. After [112], Fig. 3 (©2001 IEEE).

devices can therefore be realized, whose main limitation probably remains the higher chirp with respect to the corresponding LN electrooptic modulators.

6.11 Modulator and laser biasing

Electrooptic and electroabsorption modulators typically require a bias voltage to be superimposed onto the signal (digital or analog). In analog EOMs, biasing for maximum linearity is at $V_\pi/2$, while in digital applications, according to the signal generated by the driver, the bias point can be at 0, $V_\pi/2$, V_π, corresponding to an amplitude of the zero and one levels of $(0, V_\pi)$, $\pm V_\pi/2$, $(-V_\pi, 0)$, respectively. Several options are available for implementing the bias circuit.

In EOMs, the RF and DC modulator inputs can be *separated* by implementing an *additional DC electrode* in the phase modulation section. This solution is shown in Fig. 6.56; the device length increases, but there is no need for a bias T (see Section 3.4.1) separating the DC and the RF inputs. Note that the DC electrode has no requirements on bandwidth (and also slightly weaker constraints on the applied voltage level); thus, the length of the DC bias section can be different (typically smaller, to reduce the total device length) with respect to the length of the modulating section. Increasing the DC bias allows shortening of the DC electrode and the total device length.

Alternatively, the modulator can be driven by a single input including both the signal and the bias. In this case, a bias T is needed, which can be *external* (see Fig. 6.57), *integrated into the modulator* or, more commonly, *integrated with the driver*. The RF load should be in any case DC blocked to avoid power dissipation and damage.

Bias Ts for high-speed modulators with ultrawide bandwidths (ranging, e.g., from 30 kHz to 40 GHz or from 50 kHz to 65 GHz) are available from several manufacturers; often, such devices are also exploited for instrumentation. A sketch of a bias T is shown in Fig. 6.58. Wideband discrete bias Ts are large and expensive devices, mainly due to the need to offer enough DC blocking (between the RF and the DC inputs) and RF

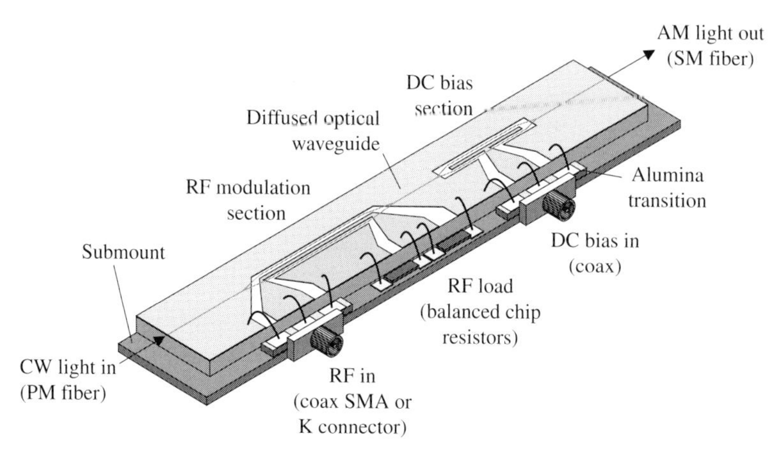

Figure 6.56 Mach–Zehnder X-cut modulator with additional DC bias electrode. For clarity the device is placed on a metal submount and unpackaged. The pigtailing sections are not shown.

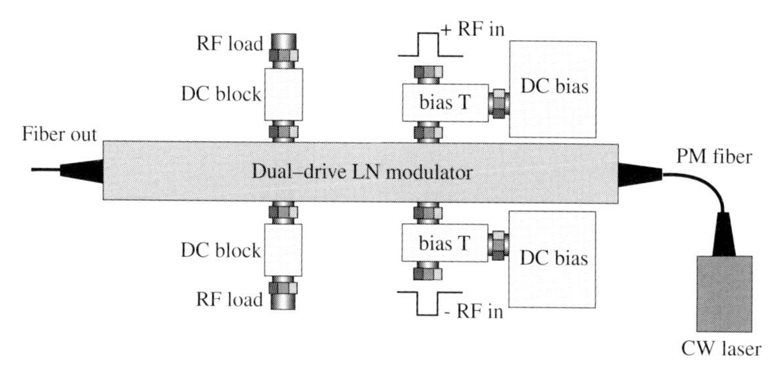

Figure 6.57 Example of dual-drive modulator setup with external bias Ts and DC-blocked load. The PM fiber keeps constant the input light polarization.

shorting (between the DC input and the DC output) at a very low frequency (e.g., 30 kHz) but with technologies able to work also at the upper frequency (e.g., 40 GHz). Wideband bias Ts often use a multiband design exploiting different component values and technologies.

Laser diodes are in principle current-driven devices, implying that a high-impedance DC current source should be used. Also, for lasers a bias T can be exploited to make the device AC-coupled only; this allows reduction of the required bias supply value. Typical laser bias currents are of the order of 10–100 mA; see [60], Ch. 8.

Due to the complex technology, wideband bias Ts are rarely integrated with the EAM, EOM or laser, and a preferred solution is integration within the device driver. Modulator drivers typically offer (for 10 Gbps or even 40 Gbps applications) peak-to-peak output signal voltages up to 3 V, with bias control ranging, for example, from −10 to 5 V. Such drivers are conceived for EAMs or for Mach–Zehnder LN modulators with optimized driving voltages (typically 3 V or less at 10 Gbps, sometimes down to 2 V). However,

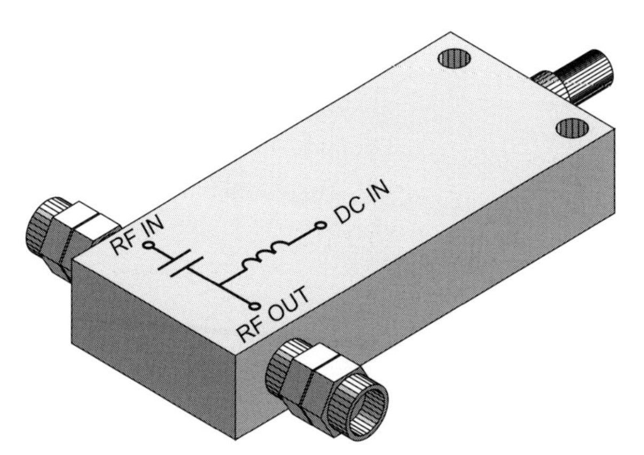

Figure 6.58 Wideband external bias T.

difficulties arise in drivers for LN 40 Gbps modulators, where typical input voltages are of the order of $V_\pi \approx 5$ V or more.

6.12 Modulator and laser drivers

The modulator or laser driving stage at large integrates a few basic functions, such as data multiplexing (MUX) from lower speed into the maximum channel speed (2.5, 10, 40 Gbps), data retiming and/or reshaping, modulator or laser biasing, and finally the *driver* – see Fig. 6.59 for a simplified example of the driving stages for a MZ modulator. The multiplexer-driver architecture is typically differential, to suppress common-mode interferers and also, sometimes, to drive differential mode modulators (such as dual-drive modulators).

The driver includes in principle three stages: (i) a *pre-driver*, often exploited as a buffer to decouple the input capacitance of the driver and to provide logical level shifting; typical pre-driver architectures include emitter- or source-follower stages; (ii) a *driver*, whose purpose is to provide the switching voltage; and (iii) the *driver amplifier*, whose output is the input data stream with a suitably large driving voltage (for modulators, 1.5–3 V_{pp} in EAMs, 3–6 V_{pp} in LN EOMs). Conventional solutions for the driver stage can readily be derived from the so-called *current-mode* logic gates, see Fig. 6.60(a), in which the driving voltage swing is able to alternately drive the two transistors of the differential pair in the off and on states, thus obtaining at the stage output two complementary voltage levels.[18] The ideal current source I_M can be practically implemented through a current mirror or other equivalent circuitry. Current-mode logic gates have a limited input and output logical swing; with a bipolar implementation the input swing could be of the order of 300 mV.

Figure 6.60(a) shows a current-mode gate loaded by a dummy load R'_D and by an EAM connected to the output through a two-port including interconnects and the driver

[18] The topology of current-mode gates is differential, similar to emitter coupled logics (ECL), see, e.g., [113].

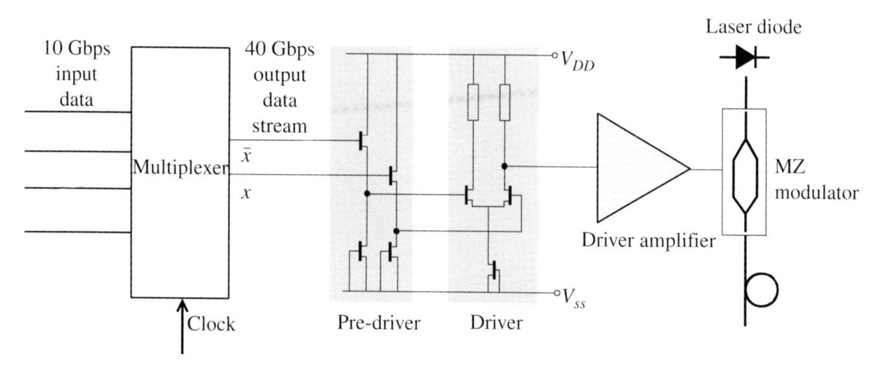

Figure 6.59 40 Gbps MZ modulator driving stage and logical circuitry (simplified). Partly adapted from [114], Fig. 1.

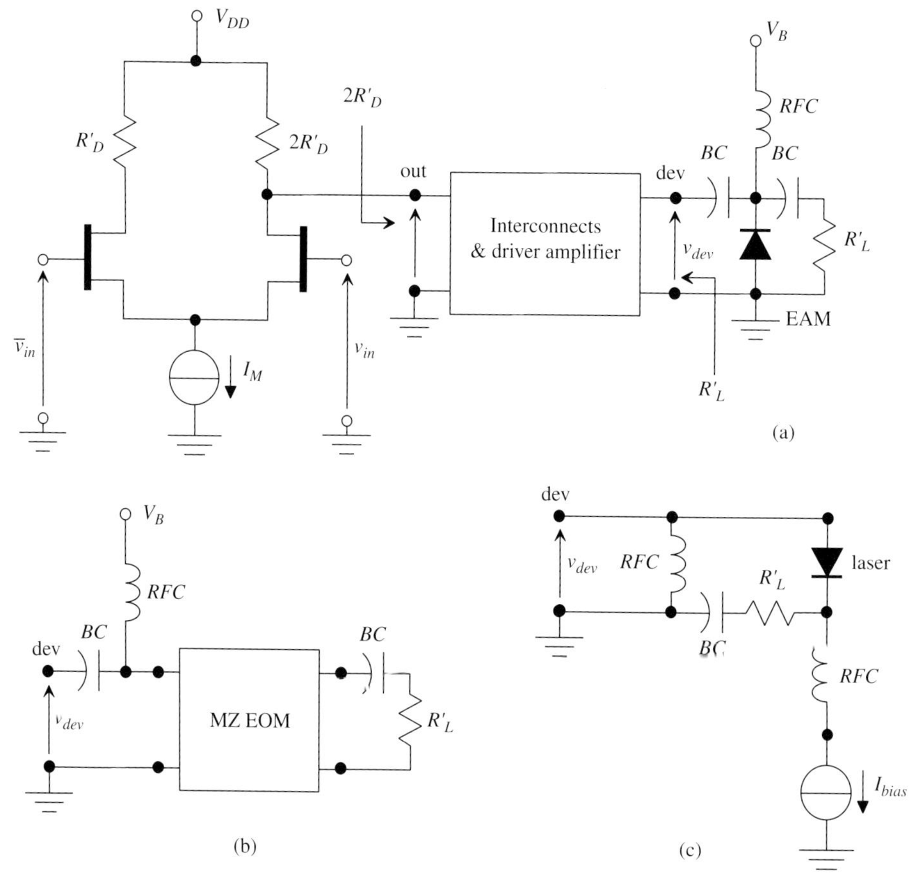

Figure 6.60 Examples of current-mode differential modulator or laser drivers. The two-port connected to the positive output (out) models interconnects and/or the *driver amplifier*. (a) Differential driver for an AC-coupled EAM. (b) AC-coupled MZ EOM. (c) AC-coupled laser. *RFC* stands for radiofrequency choke, *BC* stands for (DC) blocking capacitor.

amplifier. The dummy load is introduced for symmetry; impedance matching must be ensured (at least in the EAM bandwidth) at the "out" node, see Fig. 6.60(a), to avoid multiple reflections between this node and the device. In a high-speed implementation 50 Ω matching could be required. The EAM is shown as AC-coupled, with a bias T made of an inductive block (the RF choke RFC) and a DC blocking capacitor (BC). The frequency response of a directly coupled EAM (i.e., without a driver amplifier) can be improved by putting in series with the back termination resistor $2R'_D$ a peaking inductor. In Fig. 6.60(b) the driven device is a distributed Mach–Zehnder modulator AC coupled to an input bias T. We assume again that the device (itself a transmission line) is impedance matched at the input and output to the load R'_L in order to suppress multiple reflections leading to distortion (see Section 3.3.1). Finally, Fig. 6.60(c) shows an AC-coupled laser; the bias supply is modeled as a current source, while the loading resistor is in series (due to the low laser series impedance). Note that, due to the possibly large laser DC current (e.g., around 100 mA), the voltage drop on the back termination could be large (e.g., 5 V for a back termination of 50 Ω) and possibly incompatible with the minimum voltage drop on the laser (of the order of 1.5–2 V).

From the standpoint of the implementation, the driver chain of high-speed systems exhibits two critical points. First, the digital technology of the last MUX must be adequate for the final channel speed. 40 Gbps digital technology is limited by the material choice (SiGe, GaAs, perhaps InP) and by the logical family (typically HBT emitter coupled logic (ECL) or FET direct coupled FET logic (DCFL)). In very high-speed logic families the output logic swing is low, e.g., 300–500 mV, and therefore direct driving of the modulator or laser may be unfeasible. Secondly, the modulator driver amplifier must incorporate a number of conflicting requirements, mainly wideband operation,[19] high maximum frequency but at the same time high output voltage. The required gain depends on the modulator technology and on the output swing of the current mode driver; for 40 Gbps systems exploiting $LiNbO_3$ devices it can be as high as 27 dB, with a ±1dB flatness. The acceptable group delay is specific to the system standard and can be of the order of ±10 ps on the whole band. The electrical bandwidth can range from $0.7B_r$ to $1.3B_r$ where B_r is the bit rate; for 50 Gbps systems this means a bandwidth from 60–100 kHz to more than 50 GHz.

While at 2.5 Gbps, and perhaps 10 Gbps, Si-based ICs may still provide a suitable solution, for 40 Gbps operation the enabling technologies are SiGe, GaAs, and InP. However, SiGe HBTs, though able to cover the 40 Gbps range, exhibit decreasing device breakdown voltages with increasing cutoff frequency. An empirical rule for this application is that the cutoff frequency should be 3–4 times the maximum operation frequency; this leads to breakdown voltages of the order of 2 V for the SiGe technology; see Fig. 6.61 [6]. SiGe drivers are therefore adequate as EAM drivers, but are critical as LN MZ drivers for 40 Gbps EOMs. A considerably larger breakdown voltage, compatible with LN MZ modulator operation, is obtained with InP HBTs and,

[19] Modulator drivers typically are not DC-coupled but exhibit a lower cutoff frequency that should be low enough not to affect the bit rate. Values are standard-dependent, e.g., 64 kHz for 2.5 Gbps and 257 kHz for 10 Gbps SONET systems, respectively, see [60], Section 6.2.6.

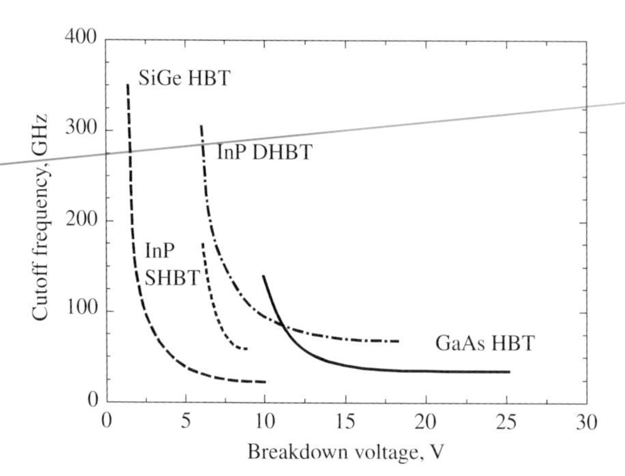

Figure 6.61 Breakdown voltage vs. cutoff frequency for a few HBT technologies. SHBT and DHBT stand for single and double heterojunction HBTs, respectively. Adapted from [6], p. 33.

even better, with GaAs-based HBTs. III-V FETs (in particular, PHEMTs) exhibit breakdown voltages in excess of 5 V (InP) and 10 V (GaAs), which allow 40 Gbps drivers to be implemented in such technologies, with GaAs as a preferred material. A promising material for high-voltage applications could also be gallium nitride (GaN), whose frequency performances are not, at least for the moment, adequate for 40 Gbps applications. In conclusion, 40 Gbps driver amplifiers for stages (like the MZ modulators on LN) requiring large peak-to-peak voltages still are technologically demanding.

6.12.1 The high-speed driver amplifier

Broadband amplifiers can be obtained through conventional open-loop or feedback circuit approaches. Resistive feedback applied to a high-gain amplifier is a simple way to achieve flat gain over a broad frequency band; however, the open loop gain of the amplifier should be suitably larger than the gain with feedback at the maximum operating frequency. This is a difficult requirement for 40 Gbps drivers, which often need alternative broadbanding approaches.[20] A number of circuit recipes for broadbanding may be based on the compensation of the load capacitance through some inductive element connected to the load (*inductive peaking*).

An interesting alternative approach, allowing broadband operation up to a frequency that can, in theory, exceed the device cutoff frequency (but is, in practice, only somewhat larger than $f_T/2$), consists in turning the amplifier into a distributed, traveling-wave structure. This solution is called the *distributed amplifier*, and will be discussed here, for simplicity, in terms of a single-ended structure. Differential implementations are also possible. The distributed amplifier is also interesting from the standpoint of the operating principle, since it is an example of an electronic amplifier exploiting a distributed or quasi-distributed interaction, yielding wideband operation.

[20] For a more complete discussion, see [60], Ch. 6.

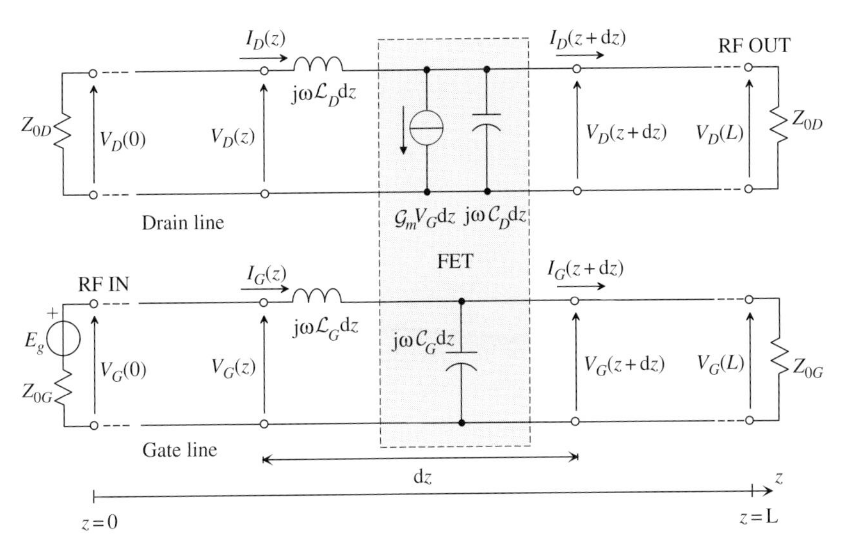

Figure 6.62 Continuous distributed amplifier.

A somewhat idealized structure for the distributed FET (e.g., PHEMT) amplifier is made of an input (gate) transmission line connecting, in a continuous way, the input of each infinitesimal device cell. The output (drain) transmission line collects the current injected by the transconductance generator. In the simplified model shown in Fig. 6.62, the input line includes only a distributed per-unit-length (p.u.l.) inductance $\mathcal{L}_G$ and a distributed p.u.l. capacitance $\mathcal{C}_G$ (mainly associated with the input gate–source capacitance). The output line includes the transconductance current generators ($\mathcal{G}_m = g_m/L$ is the p.u.l. device transconductance, g_m being the total transconductance and L the gate periphery) of p.u.l. current $\mathcal{G}_m V_G(z)$, and a p.u.l. inductance and capacitance $\mathcal{L}_D$ and $\mathcal{C}_D$.

The gate and drain lines are transmission lines with characteristic impedances

$$Z_{0G} = \sqrt{\frac{\mathcal{L}_G}{\mathcal{C}_G}}, \quad Z_{0D} = \sqrt{\frac{\mathcal{L}_D}{\mathcal{C}_D}}$$

and propagation constants

$$\beta_G = \omega\sqrt{\mathcal{L}_G\mathcal{C}_G}, \quad \beta_D = \omega\sqrt{\mathcal{L}_D\mathcal{C}_D}.$$

Line attenuations α_G and α_D are for the moment neglected. Assume that the gate line is matched at the input and output; the line input impedance will be Z_{0G} and therefore $V_G(0) = E_g/2$. Assume $V_G(0) = V_{in}$ as the line input voltage; since the line is matched, only the forward-propagating wave exists, so that $V_G(z) = V_{in}e^{-j\beta_G z}$. On the drain line, distributed transconductance current generators are present, of value $\mathcal{G}_m V_G(z)$. We can therefore write the transmission line equations for the drain line (by applying the Kirchhoff voltage and current laws to a transmission line cell of length dz) as

$$V_D(z+\mathrm{d}z) = V_D(z) - \mathrm{j}\omega\mathcal{L}_D I_D(z)\,\mathrm{d}z$$
$$I_D(z+\mathrm{d}z) = I_D(z) - \mathrm{j}\omega\mathcal{C}_D V_D(z+\mathrm{d}z)\,\mathrm{d}z - \mathcal{G}_m V_G(z)\,\mathrm{d}z,$$

i.e., in the limit $\mathrm{d}z \to 0$:

$$\frac{\mathrm{d}V_D}{\mathrm{d}z} = -\mathrm{j}\omega\mathcal{L}_D I_D(z)$$
$$\frac{\mathrm{d}I_D}{\mathrm{d}z} = -\mathrm{j}\omega\mathcal{C}_D V_D(z) - \mathcal{G}_m V_G(z) = -\mathrm{j}\omega\mathcal{C}_D V_D(z) - \mathcal{G}_m V_{in}\mathrm{e}^{-\mathrm{j}\beta_G z}.$$

Taking the derivative vs. z of the first equation and substituting into the second equation, we obtain the following second-order equation in V_D:

$$\frac{\mathrm{d}^2 V_D}{\mathrm{d}^2 z} = -\beta_D^2 V_D + \mathrm{j}\omega\mathcal{L}_D\mathcal{G}_m V_{in}\mathrm{e}^{-\mathrm{j}\beta_G z}.$$

We can express the solution as $V_D(z) = V_1 + V_2$, where V_1 is the solution of the homogeneous equation (no forcing term), V_2 is a particular solution of the forced equation. We have

$$V_1(z) = V_{D0}^+\mathrm{e}^{-\mathrm{j}\beta_D z} + V_{D0}^-\mathrm{e}^{\mathrm{j}\beta_D z}.$$

We seek V_2 in the form $K\exp(-\mathrm{j}\beta_G z)$, K to be determined. Substituting, we have

$$-\beta_G^2 K = -\beta_D^2 K + \mathrm{j}\omega\mathcal{L}_D\mathcal{G}_m V_{in}\mathrm{e}^{-\mathrm{j}\beta_G z} \to V_2 = \frac{\mathrm{j}\omega\mathcal{L}_D\mathcal{G}_m}{\beta_D^2 - \beta_G^2} V_{in}\mathrm{e}^{-\mathrm{j}\beta_G z}.$$

Therefore, the total solution is

$$V_D(z) = V_{D0}^+\mathrm{e}^{-\mathrm{j}\beta_D z} + V_{D0}^-\mathrm{e}^{\mathrm{j}\beta_D z} + \frac{\mathrm{j}\omega\mathcal{L}_D\mathcal{G}_m}{\beta_D^2 - \beta_G^2} V_{in}\mathrm{e}^{-\mathrm{j}\beta_G z}.$$

For the drain current, we have

$$I_D(z) = -\frac{1}{\mathrm{j}\omega\mathcal{L}_D}\frac{\mathrm{d}V_D}{\mathrm{d}z} = \frac{V_{D0}^+}{Z_{0D}}\mathrm{e}^{-\mathrm{j}\beta_D z} - \frac{V_{D0}^-}{Z_{0D}}\mathrm{e}^{\mathrm{j}\beta_D z} + \frac{\mathrm{j}\beta_G\mathcal{G}_m}{\beta_D^2 - \beta_G^2} V_{in}\mathrm{e}^{-\mathrm{j}\beta_G z}.$$

To derive V_{D0}^+ and V_{D0}^- we apply the boundary conditions (assuming that also the drain line is impedance matched):

$$V_D(0) = Z_{0D}(-I_D(0)), \quad V_D(L) = Z_{0D} I_D(L).$$

Separating forward and backward contributions, we obtain

$$V_{D0}^+ = -\frac{V_{in}}{2}\frac{\mathcal{G}_m(\mathrm{j}\omega\mathcal{L}_D + \mathrm{j}\beta_G Z_{0D})}{\beta_D^2 - \beta_G^2}$$
$$V_{D0}^- = -\frac{V_{in}}{2}\frac{\mathcal{G}_m(\mathrm{j}\omega\mathcal{L}_D - \mathrm{j}\beta_G Z_{0D})}{\beta_D^2 - \beta_G^2}\mathrm{e}^{-\mathrm{j}(\beta_G+\beta_D)L}.$$

The load voltage $V_D(L)$ is therefore

$$\begin{aligned} V_D(L) &= -\frac{V_{in}}{2}\frac{\mathcal{G}_m\,(\mathrm{j}\omega\mathcal{L}_D + \mathrm{j}\beta_G Z_{0D})}{\beta_D^2 - \beta_G^2}\mathrm{e}^{-\mathrm{j}\beta_D L} \\ &\quad - \frac{V_{in}}{2}\frac{\mathcal{G}_m\,(\mathrm{j}\omega\mathcal{L}_D - \mathrm{j}\beta_G Z_{0D})}{\beta_D^2 - \beta_G^2}\mathrm{e}^{-\mathrm{j}\beta_G L} + \frac{\mathrm{j}\omega\mathcal{L}_D\mathcal{G}_m}{\beta_D^2 - \beta_G^2}V_{in}\mathrm{e}^{-\mathrm{j}\beta_G L} \\ &= \frac{V_{in}}{2}\frac{\mathcal{G}_m\,(\mathrm{j}\omega\mathcal{L}_D + \mathrm{j}\beta_G Z_{0D})}{\beta_D^2 - \beta_G^2}\left[\mathrm{e}^{-\mathrm{j}\beta_G L} - \mathrm{e}^{-\mathrm{j}\beta_D L}\right], \end{aligned}$$

or, in a more convenient form:

$$V_D(L) = -\frac{V_{in}}{2}Z_{0D}\mathcal{G}_m \exp\left(-\mathrm{j}\frac{\beta_G + \beta_D}{2}L\right)\frac{\sin\left(\frac{\beta_G - \beta_D}{2}L\right)}{\frac{\beta_G - \beta_D}{2}}.$$

Thus, the voltage gain can be written as

$$|A_V| = \left|\frac{V_D(L)}{V_{in}}\right| = \left|\frac{Z_{0D}\mathcal{G}_m L}{4}\frac{\sin\left(\frac{\beta_G - \beta_D}{2}L\right)}{\frac{\beta_G - \beta_D}{2}L}\right|.$$

The amplification is maximum and becomes *frequency-independent* if the coupling between the gate and drain lines is synchronous, i.e., if

$$\beta_G = \beta_D \rightarrow \mathcal{L}_D\mathcal{C}_D = \mathcal{L}_G\mathcal{C}_G.$$

In this case, we have

$$|A_V| = \frac{Z_{0D}g_m}{4}, \tag{6.41}$$

independent of frequency. If we account for losses in the drain and gate lines through the attenuations α_D and α_G, the response becomes frequency-dependent also in case of synchronous coupling; moreover, increasing the gate periphery L increases the transconductance g_m, but this effect is ultimately countered by the increase of the total line attenuation, so that an optimum gate periphery exists:

$$L_{opt} = \frac{\log(\alpha_D/\alpha_G)}{\alpha_D - \alpha_G}.$$

In practice, distributed amplifiers are made with a number of discrete devices connected on the inputs and outputs by delay lines, as shown in Fig. 6.63. The main motivations for this design are the difficulty of achieving synchronous coupling in a continuous device (due to the fact that the device input capacitance C_{GS} typically is much larger than the output capacitance C_{DS}), and the large losses induced in the input line by the extremely thin gate electrode. With the discrete setup, on the other hand, the output line can be capacitively loaded so as to improve velocity matching (or the line length can be properly increased to compensate delays), and losses can be decreased

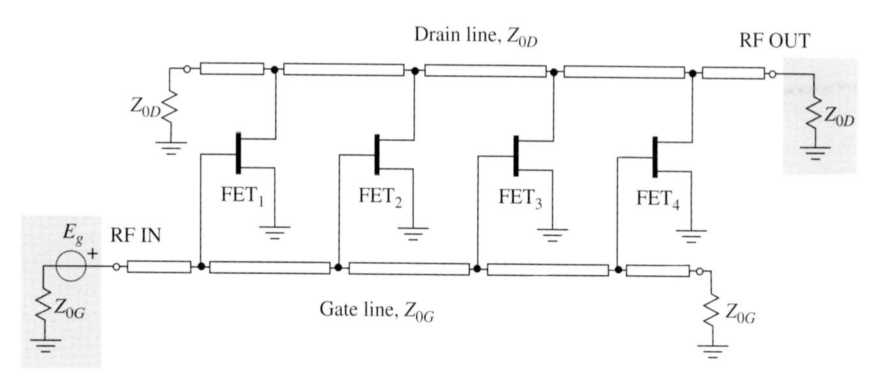

Figure 6.63 Example of discrete cell distributed amplifier with four cells.

due to the wider conductors used (typically 100 μm against less than 1 μm as in the gate fingers).

If we approximate the delay lines as LC discrete cells, the resulting gate and drain lines become quasi-distributed structures known as *artificial lines*. In such structures, the electrical behavior is similar to that of a transmission line for frequencies below a cutoff frequency given by

$$f_C = \frac{1}{\pi Z_{0D} C_D} = \frac{1}{\pi Z_{0G} C_G} = \frac{1}{\pi Z_0 C},$$

where we have assumed that the two lines are velocity matched and have the same characteristic impedance (and thus, the same p.u.l. parameters). In such conditions the low-frequency voltage gain becomes

$$|A_V(0)| = \frac{Z_0}{4} n g_m,$$

where ng_m is the total device transconductance, n being the number of cells. The gain–bandwidth product will therefore be

$$|A_V(0)| f_C = \frac{Z_0}{4} n g_m \frac{1}{\pi Z_0 C} = \frac{n}{2} \frac{g_m}{2\pi C} = \frac{n}{2} f_T;$$

in other words, the distributed structure increases the gain–bandwidth product with respect to the single cell. In practice, the bandwidth obtained so far deteriorates further because of the effect of the input and output line RF losses. Also in the discrete cell case, there is an optimum cell number; in practice, FET-based distributed amplifiers rarely exceed 10 cells.

Owing to the need for broadbanding the available devices as much as possible, a quite popular configuration in the design of distributed amplifiers for optoelectronic applications is the *cascode* cell configuration. The cascode transistor configuration (Fig. 6.64) is, for a bipolar, a common emitter stage connected to a common base stage. The same configuration can be implemented in FETs with a common source and common gate stage. The cascode configuration decreases the internal feedback capacitance and therefore improves the device stability and bandwidth when compared to the conventional

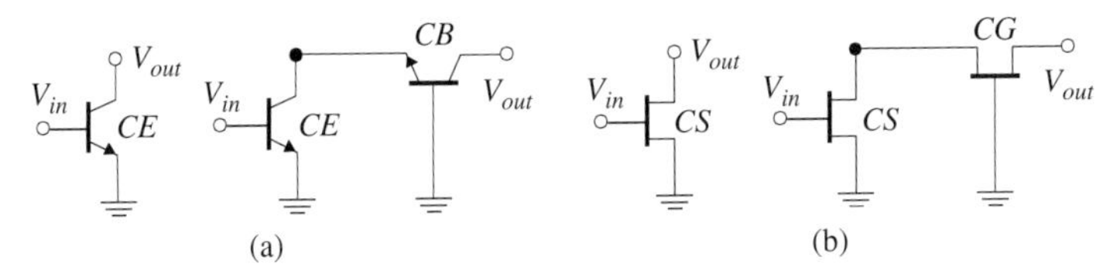

Figure 6.64 Cascode configuration of (a) bipolar transistors (right), compared to the conventional common-emitter (left) configuration; (b), same for FETs, where CE stands for common emitter, CS for common source, CB for common base, CG for common-gate.

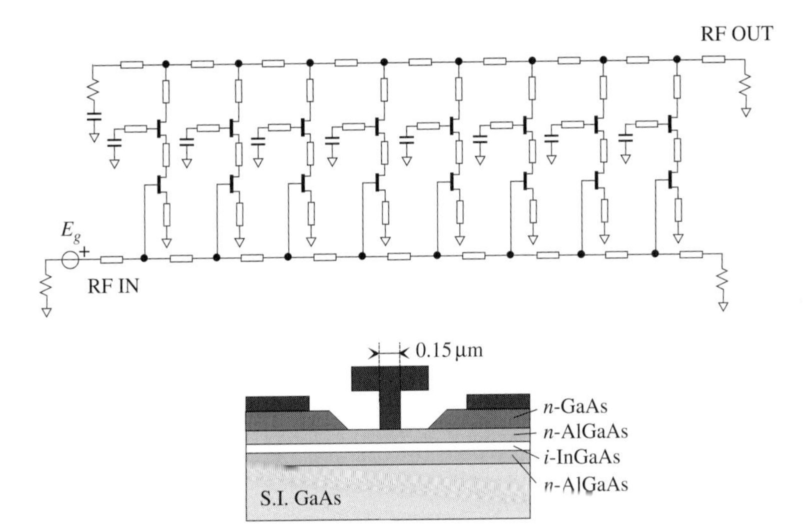

Figure 6.65 Eight-stage PHEMT based cascode distributed amplifier designed as a 40 Gbps modulator driver (above) and GaAs PHEMT cross section (below). From [114], Fig. 2 (adapted) and Fig. 7 (©2001 IEEE).

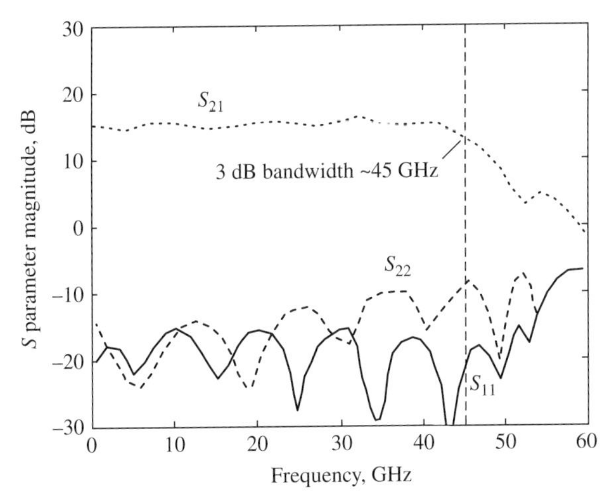

Figure 6.66 Scattering parameters of the 8-stage cascode amplifier with 6 V peak-to-peak output. From [114], Fig. 9 (©2001 IEEE).

stage. The two main consequences of the cascode configuration are the broadbanding and the more resonant response.

An example of high-speed MZ modulator driver cascode distributed amplifier for 40 Gbps systems developed by Fujitsu is shown in Fig. 6.65 [114], [115]. The technology is a 0.15 μm InGaAs/GaAs PHEMT; the amplifier is able to provide 6 V peak-to-peak output voltage with a gain around 14 dB; the 3 dB electrical bandwidth is 45 GHz. The electrical response is shown in Fig. 6.66.

In conclusion, the present transistor technology allows the development of high-voltage, broadband amplifiers to be exploited as LN modulator drivers; however, the distributed amplifier solution is expensive in terms of enabling technology, circuit cost, and power dissipation (ultrawideband amplifiers are typically class A amplifiers, with rather poor efficiency; moreover, closing the amplifier on a 50 Ω matched load, a 5 V peak-to-peak output with a square-wave behavior translates into a 250 mW average power, with a dissipated power of 2.5 W assuming overall 10% efficiency). The development of low-driving-voltage modulators is also fostered by the opportunity to reduce the requirements on this component or eliminate it altogether in favor of a direct digital driver.

6.13 Questions and problems

6.13.1 Questions

1. Explain the linear electrooptic effect in a crystal and how it can be described in a mathematical way.
2. What is the difference between an isotropic, a uniaxial and a biaxial crystal? To which class do GaAs and lithium niobate, respectively belong?
3. Explain the structure of a lumped electrooptic modulator and justify its bandwidth limitation.
4. Define the on–off voltage (V_π) in a Mach–Zehnder modulator. What happens to the on–off voltage if we increase the modulator length?
5. Discuss the structure of X-cut and Z-cut Mach–Zehnder modulators.
6. How can chirp be electronically controlled in a Mach–Zehnder modulator?
7. Explain what causes the bandwidth limitation in a traveling-wave Mach–Zehnder modulator.
8. Explain how the electrode structure of a lithium niobate electrooptic modulator can be modified in order to achieve synchronous coupling.
9. Explain why a slow-wave structure has to be exploited in the electrode design of a GaAs electrooptic modulator.
10. Quote some solutions for narrowband Mach–Zehnder modulators.
11. Explain why a distributed (traveling-wave) electrooptic modulator can overcome the RC-limited bandwidth–efficiency product of lumped EOMs.
12. Describe the limiting factors of the bandwidth–efficiency product in traveling-wave electrooptic modulators.

13. What differences in implementation are found between LN and semiconductor EOMs? Explain the need of slow-wave structures arising in semiconductor EOMs vs. LN EOMs.
14. Discuss the purpose of charge bleed layers in Z-cut LN modulators.
15. Comment on the possible advantages of polymer based and Si-based electrooptic modulators. What is the modulation mechanism in Si-based modulators?
16. Describe the process of electroabsorption in bulk semiconductors.
17. Describe the process of electroabsorption in quantum wells.
18. Compare performances of FKE- and QCSE-based modulators, highlighting the physical causes of their different behavior.
19. Explain the origin of polarization dependence in QW/MQW EAMs.
20. Describe qualitatively the effect of strain on band energies in a direct-bandgap semiconductor, and how strain can be employed to minimize polarization dependence in QCSE-based modulators.
21. Discuss the chirp effect in EAMs.
22. Explain the operation of a waveguide EAM.
23. Describe the effect of the optical detuning on the switching voltage (for a given CR) and residual transmission loss of a waveguide EAM.
24. Describe the effect of the waveguide length (L) on the switching voltage (for a given CR) and residual transmission loss of a waveguide EAM.
25. Discuss the bandwidth–efficiency trade-off in EAMs and the effect of the active region thickness and modulator length on these parameters.
26. Compare, in terms of bandwidth-efficiency product;
    (a) an open-loaded EAM;
    (b) an EAM with a resistive load equal to the driver resistance.
27. Illustrate the structure of a modulator driver, specifying the role of the multiplexer, the pre-driver, the driver, and the driver amplifier.
28. Explain why a distributed power amplifier has to be exploited to drive lithium niobate modulators at 40 Gbps, while EAMs can also be driven by a simplified stage.
29. Why are all logic signals in the driver stage typically represented through the original and negated versions of the bit (i.e., in a differential architecture)?
30. Describe the structure of a distributed amplifier and its frequency response.
31. Explain the condition of synchronism in a distributed amplifier.
32. In a discrete-cell distributed amplifier, what is the main limitation to increasing the number of cells?
33. Explain some advantages of a cascode cell in the design of distributed amplifiers.
34. Discuss the trade-off between maximum operating frequency and breakdown voltages in heterojunction bipolar transistors.
35. Explain why a matched back termination is exploited in connecting a modulator or laser to a current mode switching stage.
36. From the standpoint of DC bias, what is the basic difference between a modulator and a laser diode?

6.13.2 Problems

1. Suppose 40 Gbps operation has to be achieved in a *lossless* lithium niobate Mach–Zehnder modulator with length 15 mm. What is the allowed refraction index mismatch?
2. Suppose 40 Gbps operation has to be achieved in a *velocity-matched* lithium niobate Mach–Zehnder modulator with length 15 mm. What is the allowed line attenuation at 1 GHz?
3. A Z-cut MZ EO modulator is made with a material whose optical refraction index is $n_o = 2.2$; the electrooptic coefficient is $r_{33} = 30$ pm/V. Supposing that a lumped coplanar structure is exploited with superposition integral $\Gamma = 0.5$ and electrode gap $G = 10\,\mu$m, evaluate the modulator length L needed to obtain and ON–OFF voltage of 5 V. The operating wavelength is 1.3 μm.
4. In the above problem, the MZ modulator is implemented with a coplanar electrode structure of characteristic impedance $Z_0 = 30\,\Omega$ and RF refractive index $n_m = 3$. Assume that the modulator is connected to a 50 Ω generator and to a 30 Ω load.
   (a) Evaluate the modulation bandwidth (3 dB optical) for a modulator loaded as a lumped structure.
   (b) The same, but for a modulator loaded as a distributed structure.
5. A MZ EO modulator with symmetric optical splitter but asymmetric arms has $V_\pi =$ 4 V for the upper arm and $V_\pi = 8$ V for the lower arm.
   (a) Evaluate the total on–off voltage and the chirp parameter α_H. What happens if the two arms are exchanged? Assume the modulator is biased at half V_π.
   (b) Assuming that a triangular RF signal at 10 GHz of peak-to-peak amplitude equal to 1 V is applied to the modulator above the bias, estimate the resulting frequency and wavelength modulation of the modulated light. Assume a 1.5 μm source.
6. An EAM consists of a *pin* junction in reverse bias. The absorption (intrinsic, depleted) region thickness is $h = 0.5\,\mu$m; the absorption region width is $W = 3\,\mu$m; and the length L is to be evaluated. Suppose that the applied reverse voltage falls entirely across the depletion region, that the electric field is uniform, and that the superposition integral between the optical and RF fields is $\Gamma_o = 1$. Assume that the zero-field absorption of the material is $\alpha_0 = 10\,\text{cm}^{-1}$ and that α varies linearly with the applied field up to $\alpha_0 + \Delta\alpha = 110\,\text{cm}^{-1}$ for an electric field strength of 100 kV/cm.
   (a) Design L so that the ON–OFF voltage at 10 dB extinction is 2 V.
   (b) In these conditions, evaluate the 3dB bandwidth of the lumped modulator, assuming that the generator has 50 Ω internal impedance; assume that in the absorption region the RF permittivity is $\epsilon_r = 12$.
   (c) What is the ON–OFF generator voltage and the bandwidth if the EAM is also connected to a 50 Ω load?
7. Assume that, for a certain DC bias point, the active region absorption of an EAM varies by $\pm 20\,\text{cm}^{-1}$ for a variation in the electric field of $\pm$ 10 kV/cm. The optical refractive index at bias is $n_o = 3$.

(a) Supposing that the operating wavelength is 1.3 μm and that the chirp parameter is $\alpha_H = -1$, evaluate the variation of the optical index for an applied field of ± 1 kV/cm.

(b) Assume that the total modulator length is $L = 200$ μm; what will be the optical path difference between the bias point and an applied field of ± 1 kV/cm?

8. An EO modulator is biased at 3 V and has 6 V ON–OFF voltage. A logical driver is available with *low* logic level at −0.3 V and *high* logic level at 0.3 V. Evaluate the amount of level shift and voltage amplification needed to drive the modulator. Assuming that the driver amplifier output impedance is 50 Ω, estimate the minimum output available power needed.

Symbols

Notation

$x(t)$ scalar variable, time domain

$X(\omega)$ scalar variable, frequency domain

$\underline{x}(t)$ vector variable, time domain

$\underline{X}(\omega)$ vector variable, frequency domain

$\boldsymbol{x}(t)$ tensor (matrix) variable, time domain

$\boldsymbol{X}(\omega)$ tensor (matrix) variable, frequency domain

X_0 scalar variable, DC

X_{DC} scalar variable, DC

$\hat{x}(t)$ small-signal scalar variable, time domain

$\hat{X}(\omega)$ small-signal scalar variable, frequency domain

$\delta x(t)$ scalar variable fluctuation, time domain

$\delta X(\omega)$ scalar variable fluctuation, frequency domain

$x_n(t)$ random (noise) variable, time domain

$X_n(\omega)$ random (noise) variable, frequency domain

$\langle x(t) \rangle$ time average of deterministic function $x(t)$

$\langle x(t) \rangle$ time average of random process $x(t)$

$\overline{x(t)}$ ensemble average of random process $x(t)$

$\overline{x}$ ensemble average of random variable x

F_x Langevin random source in the rate equation for x

$S_x(\omega)$ power spectrum of x (also S_{xx})

$\overline{XX^*}$ power spectrum of x in terms of spectral average

$S_{xy}(\omega)$ correlation spectrum between x and y

$R_x(\tau)$ autocorrelation function of x (also R_{xx})

$R_{xy}(\tau)$ correlation function between x and y

∇ [m^{-1}] gradient operator

$\nabla \cdot$ [m^{-1}] divergence operator

Symbols

a [eV m^2] derivative of $E_F - E_c(0)$ vs. n_s (modulation-doped structure)

a bipolar transistor common base current gain

a	[m^2] laser differential gain
a	[m] lattice constant
a, b	[$\mathrm{W}^{1/2}$] forward and backward power waves
A_g	[A] short-circuit generator current
α	[m^{-1}] absorption
$\bar{\alpha}$	[m^{-1}] attenuation
α	[m^{-1}] attenuation
α	[J^{-1}] nonparabolic factor
α_c	[m^{-1}] conductor attenuation
α_d	[m^{-1}] dielectric attenuation
α_{fc}	[m^{-1}] absorption in cladding (electroabsorption modulators)
α_h	[m^{-1}] hole impact ionization coefficient
α_H	Henry chirp parameter (linewidth enhancement factor)
α_{loss}	[m^{-1}] absorption in cladding (lasers)
α_m	[m^{-1}] laser mirror (end) equivalent loss
α_n	[m^{-1}] electron impact ionization coefficient
α_t	[m^{-1}] laser total loss
b	bipolar transistor base transport factor
B	[Hz] bandwidth
B_r	[bps] bit rate
BER	bit error rate
β	[m^{-1}] propagation constant
β	bipolar transistor common-emitter current gain
β_m	[m^{-1}] RF propagation constant (modulators)
β_o	[m^{-1}] optical propagation constant (modulators)
β_k	spontaneous emission factor
c_0	[$\mathrm{m\,s}^{-1}$] speed of light in vacuo, $c_0 = 2.997\,924\,58 \times 10^8\ \mathrm{m\,s}^{-1}$
C_{ch}	[$\mathrm{F\,m}^{-2}$] channel capacitance per unit surface (FETs)
C_{eq}	[$\mathrm{F\,m}^{-2}$] equivalent 2DEG capacitance (HEMTs)
C_{GS}	[F] gate-source capacitance
C_j	[F] junction capacitance
$\mathcal{C}$	[$\mathrm{F\,m}^{-1}$] capacitance per unit length
$\mathcal{C}_a$	[$\mathrm{F\,m}^{-1}$] capacitance per unit length in air
D_h	[$\mathrm{m}^2\mathrm{s}^{-1}$] hole diffusivity
D_n	[$\mathrm{m}^2\mathrm{s}^{-1}$] electron diffusivity
δ	[m] skin penetration depth
$\overline{\delta}$	[rad] loss angle
δf	[Hz] laser frequency fluctuation
δn	[m^{-3}] laser population fluctuation
$\delta\phi$	[$\mathrm{rad\ s}^{-1}$] laser phase fluctuation
$\Delta\beta$	[m^{-1}] detuning vs. Bragg condition
$\Delta\beta$	[m^{-1}] detuning vs. velocity matching condition

ΔE_c [J] [eV] conduction band discontinuity
ΔE_v [J] [eV] valence band discontinuity
Δf_{ST} [Hz] Schawlow–Townes linewidth

$E(\underline{k})$ [J] [eV] dispersion relation
$\underline{E}(\omega)$ [V m^{-1}] electric field, frequency domain
E_A [J] [eV] acceptor energy level
E_c [J] [eV] conduction band edge
E_D [J] [eV] donor energy level
E_F [J] [eV] Fermi level
E_{Fh} [J] [eV] quasi-Fermi level, holes
E_{Fi} [J] [eV] intrinsic Fermi level
E_{Fn} [J] [eV] quasi-Fermi level, electrons
E_g [J] [eV] energy gap
E_g [V] generator open-circuit voltage
E_h [J] [eV] hole energy
E_n [J] [eV] electron energy
E_p [J] [eV] energy parameter associated with dipole matrix element
E_{ph} [J] [eV] photon energy
E_t [J] [eV] trap energy level
E_v [J] [eV] valence band edge
E_x [J] [eV] exciton energy level
ER modulator extinction ratio
$\underline{\mathcal{E}}$ [V m^{-1}] electric field
$\mathcal{E}_{br}$ [V m^{-1}] breakdown electric field
ϵ [F m^{-1}] dielectric permittivity
$\epsilon(\omega)$ [F m^{-1}] complex dielectric permittivity, frequency domain; $\epsilon = \epsilon'(\omega) - \mathrm{j}\epsilon''(\omega)$
ϵ [F m^{-1}] dielectric permittivity tensor
ϵ_c gain compression factor
ϵ_e [F m^{-1}] extraordinary permittivity
ϵ_{eff} effective permittivity
ϵ_o [F m^{-1}] ordinary permittivity
ϵ_r relative dielectric permittivity
ϵ_0 [F m^{-1}] vacuum dielectric permittivity, $\epsilon_0 = 8.854\,187\,817 \times 10^{-12}$ F m^{-1}
η_i injection LED quantum efficiency
η_Q detector internal quantum efficiency
η_t transmission LED quantum efficiency
η_r radiative LED quantum efficiency
η_x total LED quantum efficiency
η_x detector external quantum efficiency

f	[Hz] frequency
$f_h(E)$	occupation probability (Fermi or Boltzmann distribution), holes
$f_{\max}$	[Hz] maximum oscillation frequency
$f_n(E)$	occupation probability (Fermi or Boltzmann distribution), electrons
f_T	[Hz] cutoff frequency
f_{Tx}	[Hz] cutoff frequency, extrinsic
$f_{3\mathrm{dB}}$	[Hz] 3 dB cutoff frequency
$f_{3\mathrm{dBe}}$	[Hz] 3 dB cutoff frequency, electrical definition
$f_{3\mathrm{dBo}}$	[Hz] 3 dB cutoff frequency, optical definition
$f_{3\mathrm{dB},RC}$	[Hz] 3 dB cutoff frequency, RC limited
$f_{3\mathrm{dB},tr}$	[Hz] 3 dB cutoff frequency, transit-time limited
F	[Hz] phonon frequency
F_h	excess noise factor (SAM-APD, hole-triggered avalanche)
F_k	[$\mathrm{m}^{-3}\,\mathrm{s}^{-1}$] Langevin source in the photon density rate equation
F_n	excess noise factor (SAM-APD, electron triggered avalanche)
F_n	[$\mathrm{m}^{-3}\,\mathrm{s}^{-1}$] Langevin source in the electron density rate equation
F_o	excess noise factor (APD)
F_N	[s^{-1}] Langevin source in the electron number rate equation
F_P	[s^{-1}] Langevin source in the photon number rate equation
F_{OEIC}	[$\mathrm{A\,W^{-1}\,\Omega\,Hz}$] OEIC figure of merit
ϕ, φ	[rad] phase
ϕ	[V] electric potential
ϕ_{ch}	[V] channel potential
g	[m^{-1}] net gain
$\overline{g}$	[m^{-1}] gain
g_c	[m^{-1}] cavity gain
$g_c(E)$	[$\mathrm{J}^{-1}\,\mathrm{m}^{-3}$] conduction band density of states
g_F	[m^{-1}] gain, neglecting gain compression
g_m	[S] transconductance
g_n	[S] noise conductance (front-end amplifier)
$g_{ph}(\hbar\omega)$	[$\mathrm{J}^{-1}\,\mathrm{m}^{-3}$] photon density of states per unit energy and volume
g_{th}	[m^{-1}] laser cavity gain at threshold
$\mathrm{d}g/\mathrm{d}n$	[m^2] laser differential gain
$g_v(E)$	[$\mathrm{J}^{-1}\,\mathrm{m}^{-3}$] valence band density of states
$g_{1D}(E)$	[$\mathrm{J}^{-1}\,\mathrm{m}^{-3}$] density of states in a quantum wire
$g_{2D}(E)$	[$\mathrm{J}^{-1}\,\mathrm{m}^{-3}$] density of states in a quantum well
G_h	[$\mathrm{m}^{-3}\,\mathrm{s}^{-1}$] generation rate, holes
G_n	[$\mathrm{m}^{-3}\,\mathrm{s}^{-1}$] generation rate, electrons
G_n	[S] noise conductance
G_o	[$\mathrm{m}^{-3}\,\mathrm{s}^{-1}$] optical generation rate
$\mathcal{G}$	[$\mathrm{S\,m}^{-1}$] conductance per unit length
γ	[m^{-1}] complex propagation constant, $\gamma = \alpha + \mathrm{j}\beta$

γ	bipolar transistor emitter efficiency
γ	space-dependent noise source, Langevin approach (dimensions vary)
γ	[s^{-1}] small-signal laser damping factor
Γ	reflection coefficient
Γ_{ov}	overlap integral
Γ_{mo}	overlap integral between the microwave and optical fields
h	[J s] Planck constant, $h = 6.626\,0755 \times 10^{-34}$ J s
$\hbar$	[J s] rationalized Planck constant, $\hbar = 1.054\,572\,66 \times 10^{-34}$ J s
$\hbar\omega$	[J] [eV] photon energy
hf	[J] [eV] photon energy
i_d	[A] dark current
i_L	[A] photocurrent
i_{PD}	[A] total photodetector current
I	optical field intensity, normalized to the photon number
I_B	[A] base current
I_d	[A] DC dark current
I_C	[A] collector current
I_D	[A] drain current
I_{DSS}	[A] saturation drain current
I_E	[A] emitter current
I_G	[A] gate current
I_L	[A] DC photocurrent
I_{PD}	[A] total DC photodetector current
I_S	[A] source current
I_{ph}	[A] photocurrent (modulators)
I_{th}	[A] laser threshold current
I_0	[J] [eV] ionization
$\underline{J}_h$	[A m^{-2}] hole current density
$\underline{J}_{h,\mathrm{d}}$	[A m^{-2}] hole diffusion current density
$\underline{J}_{h,\mathrm{dr}}$	[A m^{-2}] hole drift current density
$\underline{J}_n$	[A m^{-2}] electron current density
$\underline{J}_{n,\mathrm{d}}$	[A m^{-2}] electron diffusion current density
$\underline{J}_{n,\mathrm{dr}}$	[A m^{-2}] electron drift current density
J_{th}	[A m^{-2}] laser threshold current density
$\underline{k}$	[m^{-1}] wavevector
$\underline{k}_T$	[m^{-1}] transverse wavevector
k_B	[J K^{-1}] Boltzmann constant, $k_B = 1.380\,6568 \times 10^{-23}$ J K^{-1}
k_{hn}	ratio between hole and electron ionization coefficients
k_{nh}	ratio between electron and hole ionization coefficients
$\underline{k}_{ph}$	[m^{-1}] photon wavevector

$\underline{k}_\phi$ [m^{-1}] phonon wavevector
κ [m^{-1}] Bragg grating coupling coefficient

L_α [m] absorption length
L_h [m] hole diffusion length
L_g [m] gate length
L_n [m] electron diffusion length
L_{op} modulator optical insertion loss
$\mathcal{L}$ [H m^{-1}] inductance per unit length
$\mathcal{L}_a$ [H m^{-1}] inductance per unit length in air
λ [m] wavelength
λ_0 [m] in vacuo wavelength
λ_g [m] guided wavelength
λ_B [m] Bragg wavelength
Λ [m] phonon wavelength

m_h^* [kg] hole effective mass
m_{hh}^* [kg] heavy hole effective mass
$m_{h,D}^*$ [kg] density of states hole effective mass
$m_{h,tr}^*$ [kg] transport hole effective mass
m_{lh}^* [kg] light hole effective mass
$m_{n,D}^*$ [kg] density of states electron effective mass
m_n^* [kg] electron effective mass
$m_{n,tr}^*$ [kg] transport electron effective mass
m_r^* [kg] joint density of states reduced mass
m_x^* [kg] exciton effective mass
m_0 [kg] electron mass, $m_0 = 9.109\,3897 \times 10^{-31}$ kg
$m(\omega)$ normalized laser or modulator frequency response $m(\omega) = |M(\omega)/M(0)|$
$M(\omega)$ [W V^{-1}] modulator frequency response
$M(\omega)$ [W A^{-1}] laser modulation frequency response
M_h hole multiplication factor (SAM-APD, hole triggered avalanche)
M_n electron multiplication factor (SAM-APD, electron triggered avalanche)
M_o multiplication factor (APD)
μ [H m^{-1}] magnetic permeability
μ_h [m^2 V^{-1} s^{-1}] hole mobility
μ_n [m^2 V^{-1} s^{-1}] electron mobility
μ_r relative magnetic permeability
μ_0 [H m^{-1}] vacuum magnetic permeability, $\mu_0 = 4\pi \times 10^{-7}$ H m^{-1}

n [m^{-3}] electron concentration
$n(\omega)$ complex refractive index, $n = n_r - jn_i$; sometimes $n \equiv n_{\text{eff}}$
n_e optical extraordinary index
n_{eff} effective refractive index
n_i [m^{-3}] intrinsic carrier concentration

n_m	RF refractive index (modulator)
n_o	optical refractive index (modulator)
n_{ph}	photon number
n_r	refractive index
$n_r(\omega)$	complex refractive index, $n_r = n_r' - \mathrm{j}n_r'' = n_{r1} - \mathrm{j}n_{r2}$; sometimes $n_r \equiv n_{\text{eff}}$
n_s	[C m^{-2}] QW carrier sheet concentration
n_{sp}	spontaneous emission factor
n_{th}	[m^{-3}] laser carrier concentration at threshold
$n_{th,2D}$	[m^{-2}] QW laser carrier sheet concentration at threshold
N_A	[m^{-3}] acceptor concentration
N_c	[m^{-3}] effective density of states, conduction band
$N_c(E)$	[J^{-1} m^{-3}] conduction band density of states
N_D	[m^{-3}] donor concentration
$N_v(E)$	[J^{-1} m^{-3}] valence band density of states
N_v	[m^{-3}] effective density of states, valence band
$N_{cv}(\hbar\omega)$	[J^{-1} m^{-3}] joint density of states
N_k	[m^{-3}] photon density in laser mode k
N_P	photon number in laser mode k
N	laser cavity carrier number
ω	[rad s^{-1}] angular frequency
ω_m	[rad s^{-1}] modulation frequency
ω_r	[rad s^{-1}] small-signal laser resonant angular frequency
p	[m^{-3}] hole concentration
$\underline{p}$	[kg m s^{-1}] momentum
$\underline{p}_{cv}$	[kg m s^{-1}] momentum matrix element in the dipole approximation
$\langle p_{cv}^2 \rangle$	[kg^2 m^2 s^{-2}] $\langle p_{cv}^2 \rangle = \frac{2}{3}\mathrm{p}_{cv}^2$ mean value of dipole matrix element squared, bulk
$p_{in}(t)$	[W] input (optical or electrical) power, time domain
$p_n(f)$	[W Hz^{-1}] noise available power spectral density
PRC	parameters of the Cappy noise FET model
P_{av}	[W] generator available power
$\widetilde{P}_{in}$	[W m^{-2}] input optical power density
P_k	[W] laser output power from mode k
P_{in}	[W] input (optical or electrical) power
P_{op}	[W] optical power or [W m^{-2}] optical power density
ψ, ϕ	electron or hole wavefunctions
q	[C] electron charge, $q = 1.602\,177\,33 \times 10^{-19}$ C
$q(t)$	[C] electric charge
$q_n(t)$	[C] electron charge
$q\chi$	[J] [eV] electron affinity

Q	resonator quality factor
Q_{ch}	[C m^{-2}] channel mobile charge per unit surface (FETs)
Q_n	[C] electron charge, DC
r_{ijk}	[m V^{-1}] linear electrooptic tensor
r_o^{sp}	[m^{-3} s^{-1} J^{-1}] spontaneous emission spectrum
$\overline{r}_o^{sp}$	[s^{-1}] spontaneous emission rate for a specific photon wavevector
$r_{o,\mathrm{D}}^{sp}$	[m^{-3} s^{-1} J^{-1}] spontaneous emission spectrum, degenerate semiconductor
$r_{o,\mathrm{ND}}^{sp}$	[m^{-3} s^{-1} J^{-1}] spontaneous emission spectrum, nondegenerate semiconductor
$\mathfrak{r}(\omega)$	normalized detector frequency response, $\mathfrak{r}(\omega) = \mathfrak{R}(\omega)/\mathfrak{R}(0)$
R	power reflectivity
R_g	[Ω] generator internal resistance
R_G	[Ω] generator internal resistance
R_S	[Ω] generator (source) internal resistance
RIN	laser relative intensity noise parameter
R_h	[m^{-3} s^{-1}] recombination rate, holes
R_i	[Ω] input resistance, front-end
R_n	[m^{-3} s^{-1}] recombination rate, electrons
R_n	[Ω] noise resistance
R_o^{sp}	[m^{-3} s^{-1}] total spontaneous emission rate per unit volume
$R_{o,\mathrm{D}}^{sp}$	[m^{-3} s^{-1}] total spontaneous emission rate per unit volume, degenerate
$R_{o,\mathrm{D}}^{sp}$	[m^{-3} s^{-1}] total spontaneous emission rate per unit volume
$\overline{R}_o^{sp}$	[m^{-3} s^{-1}] spontaneous recombination rate per unit volume, specific photon state
R_o^{st}	[m^{-3} s^{-1}] stimulated emission rate per unit volume
$R_{o,\mathrm{D}}^{st}$	[m^{-3} s^{-1}] stimulated emission rate per unit volume, degenerate
$R_{o,\mathrm{ND}}^{st}$	[m^{-3} s^{-1}] stimulated emission rate per unit volume, nondegenerate
$\mathcal{R}$	[Ω m^{-1}] resistance per unit length
$\mathfrak{R}$	[A W^{-1}] detector responsivity
$\mathfrak{R}(\omega)$	[A W^{-1}] small-signal detector responsivity, frequency domain
ρ	[C m^{-3}] charge density
ρ_{ph}	[m^{-3}] photon density
$\underline{S}$	[W m^{-2}] Poynting vector
$\mathbf{S}$	scattering matrix
SNR	signal-to-noise ratio
$S_i(\omega)$	[A^2 Hz^{-1}] current (i) power spectrum
$S_{i_1 i_2}(\omega)$	[A^2 Hz^{-1}] correlation spectrum between i_1 and i_2
$S_P(\omega)$	[W^2 Hz^{-1}] power spectrum of laser output power fluctuations
$S_v(\omega)$	[V^2 Hz^{-1}] voltage (v) power spectrum
$S_{v_1 v_2}(\omega)$	[A^2 Hz^{-1}] correlation spectrum between v_1 and v_2
σ	[S m^{-1}] conductivity

T	[K] absolute temperature
$T(V_{in})$	modulator transfer curve
τ_h	[s] hole lifetime
τ_{loss}	[s] external (cladding) loss lifetime
τ_m	[s] mirror (end) loss photon lifetime
τ_n	[s] electron lifetime
τ_n^{sp}	[s] electron spontaneous radiative lifetime
$\overline{\tau}_n^{sp}$	[s] electron spontaneous radiative lifetime, single photon state
$\tau_{n,\mathrm{D}}^{sp}$	[s] electron spontaneous radiative lifetime, degenerate
$\tau_{n,\mathrm{ND}}^{sp}$	[s] electron spontaneous radiative lifetime, nondegenerate
$\tau_{n,nr}$	[s] electron nonradiative lifetime, LED
τ_{nr}	[s] carrier nonradiative lifetime in laser cavity
$\tau_{n,r}$	[s] electron radiative lifetime, LED
$\tau_{n,th}^{sp}$	[s] spontaneous carrier lifetime at laser threshold
τ_n^{st}	[s] electron stimulated radiative lifetime
$\tau_{n\,\mathrm{min}}^{st}$	[s] limit value of electron stimulated radiative lifetime
$\tau_{n,th}^{st}$	[s] stimulated carrier lifetime at laser threshold
τ_{ph}	[s] photon lifetime
τ_t	[s] transit time (detectors, modulators)
τ_0	[s] spontaneous radiative lifetime (limit value)
U_h	[$\mathrm{m}^{-3}\,\mathrm{s}^{-1}$] net recombination rate, holes
U_n	[$\mathrm{m}^{-3}\,\mathrm{s}^{-1}$] net recombination rate, electrons
U^{SRH}	[$\mathrm{m}^{-3}\,\mathrm{s}^{-1}$] Shockley–Read–Hall trap-assisted recombination rate
U_0	[J] [eV] vacuum level
v_f	[$\mathrm{m\,s}^{-1}$] phase velocity
v_g	[$\mathrm{m\,s}^{-1}$] group velocity
v_h	[$\mathrm{m\,s}^{-1}$] hole drift velocity
$v_{h,\mathrm{sat}}$	[$\mathrm{m\,s}^{-1}$] hole saturation velocity
v_m	[$\mathrm{m\,s}^{-1}$] RF velocity (modulators)
v_n	[$\mathrm{m\,s}^{-1}$] electron drift velocity
$v_{n,\mathrm{sat}}$	[$\mathrm{m\,s}^{-1}$] electron saturation velocity
v_o	[$\mathrm{m\,s}^{-1}$] optical velocity (modulators)
v_{PD}	[V] photodetector voltage
V	[m^3] crystal volume, for normalization
V	[m^3] laser photon volume
V_{ac}	[m^3] laser active region volume
V_{br}	[V] breakdown voltage
V_{PD}	[V] total DC photodetector voltage
V_{SW}	[V] modulator switching voltage
V_T	[V] thermal voltage
V_{TH}	[V] threshold voltage
V_π	[V] modulator OFF voltage

w_{abs} [s^{-1}] absorption scattering rate
w_{em} [s^{-1}] emission scattering rate
w_{em}^{sp} [s^{-1}] spontaneous emission scattering rate
w_{em}^{st} [s^{-1}] stimulated emission scattering rate
W_{abs} [m] absorption region width
$W_{abs}(\hbar\omega)$ [m^{-3} s^{-1}] total absorption scattering rate per unit volume
W_{av} [m] avalanche region width
$W_{em}(\hbar\omega)$ [m^{-3} s^{-1}] total emission scattering rate per unit volume
$W_{em}^{sp}(\hbar\omega)$ [m^{-3} s^{-1}] total spontaneous emission scattering rate per unit volume
$W_{em}^{st}(\hbar\omega)$ [m^{-3} s^{-1}] total stimulated emission scattering rate per unit volume
W_k [J] energy in laser mode k
$\mathcal{W}_{abs}(\hbar\omega)$ [s^{-1}] total absorption scattering rate
$\mathcal{W}_{em}(\hbar\omega)$ [s^{-1}] total emission scattering rate
$\mathcal{W}_{em}^{sp}(\hbar\omega)$ [s^{-1}] total spontaneous emission scattering rate
$\mathcal{W}_{em}^{st}(\hbar\omega)$ [s^{-1}] total stimulated emission scattering rate

$\mathbf{Y}$ [S] admittance matrix

$\mathbf{Z}$ [Ω] impedance matrix
Z_g [Ω] generator internal impedance
Z_G [Ω] generator internal impedance
Z_i [Ω] input impedance, front-end
Z_{in} [Ω] input impedance
Z_L [Ω] load internal impedance
Z_m [Ω] TIA transimpedance
Z_{out} [Ω] output impedance
$Z_s(\omega)$ [Ω] $Z_s = R_s + jX_s$ surface impedance (resistance, reactance)
Z_S [Ω] generator (source) internal impedance
Z_0 [Ω] characteristic impedance
ζ normalized impedance

References

1. Ioffe Institute of the Russian Academy of Sciences web site on semiconductors, www.ioffe.ru/SVA/NSM/Semicond/.
2. C. Kittel, *Introduction to Solid-State Physics*, 7th edn. (Wiley, 1996).
3. M. Balkanski and R. F. Wallis, *Semiconductor Physics and Applications* (Oxford University Press, 2000).
4. J. Singh, *Semiconductor Optoelectronics, Physics and Technology* (McGraw-Hill, 1995).
5. M. Farahmand, C. Garetto, E. Bellotti, K. F. Brennan, M. Goano, E. Ghillino, G. Ghione, J. D. Albrecht, and P. P. Ruden, Monte Carlo simulation of electron transport in the III-nitride wurtzite phase materials system: binaries and ternaries. *IEEE Transactions on Electron Devices*, **48**:3 (2001), 535–542.
6. F. Schwierz, Wide bandgap and other non-III-V RF transistors: trends and prospects, *CSSER 2004 Spring Lecture Series*, ASU Tempe, 25 March 2004, www.eas.asu.edu/~vasilesk/EEE532/Talk_ASU_short.ppt.
7. T. P. Pearsall, *GaInAsP Alloy Semiconductors* (Wiley, 1982).
8. I. Vurgaftman and J. R. Meyer, Band parameters for nitrogen-containing semiconductors. *Journal of Applied Physics*, **94**:6 (2003), 3675–3696.
9. F. M. Abou El-Ela, I. M. Hamada, Electron and hole impact ionization coefficients at very high electric fields in semiconductors. *Fizika A*, **13**:3 (2004), 89–104.
10. J. D. Jackson, *Classical Electrodynamics* (Wiley, 1962).
11. S. L. Chuang, *Physics of Optoelectronic Devices* (Wiley, 1995).
12. E. Rosencher and B. Vinter, *Optoélectronique* (Dunod, 2002) [English translation: *Optoelectronics* (Cambridge University Press, 2002)].
13. W. Schäfer and M. Wegener, *Semiconductor Optics and Transport Phenomena* (Springer, 2002).
14. G. G. Macfarlane, T. P. McLean, J. E. Quarrington, and V. Roberts, Fine structure in the absorption-edge spectrum of Si. *Physical Review*, **111** (1958), 1245–1254.
15. M. D. Sturge, Optical absorption of gallium arsenide between 0.6 and 2.75 eV. *Physical Review*, **127** (1962), 768–773.
16. G. Livescu, D. A. B. Miller, D. S. Chemla, M. Ramaswamy, T. Y. Chang, N. Sauer, A. C. Gossard, and J. H. English, Free carrier and many-body effects in absorption spectra of modulation-doped quantum wells. *IEEE Journal of Quantum Electronics*, **24**:8 (1988), 1677–1689.
17. S. Gupta, P. K. Bhattacharya, J. Pamulapati, and G. Mourou, Optical properties of high-quality InGaAs/InAlAs multiple quantum wells. *Journal of Applied Physics*, **69**:5 (1991), 3219–3225.
18. G. P. Agrawal and N. K. Dutta, *Semiconductor Lasers* (Van Nostrand, 1993).

19. W. van Roosbroeck and W. Shockley, Photon-radiative recombination of electrons and holes in germanium. *Physical Review*, **94** (1954), 1558–1560.
20. N. G. Nilsson, An accurate approximation of the generalized Einstein relation for degenerate semiconductors. *Physica Status Solidi (a)*, **19**:1 (1973), K75–K78.
21. K. C. Gupta, R. Garg, I. Bahl, and P. Bhartia, *Microstrip Lines and Slotlines*, 2nd edn. (Artech House, 1996).
22. T. C. Edwards, *Foundations for Microstrip Circuit Design*, 2nd edn. (Wiley, 1991).
23. I. Wolff, *Coplanar Microwave Integrated Circuits* (Wiley, 2006).
24. W. Hilberg, From approximations to exact relations for characteristic impedances. *IEEE Transactions on Microwave Theory and Techniques*, **17**:5 (1969), 259–265.
25. G. Gonzalez, *Microwave Transistor Amplifiers: Analysis and Design*, 2nd edn. (Prentice Hall, 1996).
26. T. A. Winslow, Conical inductors for broadband applications. *IEEE Microwave Magazine*, **6**:1 (2005), 68–72.
27. A. R. Barnes, A. Boetti, L. Marchand, and J. Hopkins, An overview of microwave component requirements for future space applications. *Proceedings of 13th GAAS Symposium*, (2005), 5–12.
28. F. Schwierz and J. J. Liou, *Modern Microwave Transistors: Theory, Design, and Performance* (Wiley, 2002).
29. Kei May Lau, Chak Wah Tang, Haiou Li, and Zhenyu Zhong, AlInAs/GaInAs mHEMTs on silicon substrates grown by MOCVD. *Proceedings of IEDM 2008*, (2008), 723–726.
30. H. P. D. Lanyon and R. A. Tuft, Bandgap narrowing in moderately to heavily doped silicon. *IEEE Transactions on Electron Devices*, **26**:7 (2979), 1014–1018.
31. S. M. Sze and K. K. Ng, *Physics of Semiconductor Devices*, 3rd edn. (Wiley, 2006).
32. C-H. Huang, T-L. Lee, and H H. Lin, Relation between the collector current and the two-dimensional electron gas stored in the base-collector heterojunction notch of InAlAs/InGaAs/InAlGaAs DHBTs. *Solid-State Electronics*, **38**:10 (1995), 1765–1770.
33. T. Henderson, J. Middleton, J. Mahoney, S. Varma, T. Rivers, C. Jordan, and B. Avrit, High-performance BiHEMT HBT/E-D pHEMT integration, *CS MANTECH Conference*, May 14–17, 2007, Austin, Texas, USA, 247–250.
34. A. Papoulis and S. Pillai, *Probability, Random Variables and Stochastic Processes*, 4th revised edn. (McGraw-Hill, 2002).
35. A. Cappy, Noise modelling and measurement techniques. *IEEE Transactions on Microwave Theory and Techniques*, **36** (1988), 1–10.
36. F. Bonani and G. Ghione, *Noise in Semiconductor Devices: Modeling and Simulation* (Springer, 2001).
37. P. Bhattacharya, *Semiconductor Optoelectronic Devices*, 2nd edn. (Prentice Hall, 1997).
38. J. Michel, J. F. Liu, W. Giziewicz, D. Pan, K. Wada, D. D. Cannon, S. Jongthammanurak, D. T. Danielson, L. C. Kimerling, J. Chen, F. O. Ilday, F. X. Kartner, and J. Yasaitis, High performance Ge p-i-n photodetectors on Si, *2nd IEEE International Conference on Group IV Photonics*, Sept. 21-23 2005, 177–179.
39. Hamamatsu Photonics web site www.hamamatsu.com/.
40. K. Kato, Ultrawide-band/high-frequency photodetectors. *IEEE Transactions on Microwave Theory and Techniques*, **47** (1999), 1265–1281.
41. H. Fukano and Y. Matsuoka, A low-cost edge-illuminated refracting-facet photodiode module with large bandwidth and high responsivity. *Journal of Lightwave Technology*, **18** (2000), 79–83.

42. K. S. Giboney, M. J. W. Rodwell, and J. E. Bowers, Traveling-wave photodetectors. *IEEE Photonics Technology Letters*, **4** (1992), 1363–1365.
43. T. Chau, L. Fan, D. Tong, S. Mathai, and M. C. Wu, Long wavelength velocity-matched distributed photodetectors. *IEEE Conference on Lasers and Electro-Optics*, CLEO'98, San Francisco, CA, (1998), paper CThM3.
44. N. Shimizu, N. Watanabe, T. Furuta, and T. Ishibashi, InP-InGaAs unitraveling-carrier photodiode with improved 3-dB bandwidth of over 150 GHz. *IEEE Photonics Technology Letters*, **10**:3 (1998), 412–414.
45. H. Ito, T. Furuta, S. Kodama, and T. Ishibashi, InP/lnGaAs uni-travelling-carrier photodiode with 310 GHz bandwidth. *Electronics Letters*, **36**:21 (2000), 1809–1810.
46. R. J. McIntyre, Multiplication noise in uniform avalanche diodes. *IEEE Transactions on Electron Devices*, **13**:1 (1966), 164–168.
47. M. A. Saleh, M. M. Hayat, B. E. A. Saleh, and M. C. Teich, Dead-space-based theory correctly predicts excess noise factor for thin GaAs and AlGaAs avalanche photodiodes. *IEEE Transactions on Electron Devices*, **47**:3 (2000), 625–633.
48. R. B. Emmons and G. Lucovsky, The frequency response of avalanching photodiodes. *IEEE Transactions on Electron Devices*, **13**:3 (1966), 297–305.
49. R. B. Emmons, Avalanche-photodiode frequency response. *Journal of Applied Physics*, **38**:9 (1967), 3705–3714.
50. J. J. Chang, Frequency response of PIN avalanching photodiodes. *IEEE Transactions on Electron Devices*, **14**:3 (1967), 139–145.
51. H. Nie, K. A. Anselm, C. Hu, S. S. Murtaza, B. G. Streetman, and J. C. Campbell, High-speed resonant-cavity separate absorption and multiplication avalanche photodiodes with 130 GHz gain-bandwidth product. *Applied Physics Letters*, **70** (1997), 161–163.
52. H. Nie, K. A. Anselm, C. Lenox, P. Yuan, C. Hu, G. Kinsey, B. G. Streetman, and J. C. Campbell, Resonant-cavity separate absorption, charge and multiplication avalanche photodiodes with high-speed and high gain-bandwidth product. *IEEE Photonics Technology Letters*, **10**:3 (1998), 409–411.
53. G. S. Kinsey, J. C. Campbell, and A. G. Dentai, Waveguide avalanche photodiode operating at 1.55 μm with a gain-bandwidth product of 320 GHz. *IEEE Photonics Technology Letters*, **13**:8 (2001), 842–844.
54. T. Torikai, T. Nakata, T. Kato, and V. Makita, 40-Gbps waveguide avalanche photodiodes, *Optical Fiber Communication Conference Technical Digest*, **5** (2005), paper OFM3, 1–3.
55. A. Yariv and P. Yeh, *Photonics: Optical Electronics in Modern Communication*, 6th edn. (Oxford University Press, 2007).
56. D. Huber, R. Bauknecht, C. Bergamaschi, M. Bitter, A. Huber, T. Morf, A. Neiger, M. Rohner, I. Schnyder, V. Schwarz, and H. Jäckel, InP–InGaAs single HBT technology for photoreceiver OEIC's at 40 Gb/s and beyond. *Journal of Lightwave Technology*, **18**:7 (2000), 992–1000.
57. R. H. Walden, A review of recent progresses in InP-based optoelectronic integrated circuit receiver front-ends. *International Journal of High-Speed Electronics and Systems*, **9**:2 (1998), 631–642; also in *High-Speed Circuits for Lightwave Communications*, edn. Keh-Chung Wang (World Scientific, 1999).
58. K. Takahata, Y. Muramoto, H. Fukano, K. Kato, A. Kozen, S. Kimura, Y. Imai, Y. Miyamoto, O. Nakajima, and Y. Matsuoka, Ultrafast monolithic receiver OEIC composed of multimode waveguide p-i-n photodiode and HEMT distributed amplifier. *IEEE Journal of Selected Topics in Quantum Electronics*, **6**:1 (2000), 31–37.

59. Y. Zhang, C. S. Whelan, R. Leoni, P. F. Marsh, W. E. Hoke, J. B. Hunt, C. M. Laighton, and T. E. Kazior, 40-Gbit/s OEIC on GaAs substrate through metamorphic buffer technology. *IEEE Electron Device Letters*, **24**:9 (2003), 529–531.
60. E. Sackinger, *Broadband Circuits for Optical Fiber Communications* (Wiley, 2005).
61. T. Pinguet, B. Analui, G. Masini, V. Sadagopan, and S. Gloeckner, 40-Gbps monolithically integrated transceivers in CMOS photonics. *Silicon Photonics III, Proceedings of the SPIE*, 6898 (2008) 5–14.
62. A. Umbach, T. Engel, H-G. Bach, S. van Waasen, E. Dröge, A. Strittmatter, W. Ebert, W. Passenberg, R. Steingrüber, W. Schlaak, G. G. Mekonnen, G. Unterbörsch, and D. Bimberg, Technology of InP-based 1.55-μm ultrafast OEMMIC's: 40-Gbit/s broad-band and 38/60-GHz narrow-band photoreceivers. *IEEE Journal of Quantum Electronics*, **35**:7 (1999), 1024–1031.
63. C. Masini, L. Colace, G. Assanto, H.-C. Luan, and L. C. Kimerling, p-i-n Ge on Si photodetectors for the near infrared: from model to demonstration, *IEEE Transactions on Electron Devices*. **48**:6 (2001), 1092–1096.
64. G. Masini, S. Sahni, G. Capellini, J. Witzens, and C. Gunn, High-speed near infrared optical receivers based on Ge waveguide photodetectors integrated in a CMOS process. In *Advances in Optical Technologies* (Hindawi Publishing Corporation, 2008), Article ID 196572 (5 pages). doi: 10.1155/2008/196572.
65. H. Zimmermann and T. Heide, A monolithically integrated 1-Gb/s optical receiver in 1-μm CMOS technology. *IEEE Photonics Technology Letters*, **13**:7 (2001), 711–713.
66. R. Swoboda and H. Zimmermann, 2.5 Gbit/s silicon receiver OEIC with large diameter photodiode. *Electronics Letters*, **40**:8 (2004), 505–507.
67. G. Masini, G. Capellini, J. Witzens, and C. Gunn, A 1550 nm 10nGbps monolithic optical receiver in 130 nm CMOS with integrated Ge waveguide photodetector, *4th IEEE International Conference on Group IV Photonics*, (2007), 1–3.
68. V. Rajamani and P. Chakrabarti, A proposed ultra low-noise optical receiver for 1.55 μm applications. *Optical and Quantum Electronics*, **35** (2003), 195–209.
69. P. Chakrabarti, P. Kalra, S. Agrawal, G. Gupta, and N. Menon, Design and analysis of a single HBT-based optical receiver front-end. *Solid State Electronics*, **49**:8 (2005), 1396–1404.
70. C. A. Burrus and B. I. Miller, Small-area, double-heterostructure aluminum-gallium arsenide electroluminescent diode sources for optical-fiber transmission lines. *Optics Communications*, **4**:4 (1971), 307–309.
71. Tien-Pei Lee, C. Burrus, and B. Miller, A stripe-geometry double-heterostructure amplified-spontaneous-emission (superluminescent) diode. *IEEE Journal of Quantum Electronics*, **9**:8 (1973), 820–828.
72. M. Kicherer, A. Fiore, U. Oesterle, R. P. Stanley, M. Ilegems, and R. Michalzik, Data transmission using GaAs-based InAs-InGaAs quantum dot LEDs emitting at 1.3 μm wavelength. *Electronics Letters*, **38**:16 (2002), 906–907.
73. C. Henry, L. Johnson, R. Logan, and D. Clarke, Determination of the refractive index of InGaAsP epitaxial layers by mode line luminescence spectroscopy. *IEEE Journal of Quantum Electronics*, **21**:12 (1985) 1887–1892.
74. D. Wood, *Optoelectronic Semiconductor Devices* (Prentice-Hall, 1994).
75. Haisheng Rong, Ansheng Liu, R. Jones, O. Cohen, D. Hak, R. Nicolaescu, A. Fang, and M. Paniccia, An all-silicon Raman laser. *Nature*, **433** (2005), 292–294.
76. Chung-En Zah, R. Bhat, B. N. Pathak, F. Favire, Wei Lin, M. C. Wang, N. C. Andreadakis, D. M. Hwang, M. A. Koza, Tien-Pei Lee, Zheng Wang, D. Darby, D. Flanders, and

J. J. Hsieh, High-performance uncooled 1.3-μm $Al_xGa_yIn_{1-y-x}As/InP$ strained-layer quantum-well lasers for subscriber loop applications. *IEEE Journal of Quantum Electronics*, **30**:2 (1994), 511–523.

77. W. F. Brinkman, T. L Koch, D. V Lang, and D. P. Wilt, The lasers behind the communications revolution. *Bell Labs Technical Journal*, **5**:1 (2000), 150–167.
78. D. Bimberg, M. Grundmann, and N. N. Ledentsov, *Quantum Dot Heterostructures* (Wiley, 1999).
79. M. Sugawara, N. Hatori, M. Ishida, H. Ebe, Y. Arakawa, T. Akiyama, K. Otsubo, T. Yamamoto and Y. Nakata, Recent progress in self-assembled quantum-dot optical devices for optical telecommunication: temperature-insensitive 10 Gb s^{-1} directly modulated lasers and 40 Gb s^{-1} signal-regenerative amplifiers. *Journal of Physics D: Applied Physics,* **38**:13 (2005), 2126–2134.
80. C. H. Henry, Theory of linewidth of semiconductor lasers. *IEEE Journal of Quantum Electronics*, **18** (1982), 259–264.
81. C. H. Henry, Theory of the phase noise and power spectrum of a single-mode injection laser. *IEEE Journal of Quantum Electronics*, **19** (1983), 1391–1397.
82. K. Petermann, *Laser Diode Modulation and Noise* (Kluwer, 1988).
83. L. A. Coldren and S. W. Corzine, *Diode Lasers and Photonic Integrated Circuits* (Wiley, 1995).
84. N. K. Dutta, N. A. Ollson, L. A. Koszi, P. Besomi, R. B. Wilson, and R. J. Nelson, Frequency chirp under current modulation in InGaAsP injection lasers. *Journal of Applied Physics*, **56**:7 (1984), 2167–2169.
85. B. Mason, G. A. Fish, S. P. DenBaars, and L. A. Coldren, Widely tunable sampled grating DBR laser with integrated electroabsorption modulator. *IEEE Photonics Technology Letters*, **11**:6 (1999), 638–640.
86. E. Yamashita, K. Atsuki, and T. Mori, Application of MIC formulas to a class of integrated-optics modulator analyses: a simple transformation. *IEEE Transactions on Microwave Theory and Techniques*, **25**:2 (1977), 146–150.
87. R. Spickermann, M. G. Peters, and R. Dagli, A polarization independent GaAs-AlGaAs electrooptic modulator. *IEEE Journal on Quantum Electronics*, **32**:5 (1996) 764–769.
88. S. Y. Wang and S. H. Lin, High speed III-V electrooptic waveguide modulators at $\lambda \approx 1.3$ μm. *Journal of Lightwave Technology*, **6**:6 (1988), 758–771.
89. W. Wang, Y. Shi, D. J. Olson, W. Lin, and J. Bechtel, Push–pull poled polymer Mach–Zehnder modulators with a single microstrip line electrode. *IEEE Photonics Technology Letters*, **11** (1999), 51–53.
90. M. Pirola, F. Cappelluti, G. Giarola, and G. Ghione, Multi-sectional modeling of high-speed electro-optic modulators integrated in a microwave circuit CAD environment. *Journal of Lightwave Technology*, **21**:12 (2003), 2989–2996.
91. R. Spickermann, S. R. Sakamono, and N. Dagli, In traveling wave modulators which velocity to match? *LEOS Annual Meeting*, (1996), 2:97–98.
92. D. Erasme, D. A. Humphreys, A. G. Roddie, and M. G. F. Wilson, Design and performance of phase reversal traveling wave modulators. *Journal of Lightwave Technology*, **6**:6 (1988), 933–936.
93. R. Madabhushi, Microwave attenuation reduction techniques for wide-band Ti:$LiNbO_3$ optical modulators. *IEICE Transactions on Electronics*, **E81-C**:8 (1998), 1321–1327.

94. E. L. Wooten, K. M. Kissa, A. Yi-Yan, E. J. Murphy, D. A. Lafaw, P. F. Hallemeier, D. Maack, D. V. Attanasio, D. J. Fritz, G. J. McBrien, and D. E. Bossi, A review of lithium niobate modulators for fiber-optic communications systems. *IEEE Journal of Selected Topics in Quantum Electronics*, **6**:1 (2000), 69–82.
95. O. Mitomi, K. Noguchi, and H. Miyazawa, Broadband and low driving-voltage $LiNbO_3$ optical modulators. *IEE Proceedings-J*, **145**:6 (1998), 360–364.
96. R. G. Walker, High-speed III-V semiconductor intensity modulators. *IEEE Journal on Quantum Electronics*, **27**:3 (1991), 654–667.
97. R. A. Griffin, R. G. Walker, R. I. Johnston, R. Harris, N. M. B. Perney, N. D. Whitbread, T. Widdowson, and P. Harper, Integrated 10 Gbit/s chirped return-to-zero transmitter using GaAs/AlGaAs modulators. *Optical Fiber Communication Conference and Exhibit, 2001*, **4** (2001), PD15-1–PD15-3.
98. C. Rolland, InGaAsP-based Mach–Zehnder modulators for high-speed transmission systems. *OFC '98 Technical Digest* (1998), 283–284.
99. B. Jalali and S. Fathpour, Silicon photonics. *Journal of Lightwave Technology*, **24**:12 (2006), 4600–4615.
100. L. Liao, D. Samara-Rubio, M. Morse, A. Liu, H. Hodge, D. Rubin, U. D. Keil, T. Franck, High-speed silicon Mach–Zehnder modulator. *Optics Express*, **13**:8 (2005), 3129–3135.
101. L. Liao, A. Liu, J. Basak, H. Nguyen, M. Paniccia, D. Rubin, Y. Chetrit, R. Cohen, and N. Izhaky, 40 Gbit/s silicon optical modulator for highspeed applications. *Electronics Letters*, **43**:22 (2007), Oct. 25 2007.
102. L. Gu, W. Jiang, X. Chen, and R. T. Chen, Silicon photonic crystal waveguide modulators. *3rd IEEE International Conference on Group IV Photonics*, (2006), 43–45.
103. D. A. B. Miller, D. S. Chemla, T. C. Damen, A. C. Gossard, W. Wiegmann, T. H. Wood, and C. A. Burrus, Band edge electroabsorption in quantum well structures: the quantum-confined Stark effect. *Physical Review Letters* **53** (1984), 2173–2176.
104. S. Nojima, Electric field dependence of the exciton binding energy in $GaAs/Al_xGa_{1-x}As$ quantum wells. *Physical Review B* **37** (1988), 9087–9088.
105. J. S. Weiner, D. A. B. Miller, D. S. Chemla, T. C. Damen, C. A. Burrus, T. H. Wood, A. C. Gossard, and W. Wiegmann, Strong polarization-sensitive electroabsorption in GaAs/AlGaAs quantum well waveguides. *Applied Physics Letters*, **47**:11 (1985), 1148–1150.
106. T. H. Wood, C. A. Burrus, D. A. B. Miller, D. S. Chemla, T. C. Damen, A. C. Gossard, W. Wiegmann, High-speed optical modulation with GaAs/GaAlAs quantum wells in a p-i-n diode structure. *International Electron Devices Meeting*, **29** (1983), 486–488.
107. F. Dorgeuille and F. Devaux, On the transmission performances and the chirp parameter of a multiple-quantum-well electroabsorption modulator. *IEEE Journal of Quantum Electronics*, **30**:11 (1994), 2565–2572.
108. T. Ido, S. Tanaka, M. Suzuki, M. Koizumi, H. Sano, and H. Inoue, Ultra-high-speed multiple-quantum-well electro-absorption optical modulators with integrated waveguides. *Journal of Lightwave Technology*, **14**:9 (1996), 2026–2034.
109. G. L. Li, S. A. Pappert, P. Mages, C. K. Sun, W. S. C. Chang, and P. K. L. Yu, High-saturation high-speed traveling-wave InGaAsP-InP electroabsorption modulator. *IEEE Photonics Technology Letters*, **13**:10 (2001), 1076–1078.
110. G. L. Li, S. A. Pappert, C. K. Sun, W. S. C. Chang, P. K. L. Yu, Wide bandwidth traveling-wave InGaAsP/InP electroabsorption modulator for millimeter wave applications. *2001 IEEE MTT-S International Microwave Symposium Digest*, **1** (2001) 61–64.

111. M. Meliga, R. Paoletti, and C. Coriasso, Uncooled laser sources for plug and play transceivers for datacom and telecom applications. *SPIE Proceedings* **6020** (2005), 382–391.
112. Y. Akage, K. Kawano, S. Oku, R. Iga, H. Okamoto, Y. Miyamoto, and H. Takeuchi, Wide bandwidth of over 50 GHz travelling-wave electrode electroabsorption modulator integrated DFB lasers. *Electronics Letters*, **37**:5 (2001), 299–300.
113. M. Cooperman, High speed current mode logic for LSI. *IEEE Transactions on Circuits and Systems*, **27**:7 (1980), 626–635.
114. H. Shigematsu, N. Yoshida, M. Sato, N. Hara, T. Hirose, and Y. Watanabe, 45 GHz distributed amplifier with a linear 6-V_{pp} output for a 40 Gb/s $LiNbO_3$ modulator driver circuit. *Gallium Arsenide Integrated Circuit (GaAs IC) Symposium Technical Digest*, (2001), 137–140.
115. H. Shigematsu, M. Sato, T. Hirose, and Y. Watanabe, A 54-GHz distributed amplifier with 6-V_{pp} output for a 40-Gb/s $LiNbO_3$ modulator driver. *IEEE Journal of Solid-State Circuits*, **37**:9 (2002), 1100–1105.

Index